T0187420

Book 2

Plumbing

SECOND EDITION

LEVEL 3 APPRENTICESHIP (9189)
LEVEL 3 ADVANCED TECHNICAL DIPLOMA (8202)
LEVEL 3 DIPLOMA (6035)
T LEVEL OCCUPATIONAL SPECIALISMS (8710)

The City & Guilds textbook

Stephen Lane
Peter Tanner

HODDER EDUCATION
AN HACHETTE UK COMPANY

Orders: please contact Hachette UK Distribution, Hely Hutchinson Centre, Milton Road, Didcot, Oxfordshire, OX11 7HH. Telephone: +44 (0)1235 827827. Email education@hachette.co.uk Lines are open from 9 a.m. to 5 p.m., Monday to Friday.

You can also order through our website: www.hoddereducation.co.uk

ISBN: 978 1 3983 6162 1

© The City and Guilds of London Institute and Hodder & Stoughton Limited 2019

First published in 2019

This edition published in 2022 by
Hodder Education,
An Hachette UK Company
Carmelite House
50 Victoria Embankment
London EC4Y 0DZ

www.hoddereducation.co.uk

Impression number   10 9 8 7 6 5 4

Year   2026 2025 2024

Cover photo © vladdeep - stock.adobe.com

City & Guilds and the City & Guilds logo are trade marks of The City and Guilds of London Institute. City & Guilds Logo © City & Guilds 2022

Typeset by Integra Software Services Pvt. Ltd., Pondicherry, India

Produced by DZS Grafik, Printed in Slovenia

A catalogue record for this title is available from the British Library.

Contents

About your qualification

INTRODUCTION TO THE PLUMBING QUALIFICATIONS

You are completing one of the following qualifications:

- Level 3 Advanced Technical Diploma in Plumbing (8202-35)
- Level 3 Diploma in Plumbing Studies (6035-03)
- Level 3 Diploma in Plumbing and Domestic Heating (9189).
- T Level Technical Qualification in Building Services Engineering for Construction (8710).

The Level 3 Advanced Technical Diploma and Level 3 Diploma in Plumbing Studies are for learners who are interested in developing the specific technical and professional skills that can support development towards becoming a plumber, progressing from Level 2 qualifications.

The Level 3 Diploma in Plumbing and Domestic Heating is the on-programme qualification for the Plumbing and Heating Technician Apprenticeship and is designed to provide the apprentice with the opportunity to develop the knowledge, skills and core behaviours that are expected of a competent Plumbing and Domestic Heating Technician operating in a number of regulated areas.

T Levels are new Level 3 vocational qualifications available to learners following the completion of GCSEs; they are the same size as three A Levels and you will sit them across two years. They offer a mixture of classroom, workshop and on-the-job work experience through industrial placements. The Plumbing Engineering (356) and Heating Engineering (355) occupational specialisms will offer the knowledge and experience needed to open the door to skilled employment or further study.

HOW TO BECOME A PLUMBING AND HEATING TECHNICIAN

To become a fully recognised plumber, you must complete the following:

- Plumbing and Heating Technician Apprenticeship (9189).

The 8202 Advanced Technical Diploma and 6035 Level 3 Diploma provide the knowledge and practical skills to prepare you for an apprenticeship.

The apprenticeship and 9189 Level 3 Diploma will give you an understanding of suitable on-site skills and further knowledge required to work in the plumbing industry. Once qualified, there are many specialist qualifications available, such as environmental technology systems and designing and planning complex water systems.

How to achieve your qualification

The requirements for successfully obtaining your qualification depend on which programme you are enrolled on.

6035

The 6035 diploma is assessed by a range of multiple choice exams, assignments and practical tests. You will be assessed, by one of these methods, at the end of each unit.

For details on which assessments will follow which units, you should consult the City & Guilds qualification handbook. For details on when you will complete your assessments, consult your tutor.

8202

The 8202 qualification is assessed using one multiple choice examination and one practical synoptic assignment.

For the synoptic assignment, a typical brief might be to install a cold water supply and hot water distribution pipework connected to all sanitary appliances. You will need to draw on skills and understanding developed across the qualification content in order to consider the specific requirements of the particular system and related plumbing principles, and carry out the brief. This includes the ability to plan tasks, such as plant, materials and equipment for an installation, and apply the appropriate practical and hand skills to carry them out using appropriate tools and equipment.

You will also demonstrate that you are following health and safety regulations at all times by drawing upon your knowledge of legislation and regulations.

The exam draws from across the content of the qualification, using multiple choice questions to:
- confirm breadth of knowledge and understanding
- test applied knowledge and understanding – giving the opportunity to demonstrate higher-level integrated understanding through application, analysis and evaluation.

9189

Level 3 is assessed using multiple choice tests and practical assignments. These will happen at the end of each phase of learning, with there being four phases in total. Learners will also be expected to keep a work log for the duration of the programme.

The apprenticeship is assessed separately to the on-programme qualification and is assessed by an end-point assessment (EPA). In order to progress through the end-test gateway to end-point assessment, you must complete the following:
- Level 3 Diploma in Plumbing and Domestic Heating qualification (9189)
- Level 2 Maths
- Level 2 English.

The graded EPA will be comprised of the following assessment methods:
- multiple choice test
- design project

- practical installation test
- practical application test
- professional discussion.

T Level (8710)

This Level 3 course, which runs alongside the apprenticeship programme, offers the opportunity to gain essential skills that will enable you to enter full-time employment within the plumbing and heating sector.

The course is a two-year programme. All learners studying a Building Services Engineering for Construction T Level will complete the core component (350), which introduces the foundational industry principles. This is assessed by two written exams and an employer-set project. This core component is covered in another Hodder Education textbook: *Building Services Engineering for Construction T Level: Core*.

You will also choose one or two occupational specialisms. These include:
- 355 Heating engineering
- 356 Plumbing engineering.

Although these specialisms involve practical work (which you will cover with your tutor in the workshop, and which will be assessed by observation of practical tasks), the key underpinning plumbing and heating content needed for these specialisms is covered across this book and *The City & Guilds Textbook: Plumbing Book 1* (also published by Hodder Education).

How to use this book

Throughout this book you will see the following features:

Industry tips and **Key points** are particularly useful pieces of advice that can assist you in your workplace or help you remember something important.

> ### INDUSTRY TIP
>
> Sedimentation tanks require cleaning when their performance begins to deteriorate, and a 12-month period between cleaning operations would normally be sufficient.

> ### KEY POINT
>
> It is vital that fuels are kept dry and that they are delivered in good condition for optimum combustion efficiency to occur.

Key terms in bold purple in the text are explained in the margin to aid your understanding. (They are also explained in the Glossary at the back of the book.)

> ### KEY TERM
>
> **Thermal shock:** the rapid cooling or heating of a substance that can lead to failure of the material.

Health and safety boxes flag important points to keep yourself, colleagues and clients safe in the workplace. They also link to sections in the health and safety chapter for you to recap learning.

> ### HEALTH AND SAFETY
>
> It is dangerous to use a disposable knife blade when stripping cables as the blade can break, creating sharp shards.

Activities help to test your understanding and learn from your colleagues' experiences.

> ### ACTIVITY
>
> Research reed bed filtration systems to assess the range of systems available in the UK.

Values and behaviours boxes provide hints and tips on good workplace practice, particularly when liaising with customers.

> ## VALUES AND BEHAVIOURS
>
> Do not forget to keep the householder/responsible person informed of the areas that are going to be isolated during maintenance tasks and operations.

Improve your maths items provide opportunities to practise or improve your maths skills.

Improve your English items provide opportunities to practise or improve your English skills.

At the end of each chapter there are some **Test your knowledge** questions. These are designed to identify any areas where you might need further training or revision. Answers are provided at www.hoddereducation.co.uk/construction.

Acknowledgements

This book draws on several earlier books that were published by City & Guilds, and we acknowledge and thank the writers of those books:

- Michael Maskrey
- Neville Atkinson
- Eamon Wilson
- Andrew Hay-Ellis
- Trevor Pickard.

We would also like to thank everyone who has contributed to City & Guilds photoshoots. In particular, thanks to: Jules Selmes and Adam Giles (photographer and assistant); Martin Biron and the staff at the College of North West London and the following models: Vivian Chioma, Jennifer Close, Peko Gayle-Reveault, Adam Giles, Michael Maskrey, Nahom Sirane, Zhaojie Yu; Michael Maskrey and the staff at Stockport College and the following models: Michael Maskrey, Jordan Taylor; Jocelynne Rowan, Steve Owen, Mick Gibbons, Diane Whinney and Dave Driver/Baxi Training Centre; Paul Morgan/Cotherm; Jamie Purser, Graham Fleming, John Pierce and Sabir Ahmed/Hackney Community College; David Simoes/Honeywell; Jonathan Madden, Andrew Patterson and Carl Spalding/Heatrae Sadia; Rob Wellman/National Skills Academy, and models Anup Chudasama, Michaela Opara and Sami Simela; Mykal Trim and Sam, CHS Gas Assessment Centre Norwich.

Contains public sector information licensed under the Open Government Licence v3.0.

Permission to reproduce extracts from British Standards is granted by BSI Standards Limited (BSI). No other use of this material is permitted. British Standards can be obtained in PDF or hard copy formats from the BSI online shop: https://shop.bsigroup.com/

Picture credits

Every effort has been made to trace and acknowledge ownership of copyright. The publishers will be glad to make suitable arrangements with any copyright holders whom it has not been possible to contact.

Fig.1.1 © WRAS; Fig.1.2 © Kiwa; Fig.1.7 Aquatech Pressmain product image courtesy of Aquatronic Group Management Plc, www.agm-plc.co.uk; Fig.1.9 © Brimar Plastics Limited; Fig.1.10 © Dewey Waters Ltd; Fig.1.17 Image reproduced with permission from Keraflo Ltd; Fig.1.18 © Danfoss; Fig.1.24 © BD|SENSORS GmbH; Fig.1.25 Aquatech Pressmain product images courtesy of Aquatronic Group Management Plc, www.agm-plc.co.uk; Fig.1.26 © KSB SE & Co. KGaA; Fig.1.29 Image provided courtesy of Dales Water Services Ltd – Water Well Drillers and Private Water Supply Engineers; Fig.1.42 © Silverline; Fig.1.47 © KSB SE & Co. KGaA; Fig.1.48 © Whisper Pumps Ltd, www.whisperpumps.com; Fig.1.51 © Audrius Merfeldas/stock.adobe.com; Figs.1.52 & 1.53 © Ultra Finishing; Fig.1.54 © Methven Limited; Fig.1.55 © hiv360/Adobe Stock; Fig.1.57 Image reproduced with permission from Cistermiser Ltd; Fig.1.59 Monsoon pump image courtesy of Stuart Turner Limited © 2019; Fig.1.60 © Toolstation; Fig.1.61 © Ultra Finishing; Fig.1.63 Monsoon pump image courtesy of Stuart Turner Limited © 2019; Fig.1.64 Image courtesy of Stuart Turner Limited © 2019; Fig.1.77 © Paxton Agricultural; Figs.1.80 & 1.88 © Arrow Valves; Fig.1.91 © Hans Sasserath GmbH & Co. KG; Fig.1.94 © Arrow Valves; Fig.1.96 top image © & courtesy of Valves Online – valvesonline.co.uk, bottom © Screwfix Direct Ltd; Fig.1.103 © GARDENA Multi-Purpose Spray Gun, www.gardena.com/UK; Fig.1.107 © HOLLANDGREEN; Fig.1.108 © Thaifairs/stock.adobe.com; Fig.1.121 Image kindly provided by Monument Tools Ltd, Hackbridge, Surrey, UK; Fig.1.124 Photograph by kind permission of ROTHENBERGER UK; Fig.1.126 © SUKU - Druck- und Temperaturmesstechnik GmbH; Fig.2.10 © Crown Water Heaters Ltd; Fig.2.15 © GDC Group Limited; Fig.2.18 © Sebastian Kaulitzki/Shutterstock.com; Fig.2.19 © Horne Engineering Ltd; Fig.2.23 © Grundfos Pumps Ltd; Fig.2.26 © Toolstation; Fig.2.32 © RZ/stock.adobe.com; Fig.2.39 © Zilmet UK Limited; Fig.2.41 Courtesy of Flamco Group, part of Aalberts; Fig.2.43 © Anton_antonov/stock.adobe.com;

Fig.2.44 © Upperplumbers Ltd; Fig.2.51 © Fefufoto/stock.adobe.com; Fig.2.60 City & Guilds; Fig.2.61 © Ddukang/stock.adobe.com; Fig.2.62 Michael Maskrey; Fig.3.3 © Danfoss; Fig.3.10 © Bosch Thermotechnology Ltd; Figs.3.16 & 3.17 © Grundfos Pumps Ltd; Fig.3.20 © Borders Underfloor Heating Ltd; Fig.3.21 © Rigamondis/Shutterstock.com; Fig.3.22 © Beggs & Partners; Fig.3.31 © M H Mear & Co. Huddersfield, England (www.mhmear.co.uk); Fig.3.41 © Anton_antonov/stock.adobe.com; Fig.3.42 © Toolstation; Fig.3.51 © Megger; Figs.3.52–3.54 & 3.56 © Fernox; p.286 © Grundfos Pumps Ltd; Fig.4.1 © kurhan/Shutterstock.com; Fig.4.2 Neville Atkinson; Fig.4.3 © CIPHE - Chartered Institute of Plumbing and Heating Engineering; Fig.4.4 © Artazum/Shutterstock.com; Figs.4.17–4.19 & 4.21 © Kingspan Group; Fig.4.22 © Saniflo Ltd; Figs.4.32, 4.33 & 4.35, 4.36 © Grundfos Pumps Ltd; Fig.4.37 Neville Atkinson; Fig.4.38 © Contactum Limited; Fig.4.39 © zstock/Shutterstock.com; Fig.4.48 © Astroflame (Fire Seals) Ltd; Figs.5.3, 5.4 & 5.10 City & Guilds; Fig.5.11 Neville Atkinson; Figs.5.16 & 5.17 © Contactum Limited; Fig.5.18 © Toolstation; Fig.5.23 City & Guilds; Fig.5.25 top left © Martindale Electric Co. Ltd, top right & bottom City & Guilds; Fig.5.26 City & Guilds; Fig.5.27 © Axminster Tool Centre Ltd; Fig.5.28 City & Guilds; Fig.5.29 © Axminster Tool Centre Ltd; Figs.5.33–5.48 City & Guilds; p.380 City & Guilds; Fig.5.49 Andrew Hay-Ellis; Figs.5.50–5.54 City & Guilds; p.384 @ City & Guilds; Figs.5.55–5.73 City & Guilds; Figs.5.74–5.76 Andrew Hay-Ellis; Figs.5.77–5.95 City & Guilds; Fig.5.96 © Health and Safety Executive (HSE); Figs.5.97 & 5.98 Neville Atkinson; Figs.5.101–5.103 © Megger; Figs.5.104 & 5.105 City & Guilds; Fig.5.107 © Megger; Fig.5.108 City & Guilds; Figs.5.109 & 5.110 © Megger; Figs.5.113 & 5.114 City & Guilds; p.405 City & Guilds; Figs.5.122 & 5.123 City & Guilds; p.408 City & Guilds; Fig.5.124 City & Guilds; p.410 City & Guilds; Figs 5.129–5.131 City & Guilds; Figs.5.132 & 5.133 Neville Atkinson; Fig.5.134 © Megger; Figs.5.135, 5.138, 5.141–5.144, 5.147 & 5.148 City & Guilds; Fig.5.149 Neville Atkinson; pp.420–21 City & Guilds; Figs.5.150–5.151 City & Guilds; Figs.5.152–5.154 © Contactum Limited; Figs.5.155–5.159 Neville Atkinson; Fig.5.160 City & Guilds; Figs.5.161 & 5.162 Neville

CHAPTER 1

COLD WATER SYSTEMS, PLANNING AND DESIGN

This chapter provides learning in the application of design techniques, installation, maintenance, diagnostics and rectification of faults and commissioning procedures, along with the backflow protection in plumbing systems to comply with current legislation and regulations. The chapter also covers systems in multi-storey dwellings with water supplied from the water undertaker and private water supplies.

By the end of this chapter, you will have knowledge and understanding of the following:
- the legislation relating to the installation and maintenance of cold water systems
- cold water system layouts
- backflow protection in plumbing systems
- design techniques for cold water systems
- diagnosing and rectifying faults in cold water systems and components
- commissioning cold water systems and components
- servicing and maintenance of cold water systems.

Return to Book 1, Chapter 5, Cold water systems, which covered the following topics:
- the sources and properties of water
- the types of water supply to dwellings
- the treatment and distribution of water
- the sources of information relating to cold water systems
- the water service pipework to dwellings
- selecting cold water systems
- backflow protection
- installing cold water systems and components
- replacing or repairing defective components
- decommissioning cold water systems.

1 THE LEGISLATION RELATING TO THE INSTALLATION AND MAINTENANCE OF COLD WATER SYSTEMS

In this, the first section of this chapter, we will look at some of the many pieces of legislation that govern the installation and maintenance of cold water systems.

Cold water system legislation

The water industry in England and Wales is regulated by the Water Industry Act 1991 (the Act), as amended by the Water Act 2003 (Commencement No. 11) Order 2012, the Private Water Supplies Regulations 2016

HEALTH AND SAFETY
Return to Book 1, Chapter 1, Health and safety practices and systems.

and the Water Supply (Water Fittings) Regulations 1999. These three documents have specific roles to play within the plumbing industry. We will look at the impact of each of these documents separately.

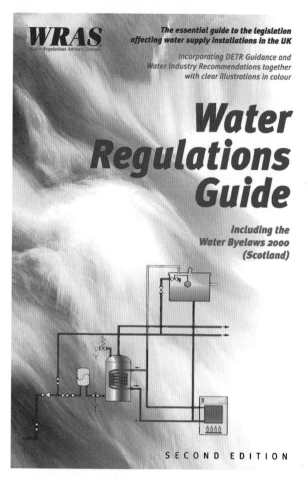

▲ Figure 1.1 Water Regulations Guide

The Water Act 2003 (Water Industry Act 1991)

The Water Act 2003 amalgamates and amends two previous pieces of legislation; the Water Industry Act 1991 and the Water Resources Act 1991. The Water Act 2003 introduced changes to the regulation of the water industry in England and Wales originally made under the Water Industry Act 1991. They are enforced by the Environment Agency and deal with such matters as:

- the appointment and regulation of water and sewerage companies and licensed water suppliers by the Water Services Regulation Authority (OFWAT)
- water supply and sewage disposal powers and duties of the water companies and suppliers
- the obligations of the water companies and licensed water suppliers to supply water that is fit for human consumption and the enforcement of those obligations by the Department of the Environment, Food and Rural Affairs (DEFRA) and the Drinking Water Inspectorate

- charging powers of water companies and suppliers and the control of those charges by OFWAT
- protection of customers and consumers by OFWAT and the Consumer Council for Water.

Under the provisions laid down by the Water Act 2003, the UK Government introduced two documents that regulate how plumbers install, commission and maintain water supplies within domestic buildings. These are:

- The Water Supply (Water Fittings) Regulations 1999
- The Private Water Supplies Regulations 2016.

The Water Supply (Water Fittings) Regulations 1999

Before 1999, each water authority had its own set of water by-laws that were based upon the 101 model water by-laws issued by the UK Government in 1986. The problem was that each **water undertaker** had local variations, which caused much confusion, as there was no 'common' standard throughout the UK.

KEY TERM

Water undertaker: a water authority or company that supplies clean, cold wholesome water under Section 67 of the Water Act 1991.

On 1 July 1999, the Office of the Deputy Prime Minister issued the first ever water regulations to be enforced in the UK. They are known as 'The Water Supply (Water Fittings) Regulations 1999' and they offer a common practice throughout the United Kingdom. **BS 8558** is the lead British Standard on the design, installation, testing and maintenance of services supplying water for domestic use within buildings and their curtilages. **BS EN 806** contains specifications for installations inside buildings conveying water for human consumption.

BS EN 806 is a relatively new British Standard that came into force in 2011. It is divided into five parts:

- general recommendations
- design
- pipe sizing
- installation
- operation and maintenance.

Both **BS 8558** and **BS EN 806** replace the old British Standard, **BS 6700**, Design, installation, testing and maintenance of services supplying water for domestic use within buildings and their curtilages. However, parts of **BS 6700** that are not covered in either **BS EN 806** or **BS 8558** will be retained.

> ### INDUSTRY TIP
>
> For more information go to the Water Regulations Advisory Scheme (WRAS) website at: www.wras.co.uk

Simply put, the Water Supply (Water Fittings) Regulations were made under Section 74 of the Water Industry Act 1991 and have been put in place to ensure that the plumbing systems we install and maintain prevent:

- contamination of water
- wastage of water
- misuse of water
- undue consumption of water
- **erroneous** metering of water.

An important factor here is that these regulations ONLY cover installations where the water is supplied from a water undertaker's water main and are enforced by the water undertaker in your area. They are **NOT**, however, enforceable where the water is supplied from a private water source.

> ### KEY TERM
>
> **Erroneous:** wrong or incorrect.

> ### INDUSTRY TIP
>
> A free copy of the Water Supply (Water Fittings) Regulations 1999 can be downloaded from: www.legislation.gov.uk/ uksi/1999/1148/contents/made

The Private Water Supplies Regulations 2016

A private water supply is defined as any water supply, which is **NOT** provided by a water undertaker. It is not connected to any part of the water mains network and, as such, water rates are not charged, although the owner of any such supply may make a charge for any water used. Private supplies are commonly used in rural areas where connection to the water mains is difficult. A private supply may serve one property or many properties on a private network. The water may be supplied from a borehole, spring, well, river, stream or pond.

Under the Water Act 2003, the local authority in the area where the private water supply is located is responsible for the inspection and testing of the water supply to ensure that it is maintained to a quality that is fit for human consumption. These inspections and tests are made in accordance with the Private Water Supplies Regulations 2016. Generally speaking, the more people that use the supply, the more detailed the tests and the more regular the inspections have to be. Supplies for commercial properties and activities or food production and preparation have to be tested more frequently and meet more stringent requirements than domestic supplies.

The Private Water Supplies Regulations 2016 stipulate that a risk assessment must be made of **ALL** private water supplies including the source, storage tanks, any treatment systems and the premises using the water supply.

> ### INDUSTRY TIP
>
> A free copy of the Private Water Supplies Regulations 2016 can be downloaded from: www.legislation.gov.uk/ uksi/2016/618/contents

Installer and user responsibilities under the Water Supply (Water Fittings) Regulations 1999

Plumbers who install plumbing systems and fittings, users, owners and occupiers have legal responsibilities under the Water Supply (Water Fittings) Regulations 1999 to ensure that any installation and the materials and fittings used comply with the regulations. In most cases, as we have seen, advanced notification of the proposed installation must be given and so architects, building developers and plumbers have to abide by the regulations on behalf of their client. These regulations are not **retrospective**, so systems installed BEFORE the regulations came into force can still be used without the need to update to current standards, even if that installation would not meet current requirements. However, any alteration or extension completed on an existing installation must comply with the regulations in force at the time of the installation date. Regulation 3 of the Water Supply (Water Fittings) Regulations 1999 makes this very clear.

KEY TERM

Retrospective: looking back, in this case, at previous installations.

Regulation 3 states:

1 No person shall:
   a Install a water fitting to convey or receive water supplied by a water undertaker, or alter, disconnect or use such a water fitting; or
   b Cause or permit such a water fitting to be installed, altered, disconnected or used, in contravention of the following provisions of this Part.

2 No water fitting shall be installed, connected, arranged or used in such a manner that it causes or is likely to cause:
   a Waste, misuse, undue consumption or contamination of water supplied by a water undertaker; or
   b The erroneous measurement of water supplied by a water undertaker.

3 No water fitting shall be installed, connected, arranged or used which by reason of being damaged, worn or otherwise faulty, causes or is likely to cause:
   a Waste, misuse, undue consumption or contamination of water supplied by a water undertaker; or
   b The erroneous measurement of water supplied by a water undertaker.

The following points are worth remembering:

- Plumbers should obtain a copy of the regulations, any amendments and guidance notes from Her Majesty's Stationery Office (HMSO) or a copy of the Water Regulations Guide from the Water Regulations Advisory Scheme (WRAS) to ensure that any plumbing work complies with the regulations. They must ensure that any fittings and materials are of a sufficient quality and that any installation is installed in a workmanlike manner to an approved installation standard requirement. Membership of a competent person scheme is advisable.

- Users, owners and occupiers must ensure that the person employed to undertake the proposed work is aware of the regulations and that any work completed is done so in accordance with the regulations. A certificate of compliance MUST be obtained for the work and retained for future reference by the user, owner or occupier. Regulation 3 also makes it clear that the user, owner or occupier is responsible for ensuring that waste, misuse, undue consumption, contamination or erroneous metering of the water supply does not occur during usage.

- The Government requires the water undertakers enforce the regulations within their area of supply. They will undertake inspections of new and existing installations to check that the regulations are being complied with. Where breaches of the regulations are found, they must be remedied as soon as practicable. Where breaches present a significant health risk, the water supply to the premises may be isolated or, in severe cases, disconnected immediately to protect the health of occupants and/or others fed from the same public supply. The Government has deemed that it is a criminal offence to breach the regulations, and offenders, including users, owners and occupiers in cases where the original installer cannot be traced, may face prosecution.

Notification requirements

Now we will investigate the requirements of notification and consent with regard to cold water installations. We need to consider:

- Who needs to notify?
- Which types of installations require notification and consent?
- Why do we need to notify?
- Who do we notify?

The Water Supply (Water Fittings) Regulations requirements

Regulation 5 of the Water Supply (Water Fittings) Regulations requires that the water undertaker must be notified before work is commenced on most types of plumbing installations and anyone installing or using the installation without the water undertaker's written consent could be committing a criminal offence.

> ## VALUES AND BEHAVIOURS
>
> Being aware of the regulations that may apply to any job you commence will ensure you are working within the law.

Notification and consent are also legally required where water fittings are to be installed on any water or waste water plants. This requirement applies irrespective of whether the plant is owned/operated by the same organisation as the enforcing water undertaker.

Consent is necessary for the installation of fittings in new buildings and dwellings, extensions and alterations of water systems in existing non-domestic premises, where there is a material change of use of a building and for the installation of certain specified items. These include:
- a bidet with an ascending spray or flexible hose
- a bath larger than 230 litres (measured to the centre of the overflow)
- a shower unit of a type specified by the regulator
- a pump or booster drawing more than 12 litres per minute
- a reverse osmosis unit
- a water treatment unit producing a waste water discharge or requiring water for regeneration or cleaning
- a reduced pressure zone (RPZ) valve or other mechanical device for protection against backflow in fluid category 4 or 5
- a garden watering system unless designed to be a hand-operated one
- any water system laid outside a building and either less than 750 mm or more than 1350 mm below ground level

- construction of an automatically-replenished pond or swimming pool of more than 10,000 litres.

When notifying the water undertaker, the following information must be sent:
- the name and address of the person giving notice and, if different, of the person to whom the consent should be sent
- a description of the proposed work and any related change of use of premises
- the location of the premises and their use or intended use
- the plumbing contractor's name and address, if an approved plumber is to do the work.

Consent cannot be withheld unless there are reasonable grounds to do so, and may be granted subject to conditions, which must be followed. If written approval is not given within ten working days it can be assumed that consent has been granted but this does not alter the obligation upon the installer and the owner or occupier to ensure that the regulations have been complied with.

Notification and consent requirements: the Building Regulations Approved Document G requirements

From 6 April 2010 an updated and extended version of Part G was implemented, bringing in a number of new areas under Building Regulations control. As such, under these new areas of control, the installation of the systems mentioned in Regulation 5 of the Water Supply (Water Fittings) Regulations are notifiable to the local authority building control. In general terms, notification must take place prior to the work starting and within five working days of the work being completed.

> **INDUSTRY TIP**
>
> A free copy of the Building Regulations Approved Document G can be downloaded from: www.gov.uk/government/publications/sanitation-hot-water-safety-and-water-efficiency-approved-document-g

The new regulation Approved Document G is broken down into six parts:

- G1 Cold water supply – new requirements on supply of wholesome water for purposes of drinking, washing or food preparation. Also for the provision of water of a suitable quality to sanitary conveniences fitted with a flushing device.
- G2 Water efficiency – G2 and Regulation 17K of the Building Regulations 2000 set out new requirements on water efficiency in dwellings.
- G3 Hot water supply and systems – sets out enhanced and amended provisions on hot water supply and safety, applying safety provisions to all types of hot water systems and a new provision on scalding prevention.
- G4 Sanitary conveniences and washing facilities – sets out requirements for sanitary conveniences and hand washing facilities.
- G5 Bathrooms – sets out requirements for bathrooms, which apply to dwellings and to buildings containing one or more rooms for residential purposes.
- G6 Kitchens and food preparation areas – contains a new provision requiring sinks to be provided in areas where food is prepared.

Certification of plumbing installations

As a direct result of the introduction of the Water Supply (Water Fittings) Regulations in 1999, the water undertakers began to recognise competent persons schemes, approved plumber schemes and self-certification of certain plumbing installations. These are administered and regulated by the local water undertaker.

Under Water Supply Regulation 5, if a person proposes to carry out plumbing work on any building other than a house or dwelling, they can either:

- notify their local water undertaker, or
- use an **approved plumber**/contractor who will issue a Water Regulation Compliance Certificate. This certificate confirms that the work has been carried out in accordance with the regulations and that the fittings and materials used meet the strict requirements of the regulations.

▲ Figure 1.2 An example of a certificate produced by KIWA, which proves quality, credibility and compliance

KEY TERM

Approved plumber: a plumber that has undertaken specific Water Regulations training and is recognised as competent by the water undertaker.

What are competent persons schemes and approved plumber schemes?

A plumber who can prove, via an assessment, a substantial knowledge of the Water Supply (Water Fittings) Regulations, can join a competent persons scheme (CPS). The main benefit for members of these schemes is that they can self-certify plumbing supply work (domestic commercial and industrial) that would otherwise have to wait for water company approval before work could commence.

The first of these schemes to emerge was the water industry approved plumber scheme (WIAPS), set up by most of the water undertakers in England and Wales.

2 COLD WATER SYSTEM LAYOUTS

In this section, we will take a look at the component layout features and functions for cold water systems and the methods of providing water supplies.

Layout features for multi-storey dwellings

In plumbing systems, the phrase '**multi-storey**' applies to buildings that are simply too tall to be supplied totally using just the pressure of the water main. Because of their design, these buildings have particular cold water system requirements that can only be satisfied by pumping or 'boosting' the cold water supply either in part or in total.

> **KEY TERM**
>
> **Multi-storey building**: a building having more than three floors.

Most cold water supplies that are delivered from the mains cold water supply arrive at a building at a three to seven bars pressure (30–70 metres head). A 30 m head is equivalent to around eight storeys in height. When taking into account a two-storey margin to allow for frictional losses, it becomes obvious that the height of the building will often outstrip the head of pressure available. In some parts of the UK, it is not unusual to find premises with pressures lower than two bars and flow rates of below 15 litres/minute. In these cases, the water undertaker should be consulted as to where supply pressures can be relied upon to ensure the correct operation of the cold water system.

If the public supply is inadequate or the building too high, then the water supply within the building must be boosted. There are several ways that this can be achieved and these can be divided into:

- direct boosting systems, direct from the cold water mains supply
- indirect boosting systems, from a break cistern.

Indirect systems are the most common, as direct boosting systems are often forbidden by the water undertakers because they can often reduce the mains pressure available to other consumers in the locality and can increase the risk of contamination by backflow.

However, where insufficient water pressure exists and the demand is below 0.2 litres/second, then drinking water may be boosted directly from the supply pipe provided that the water undertaker agrees. With indirect systems, a series of float switches in the break cistern starts and stops the pumps depending upon the water levels in the cistern.

Boosting pumps can create excessive aeration of the water, which, although causing no deterioration of water quality, can cause concern to the consumer because of the opaque, milky appearance of the water. There are several common examples of these systems:

- direct boosting systems
- direct boosting to a drinking water header and duplicate storage cisterns
- indirect boosting to a storage cistern
- indirect boosting with a pressure vessel.

Direct boosting systems

Where permission from the water undertaker has been granted, pumps can be directly fitted to the incoming supply pipe to enable the head of pressure to be increased.

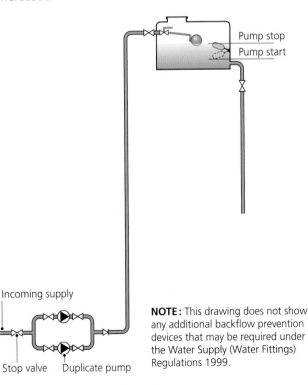

Pump stop
Pump start

Incoming supply

Stop valve   Duplicate pump

NOTE: This drawing does not show any additional backflow prevention devices that may be required under the Water Supply (Water Fittings) Regulations 1999.

▲ Figure 1.3 Direct boosting system

A float switch, or some other no less effective device, situated inside the high-level cistern controls the pumps. The pumps either switch on or off depending upon the water level in the cistern. The pumps are activated when the water drops to a depth normally equal to about half cistern capacity and switch off again when the water level reaches a depth approximately 50 mm below the shut off level of the float operated valve.

If the cistern is to be used for drinking water, then it must be of the protected type.

Direct boosting to a drinking water header and duplicate storage cisterns

This system is mainly used for large and multi-storey installations. With this system, the cisterns at high level are for supplying non-drinking water only, a drinking water header sited on the boosted supply pipe provides limited storage of 5–7 litres drinking water to sinks in each dwelling when the pump is not running. Excessive pressure should be avoided as this can lead to an increase in the wastage of water at the sink taps along with the nuisance of excessive splashing.

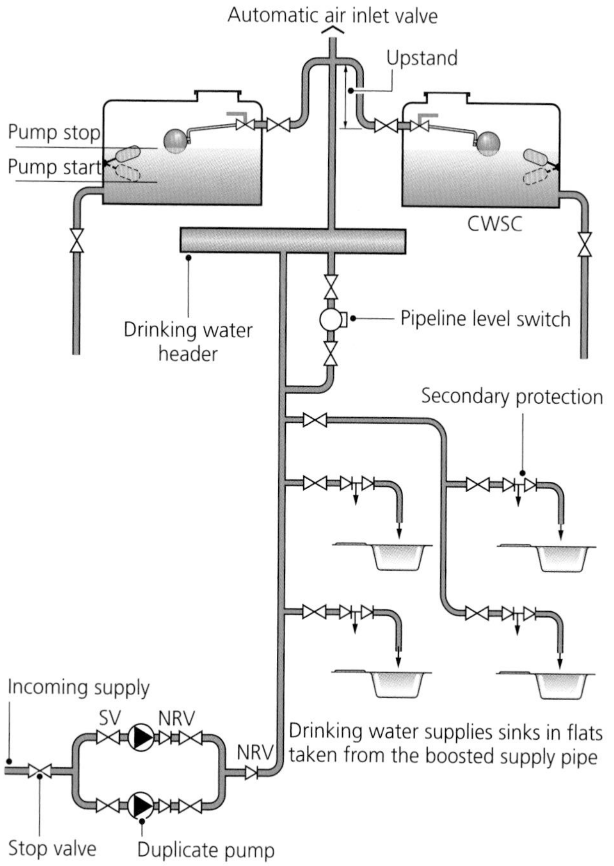

A pipeline switch on the header by-pass starts the pumps when the water level falls to a predetermined level. The pumps can be time controlled or activated to shut down by a pressure switch. When filling the cisterns, the pumps should shut down when the water levels in the cisterns are approximately 50 mm below the shut off level of the float operated valve.

Secondary backflow devices may be required at the drinking water outlets on each floor.

Indirect boosting to a storage cistern

This system incorporates a break cistern to store the water before it is pumped via a boosting pump (known as a booster set) to a storage cistern at high level. The pumps should be fitted to the outlet of the break cistern. The capacity of the break cistern needs careful consideration and will depend upon the total water storage requirements and the cistern's location within the building, but it should not be less than 15 minutes of the pumps maximum output. However, the cistern must not be oversized as this may result in water stagnation within the cistern.

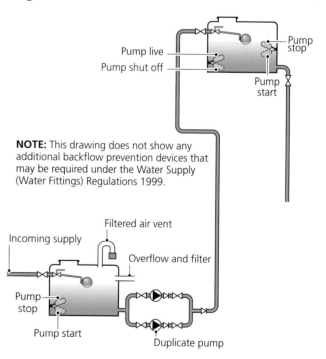

NOTE: This drawing does not show any additional backflow prevention devices that may be required under the Water Supply (Water Fittings) Regulations 1999.

▲ Figure 1.5 Indirect boosting to a storage cistern

The water level in the storage cistern (or cisterns) is usually controlled by means of water level switches which control the pumps. When the water drops to a predetermined level the pumps start filling the storage

▲ Figure 1.4 Direct boosting to drinking header

cisterns. The pumps are then switched off when the water level reaches a point about 50 mm from the shut off level of the float operated valve. A water-level switch should also be positioned in the break cistern to automatically shut off the pumps if the water level drops to within 225 mm of the suction connection near the bottom of the break cistern. This is simply to ensure the pumps do not run dry.

Indirect boosting with a pressure vessel

This rather complicated system is mainly used in buildings where a number of storage cisterns are fed at various floor levels making it impractical to control pumps by water level switches. It utilises a pneumatic pressure vessel to maintain the pressure boost to the higher levels of the building.

The pneumatic pressure vessel comprises of a small water reservoir with a cushion of compressed air. The water pumps and the compressed air operate intermittently. The pumps replenish the water level and the pressure vessel maintains the system pressure. Since the system may be supplying drinking water, the vessel capacity is kept purposely low to ensure a rapid and regular turnover of water. The compressed air must be filtered to ensure that dust and insects are eliminated.

▲ Figure 1.7 A typical booster set with pressure vessel and control boards

Normally, the controls including the pressure vessel, pumps, air compressor and control equipment are purchased as a package, although self-assembled booster sets are available.

As can be seen in Figure 1.8, some of the floors below the limit of the mains cold water supply pressure are supplied un-boosted directly from the cold water main with the floors above the mains pressure limit being supplied via the break cistern and booster set. Drinking water supplies must be from a protected cistern.

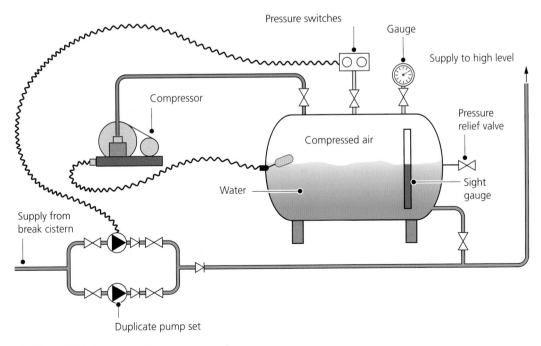

▲ Figure 1.6 Auto pneumatic pressure vessel

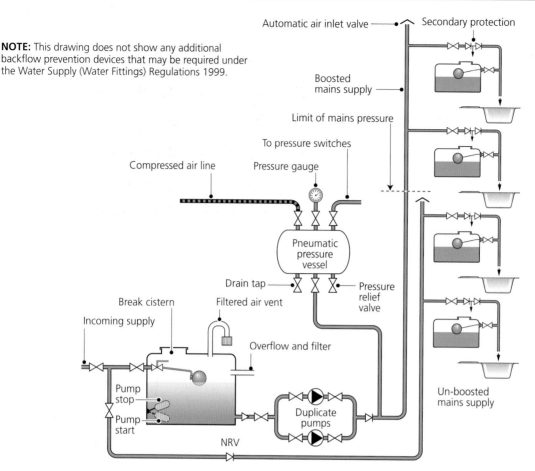

NOTE: This drawing does not show any additional backflow prevention devices that may be required under the Water Supply (Water Fittings) Regulations 1999.

▲ Figure 1.8 Indirect boosting with pressure vessel

Layout features for large-scale storage cisterns used in multi-storey cold water systems

The installation of large-scale cisterns differs somewhat from the cisterns that were introduced in Chapter 5 of Book 1. Large cisterns must be installed in accordance with the Water Supply (Water Fittings) Regulations 1999 (and the Scottish Water Byelaws 2014). Regulation 5 states that the water undertaker must be notified before the installation of large cisterns begins and it is important to remember that the correct backflow protection must be present in relation to the fluid category of the contents of the cistern.

In this section of the chapter, we will look at the general requirements of large-scale cisterns.

Materials for large-scale cisterns

Large cisterns can be made from several materials and can be either one piece or sectional.

Sectional cisterns are constructed, usually on site, from 1 m² sections, which are bolted together and can be made to suit literally any capacity and tailored to fit any space. Sectional cisterns can be internally or externally flanged and are bolted together with stainless steel bolts.

The main materials used for:

- One-piece cisterns:
  - glass-reinforced plastic (GRP) **BS EN 13280:2001**
  - plastic **BS 4213:2004** and **BS EN 12573.1:2000**:
    - polyproylene (PP)
    - polyethylene (PE)
    - polyvinyl chloride (PVC)
- Sectional cisterns:
  - glass-reinforced plastic (GRP) **BS EN 13280:2001**
  - steel with protection against corrosion and subsequent water contamination in the form of:
    - paint protected with a paint that is listed in the Water Materials and Fittings Directory

- glass-coated
- galvanised
- rubber-lined.
- aluminium – rubber-lined.

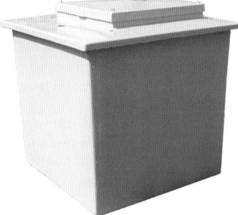

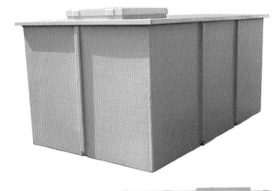

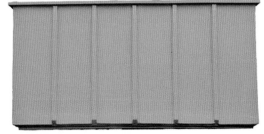

▲ Figure 1.9 One-piece GRP cisterns

▲ Figure 1.10 A sectional cistern

Overflow and warning pipe requirements of large-scale cisterns

Overflows for large cisterns are quite different from those fitted to cisterns for domestic purposes. The objective is the same – to warn that the float operated valve is malfunctioning and to remove water that may otherwise damage the premises. But with larger cisterns, the potential for water wastage and water damage is far greater. Therefore, the layout is different.

The overflow/warning pipe on large-scale cisterns must:
- contain a vermin screen to prevent the ingress of insects and vermin
- be capable of draining the maximum inlet flow without compromising the inlet air gap
- contain an air break before connection to a drain
- not be of such a length that it will restrict the flow of water, causing the air gap to be compromised
- discharge in a visible, conspicuous position.

The warning pipe invert needs to be located a minimum of 25 mm above the maximum water level of the cistern and the air gap not less than 20 mm or TWICE the internal diameter of the inlet pipe, whichever is the greater.

The general features of larger cisterns are:
- Cisterns with an actual capacity of 1000 litres to 5000 litres:
  - The discharge level of the inlet device must be positioned at least twice the diameter of the inlet pipe above the top of the overflow pipe.

- The overflow pipe invert must be located at least 25 mm above the invert of the warning pipe (or warning level if an alternative warning device is fitted).
- The warning pipe invert must be located at least 25 mm above the water level in the cistern and must be at least 25 mm diameter.
- Cisterns with an actual capacity greater than 5000 litres:
  - The discharge level of the inlet device must be positioned at least twice the diameter of the inlet pipe above the top of the overflow pipe.
  - The overflow pipe invert must be located at least 25 mm above the invert of the warning pipe (or warning level if an alternative warning device is fitted).
  - The warning pipe invert must be located at least 25 mm above the water level in the cistern and must be at least 25 mm diameter.

- Alternatively, the warning pipe may be discarded provided a water level indicator with an audible or visual alarm is installed that operates when the water level reaches 25 mm below the invert of the overflow pipe.

INDUSTRY TIP

References to the 'actual capacity' of a cistern simply means the maximum volume which it could hold when filled to its overflowing level. 'Nominal capacity' is the total volume it could hold when filled to the top of the cistern.

In both cases, the size of the overflow pipe will depend upon the type of air gap incorporated into the cistern (we will look at air gaps and backflow protection a little later in this chapter) and this will depend upon the fluid category of the cistern contents. It must be remembered that:

- if a type AG air gap (fluid category 3) is fitted, the overflow diameter shall be a minimum of twice the inlet diameter
- if a type AF air gap (fluid category 4) is fitted, the minimum cross-sectional area of the overflow pipe must be, throughout its entire length, four times the cross-sectional area of the inlet pipe.

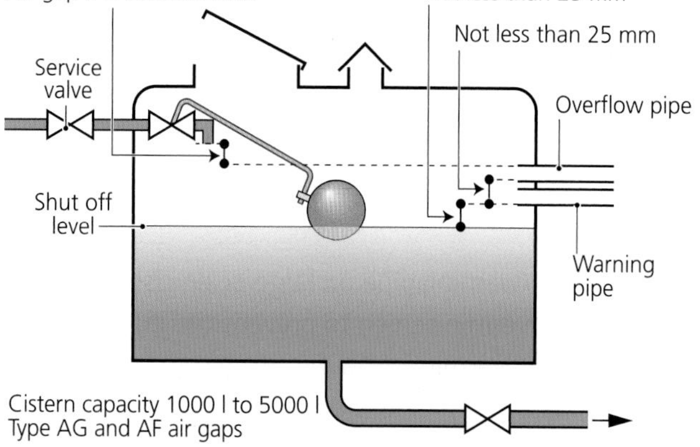

▲ Figure 1.11 Cisterns 1000–5000 litres

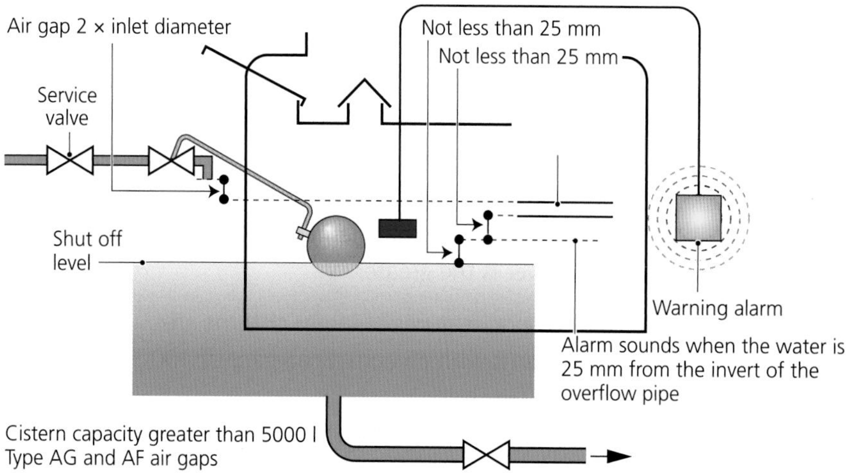

▲ Figure 1.12 Cisterns greater than 5000 litres

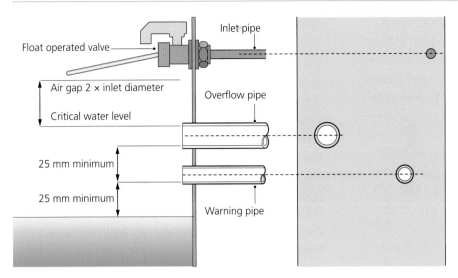

▲ Figure 1.13 Relative positions of inlet, warning pipe and overflow for type AG and AF air gaps

For all cisterns greater than 1000 litres, the invert of the overflow must not be less than 50 mm above the working level of the cistern.

Methods of filling large-scale cisterns

Section 7, Schedule 2, Paragraph 16 of the Water Supply (Water Fittings) Regulations states:

> Every pipe supplying water connected to a storage cistern shall be fitted with an effective adjustable valve capable of shutting off the inflow of water at a suitable level below the overflowing level of the cistern.

In domestic cistern installations up to 1000 litres, this is usually a float operated valve to **BS 1212**. In large-scale cisterns, however, other means of filling the cistern are available:

- float operated valves
- solenoid valves.

Float operated valves BS 1212

There are basically four types of float operated valve (FOV) that can be installed on large-scale cisterns. These are:

- **BS 1212** Parts 1 and 2 float operated valves
- equilibrium float operated valves
- pressure operated float valves
- Keraflo delayed action float valves.

We will look at the merits of each valve in turn.

BS 1212 Parts 1 and 2

These are the most common types of FOV. The main problem with this type of FOV is that they are very restrictive to water flow and incur a much greater pressure (head) loss than other types of FOVs, making cistern filling a long process. Wear on the washer and orifice can also be problematic when the valve is in constant use. They are, however, satisfactory when intermittent use is anticipated. These were looked at in detail in Book 1.

Equilibrium float operated valves

The **equilibrium** FOV offers a greater flow rate and lower pressure loss than the **BS 1212** type and are especially beneficial for large cisterns with a high-pressure inlet cold water supply.

> **KEY TERM**
>
> **Equilibrium**: in perfect balance. In this case, the pressure is balanced both sides of the valve.

Unlike other FOVs, the equilibrium type does not rely solely on the float to successfully close the valve. Instead, the closing operation is aided by the water pressure of the incoming mains cold water supply allowing a smaller float to be used.

As can be seen in Figure 1.14, the piston has a hole running through its length. This allows water to pass through to the back of the piston, which has the effect of pushing the piston towards valve shut off whilst the water at the front of piston tends to push it away, equalising the pressure both sides of the barrel. The float now only has to lift the arm to close the valve greatly reducing the effort required to stop the flow of water.

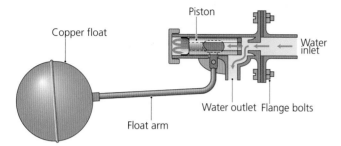

▲ Figure 1.14 An equilibrium float operated valve

Equilibrium valves are an advantage where the water pressure is high and water hammer may be a concern. Almost all FOVs above 54 mm are of this type.

Pressure operated valves

The pressure operated FOV uses the pressure of the cold water main to assist the valve closure by the use of a pilot valve controlled by the lever and the float. Often called the pilot operated valve, the advantage here is that the variations in water tank levels between fully open and fully closed are greatly reduced. Although the head loss is greater than with the equilibrium type, the pressure operated valve is particularly suited to large cisterns with a high-pressure supply.

How do pressure operated valves work?

At zero pressure, the valve is closed. As water enters the valve inlet and the pressure increases, the valve opens to allow water to flow to the cistern. When the water has reached its shut off level, the pilot valve, operated by the float and lever, closes. This causes the pressure within the diaphragm chamber to increase, thereby closing the water inlet and stopping the flow of water. As the water level drops, the float operated pilot valve opens, releasing the pressure of the diaphragm chamber. Water pressure then re-opens the inlet valve and the cistern fills again to its shut off level.

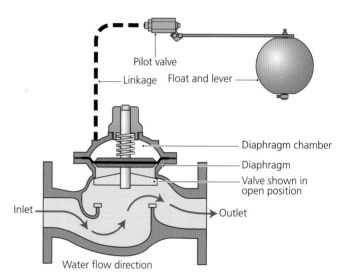

▲ Figure 1.15 The pressure operated valve

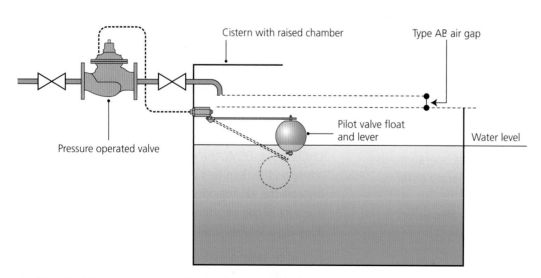

▲ Figure 1.16 The pressure operated valve cistern layout

Keraflo delayed action float valve

An alternative approach to cistern filling is the Keraflo delayed action float valve. This type of valve only opens when the water falls to a pre-set level in the cistern, opening fully to achieve a fast cistern fill. The benefits here are that not only does the cistern fill quickly but also that the velocity of the water entering the cistern means that the water will mix with the cistern contents preventing **stratification**.

The valve opens at a predetermined water level, opening and closing fully eliminating water hammer and unwanted system noise. An adjustable water level enables the levels to be set based upon water usage or, in the case of large domestic installations, occupancy levels.

The design of the valve means that boosting pumps are used less frequently. With a conventional FOV, pumps activating every few minutes wastes energy and increases pump wear. A delayed action float valve eliminates this by allowing the pumps to be activated only once every few hours when the water level has fallen sufficiently.

KEY TERM

Stratification: describes how the temperature of the water varies with its depth. The nearer the water is to the top of the cistern, the warmer it will be. The deeper the water, the colder it will be. This tends to occur in layers, whereby there is a marked temperature difference from one layer to the next. The result is that water quality can vary, the warmer water near the top being more susceptible to biological growth such as *Legionella pneumophila* (Legionnaires' disease).

Solenoid valves

A solenoid valve is an electromechanical valve that controls the flow of water into the cistern. The solenoid is an electromagnet that operates when an electrical current runs through the coil. When the coil is not energised, a spring keeps the valve shut.

Most solenoid valves used on large cisterns are of the servo-type (also called the pilot type solenoid valve). With this type, the electromagnet operates a plunger, which opens and closes a pilot orifice. The incoming water pressure, which is fed through the pilot orifice, opens the valve seal allowing water to flow through the valve. As the pilot valve closes, the pressure on the valve seal decreases and a spring closes the valve.

Although very rarely used with modern systems, a solenoid valve discharging through an open-ended pipe when used in conjunction with a float switch to activate the solenoid valve is an acceptable alternative method of filling large-scale cisterns (we will look at float switches a little later in the chapter). They are generally associated with boosted cold water systems.

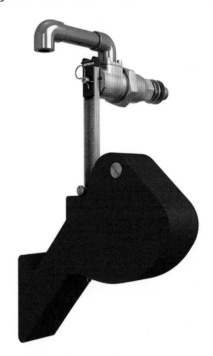

▲ Figure 1.17 The Keraflo 'Aylesbury' type delayed action float valve

▲ Figure 1.18 A servo-type solenoid valve

Multiple cistern installations

Where large quantities of water are required but space is limited, then cisterns can be interlinked, provided the cisterns are the same size and capacity. Problems can occur if the cisterns are not linked correctly, especially where the cisterns are to supply drinking water. Stagnation of the water in some parts of the cistern may cause the quality of the water to deteriorate. It should be remembered that the number of cisterns to be linked should be kept to a minimum.

Stagnation can be avoided by following some basic rules. Connection must be arranged to encourage the flow of water through each cistern. This can be achieved by:

● keeping the cistern volumes to a minimum to ensure rapid turnover of water to prevent stagnation
● connecting the cisterns in parallel wherever possible
● connecting the inlets and the outlets at opposite ends of the cistern
● using delayed action float operated valves to limit stratification.

Where it is not possible to connect cisterns in parallel, cisterns may be connected in series.

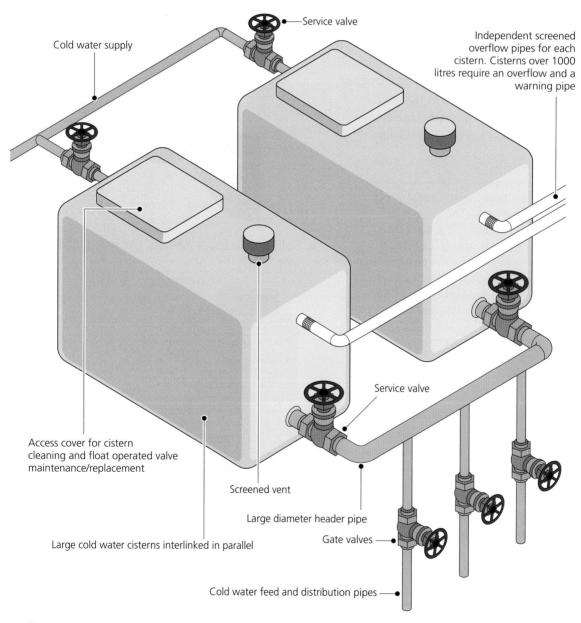

Service valve

Cold water supply

Independent screened overflow pipes for each cistern. Cisterns over 1000 litres require an overflow and a warning pipe

Service valve

Access cover for cistern cleaning and float operated valve maintenance/replacement

Screened vent

Large diameter header pipe

Large cold water cisterns interlinked in parallel

Gate valves

Cold water feed and distribution pipes

▲ Figure 1.19 Cisterns in parallel

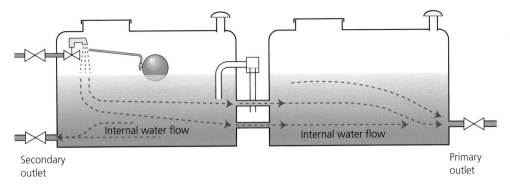

▲ Figure 1.20 Cisterns in series

In practice, cisterns in series should be **interconnected** to allow free movement of water from one cistern to the other. They should be connected at the bottom AND the middle so that water passes evenly through them. The primary outlet connection should be made on the opposite cistern to the FOV to encourage water movement with the secondary connection made on the cistern with the FOV installed. The overflow/warning pipe should be fitted on to the same cistern as the FOV. Both cisterns must be of the same size and capacity.

KEY TERM

Interconnected: connected together to form one cistern.

When connecting two or more cisterns together, care should be taken to ensure that the water movement is regular and even across all cisterns. In this situation, it is a good idea to install FOVs on **ALL** cisterns with appropriate service valves as detailed in the Water Regulations:

> Every float operated valve must have a service valve fitted as close as is reasonably practicable.

Wherever an FOV is fitted, then an overflow/warning pipe must accompany it. These should terminate in a conspicuous, visible position outside the building. On no account should they be coupled together.

There should be service/gate valves positioned to allow for isolation and maintenance of the cisterns without interrupting the supply. In Figure 1.21, you will see that any two of the four cisterns can be decommissioned, leaving two in operation. This ensures continuation of supply.

Every cylinder has a float operated valve to allow movement of water in every cylinder
Each FOV is fitted with a service valve as detailed in the Water Supply (Water Fittings) Regulations

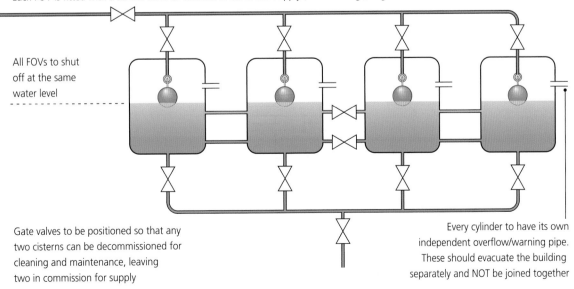

All FOVs to shut off at the same water level

Gate valves to be positioned so that any two cisterns can be decommissioned for cleaning and maintenance, leaving two in commission for supply

Every cylinder to have its own independent overflow/warning pipe. These should evacuate the building separately and NOT be joined together

▲ Figure 1.21 Installing three or more cisterns

Break cisterns

Break cisterns (often called break tanks) are used in large cold water installations in order to supply the system with water via a set of boosting pumps when the mains supply is insufficient. They provide a 'break' in the supply between the mains supply and the installation. This has several advantages over pumping direct from the mains supply:

- Using break cisterns ensures that there is no surge on the mains supply when the boosting pumps either start or stop.
- Break cisterns ensure that contamination of the mains cold water supply from multi-storey installations does not occur.
- Break cisterns ensure that there is sufficient supply for the installation requirements at peak demand.
- Break cisterns safeguard the water supply to other users by not drawing large amounts of water from the mains supply through the boosting pumps.

Break cisterns are often used in very tall buildings as intermediate cisterns on nominated service floors, thus dividing the system into a number of manageable pressure zones. The break cisterns provide water to both user outlets and other break cisterns higher up where the water is then boosted to other pressure zones further up the building.

As with all cistern installations, break cisterns must be fitted with an appropriate air gap that ensures zero backflow into any part of the system.

Components of systems in multi-storey dwellings

Float switches, transducers and temperature sensors

Float switches, transducers and temperature sensors play a vital part in modern boosted large-scale cold water systems. The problems encountered are not just those of how to install them but also where to install them. Installations of large cisterns are often undertaken in tight and restricted spaces. Difficulties arise in positioning these components whilst providing access for maintenance and inspection.

We will look at these important components.

Float switches

Float switches, often called level switches, provide detection of water levels within the cistern to activate various other pieces of remote equipment such as start/stop functions on boosting pumps, open/close functions on solenoid valves, water level alarms, and water level indicators.

There are many different types of float switches available and these can vary in design from simple magnetic toggle switches to ultrasonic and electronic types.

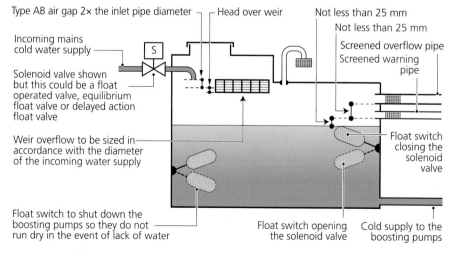

▲ Figure 1.22 The layout of a break cistern with a raised chamber

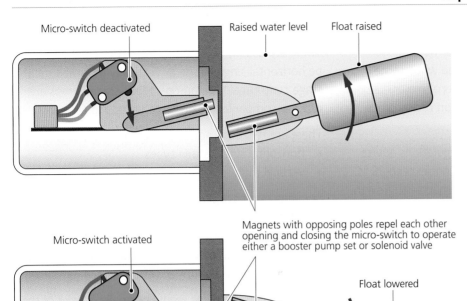

Micro-switch deactivated   Raised water level   Float raised

Micro-switch activated

Magnets with opposing poles repel each other opening and closing the micro-switch to operate either a booster pump set or solenoid valve

Float lowered

Lowered water level

▲ Figure 1.23 Magnetic toggle float switch – how it works

Popular types include:
- magnetic toggle – a simple float switch that uses the opposing forces of magnets to activate a micro-switch
- sealed float
- pressure activated diaphragm
- electronic
- ultrasonic.

Transducers

A transducer is an electronic sensor that converts a signal from one form to another. In large-scale, multi-storey water systems it senses system pressure variations and converts a pre-set low pressure into voltage to activate either the boosting pumps or the compressor feeding the pressure vessel to boost the pressure to normal operating pressure. They may also be used to sense over-pressurisation.

▲ Figure 1.24 Water pressure transducer with pressure gauge

Temperature sensors

Temperature sensors are often used to monitor the temperature of large volumes of stored wholesome, **potable** water where the installation is of major importance, such as in a hospital, prison or any place where there is a **duty of care**.

The Water Supply (Water Fittings) Regulations advise that stored wholesome water should not exceed 20 °C in order to minimise the risk of micro-bacterial growth.

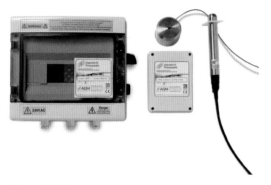

▲ Figure 1.25 Water temperature sensor

KEY TERMS

Potable: pronounced poe-table, from the French word 'potable' meaning drinkable.

Duty of care: in British law, this is a moral and legal obligation imposed on an organisation or an individual, which necessitates that a standard of reasonable care is adhered to. If an action does not meet the standard of care, the actions are considered to be negligent and damages may be claimed for in a court of law.

Boosting pumps

These are usually sited in the plant room and can either be horizontal single-stage or vertically mounted multi-stage centrifugal type pumps. Multi-stage pumps are often accompanied by pressure vessels to aid boosting.

There are two common types of boosting pump:
- horizontal single-stage types
- vertical multi-stage types.

Boosting pumps are available either as single components or as packaged units containing all the necessary equipment pre-fitted. The latter are the easiest to install and only require the final plumbing and electrical connections.

A typical pump package would normally consist of the following components:
- the pump
- a transducer to sense pressure and flow
- a control box to monitor pressure differentials and flow rate
- an accumulator to assist in providing sufficient system pressure for the installation
- a float switch to prevent the pumps running dry.

Water feed to property

Control box to monitor pressure and flow rate and to vary the pump speed if necessary

Pressure and flow rate LCD read out

Water supply in

Single-stage pump

▲ Figure 1.26 Components of a horizontal pump

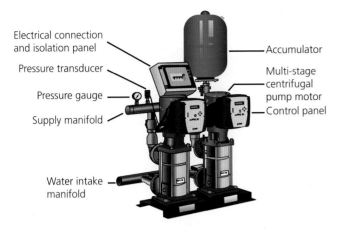

▲ Figure 1.27 Components of a vertical multi-stage pump set

The accumulator

The accumulator is a pressurised vessel that holds a small amount of water for distribution within the installation. They are designed to maintain mains operating pressure when the pump is not working and to reduce pump usage. Small accumulators can also be used to suppress water hammer.

Small domestic installations use bladder type accumulators. These consist of a synthetic rubber bladder or bag within a coated steel cylinder or vessel.

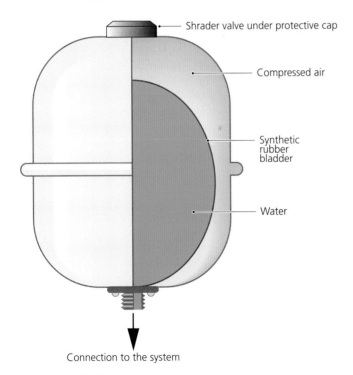

▲ Figure 1.28 A bladder type accumulator

Water for single occupancy dwellings

A private water supply (PWS) can be described as any water supply that is not provided by a water undertaker. The responsibility for the maintenance and repair of the supply lies solely with the person that owns or uses it. Private water supplies are regulated by the Private Water Supplies Regulations 2016.

> **INDUSTRY TIP**
>
> Access the Private Water Supplies Regulations 2016 at: www.legislation.gov.uk/uksi/2016/618/contents/made

A private water supply can serve a single occupancy dwelling providing less than 1000 litres (1 m³) per day or it may supply many properties or commercial/industrial buildings where the extraction rate may be in excess of 1,000,000 litres (1000 m³) per day.

Sources of water vary from boreholes, wells, springs and streams to rivers and lakes, and each one needs to be assessed for its water quality and suitability. The monitoring requirements of the Drinking Water Directive will vary according to the source and the size of the supply. In addition to this, the volume of water produced and the population it is serving should also be classified by the nature of the supply and whether the supply is:

- serving a single occupancy dwelling
- for domestic use for the persons living in the dwelling
- supplying water for commercial food production.

The issue of private water supply is an especially important part of a plumber's work as the Water Supply (Water Fittings) Regulations do not apply in this instance. The special regulations mentioned previously need to be followed with respect of cleansing, sterilising and testing the water to ensure that it is fit for human consumption.

The methods of providing private water supplies to single occupancy dwellings

There are several methods of water extraction that can be used to supply single occupancy dwellings. In this, the first of the assessment criteria for this section, we will investigate these methods and the equipment that is used. We will look at:

- water that is pumped from wells and boreholes
- water that is collected from surface water sources such as streams and springs
- the use of externally sited break cisterns.

Water that is pumped from wells and boreholes

Many small private drinking water supplies are extracted from boreholes and wells. Wells are usually large and circular, and not less than 1 m diameter, often dug by hand and occasionally by mechanical excavators. Boreholes, however, are smaller in diameter and are drilled by specialist companies using a variety of methods, including percussion and rotary drilling.

The quantity of available water will largely depend upon the type of **aquifer** that the borehole is to access. Obviously, the bigger the aquifer, the more water will be available for extraction. Estimated amounts can be calculated by test pumping after the borehole has been sunk. Perched aquifers are the most unreliable as these may well dry up after long periods without rain.

▲ Figure 1.29 Water pumped from boreholes

KEY TERM

Aquifer: a type of rock that holds water like a sponge. There are three basic types:
- confined aquifers – these are aquifers that have a confining layer between the water level and ground level. A confining layer is a layer of material that is impermeable with little or no porosity
- unconfined aquifers – these are aquifers that have no confining layers between the water level and ground level
- perched aquifers – these are aquifers that have a confining impermeable layer below the water bearing strata. They sit above the main water table.

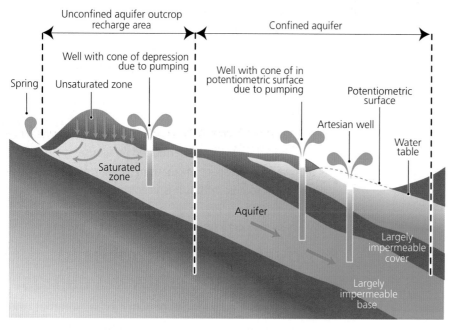

▲ Figure 1.30 Aquifers

Normally, a well-sunk and properly constructed borehole will comfortably be able to supply a single occupation domestic dwelling with water pumped directly from the aquifer, after appropriate treatment, by either:

- **a submersible pump** – this is a centrifugal pump that has a hermetically sealed motor and is, therefore, waterproof. The whole assembly is submerged in the water. The main advantage with this type of pump is that it prevents water cavitation. Submersible pumps are specifically designed to push the water from the well or borehole rather than pull water from it. They are more efficient than surface pumps, which have to pull the water upwards from the borehole.
- **a surface mounted pump** – usually sited in a pump house, these can either be horizontal single-stage or vertically mounted multi-stage centrifugal-type pumps. Multi-stage pumps are often accompanied by pressure vessels to aid boosting, especially where the water supplies are unreliable or inefficient.

▲ Figure 1.32 A typical surface pumping set using multi-stage centrifugal pumps and pressure vessels

Water taken from deep wells and boreholes may have travelled from catchments several kilometres away. If the water supply is extracted from a sand and gravel type aquifer, then the water will be very clean having gone through a thorough many stages of filtration through the sand/gravel geology. If the aquifer is predominantly limestone, the water will have travelled through fissures in the rock itself and will generally not be as clean as sand/gravel aquifers. Although ground waters such as aquifers are usually of good quality, some may contain high levels of iron and manganese and others may be contaminated with nitrates and other chemicals from farming and agricultural activities.

Boreholes extract the water from far deeper than any other source of private water supply. Boreholes, often 100 mm to 150 mm in diameter can be drilled as deep as 50 m.

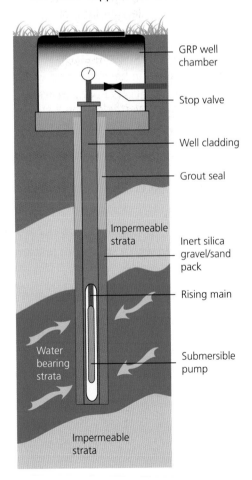

▲ Figure 1.31 A typical borehole installation

Water that is collected from surface water sources

There are several water sources that can be classified as surface water sources.

Rivers

Rivers offer greater, more reliable yields than boreholes, but can be susceptible to pollution and often show a variable quality of water. Pollution often depends on the catchment area and the activities in the general vicinity. Water that is derived from ground where there is little peat and agricultural activities is usually of good quality. Peaty ground tends to yield acidic water because of the concentration of CO_2, and this can lead to high concentrations of dissolved metals such as lead.

Lowland water is most likely of poor quality and may show seasonal variations in terms of quality and colour at different times of the year, with late autumn being the peak of colour changes. Microbiological contamination may also be high during periods of heavy rainfall.

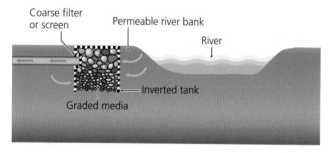

Coarse filter or screen

Permeable river bank

River

Inverted tank

Graded media

▲ Figure 1.33 River and stream water collection

Because of the potential problems associated with rivers, a surface water source should only be considered as a potential drinking water supply where no ground water source exists. The water will require a minimum of filtration and disinfection treatments. This must be designed for a worst-case scenario when the water is raw, especially after snow-melt and heavy rainfall. Installing a small tank or reservoir will help the settlement process and will do much to reduce the variations in water quality. The tank will require regular inspections and cleaning because of the solid matter within the water.

Streams

Small streams often show a variable quality to the water because of animal and human activity within the streams' catchment area and may show varying colour changes due to the levels of humic and fulvic acids that are used in agriculture as soil supplements.

Pollution and natural variations in the quality of the water are the most common problems that can occur with both stream and river water sources and these need to be considered carefully when siting the water supply intake point. Water can be pumped directly from the stream or collected from the ground in the immediate vicinity of the stream or riverbank. This is desirable in certain situations where the geology allows a natural filtration process, the water therefore being cleaner than directly from the river itself. The intake should not be positioned in an area where water turbulence may be created, especially during periods of heavy rainfall (for example, on the bend of a river or at sudden changes in water level). Water intakes must be protected by a strainer to prevent the ingress of fish, vermin and debris and the inlet pipe must feed a settlement tank that allows particulate matter within the water to settle.

The outlet of the tank should be situated above the floor of the tank and be fitted with a strainer to prevent sediment contamination.

The tank must be constructed of a material that will not contaminate or impair the quality of the water and must be designed in such a way as to prevent the ingress of vermin and debris.

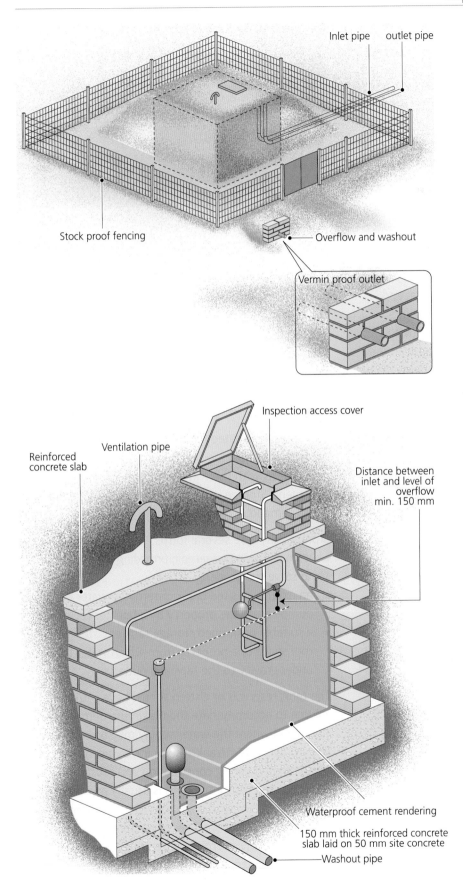

Inlet pipe   outlet pipe

Stock proof fencing

Overflow and washout

Vermin proof outlet

Inspection access cover

Ventilation pipe

Reinforced concrete slab

Distance between inlet and level of overflow min. 150 mm

Waterproof cement rendering

150 mm thick reinforced concrete slab laid on 50 mm site concrete

Washout pipe

▲ Figure 1.34 Concrete reservoirs and tanks

Springs

Where the **water table** and the surface coincide, then a spring is formed. The presence of fissures in the Earth's surface usually dictates where natural springs occur. The most reliable of these are from deep aquifers, whereas those aquifers nearer the surface are susceptible to drying up after a short period without rain, especially if the water is flowing from fissured limestone or granite.

KEY TERM

Water table: the point where the earth below ground becomes saturated with water causing water to pool.

Spring water is usually of good microbiological and chemical quality although, again, shallow aquifers may suffer from variations in water quality due to surface contamination. The probability of agricultural contamination must be carefully considered especially when the aquifer evacuates the surface. It must also be remembered that some shallow 'springs' may actually be land surface drains. Here, the water quality is likely to be unacceptable.

The treatment of spring water is usually much simpler than river or stream water because there is much less suspended solid matter.

Spring water must be protected from surface contamination once it reaches ground level. It is necessary to consider leakage from septic tanks and agricultural activity. A small chamber built over the spring will protect the source from surface contamination and also serve as a collection source and header tank. It should be constructed so the water enters from the base or the side. The chamber top must be above ground level and be fitted with a lockable, watertight access cover. An overflow must be provided and sized to accommodate the maximum flow of the spring. The outlet should be fitted with a strainer and positioned above the floor of the chamber.

The chamber must be constructed of a material that will not contaminate or impair the quality of the water and must be designed in such a way as to prevent the ingress of vermin and debris.

The land in the immediate vicinity of the chamber must also be fenced off and small ditch dug upslope to intercept and divert surface water run-off.

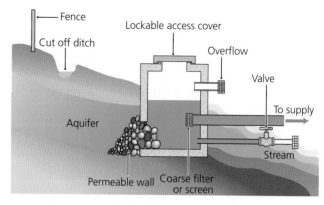

▲ Figure 1.35 Spring water collection

KEY POINT

Remember that all water sources should be protected from the ingress of vermin and surface-water contamination.

Storing treated water

Plumbing systems in domestic dwellings that use water from private water supplies need to include a method of storing the treated water to provide a water reserve in the event of planned or emergency maintenance, or problems with the water source or the water treatment. Storage will also cater for fluctuations in water demand.

Storage may take several forms:
- a small covered reservoir, providing sufficient head to serve the property
- an externally sited break cistern, the water being pumped into the property by a booster set
- a suitably positioned storage tank or cistern in the roof space of a property from which water flows under gravity to the taps and outlets.

The tank, reservoir or cistern should have sufficient volume to accommodate the maximum demand and any period where the water supply may be interrupted. The storage cistern/tank/reservoir may possibly be contaminated during construction and must, therefore be disinfected before use. This is usually achieved by filling the system with a solution of water/chlorine at 20 mg/l of chlorine and leaving to stand for several hours, preferably overnight, after which the chlorine should be drained and the system thoroughly flushed using treated water.

External break cisterns and cisterns in the roof space must be protected against contamination and insulated against freezing in cold weather and undue warming in warm weather. Cisterns and tanks should also be fitted with a lockable, well-fitting but not airtight lid to prevent the ingress of insects and vermin, and overflows and warning pipes must be protected by a mesh screen.

Storage tanks must be inspected every six months and cleaned, if necessary, to prevent the build-up of silt and debris. This should be followed by disinfection.

Water treatment processes for private water supplies

Larger water supplies served from a private water source (for example, those that serve many properties or commercial/industrial establishments) are often treated by 'point of entry' treatment methods. These are very similar to those that are used by the local water undertaker, and were discussed in detail in Book 1. It should be remembered that:

- the design of the treatment process should be based on a full investigation of site conditions

- the chemical and microbiological content of the water must be established and tests performed to determine the effectiveness of any treatment process and the chemical dosing requirements.

For small supplies, to a single dwelling, for instance, the treatment is often precautionary and should include disinfection. The disinfection stage should only be discarded if it can be shown without reasonable doubt by risk assessment and frequent testing that the water supply is likely to be consistently **pathogen**-free.

KEY TERM

Pathogen: a germ or bacteria.

In this section of the chapter we will look at the different methods of water treatment that are often used with private water supplies. Table 1.1 shows the methods of water treatment available and their effectiveness. We will look at the findings of this table method by method.

▼ Table 1.1 Water treatment methods

	Bacteria	Viruses	Algae	Coarse particles	Turbidity	Colour	Aluminium	Ammonia	Arsenic	Iron and manganese	Nitrate	Pesticides	Solvents	Taste and colour
Coagulation and flocculation[1]	+	+	+	++	++	++	++		+	++				
Sedimentation				++	+		+			+				
Gravel filter/screen			+	++	+		+			+				
Rapid sand filtration	+	+	+	++	+		+			+				
Slow sand filtration	++	++	++	++	++		+			+				
Chlorination	++	++	+			+		++						
Ozonation	++	++	++			+						++		++
Ultraviolet (UV)	++	++	+											
Activated carbon						+						+	+	++
Activated alumina									++					
Ceramic filter	++		++	++	++									
Ion exchange								+	+		++			
Membranes	++	++	++	++	++	++	++		+	++	++	++		++

+ partly effective

++ preferred technique/effective

1 pre-oxidation may be required for effective removal of aluminium, arsenic, iron and manganese

Without exception, all sources of water will need treatment before it is acceptable for human consumption. The health risks presented by poor quality water can be due to microbiological or chemical contaminants. Microbiological contamination is the most important issue as this can lead to infectious diseases such as Legionnaires' disease and cholera. Chemical contamination often leads to more long-term health risks. Substances that affect the appearance, odour or taste often make the water unpalatable to the consumers. Particulates in the water may also present a health risk as these could also be contaminated with microbiological organisms. In these circumstances, disinfection becomes more difficult. Final disinfection must always be included in any treatment system to effectively kill off any remaining micro-organisms. Disinfection solutions containing chloride provide a residual that will act to preserve the quality of the water during storage and distribution (in larger systems).

Treatment of the water is based upon the physical removal of contaminants through:

- filtration
- settling, often assisted by the addition of chemicals, or
- the biological removal of micro-organisms.

This usually consists of a number of key stages:

- initial pre-treatment by settling
- pre-filtration through a coarse medium
- sand filtration
- disinfection or chlorination.

This process is known more commonly as the 'multiple barrier principle' and is designed to provide an effective water treatment by not relying on a single, less effective process or the failure of one stage in the process. For example, if a system consists of coagulation/flocculation, sedimentation, sand filtration and finally chlorination, failure of say, the rapid sand filter does not mean that untreated water will be supplied to the property. Other processes will remove the majority of the suspended particles and, therefore, many of the microbiological contaminants. Disinfection will remove those remaining. Provided the sand filter is repaired quickly, there will be little by the way of deterioration in water quality.

▼ Table 1.2 The Private Water Supplies Regulations 2016 – water quality parameters

Parameter	Unit of measurement	Concentration or value (maximum unless otherwise stated)
Colour	mg/1 Pt/Co	20
Turbidity	FTU	4
Odour (including hydrogen sulphide)	Dilution no.	3 at 25 °C
Taste	Dilution no.	3 at 25 °C
Temperature	°C	25
Hydrogen ion	pH value	9.5 (min. 5.5)
Sulphate	mg SO_4/1	250
Magnesium	mg Mg/1	50
Sodium	mg Na/1	150
Potassium	mg K/1	12
Nitrite	mg NO_2/1	0.1
Nitrate	mg NO_3/1	50
Ammonia	mg NH_4/1	0.5
Silver	µg Ag/1	10
Fluoride	µg F/1	1500
Aluminium	µg Al/1	200
Iron	µg Fe/1	200
Copper	µg Cu/1	3000
Manganese	µg Mn/1	50
Zinc	µg Zn/1	500
Phosphorus	µg P/1	2200
Arsenic	µg As/1	50
Cadmium	µg Cd/1	5
Cyanide	µg CN/1	50
Chromium	µg Cr/1	50
Mercury	µg Hg/1	1
Nickel	µg Ni/1	50
Lead	µg Pb/1	50
Pesticides	µg/1	0.1
Conductivity	µS/cm	1500 at 20 °C
Chloride	mg Cl/1	400
Calcium	mg Ca/1	250
Total hardness	mg Ca/1	min. 60
Alkalinity	mg HCO_3/1	min. 30
Total coliforms	number/100 ml	0
Faecal coliforms	number/100 ml	0
Faecal streptococci	number/100 ml	0

Coagulation and flocculation

These processes are used to remove colour, **turbidity**, algae and other micro-organisms from surface water. It involves the addition of a chemical coagulant, which encourages the formation of a 'precipitate' or floc. This entraps the impurities. In certain conditions, iron and aluminium can also be removed in this way. The floc is then removed from the water by sedimentation and filtration.

KEY TERM

Turbidity: refers to how clear or cloudy the water is, due to the amount of total suspended solids it contains. The greater the amount of total suspended solids (TSS) in the water, the cloudier it will appear. Cloudy water can therefore be said to be turbid.

Sedimentation

Sedimentation tanks are designed to slow down the water velocity to allow the solids that the water contains to sink to the bottom and settle under gravity. Simple sedimentation may also be used to reduce turbidity.

Sedimentation tanks are usually rectangular in shape with a length-to-width ratio of 2:1 and are usually 1.5 to 2 m deep. The inlet and outlets must be on opposite sides of the tank and the inlet designed to distribute the incoming flow as evenly across the tank as possible. The outlet should be designed to collect the cleared water across the entire width of the tank. The tank will also require covering to prevent external contamination.

INDUSTRY TIP

Sedimentation tanks require cleaning when their performance begins to deteriorate, and a 12-month period between cleaning operations would normally be sufficient.

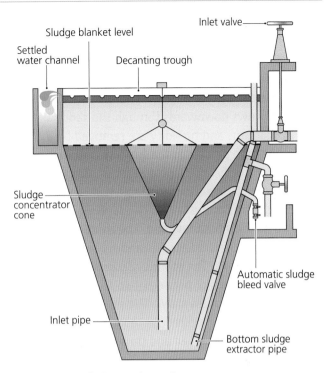

▲ Figure 1.36 Sedimentation tank

Filtration

Filtration is an important process that removes turbidity and algae from the raw, untreated water. There are many different types, including screens, gravel filters, slow sand filters, rapid sand filters and pressure filters. We will concentrate on five of these:

- slow sand filters
- rapid sand filters
- pressure filters
- absorption filters
- reverse osmosis.

Note that the difference between slow sand filters and rapid sand filters is not just a matter of the speed of the filtration process, but the underlying principle of the method. Slow sand filtration is a biological process and rapid sand filtration is a physical treatment process.

Slow sand filters

Slow sand filters are often preceded by micro-straining or coarse filtration; these filters are used primarily to remove micro-organisms, algae and turbidity. It is a slow but very reliable method of water treatment often suited to small supplies, provided that there is sufficient area to properly construct the filtration tanks.

Slow sand filters consist of tanks containing sand with a size range of 0.15 to 0.30 mm and to a depth of around 0.5 to 1.5 m. For single dwellings, circular modular units, usually used in tandem, are available. These have a diameter of around 1.25 m. As the raw water flows downwards through the sand, micro-organisms and turbidity are removed by a simple filtration process in the top few centimetres of sand. Eventually, a biological layer of sludge develops, which is extremely effective at removing micro-organisms in the water. This layer of sludge is known as the **schmutzdecke**. The treated water is then collected in underdrains and pipework at the bottom of the tank. The schmutzdecke will require removing at periods of between two to ten weeks as the filtration process slows. The use of tandem filters means that one filter can remain in service whilst the other is cleaned and time allowed for the schmutzdecke to re-establish.

Slow sand filters should be sized to deliver between 0.1 and 0.3 m³ of water for every 1 m² of filter per hour.

KEY TERM

Schmutzdecke: a layer of mud that is saturated with friendly, water-cleansing bacteria.

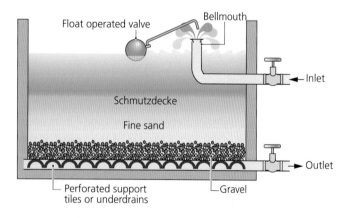

▲ Figure 1.37 Slow sand filter

Rapid (gravity) sand filters

Rapid sand filters are predominantly used to remove the floc from coagulated water, but they can be successfully used to remove algae, iron, manganese and water turbidity from raw water. Activated carbon in granular form is used to remove any organic compounds. Some filters also incorporate an alkaline medium to increase the pH value of acidic water.

Rapid sand filters are usually constructed from rectangular tanks containing coarse silica sand with a size range of 0.5 to 1 mm, laid to a depth of between 0.6 and 1 m. As the water flows downwards through the filter, the solids remain in the upper part of the sand bed where they become concentrated. The treated water collects at the bottom of the filter and flows through nozzles in the floor. The accumulated solids are removed either manually every 24 hours or automatically when the headloss reaches a predetermined level. This is achieved by backwashing.

A variety of proprietary units are available containing filtering media of different types and sizes. In some filters, the water flows upwards, improving the efficiency.

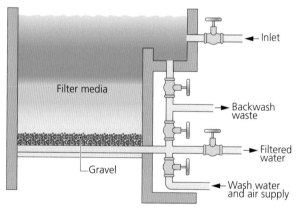

▲ Figure 1.38 Rapid (gravity) sand filter

Pressure filters

Pressure filters are sometimes used where it is important to maintain a head of pressure to remove the need to pump the water into the supply. The filter bed is enclosed in a cylindrical pressure vessel. Some small pressure filters are capable of delivering as much as 15 m³/h. The cylinder is typically made of specially coated steel, and smaller units can be manufactured from glass-reinforced plastic. They operate in a similar way to the rapid sand filter.

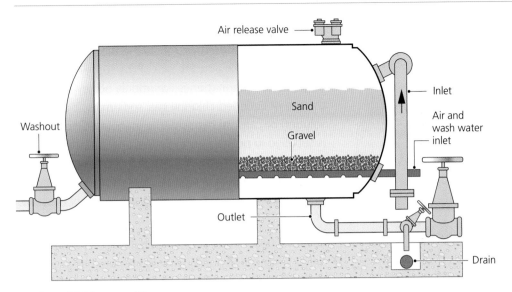

▲ Figure 1.39 Pressure filter

Absorption filters

Absorption filters fall into two distinct categories:

● **Activated carbon** – these filters remove contaminants by the process of physical absorption. Granular activated carbon (GAC) is the most commonly used medium, although block carbon and powdered carbon can also be used. Most filters use replaceable cartridges that can easily be changed when the old one is exhausted. Activated carbon filters will remove suspended solids, chlorine and some organic contaminants, including pesticides. They will also remove some humic acids, which are responsible for giving water derived from peat its brown appearance. Unfortunately, activated carbon provides a good medium for the development of micro-organisms and there is some concern that this can cause health problems should the bacteria be reintroduced into the water. Water inhaled in the form of an aerosol during activities such washing is also a concern. Because activated carbon removes chlorine, bacterial growth has also been found on filters treating chlorinated water. Some manufacturers state that activated carbon should not be used if the water is of unknown quality or contains microbiological organisms.

● **Activated alumina** – these filters can be used where the water contains contaminants, such as arsenic, and other chemicals, such as fluoride. They are manufactured using aluminium hydroxide.

Reverse osmosis

Large **reverse osmosis** (RO) units have been used for many years for producing good drinking (potable) water from low-quality water. They can also be used to produce drinking water from saline (salt) water where the supplies of fresh water are inadequate. RO will successfully remove a wide range of organic and chemical contaminants, such as sodium, calcium, nitrates, fluoride, pesticides and solvents.

KEY TERM

Reverse osmosis: a method of purifying water.

Reverse osmosis units work by forcing the water under pressure through a semi-permeable membrane. The membrane is usually manufactured from polyamide. This material is preferred, as some membranes, such as cellulose, actively support the growth of bacteria. The filters do not need replacing, but the membrane may require periodic chemical cleaning and de-scaling.

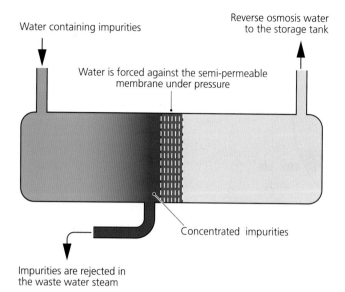

Water containing impurities

Reverse osmosis water to the storage tank

Water is forced against the semi-permeable membrane under pressure

Concentrated impurities

Impurities are rejected in the waste water steam

▲ Figure 1.40 The principle of reverse osmosis

The flow rate through the units tends to be low in domestic RO units, with the treated water being collected in a storage tank to cushion supply and demand. A level sensor within the storage tank itself usually controls the RO unit. One disadvantage with reverse osmosis is that it produces a lot of wasted water. For every litre of fresh drinking water, three litres of wasted water are produced. An alternative would be to use the wasted water for non-potable supplies such as WC flushing.

The water treated by reverse osmosis tends to be soft. For this reason, some units contain a re-hardener to increase both hardness and alkalinity, thus making it less aggressive on metals such as lead.

Disinfection

The greatest danger to water for human consumption (potable water) is from contamination by human and animal faeces. These contain the organisms of many communicable diseases, and so the use of disinfection to kill or inactivate their pathogenic organisms is vital.

There are several disinfection methods available for treating water from a private source. Disinfection by the use of chlorine is the most widely used for large water users, but is much less common for smaller, single-dwelling supplies. Smaller water supplies commonly use ultraviolet irradiation and ozonation.

Here, we will investigate these three very different methods of disinfection.

Chlorination

There are several methods of chlorination that can be used, depending on the size of the supply being disinfected. Chlorination is commonly achieved by the use of liquefied chlorine gas, sodium hypochlorite solution or calcium hypochlorite granules. Chlorine gas is supplied in pressurised cylinders and is extremely dangerous, and so great care must be taken of how it is stored and handled. The gas is taken from the cylinder by a chlorinator, which measures and controls the flow of gas.

Sodium hypochlorite solution is delivered in drums, and only a one-month supply should be stored at any one time. Exposure to sunlight can result in a decrease in the disinfection capabilities and results in the loss of available chlorine and an increase in the concentration of chlorate relative to chlorine. There are many different methods of dosing the water with sodium hypochlorite. A simple gravity-fed system where the solution is drip-fed at a constant rate has proven very successful, provided that the rate of flow needed and the water quality and composition remain constant.

Calcium hypochlorite can be supplied in a powder form, tablets or granules. Calcium hypochlorite is very stable provided it is kept dry, and several months' supply may be stored.

Dosing via a calcium hypochlorite doser is very simple and involves dissolving a measured amount of calcium hypochlorite in a measured volume of water, which is then introduced into the supply. Tablet form is preferred, as the rate of dissolve is very predictable.

With all systems, the resultant free residual chlorine after disinfection should remain within the range of 0.2 to 0.5 mg/l, and a contact time of 30 minutes is recommended. The design of the disinfection system is most important and must evenly distribute the chlorine within the water supply. It must not allow chlorine concentrations to build up in dead zones.

HEALTH AND SAFETY
Care should be taken to avoid contact between calcium hypochlorite and moisture, as this then reacts to form chlorine gas.

Ozonation

This system uses ozone gas to disinfect the water supply. Ozone is a powerful disinfectant and oxidant that completely kills bacteria and viruses. It cannot, however, be relied upon solely as the means of disinfection where the water contains **cryptosporidium**. Ozone may also help to reduce the levels of taste, colour and odour within the water.

Ozone is a gas that is produced by discharging alternating current through dry air. Small proprietary units use a 230 V single-phase AC current, but larger units run on 400 V three-phase supplies. The ozone containing air is then mixed with water in the contact column. To be completely effective, the column should give at least four minutes contact time, giving a residual ozone content of 0.4 mg/l. Unless the water is used quickly, the ozone will decompose rapidly and so it is recommended that the disinfectant process be reinforced by using a small amount of chlorine.

KEY TERM

Cryptosporidium: a microscopic protozoan parasite that affects humans and cattle and presents itself as severe diarrhoea. It usually affects children between the ages of 1 and 5, but it can affect anyone and the symptoms can be very severe in people with low immune systems.

Small-scale ozone units are available for single-dwelling private water supplies, but they are not widely used. This is generally because of the high power usage and complexity of the equipment.

Ultraviolet irradiation

This is the preferred method of disinfection of private water supplies for small single domestic dwellings. They use low-frequency UV light to change the cellular structure of microbiological organisms, effectively destroying them.

UV disinfection is affected by the water quality, and its flow rate and the water must be of good quality with low turbidity and colour. Pre-filtration is necessary with this method.

Special low mercury (Hg) lamps are used to generate the ultraviolet radiation in an enclosed chamber, usually manufactured from stainless steel. The lamps look very similar to fluorescent tubes, but are made from UV transparent quartz instead of phosphor-coated glass. The lamps generate UV radiation at a wavelength of 254 nanometres (nm). The optimum germicidal wavelength is between 250 and 265 nm. The temperature of the lamp is around 40°C. The lamp is separated from the water by a sleeve to prevent it being cooled by the water. The UV effects of the lamps deteriorate with age, and after 10 to 12 months the efficiency is down to 70 per cent of that of a new lamp. Therefore, lamp replacement is recommended every 12 months.

As with all systems, disinfection will only be effective provided a sufficient dose of UV is applied, and a dosage of 16 to 40 mWs/cm² (milliwatt seconds per cm²) is recommended, but this depends on factors such as the ability of the micro-organisms to withstand UV light.

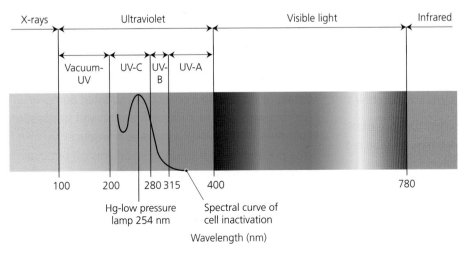

▲ Figure 1.41 The UV spectrum

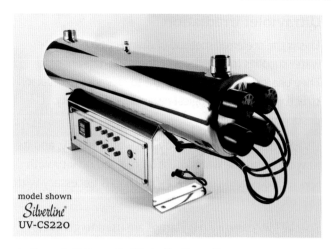

▲ Figure 1.42 Domestic UV sterilisation unit

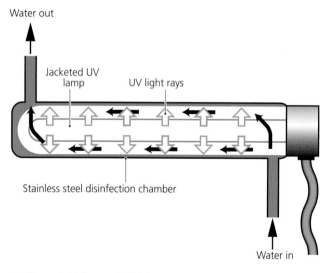

▲ Figure 1.43 Domestic UV chamber

The system layout features of cold water systems fed from private water supplies

The United Kingdom has over 50,000 people whose only source of potable drinking water is from a private supply. These, as we have already seen, fall into four main categories of wells, springs, streams and boreholes.

Occasionally, properties may have access to more than one supply and in some cases a licence may be required to extract the water from a given source.

The delivery of feed waters into a domestic property can be one of two methods. These are:

- pumped supply direct from the borehole or well, or
- gravity supply from a catchment tank in a spring or stream.

There are no hard and fast rules as to which method is the best to use in any given situation. If the water is being delivered from a borehole then, obviously, a pumped method of supply is the one that will be used because of the location of the supply. A gravity supply may be used where the water source is higher than the property, the water flowing by gravity from the catchment tank in the water source (for example, a spring) to either an external storage/break cistern or tank before being pumped into the property or direct to a storage cistern located within the property, with the water being distributed to all outlets from the storage cistern by gravity supply.

Pumped supplies

There are two methods of pumped supply from a well or a borehole:

- pumped supply with pressure control
- pumped supply with level control.

We will look at each of these methods separately.

Pumped supply with pressure control

This type of system provides directly drawn water at the point of use. Pressure is maintained within the system by the use of an accumulator (often called a pressure vessel) and a pump. The accumulator is a vessel that contains air under pressure and water. The water is contained within a neoprene rubber bag inside the accumulator, which expands when water is pumped into it under pressure. The air is then compressed and the pressure rises. As the water within the accumulator is used, the pressure will drop. At a predetermined pressure, the pump will start and the accumulator is refilled, raising the pressure to its operating pressure. These systems generally operate at 1.5 to 3 bar. This system is preferred when water treatment is being considered.

Control of the system is automatic. The system contains a submersible or surface mounted pump to bring the water to the surface, filtration and sterilisation equipment (usually UV), a pressure transducer to sense pressure drop across the installation, a pressure gauge and an accumulator. The kitchen sink is usually installed with water under pressure directly from the accumulator. All other outlets are supplied from a low-pressure supply from a storage cistern situated in the roof space. A non-return or check valve must be fitted upstream of the accumulator.

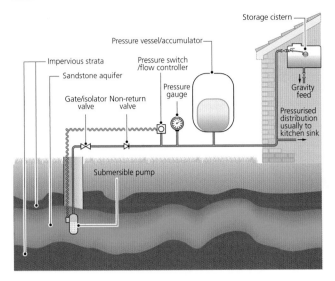

▲ Figure 1.44 A typical borehole installation with pressure control

Pumped supply with level control

This system uses a float switch to monitor the level of the water in a storage cistern. The storage cistern is normally situated in the roof space of a dwelling. The float switch operates a surface mounted pump, which fills the tank until the level of the float switch is reached. All water for the dwelling passes through the storage cistern and this supplies all outlets with a low-pressure supply. Water fed directly from the borehole to a kitchen sink under pressure is not possible with this installation.

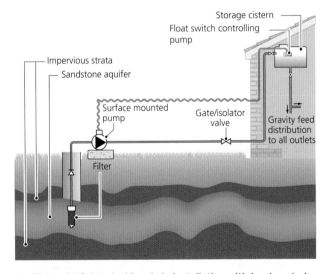

▲ Figure 1.45 A typical borehole installation with level control

Because all of the water for the dwelling is supplied at low pressure, this system can also be used with supplies that are fed via a catchment tank in a stream or spring via an external break/storage cistern. It is also possible

to use water directly from a catchment tank without the use of a pump provided that the source of water is higher than the dwelling. It must be remembered, however, that some form of filtration and sterilisation of the water is necessary. A non-return or check valve must be fitted upstream of the pump.

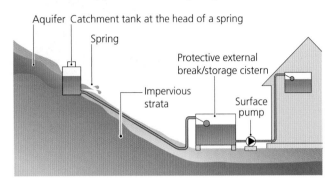

▲ Figure 1.46 A typical spring catchment tank installation with level control

The components used in boosted (pumped) cold water supply systems from private sources for single occupancy dwellings

In this, the final part of the chapter dealing with private water supplies, we will investigate the components used with private water supplies to single domestic dwellings:

- small booster pump sets which incorporate all controls and components
- boosted system with separate controls and components
- use of accumulators in increasing system flow rate.

Vertical, horizontal and submersible pumps

As we have already seen earlier in the chapter, there are two different types of pump that can be used with private water supplies and, more specifically, boreholes and springs:

- submersible pumps
- surface pumps:
  - horizontal single-stage types
  - vertical multi-stage types.

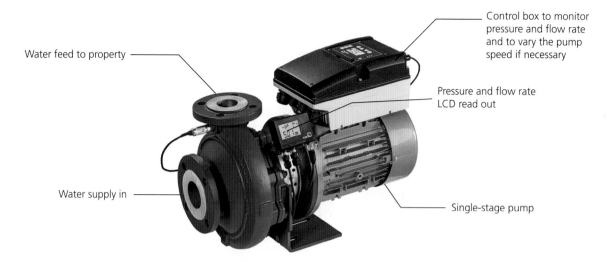

Water feed to property

Control box to monitor pressure and flow rate and to vary the pump speed if necessary

Pressure and flow rate LCD read out

Water supply in

Single-stage pump

▲ Figure 1.47 Components of a horizontal pump

Surface pumps for private water supplies are available either as single components or as packaged units containing all the necessary equipment pre-fitted. The latter are the easiest to install and only require the final plumbing and electrical connections.

Submersible pumps may be purchased as separate components or in 'pack form' whereby all the separately matched equipment is supplied ready to assemble.

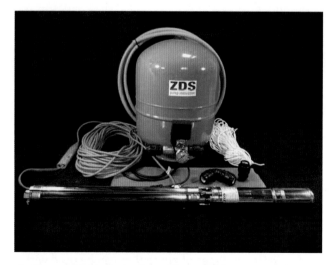

▲ Figure 1.48 A typical submersible pump kit

A typical pump package would normally consist of the following components:
- the pump
- a transducer to sense pressure and flow
- a control box to monitor pressure differentials and flow rate

- an accumulator to assist in providing sufficient system pressure for the installation
- a float switch to prevent the pumps running dry.

Electrical connection and isolation panel

Pressure transducer

Pressure gauge

Supply manifold

Water intake manifold

Accumulator

Multi-stage centrifugal pump motor

Control panel

▲ Figure 1.49 Components of a vertical multi-stage pump set

The accumulator

The operation of an accumulator can be broken down into three stages:

1 When the pump operates it forces water into the accumulator bladder compressing the air surrounding it to a pressure greater than the vessel's pre-charge pressure. This is the source of the stored energy.

2 When the bladder expands due to water being forced in by the pump, it deforms in shape and the pressure within the accumulator increases. Bladder deformation stops when the water and the now-compressed air charge become balanced.

3 When a tap is opened, the pressure within the system drops and the compressed air forces the water out of the accumulator. When all of the water inside the accumulator is used and the pressure falls to a predetermined level, the pump energises to recharge the accumulator water storage and pressure and the cycle begins again.

Probably the most important consideration when applying an accumulator is calculating the correct pre-charge pressure. There are three points to be considered:

- the type of accumulator being used
- the work to be done, and
- the system operating limits.

The pre-charge pressure is usually 80–90 per cent of the minimum system cut-in pressure (the pressure at which the pump energises) to allow a small amount of water to remain in the vessel at all times. This prevents the bladder from collapsing totally.

IMPROVE YOUR MATHS

To calculate the pre-charge pressure, follow this simple procedure.

If the minimum working pressure of a cold water system is 2 bar, then:

$2 \times 0.9\ (90\%) = 1.8$ bar

Pre-charge pressure $= 1.8$ bar

The accumulator air charge must be lower than the mains pressure for water to enter the vessel, and on average, a pressure differential of around 1.5 bar lower than the supply pressure would be acceptable (but no more than 2 bar and no less than 0.8 bar). This means that if the supply pressure is 3.5 bar, then the air charge within the accumulator must be around 2 bar, a supply pressure of 4.5 bar would require a 3 bar air charge, and so on. Air pressure can be checked and topped up as necessary at the Schrader valve situated at the top of the accumulator.

Specialist components for cold water systems

Our work as plumbers covers a multitude of various installations, systems and components. Occasionally, we may be asked to install specialist components that we may only come into contact with on a limited number of occasions. Even so, it is important that we become familiar with these 'specialist' components to ensure that we position and install them correctly and according to the manufacturer's instructions and in line with any regulations or recommendations.

In this part of the chapter, we will look at a selection of components that may be unfamiliar to you. We will investigate how they operate and the best ways of installing them in accordance with the recommendations in place.

Infrared operated taps

Low voltage (6 V DC current) infrared operated outlets are becoming popular for use in public conveniences, hotels and public buildings. They use infrared sensors to operate solenoid valves. The solenoid valves open for a defined length of time to allow a certain quantity of water to flow through them. They are frequently used to flush WCs and urinals, and to operate taps and shower fitments. The use of infrared operated outlets has several advantages over standard taps and outlets:

- They are easy to operate.
- They stop the spread of germs and bacteria.
- They can help with water conservation.
- They can prevent scalding injury.

How do they work?

The method of operation of infrared operated outlets is quite simple:

1 The infrared sensor eye emits an infrared beam that is approximately 200 to 260 mm wide. When an object, such as a hand, is within range of the infrared sensing zone, the infrared beam is interrupted and a signal wire transfers an electronic signal to a solenoid valve.
2 The solenoid valve acts as a latching mechanism that allows a restricted flow of water to flow through it. As soon as the valve receives the electronic signal, it snaps open, allowing the water to flow to the outlet.
3 When the object leaves the sensing zone, the infrared beam returns to normal, the electronic signal ceases and the valve closes.

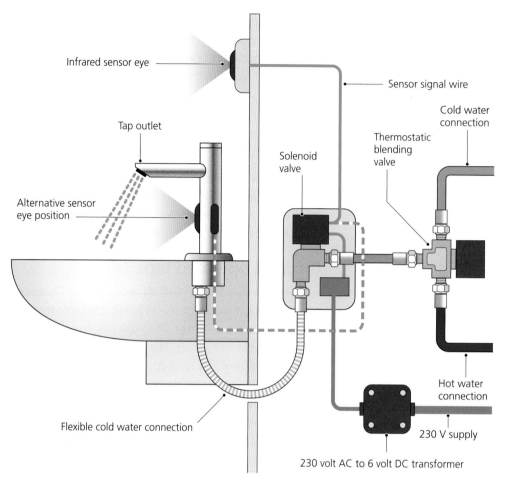

▲ Figure 1.50 The operation of an infrared tap

Non-concussive (self-closing) taps

Non-concussive taps are self-closing taps that are usually used in public washrooms, hotels and places of work where there is a risk that the tap may either be left open or there is a high risk of vandalism. They are operated by pressing the tap head downwards, which operates a spring-loaded plunger to open the tap. After a period of time, the spring then lifts the plunger to close the tap. The time the tap is open can be adjusted up to a maximum of about 20 seconds of water flow.

Most non-concussive taps use an internal cartridge system so repairs and maintenance are fairly simple. When the tap requires maintenance, the cartridge can be replaced easily by removing the tap head and withdrawing the cartridge.

Non-concussive taps are available as single taps or as thermostatic mono-bloc mixer taps.

▲ Figure 1.51 Non-concussive type thermostatic mixer tap

▼ Table 1.3 The advantages and disadvantages of non-concussive taps

Advantages	Disadvantages
They are self-closing so water is not wasted due to the tap being left open.	The rapid closing of the tap can cause water hammer and pipework reverberation.
Most models are vandal-proof.	They require regular maintenance.
They can help in saving money on water costs if a water meter is fitted.	They should not be used where there is a risk of fouling by grease or dirt.
	The tap may block with scale deposits in hard water areas.
	Because of the restricted amount of water released when the tap is used, waste pipes may not reach a self-cleansing velocity and may block with residue.

Combination bath tap and shower head

This type of tap is more commonly known as a bath/shower mixer tap. There are many different styles including:

- pillar type – a traditional style predominantly used in period bathrooms
- deck mixer type – there are many types to suit all types of modern bathroom styles
- thermostatic type – has the benefit of thermostatic control by the inclusion of a temperature sensitive wax cartridge
- wall mounted – specifically designed for baths without tap holes.

▲ Figure 1.53 Deck type bath/shower mixer tap

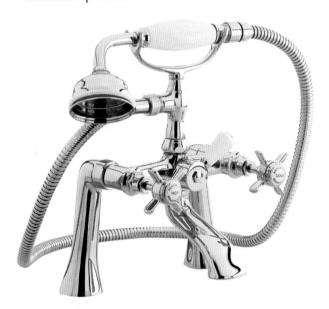

▲ Figure 1.52 Pillar type bath/shower mixer tap

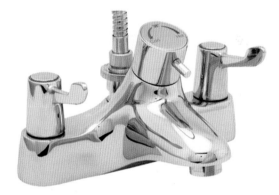

▲ Figure 1.54 Thermostatic type bath/shower mixer tap

▲ Figure 1.55 Wall mounted bath/shower mixer tap

Bath/shower mixer taps are designed for use where the pressures of the hot and cold water are equal. They should not be fitted where the cold water is direct from the mains supply and the hot water is fed from a vented, low-pressure hot water storage cylinder. This type of installation creates an imbalance of water supply and correct mixing of both hot and cold water cannot take place. It can also cause the hot water to be pushed back into the cylinder by the high pressure of the cold water supply.

Where there is a risk that the bath water can be siphoned back into the water undertaker's mains cold water supply through the shower hose of the mixer tap, the hot and cold water connections to the bath/shower mixer should be fitted with double check valves or, as an alternative, the shower hose should be fixed by a retaining ring so that the head cannot be placed below the overspill level of the bath.

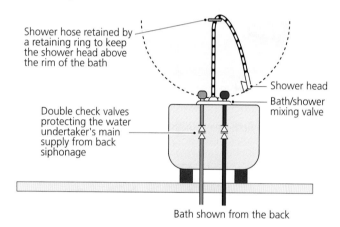

Shower hose retained by a retaining ring to keep the shower head above the rim of the bath

Shower head

Bath/shower mixing valve

Double check valves protecting the water undertaker's main supply from back siphonage

Bath shown from the back

▲ Figure 1.56 Bath/shower mixer tap installation

Most modern bath/shower mixer taps have a type HC diverter with automatic return backflow protection device built into them as part of the design of the tap (see page 63).

Flow limiting valves

Flow limiting valves are designed to limit the flow of water to appliances irrespective of the pressure upstream. Most are designed with an integral service valve and interchangeable cartridges that allow the

same valve body to be used for differing flow rates depending on the appliance that it is serving. The flow rate can be changed by using a different cartridge insert. They can be used on both hot and cold water supplies and are ideal for limiting the flow to:

- washbasins
- baths and sinks
- bidets
- WCs (when used with a strainer cartridge as required by Water Supply (Water Fittings) Regulation R25.6).

The flow limiting cartridges are capable of delivering a flow rate of between 0.07 and 0.43 l/s. By limiting the maximum flow rate, the guess-work is removed from designing the system, such as pipe sizing, cistern capacities and pump size. The use of flow limiting valves also helps to:

- balance the flow rates between hot and cold supplies
- assist with balancing thereby preventing some appliances (for example, a shower) from consuming all available water while other appliances at higher level or further downstream are starved of water
- save money on water and energy costs.

Spray taps

Spray taps are designed to reduce the flow of water from the outlet of the tap to a fine spray. This helps to conserve water, often reducing the water consumption of the tap by as much as 20 per cent. There are many different types of spray taps for use in both bathrooms and kitchens. Some taps use a special insert to make the water appear bubbly. This is known as an aerator, and simply introduces air into the water flow as the tap is running. Kitchen spray taps often have a pull-out nozzle for rinsing plates and dishes.

Urinal – water conservation controls

Many urinal installations do not have any form of water control and so they flush continuously even during those periods when the building is unoccupied. Quite often, the flow rate is higher than that specified by the

Water Supply (Water Fittings) Regulations. Under the regulations, a urinal should use no more than 7.5 litres per bowl/hour and 10 litres/hour for a single bowl. The urinal should also have some form of limiting device to prevent unnecessary flushing during periods when the building is not used. In practice, flow rates from urinals are not measured and often deliberately increased in an attempt to control unwanted odours.

There are many designs of flush controller available. These can use:

- a timer to match the hours of occupancy
- infrared detectors to detect the presence of people to open a solenoid valve for a short period of time
- a mechanical means to detect reduced hydraulic water pressure caused by taps being opened to open a valve controlling water to a urinal cistern
- some controls even allow a cistern to fill only very slowly if no movement is detected for a pre-set period of time.

Where a large number of urinals are installed, separate controllers may be necessary to prevent all of the urinals flushing simultaneously when only one person enters the room.

▲ Figure 1.57 Mechanical urinal flushing system

> **KEY POINT**
>
> Whichever system of water control is used, it must be correctly set up to both the manufacturer's instructions and the requirements of the regulations.

Shower pumps – single and twin impeller

Shower booster pumps are used to give a high flow rate shower and are used in conjunction with shower mixing valves. There are two types of shower booster pump:

- **Twin impeller pump on the inlet to the mixer valve.** The pump increases the pressure of the hot and cold water supplies to the mixer valve independently. The water is then mixed to the correct temperature in the valve before flowing to the shower head.

  Care must be exercised when making the hot connection to the cylinder. There are two ways in which this can be done. The first method involves installing the hot water draw-off from the cylinder at an angle of between 30 and 60°, with the hot shower pump connection being made at an angle of 90° with a tee piece (see Figure 1.58). This allows any air in the system to filter up to the vent and away from the hot shower pump inlet.

  The second method involves making a direct connection to the cylinder using a special fitting called an Essex flange. With this method, the hot water is taken directly from the hot water storage vessel avoiding any air problems which may occur.

- **Single impeller pump off the outlet from the mixer valve.** These boost the water AFTER it has left the mixer valve. They are usually used with concealed shower valves and fixed 'deluge' type, large water volume shower heads.

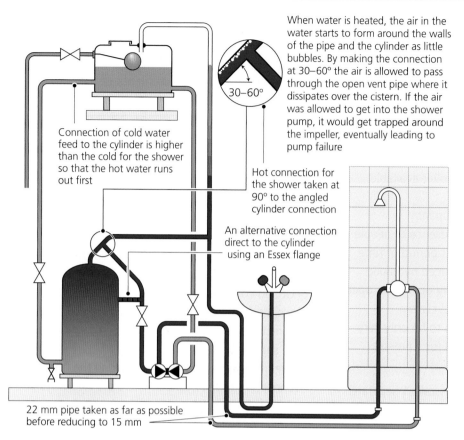

When water is heated, the air in the water starts to form around the walls of the pipe and the cylinder as little bubbles. By making the connection at 30–60° the air is allowed to pass through the open vent pipe where it dissipates over the cistern. If the air was allowed to get into the shower pump, it would get trapped around the impeller, eventually leading to pump failure

30–60°

Connection of cold water feed to the cylinder is higher than the cold for the shower so that the hot water runs out first

Hot connection for the shower taken at 90° to the angled cylinder connection

An alternative connection direct to the cylinder using an Essex flange

22 mm pipe taken as far as possible before reducing to 15 mm

▲ Figure 1.58 Pump-assisted shower installation with twin impeller, inlet shower booster pump

▲ Figure 1.59 Twin impeller, inlet shower booster pump

▲ Figure 1.60 An Essex flange

▲ Figure 1.61 A large water volume 'deluge' shower head

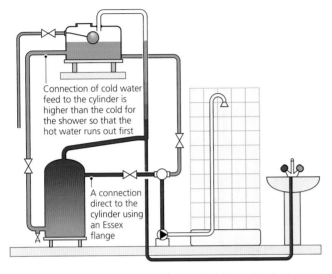

Connection of cold water feed to the cylinder is higher than the cold for the shower so that the hot water runs out first

A connection direct to the cylinder using an Essex flange

▲ Figure 1.62 Pump-assisted shower installation with single impeller, outlet shower booster pump

▲ Figure 1.63 Single impeller shower booster pump

In both of these installations, the pump increases the pressure of the water, which means that the minimum 1 m head is not necessary. However, a minimum head of 150 mm is required to lift the flow switches as these switch the pump on.

Using shower pumps in negative head situations

With some installations it is possible to install the pump where a negative head exists. A negative head is where the cistern is lower than the pump. In this instance a means of starting the pump must be in place.

▲ Figure 1.64 A negative head shower installation

Negative head shower pumps have the inclusion of a small accumulator or pressure vessel that sits on top of the pump. They work by sensing a sudden drop in pressure. When the shower valve is opened, the pressure within the system, caused by the pumping power of the shower pump, suddenly drops. The accumulator then immediately forces a small amount of water out via a small plastic tube, to activate the flow switch located inside the pump casing. Once the flow switch is activated the shower pump will run at negative head. When the shower is turned off, the system remains charged at the shower pump pressure until the shower valve is opened again and the sudden pressure drop occurs, starting the pump once more.

Some shower pumps also use a small pressure transducer (pressure switch), which activates the shower pump directly, rather than using a flow switch.

43

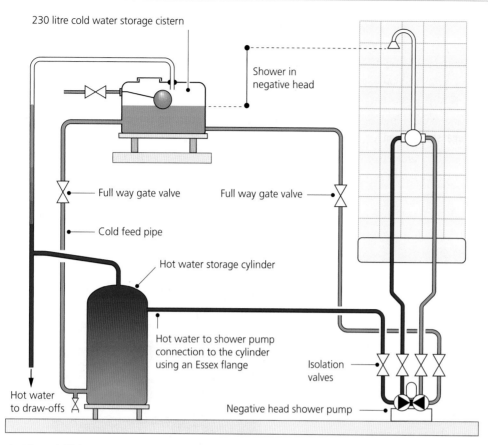

230 litre cold water storage cistern

Shower in negative head

Full way gate valve

Full way gate valve

Cold feed pipe

Hot water storage cylinder

Hot water to shower pump connection to the cylinder using an Essex flange

Isolation valves

Hot water to draw-offs

Negative head shower pump

▲ Figure 1.65 A negative head shower booster pump

Pressure reducing valves

The water pressure in England and Wales varies considerably. Older properties that have not had their mains cold water service updated for many years could have less than 1 bar pressure. This is insufficient to run even the most basic of modern appliances such as combination boilers and even some electric showers. At the extreme, some areas where a new water main has been laid could receive up to 10 bar pressure. High water pressures (anything above 4 bar pressure) although very good for firefighting, can be damaging to domestic plumbing systems. It can cause erosion corrosion where the flow of the water wears away pipes and fittings, especially where changes of direction occur. It can also cause leaks to water heaters, increased noise within the system, dripping taps and water hammer. Therefore, any pressures above those needed to provide sufficient flow to water fittings and appliances becomes damaging, wasteful and reduces considerably the life expectancy of the system as a whole. This adds to the cost of water due to water wastage and increases energy usage.

INDUSTRY TIP

The average water pressure in the UK tends to be in the region of 3 bar during the day (although most water undertakers will only guarantee at least 1 bar at all times, including times of peak usage), increasing to around 4.5 to 5 bar during the night as the pressure rises due to lack of usage.

Pressure reducing valves (PRVs) are used to reduce a high upstream pressure to a lower downstream pressure. They act as a buffer between the supply pressure and the systems or appliances they are fitted to during both flow and non-flow conditions. They perform two functions:

- They reduce high supply pressures upstream to a lower, more functional pressure for distribution.
- They maintain a set pressure ensuring that the pipework and appliances are not subjected to excessive stress and operate at a more moderate, acceptable pressure.

How do PRVs work?

A pressure reducing valve takes high pressure water and reduces it to a lower pressure under flow and no-flow conditions, which means that the valve effectively stops the water pressure from creeping up when there is no water flow. This type of control is known as 'drop tight'. Most PRVs use a balanced spring and a diaphragm to control downstream water pressure. They work by sensing water pressures either side of the diaphragm.

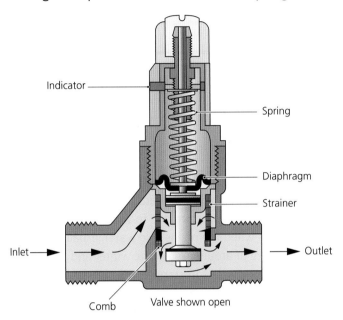

▲ Figure 1.66 A pressure reducing valve cut-away

The diaphragm separates all of the parts in contact with the water from the control spring and the valve's mechanism. The valve body is internally protected from debris by a stainless steel mesh strainer.

● Under no-flow conditions, the downstream pressure applies back pressure on the valve seat and the diaphragm. This overcomes the pressure of the spring and forces the seat against the diaphragm to prevent the downstream pressure from increasing.

● Under flow conditions, the back pressure is reduced and the spring forces the valve open to allow water flow.

Shock arrestors/mini expansion vessels

Before we look at shock arrestors (or water hammer arrestors as they are more commonly known), it may help to understand why some systems may need to include them.

Water hammer

Water hammer (or fluid hammer) is a pressure surge or wave caused when the fluid (both liquid and gas) is suddenly forced to stop or change direction. This can occur in plumbing systems when a valve is closed very quickly at the end of a pipeline. When the valve is closed, a pressure wave propagates within the pipework. This is known as hydraulic shock. The shock wave travels in the opposite direction to the water flow, often with disastrous consequences. The shock wave can cause major problems with repeated water hammer, vibration and noise, which can lead to joint failure and pipework damage. To put this into perspective, a system that is normally operating at 3 bar, can suffer shock waves equalling twice this pressure. If the pipework is installed poorly or has not been clipped correctly, the noise and vibration, and subsequent damage can often be much worse. A shock arrestor can help prevent water hammer by cushioning the effects of the shock wave.

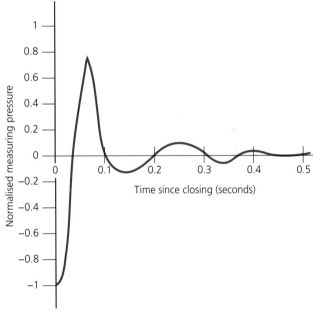

▲ Figure 1.67 The hydraulic shock wave

Shock arrestors

Shock arrestors are often manufactured from corrosion resistant brass, copper or stainless steel. They contain a piston or a diaphragm, which is cushioned by a calculated amount of inert gas or air. When the shock wave hits the arrestor, the piston (or diaphragm) moves with the shock wave (or hydraulic impulse) to dampen its effects by absorbing the kinetic energy. This allows the shock wave to dissipate safely without damaging the pipework and fittings.

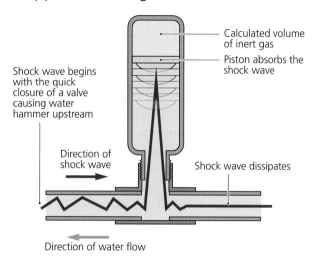

▲ Figure 1.68 How the shock arrestor works

Using the shock arrestor as a mini expansion vessel

Some types of shock arrestors can be used successfully as mini expansion vessels. These are usually required where there is a small amount of water expansion associated with hot water pipework on mains-fed hot water supplies (unvented hot water storage systems, combination boilers or multi-point instantaneous hot water heaters). The expanded water increases the internal pressure within the pipework and this can cause damage to terminal fittings such as shower mixing valves and backflow protection devices. A mini

expansion vessel/shock arrestor allows the expansion to take place within the vessel, thereby protecting systems and equipment from internal damage. They are a requirement of most major shower manufacturers.

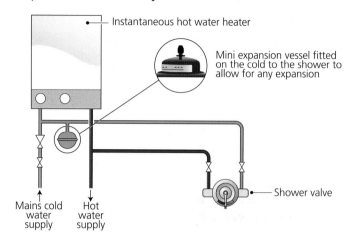

▲ Figure 1.69 The use of a mini expansion vessel

Mini expansion vessels can also be fitted on cold water installations to counteract the problems of increased system pressure due to water expansion. This occurs when the mains cold water, usually at a temperature of around 4°C in the winter and 16°C in the summer, is left to stand in the pipework for a period of time. The ambient air temperature in the building can be considerably higher, which causes the water to expand slightly. This expansion can damage backflow protection devices such as single check valves, and terminal fittings such as ceramic disc type taps, especially where the run of pipework is extensive.

③ BACKFLOW PROTECTION IN PLUMBING SYSTEMS

The Water Supply (Water Fittings) Regulations 1999 were implemented to harmonise the working practices of plumbers in England and Wales.

Simply put, the Water Supply (Water Fittings) Regulations have been put in place to ensure that the plumbing systems we install and maintain prevent:
- contamination of water
- wastage of water
- misuse of water

- undue consumption of water
- erroneous metering of water.

Of these, by far the biggest issues surround contamination.

Up until 1999, there were three classes of fluid. At that time, water was either considered wholesome, suspicious or dangerous. With the implementation of the 1999 Water Regulations, these three fluid categories became five to bring the UK into line with the rest of Europe.

In this part of the chapter, we will investigate the different fluid categories as defined by the Water Supply (Water Fittings) Regulations 1999 and the methods, both physical and mechanical, we can employ to prevent contamination of domestic cold water installations by back pressure and back siphonage.

The fluid categories

Any water that is not cold wholesome drinking water supplied by a water undertaker can be classed as a potential hazard. The Water Supply (Water Fittings) Regulations 1999 lists five fluid categories.

Fluid category 1

Fluid category 1 is wholesome water supplied by a water undertaker, complying with the Water Quality Regulations made under Section 67 of the Water Industry Act 1991. It must be wholesome, clean, cold and potable. All water undertakers have a duty to supply water that conforms to these regulations which ensures wholesome water suitable for domestic use or food production purposes. Whenever practicable, water for drinking water purposes should be supplied direct from the water undertaker's mains without any intervening storage.

Fluid category 2

Fluid category 2 is water that would normally be classified as fluid category 1 but whose aesthetic quality has been impaired because of:

- a change in temperature
- a change in appearance, taste or odour owing to the presence of substances or organisms.

These changes are aesthetic only and do not constitute a health risk. Typical situations where this may occur in domestic properties are:

- water heated in a hot water secondary system
- mixed fluid categories 1 and 2 water discharged from combination taps or showers.

Fluid category 3

Fluid category 3 is water that constitutes a slight health hazard because of the concentration of low toxicity substances. Fluids in this category are not suitable for drinking or any other domestic purpose or application. This includes:

- ethylene glycol (anti-freeze), copper sulphate or similar chemical additives such as heating inhibitors, cleansers and de-scalers used in domestic properties
- sodium hypochlorite and other common disinfectants.

Typical fluid category 3 situations are:

- in houses, apartments and other domestic dwellings:
  - water in the primary circuits of heating systems whether chemicals have been administered or not
  - water in washbasins, baths and shower trays
  - clothes and dishwashing machines
  - home dialysis machines

- hand-held garden hoses with a flow-controlled spray or shut off valve
- hand-held fertilisers
- in premises other than single occupancy domestic dwellings:
  - domestic fittings and appliances such as washbasins, baths, or showers installed in commercial, industrial or other premises may be regarded as fluid category 3. However, if there is a potential for a higher risk, such as a hospital, medical centre or other similar establishment, then a higher fluid category risk should be applied in accordance with the regulations
- house garden or commercial irrigation systems without insecticides.

Fluid category 4

Fluid category 4 is water that constitutes a significant health hazard because of the concentration of toxic substances, which can include:

- chemical, carcinogenic substances or pesticides (including insecticides and herbicides)
- environmental organisms of potential health significance.

Typical fluid category 4 situations are:

- general:
  - primary circuits of heating systems in properties other than a single occupancy dwelling (commercial systems)
  - fire sprinkler systems using anti-freeze chemicals
- house gardens:
  - mini irrigation systems without fertiliser or insecticides, including pop-up sprinkler systems and permeable hoses
- food processing:
  - food preparation
  - dairies
  - bottle washing plants
- catering:
  - commercial dishwashers
  - refrigerating equipment
- industrial and commercial installations:
  - dyeing equipment
  - industrial disinfection equipment
  - photographic and printing applications
  - car washing and degreasing plant
  - brewery and distilling processes

- water treatment plant or softeners that use other methods than salt
- pressurised fire-fighting systems.

Fluid category 5

Fluid category 5 represents a serious health risk because of the concentration of pathogenic organisms, radioactive material or very toxic substances. These include water that contains:

- faecal material or any other human waste
- butchery or any other animal waste
- pathogens from any source.

Typical fluid category 5 situations are:

- general:
  - industrial cisterns and tanks
  - hose union bib taps in a non-domestic installation
  - sink, WC pans, urinals and bidets
  - permeable pipes in any non-domestic garden whether laid at or below ground level
  - grey water recycling systems
- medical:
  - laboratories
  - any medical or dental equipment with submerged inlets
  - bedpan washers and slophoppers
  - mortuary and embalming equipment
  - hospital dialysis machines
  - commercial clothes washing equipment in care homes and similar premises
  - baths, washbasins, kitchen sinks and other appliances that are in non-domestic installations
- food processing:
  - butchery and meat trade establishments
  - slaughterhouse equipment
  - vegetable washing
- catering:
  - dishwashing machines in healthcare premises and similar establishments
  - vegetable washing
- industrial/commercial:
  - industrial and chemical plants
  - laboratories
  - any mobile tanker or gulley cleaning vehicles
- sewerage treatment works and sewer cleaning:
  - drain cleaning plant
  - water storage for agricultural applications
  - water storage for firefighting systems

- commercial agricultural:
  - commercial irrigation outlets below or at ground level and/or permeable pipes, with or without chemical additives
  - insecticide or fertiliser applications
  - commercial hydroponic systems.

INDUSTRY TIP

The list of examples of applications shown for each fluid category is not exhaustive.

Backflow and back siphonage risks in the home

There are many instances in the home where backflow and back siphonage could present contamination risks. These will need to be considered during any planning, design and installation of hot and cold water supplies and central heating systems. Let us look first at some of the appliances and systems we use and consider the risks. This will give you some idea of how the fluid categories occur. Table 1.4 is designed to give

a brief overview of how and where fluid categories occur in the home and should not be viewed upon as exhaustive.

As you can see from the table, there are many potential contamination risks in every dwelling and the bigger the building then the more risks there are likely to be.

KEY POINT

The distinction between fluid category 4 and fluid category 5 is often difficult to interpret. In general we can assume that fluid category 4 is such that the risk to health, because of the level of toxicity or the concentration of substances, is such that harm will occur over a prolonged period of days to weeks to months whereas the risk from fluid category 5, because of the high concentration of substances or the level of toxicity, is such that serious harm could occur after a very short exposure of minutes to hours to days or even a single exposure.

We must remember that fluid category 1 is clean, cold, wholesome water direct from the water undertaker's main and no other fluid category must come into contact with it or contamination may occur.

▼ Table 1.4 Appliances and fluid category risk

Appliance or system	Content of the water	Risk
Kitchen sink	May contain animal remains from food preparation	Fluid cat. 5
WC	Contains human waste	
Bidet (over rim type)	May contain human waste	
Grey water and rainwater harvesting systems	May contain bacteria and disinfectants	
Washing machines and dish washers	Contains soap and other detergents and chemicals from dish washing and clothes cleaning	Fluid cat. 3
Bath	May contain soap and other detergents from personal hygiene	
Wash hand basin		
Shower valves and instantaneous showers	At risk from soap and other detergents from personal hygiene	
Hose union bib taps (outside tap)	At risk from gardening and other activities such as watering, weed killing, car washing, irrigation etc.	
Combination boilers	The water in the heating system is often contaminated with dissolved metals, flux and some form of chemical inhibitor	Fluid cat. 3 or 4 (depending on boiler size)
Hot water system	Contains hot water	Fluid cat. 2

Whole site, zone and point of use protection

There are many commercial and industrial processes where the whole or part of a plumbing system can present a high risk of backflow to other parts of the installation or even the water undertaker's mains supply despite the fact that the installation is installed to the required standards. In these circumstances whole site or zone protection must be installed on those parts that are deemed to be high risk.

Whole site protection

The term 'whole site protection' simply means that the water undertaker's main is protected at all times from backflow or back siphonage from any fluid category that is not fluid category 1 by a suitable backflow device. Protection should be at the point of entry of the cold water supply.

If whole site protection is required, it is important that the water undertaker is informed at the application/notification for water supply stage. They will assess the application for a water supply and advise on what fluid category of backflow protection device must be installed to comply with the Water Supply (Water Fittings) Regulations. The backflow protection device MUST be installed before the system is commissioned.

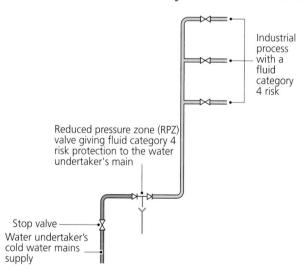

▲ Figure 1.70 Whole site protection

Zoned protection

Zoned backflow protection simply means that where different fluid categories exist within the same building,

premises or complex, these have their own backflow protection devices to protect any part of the system that is fluid category 1. Zoned protection is also required where any water supply pipe is supplying more than one separately occupied premises.

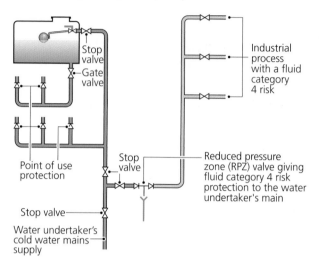

▲ Figure 1.71 Zoned protection

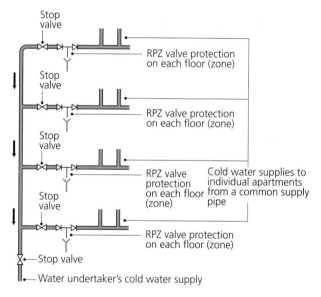

▲ Figure 1.72 Zoned protection for domestic premises

Point of use protection

This is the simplest form of backflow protection. Point of use backflow protection devices are used to protect an individual fitting or outlet against backflow and are usually located close to the fitting which they protect, such as a single check valve on a mixer tap to protect against fluid category 2 or a double check valve on a domestic hose union bib-tap as protection against fluid category 3.

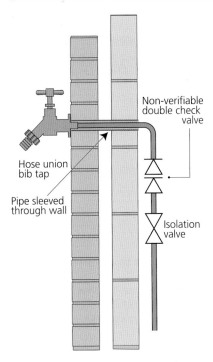

▲ Figure 1.73 Point of use protection

Labels in figure: Non-verifiable double check valve; Hose union bib tap; Pipe sleeved through wall; Isolation valve

Eliminating the risk of contamination of wholesome water

The Water Regulations and, more specifically, the Water Regulations Guide, can help us to choose the right course of action based upon the risk. The manufacturers too, help in this regard by designing and manufacturing their appliances, taps and valves to conform to the Water Regulations. For example, most kitchen and bidet taps are designed and made with fluid category 5 risk in mind and most bath and wash basin taps are designed and made with fluid category 3 in mind.

In most cases, where baths, washbasins, bidets and kitchen sinks are concerned, a simple air gap will protect the mains cold water supply. The size of the air gap, however, is dependent on the size of the tap, appliance type and its likely contents.

Air gaps used as a method of backflow prevention

An air gap is simply a physical unrestricted open space between the wholesome water and the possible contamination, the greater the air gap, the greater the level of protection that is offered. It does not require the use of a mechanical backflow prevention device. Here, we will consider the most important air gaps and how we can apply them (see Table 1.5).

▼ Table 1.5 Schedule of non-mechanical backflow prevention arrangements and their respective fluid category protection

	Type	Description of backflow prevention arrangements and devices	Suitable for protection against fluid category	
			Back pressure	Back siphonage
a	AA	Air gap with unrestricted discharge above spillover level	5	5
b	AB	Air gap with weir overflow	5	5
c	AD	Air gap with injector	5	5
d	AG	Air gap with minimum size circular overflow determined by measure or vacuum test	3	3
e	AUK1	Air gap with interposed cistern (e.g. a WC suite)	3	5
f	AUK2	Air gaps for taps and combination fittings (tap gaps) discharging over domestic sanitary appliances, such as a washbasin, bidet, bath or shower tray shall not be less than the following: <table><tr><td>Size of tap or combination fitting</td><td>Vertical distance of bottom of tap outlet above spillover level of receiving appliance</td></tr><tr><td>Not exceeding G ½</td><td>20 mm</td></tr><tr><td>Exceeding G ½ but not exceeding G ¾</td><td>25 mm</td></tr><tr><td>Exceeding G ¾</td><td>70 mm</td></tr></table>	X	3

▼ Table 1.5 Schedule of non-mechanical backflow prevention arrangements and their respective fluid category protection (continued)

	Type	Description of backflow prevention arrangements and devices	Suitable for protection against fluid category	
			Back pressure	Back siphonage
g	AUK3	Air gaps for taps or combination fittings (tap gaps) discharging over any higher risk domestic sanitary appliances where a fluid category 4 or 5 is present, such as: ● any domestic or non-domestic sink or other appliance, or ● any appliances in premises where a higher level of protection is required, such as some appliances in hospitals or other health care premises shall be not less than 20 mm or twice the diameter of the inlet pipe to the fitting, whichever is the greater.	X	5
h	DC	Pipe interrupter with permanent atmospheric vent	X	5

X indicates that the backflow prevention arrangement or device is not applicable or not acceptable for protection against back pressure for any fluid category within water installations in the UK.

Arrangements incorporating type DC devices shall have no control valves on the outlet of the device; they shall be fitted not less than 300 mm above the spillover level of a WC pan, or 150 mm above the sparge pipe outlet of a urinal, and discharge vertically downwards.

Overflows and warning pipes shall discharge through, or terminate with, an air gap, the dimension of which should satisfy a type AA air gap.

Each of the air gaps here will have two fluid categories attached to it, one for back pressure and one for back siphonage. The difference between the two is simple to explain:

- **Back pressure** – caused when a **downstream** pressure is greater than the **upstream** or supply pressure in the water undertakers main or the consumer's potable water supply. Back pressure can be caused by:
  - a sudden loss of upstream pressure, such as a burst pipe on a water undertaker's mains supply
  - an increase in downstream pressure caused by pumps or expansion of hot water
  - a combination of both of the above.
- **Back siphonage** – backflow caused by a negative pressure creating a vacuum or partial vacuum in the water undertaker's mains cold water supply. It is similar to drinking through a straw. If a sudden loss of pressure on the mains supply was to occur whilst a submerged outlet was flowing, then water would backflow upwards through the submerged outlet and down into the water undertaker's main.

KEY TERMS

Downstream: in water systems, downstream means travelling away from the point of supply.

Upstream: in water systems, upstream means travelling toward the point of supply.

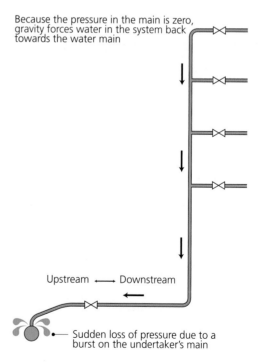

Because the pressure in the main is zero, gravity forces water in the system back towards the water main

Upstream ←→ Downstream

Sudden loss of pressure due to a burst on the undertaker's main

▲ Figure 1.74 Back pressure

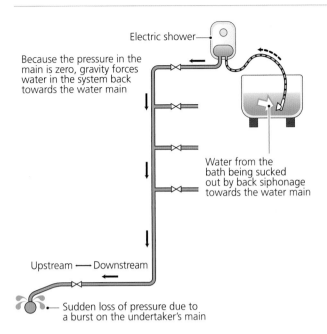

Because the pressure in the main is zero, gravity forces water in the system back towards the water main

Electric shower

Water from the bath being sucked out by back siphonage towards the water main

Upstream —— Downstream

Sudden loss of pressure due to a burst on the undertaker's main

▲ Figure 1.75 Back siphonage

Type AA – air gap with unrestricted discharge above spillover level

This gives protection against fluid category 5 and is a non-mechanical backflow prevention arrangement of water fittings where water is discharged through an air gap into a cistern, which has, at all times, an unrestricted spillover to the atmosphere. The air gap is measured vertically downwards from the lowest point of the inlet discharge orifice to the spillover level. It should be remembered that:

- the type AA air gap is suitable for all fluid categories
- the size of the air gap is subject to the size of the inlet (see Table 1.6)
- the flow from the inlet into the cistern must not be more than 15° from the vertical.

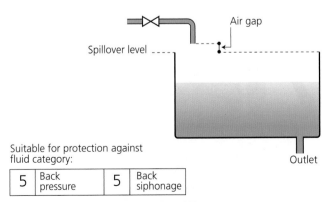

Air gap

Spillover level

Suitable for protection against fluid category:

5	Back pressure	5	Back siphonage

Outlet

▲ Figure 1.76 Type AA air gap with unrestricted discharge above spillover level

▼ Table 1.6 Air gaps at taps, valves, fittings and cisterns

Situation	Nominal size of inlet, tap, valve or fitting	Vertical distance between tap or valve outlet and the spillover level of the receiving appliance or cistern
Domestic situation with fluid categories 2 and 3 (AUK2)	Up to and including G ½	20 mm
	Over G ½ and up to G ¾	25 mm
	Over G ¾	70 mm
Non-domestic situation with fluid categories 4 and 5 (AUK3)	Any size of inlet pipe	Minimum diameter of 20 mm or twice the diameter of the inlet pipe, whichever is the greater of the two

A good example for the use of a type AA air gap is in the form of animal drinking troughs where the discharge of water into the trough is in a raised housing on the edge of the trough. The housing is covered to prevent the animals from having access to the water supply.

▲ Figure 1.77 Animal trough

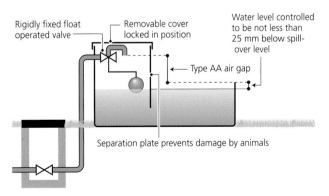

Rigidly fixed float operated valve

Removable cover locked in position

Water level controlled to be not less than 25 mm below spillover level

Type AA air gap

Separation plate prevents damage by animals

▲ Figure 1.78 Animal trough schematic

Type AB – air gap with weir overflow

This gives protection against fluid category 5. It is a non-mechanical backflow prevention arrangement of water fittings complying with type AA, except that the air gap is the vertical distance from the lowest point of the discharge orifice which discharges into the receptacle, to the critical level of the rectangular weir overflow.

The type AB air gap is suitable for high risk fluid category 5 situations and is particularly suited to installations where the contents of the cistern need to be protected from contaminants such as insects, vermin and dust. A good example of this is feed and expansion cisterns in industrial/commercial installations or where high-quality water is required, such as in dental surgeries.

The size of the weir needs to be calculated based upon the inlet size. This is usually completed using a weir overflow calculator.

IMPROVE YOUR MATHS

An example of a weir calculator can be found at: www.airgapcalculator.co.uk/inletcalc/index.html

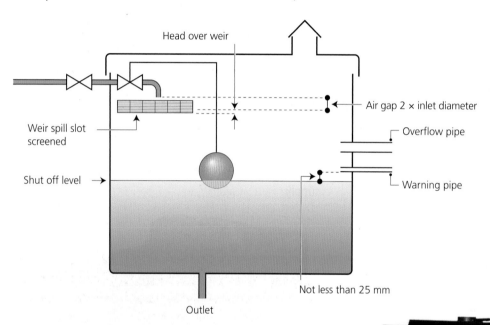

Suitable for protection against fluid category:

5	Back pressure	5	Back siphonage

▲ Figure 1.79 Type AB air gap with weir overflow

▲ Figure 1.80 Type AB air gap with weir overflow on a cistern

Type AD – air gap with injector

This is defined as a non-mechanical backflow prevention arrangement of water fittings with a horizontal injector and a physical air gap of 20 mm or twice the inlet diameter, whichever is the greater. It gives protection against back pressure and back siphonage up to fluid category 5. This device is commonly known as a 'jump jet'.

The principal uses of this type of air gap arrangement are in commercial clothes washing and dish washing machines. It also has the potential to be used in catering equipment such as steaming ovens.

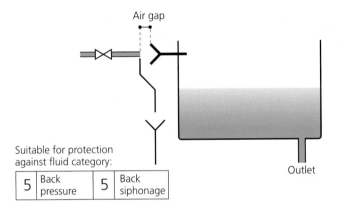

Suitable for protection against fluid category:

| 5 | Back pressure | 5 | Back siphonage |

▲ Figure 1.81 Type AD air gap with injector

Type AG – air gap with minimum size circular overflow determined by measure or vacuum test

This means a non-mechanical backflow prevention arrangement of water fittings with an air gap; together with an overflow, the size of which is determined by measure or a vacuum test. This arrangement gives protection against fluid category 3.

The type B air gap fulfils the requirements of **BS EN 14623:2005**, Devices without moving parts for the prevention of contamination of water by backflow; Specification for type B air gaps. In a cistern that is open to the atmosphere, the vertical distance between the lowest point of discharge and the critical water level should comply with one of the following requirements:

- It should be sufficient to prevent back siphonage.
- It should not be less than the distances specified in Table 1.6, depending on cistern type.

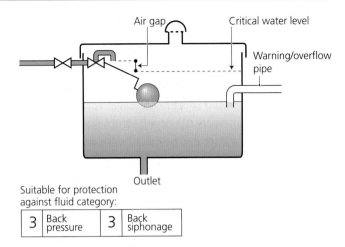

Suitable for protection against fluid category:

| 3 | Back pressure | 3 | Back siphonage |

▲ Figure 1.82 Type AG gap with minimum size circular overflow determined by measure

The following points about type AG air gaps should be noted:

- The air gap is related to the size of the inlet supply and is minimum vertical distance between the critical water level and the lowest part of the discharge outlet of the float operated valve (FOV) as specified in Table 1.6.
- The critical water level is the level that is reached when the FOV has failed completely and the water is running freely at maximum full-bore flow rate and pressure.
- Type A unrestricted air gaps must comply with the requirements of **BS EN 13076:2003**.

Where storage cisterns are installed, it is likely that the critical water level would differ from installation to installation because of varying flow rates and pressures of the incoming supply and the differing lengths and gradients of the overflow pipe. With this type of installation, the type AG air gap is not practical because the critical water level cannot be accurately calculated. It is the critical water level that would determine the position on the cistern of the float operated valve and the distance between the FOV and the overflow.

Type AUK1 – air gap with interposed cistern

This is a non-mechanical backflow prevention arrangement consisting of a cistern incorporating a type AG overflow and an air gap. The spillover level of the receiving vessel is located not less than 300 mm below the overflow pipe and not less than 15 mm below the lowest level of the interposed cistern. It is suitable for protection against fluid categories 5 for back siphonage and 3 for back pressure.

This arrangement is most commonly found on WC installations with the WC pan being the receiving vessel containing fluid category 5 water. A conventional domestic WC suite consists of a 6 litre/4 litre dual flushing cistern, a Part 2, 3 or 4 FOV with an AG air gap and overflow arrangement. This creates an AUK1 interposed cistern or, in other words, a cistern that can be supplied from a mains supply or another protected cistern without the need for additional backflow protection.

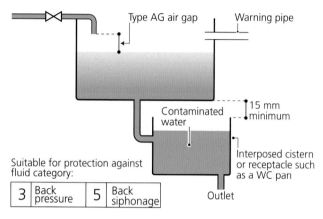

▲ Figure 1.83 AUK1 air gap with interposed cistern

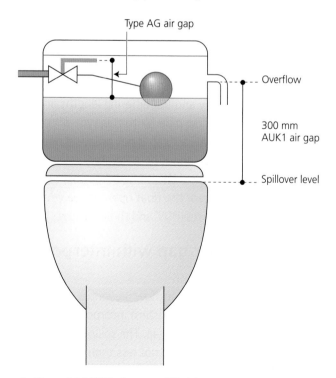

▲ Figure 1.84 AUK1 air gap on WC cisterns

Type AUK2 – air gaps for taps and combination fittings (tap gaps) discharging over domestic sanitary appliances

This means the height of air gap between the lowest part of the outlet of a tap, combination fitting, shower head or other fitting discharging over a domestic sanitary appliance or other receptacle, and the spillover level of that appliance, where a fluid category 2 or 3 risk is present downstream. An AUK2 air gap is only suitable for back siphonage up to fluid category 3 and must comply with the distances stated in Table 1.6.

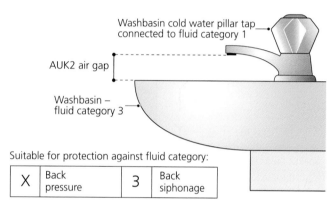

▲ Figure 1.85 AUK2 air gap (tap gaps)

Type AUK3 – air gaps for taps or combination fittings (tap gaps) discharging over any higher risk domestic sanitary appliances where a fluid category 4 or 5 is present

This means the height of an air gap between the lowest part of the outlet of a tap, combination fitting, shower head or other fitting discharging over any appliance or other receptacle, and the spillover level of that appliance, where a fluid category 4 or 5 risk is present downstream.

In a domestic dwelling, AUK3 air gaps are most common at the kitchen sink in the form of high-necked pillar taps, sink mixer taps or sink monobloc taps. Sink mixers and monoblocs have a swivel spout. If a cleaner's sink, Belfast sink or London sink is being installed, it is important that any bib taps installed are positioned so as to maintain an AUK3 air gap.

Taps and combination fittings discharging on non-domestic appliances and any appliances in premises where a higher level of protection is required, such as appliances in hospitals or other healthcare premises require a type AUK3 tap gap.

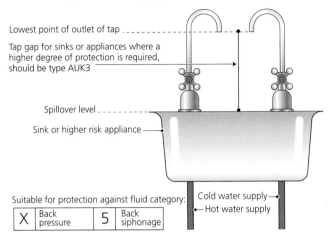

Lowest point of outlet of tap

Tap gap for sinks or appliances where a higher degree of protection is required, should be type AUK3

Spillover level

Sink or higher risk appliance

Cold water supply
Hot water supply

X	Back pressure	5	Back siphonage

Suitable for protection against fluid category:

▲ Figure 1.86 AUK3 air gap (higher risk tap gaps)

Type DC – pipe interrupter with a permanent atmospheric vent

This means a non-mechanical backflow prevention device with a permanent unrestricted air inlet, the device being installed so that the flow of water is in a vertical downward direction. They are used where there is a threat of back siphonage from a fluid category 5.

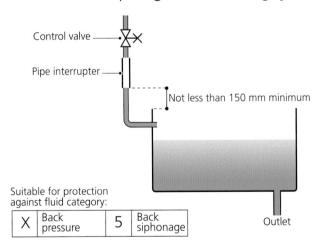

Control valve

Pipe interrupter

Not less than 150 mm minimum

Suitable for protection against fluid category:

X	Back pressure	5	Back siphonage

Outlet

▲ Figure 1.87 A DC pipe interrupter

The idea behind the DC pipe interrupter is to create an air inlet should a back siphonage situation occur. When water begins to backflow upwards due to back siphonage, the DC pipe interrupter allows air into the system to break the siphonic action, thus preventing contamination.

▲ Figure 1.88 A typical DC pipe interrupter

Type DC pipe interrupter must be fitted with the lowest point of the air aperture not less than 150 mm above the free discharge or spillover level of an appliance and have no valve, flow restrictor or tap on its outlet

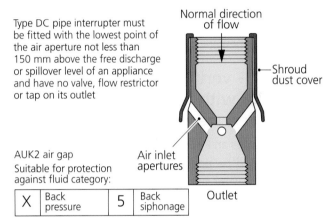

Normal direction of flow

Shroud dust cover

AUK2 air gap
Suitable for protection against fluid category:

Air inlet apertures

X	Back pressure	5	Back siphonage

Outlet

▲ Figure 1.89 A schematic drawing of a DC pipe interrupter

The DC pipe interrupter is a non-mechanical fitting. It does not contain any moving parts. It is manufactured from corrosion-resistant brass. Typical uses include WCs and urinal installations. The following points should be noted:

- The valve should be fitted in the vertical position, discharging downwards.
- It must be installed at least 300 mm above the overflowing level or 150 mm if fitted above a urinal.
- No tap or valve should be installed downstream of the interrupter.
- Pipe size reductions downstream of the interrupter are not allowed.
- The length of the pipe downstream after the interrupter should be as short as possible.
- The interrupter should be accessible for replacement and repair.
- DC pipe interrupters must comply with **BS EN 14453:2005**.

Mechanical backflow prevention devices

An air gap is the most effective method of preventing contamination of the water supply and most installers will try to achieve this within their installations and designs but there are many cases where air gaps are not practical as a method of protection. In these instances, installers may opt to install a mechanical backflow prevention device. These provide a physical barrier to backflow. However, it must be remembered that mechanical backflow prevention devices have limitations and can be subject to failure.

In this section of the chapter, we will look at some of the more common mechanical backflow prevention devices and where we can install them (see Table 1.7).

Type BA – verifiable backflow preventer with reduced pressure zone (a reduced pressure zone valve)

Better known as an RPZ valve, this is a mechanical, verifiable, backflow prevention device, offering protection to water supplies up to and including fluid

category 4. Verifiable simply means that the valve can be checked via test points to see if it is working correctly (verified).

> **KEY TERM**
>
> **Verifiable:** able to be checked.

Most RPZ valves consist of three separate elements:
- two check valves
- a differential relief valve
- three test points.

The first check valve is spring loaded to generate a specific pressure drop across this part of the valve. This creates a reduced pressure zone downstream in the middle chamber of the valve and on the downstream side of the differential relief valve. The incoming mains supply maintains supply pressure on the upstream side of the differential valve and, as long as the mains pressure is higher, the differential relief valve will remain closed.

▼ Table 1.7 Schedule of mechanical backflow prevention arrangements and fittings and their respective fluid category protection

	Type	Description of backflow prevention arrangements and devices	Suitable for protection against fluid category	
			Back pressure	Back siphonage
a	BA	**Verifiable** backflow preventer with reduced pressure zone	4	4
b	CA	Non-verifiable disconnector with difference between pressure zones not greater than 10%	3	3
c	DB	Pipe interrupter with atmospheric vent and moving element	X	3
d	EA/EB	Verifiable and non-verifiable single check valves	2	2
e	EC/ED	Verifiable and non-verifiable double check valves	3	3
f	HA	Hose union backflow preventer. Only permitted for use on existing hose union bib tap in house installations	2	3
g	HUK1	Hose union bib tap incorporating a double check valve arrangement. Only permitted as a replacement for existing bib taps in house installations	3	3
h	HC	Diverter with automatic return (normally integral with some domestic appliance applications only)	X	3

X indicates that the backflow prevention device is not acceptable for protection against back pressure for ANY fluid category.

Arrangements incorporating a type DB device shall have no control valves on the outlet of the device. The device shall not be fitted less than 300 mm above the spillover level of an appliance and must discharge vertically downwards.

Relief ports from BA and CA devices should terminate with an air gap, the dimension of which should satisfy a type AA air gap.

The following applies:

- If, under static conditions, the mains pressure reduces where it is just 0.14 bar above the pressure in the reduced pressure zone, the differential relief valve will open and release the contents of the middle chamber to drain.
- Should backflow occur past the first check valve element, the pressure on both sides of the differential valve will equalise and the differential relief valve will open to discharge the water.
- If complete mains failure occurs, the contents of the middle chamber are discharged to drain, providing that both check valve elements are functioning correctly. However, should the upstream check valve become faulty, the pressure in the middle chamber will equalise to that of mains pressure and the differential relief valve will open and continuously discharge water at a steady rate.
- If the downstream check valve fails under zero mains pressure conditions, the differential relief valve will open and water will discharge from the downstream side of the system until the pressure here also becomes zero.

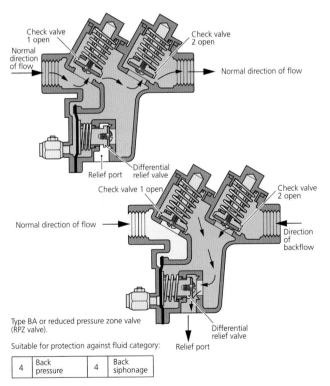

Type BA or reduced pressure zone valve (RPZ valve).

Suitable for protection against fluid category:

4	Back pressure	4	Back siphonage

▲ Figure 1.90 Use of a type CA backflow preventer

INDUSTRY TIP

Testing, commissioning, maintenance and annual inspection can ONLY be carried out by a trained and approved installer. Anyone who tests RPZ valves must be certificated. Specialist training is available from various test centres across the UK. Further recommended reading is the Water Regulations Advisory Scheme Information and Guidance Note No. 9-03-02.

Type CA – non-verifiable disconnector with difference between pressure zones not greater than 10%

These are very similar to BA devices (RPZ valves) in that they provide a positive disconnection chamber between the downstream water and the upstream water. The disconnection area between the two main check valves is open to the atmosphere under fault conditions thereby maintaining an air gap should a loss of upstream pressure occur. Like the RPZ valve, any water discharged would run to drain via a tundish. They are suitable for fluid category 3.

A typical use of a type CA disconnector is a permanent connection between a sealed central heating system and the water undertaker's cold water supply.

▲ Figure 1.91 A type CA backflow preventer

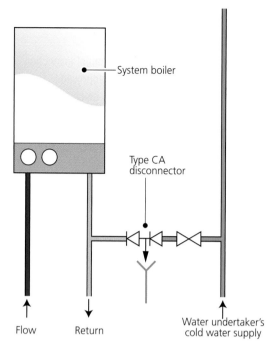

▲ Figure 1.92 Use of a type CA backflow preventer

Type DB – pipe interrupter with atmospheric vent and moving element

The type DB pipe interrupter is a backflow prevention device specifically designed for fluid category 4 applications. The concept of the DB interrupter is very simple. Water enters a tube which has one end blanked off. Around the tube are a series of small holes over which a flexible rubber membrane is stretched. As the water flows into the tube, it is forced through the holes and this flexes the rubber membrane to allow water to flow. If the supply pressure suddenly stops, then the membrane contracts against the holes to effectively prevent backflow. Any backflowing water is then released to atmosphere through another series of holes in the outer casing of the device. They are approved for use as protection against back siphonage but not back pressure.

DB pipe interrupters are generally used externally as attachments to hose union bib taps and must not be used on appliances that have a control valve restriction, such as a clothes washing machine. They are resistant to frost damage. They must be fitted vertically and have no valves fitted downstream of the device.

Some DB interrupters are manufactured with bayonet type attachments for domestic garden perforated hose irrigation systems.

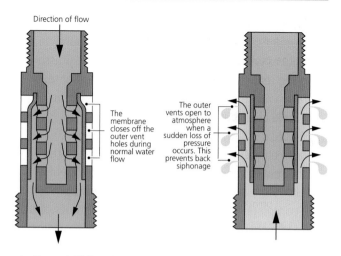

▲ Figure 1.93 Type DB pipe interrupter schematic

▲ Figure 1.94 Type DB pipe interrupter

Type EA and Type EB – verifiable and non-verifiable single check valves

These two valves are the most simple of all mechanical backflow prevention devices and can be used to protect against fluid category 2 for both back pressure and back siphonage. Generally regarded as point of use protection, they consist of a spring-loaded one-way valve that will allow water to flow from upstream to downstream only. If back siphonage or back pressure occurs, the valve will shut to prevent a reverse water flow. When no water is flowing, the valve remains in the closed position. Both types are almost identical in appearance. The difference between them is:

- Type EA device – has a test nipple situated on the upstream side of the valve so that it can be tested whilst in position to verify that it is working correctly

- Type EB non-verifiable single check valve – does not have a test point but can be used in the same way as the type EA single check valve.

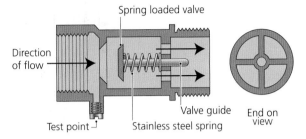

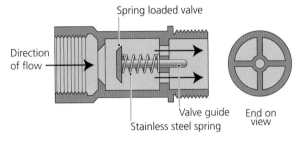

Suitable for protection against fluid category:

2	Back pressure	2	Back siphonage

▲ Figure 1.95 Type EA verifiable single check valve and type EB non-verifiable single check valve

Both valves are manufactured from **de-zincification resistant (DZR)** resistant brass and have either type A compression fittings or female BSP threads for connection to the pipework. The valves should conform to **BS EN 13959:2004** for use in hot or cold water systems up to 90 °C.

> **KEY TERM**
>
> **De-zincification resistant (DZR):** a type of brass that resists electrolytic corrosion.

In domestic premises the risk from fluid category 2 generally occurs where the hot and cold supplies are taken to a single terminal fitting such as mixer taps or shower valves. This is known as a cross connection. However, care must be taken when installing single check valves to hot water supplies as the expansion of

the water can cause excessive pressure on the check valve causing it to fail. Other uses include the cold water connections to drinks machines.

▲ Figure 1.96 Verifiable and non-verifiable single check valves

Type EC and Type ED – verifiable and non-verifiable double check valves

These are mechanical backflow prevention devices consisting of two single check valves in series, which will permit water to flow from upstream to downstream but not in the reverse direction. They are used primarily to protect against fluid category 3 for both back pressure and back siphonage.

The type EC verifiable double check valve has two test nipples, one on the upstream side of the first check valve and another in the chamber between the first and second check valve. These are used to verify that the valve is working correctly. The type ED non-verifiable double check valve does not have a test point but can be used in the same way as the type EA single check valve.

Suitable for protection against fluid category:

3	Back pressure	3	Back siphonage

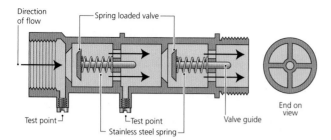

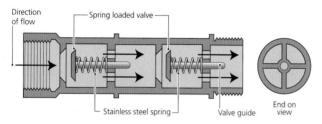

▲ Figure 1.97 Type EC verifiable double check valve and type ED non-verifiable double check valve

Typical uses in domestic installations include garden hose union bib-taps and sealed heating systems fitted in conjunction with a temporary filling loop. When used with sealed heating systems, the double check valve MUST be fitted to the cold water supply connection to the filling loop and NOT to the sealed heating connection.

Type HA – hose union backflow preventer (only permitted for use on existing hose union bib tap in house installations)

As the name suggests, this mechanical backflow prevention device screws on to the outlet thread of a hose union bib tap. It is specifically for use with existing hose union bib taps that do not have any form of backflow protection. It is used to protect against back pressure at fluid category 2 and back siphonage at fluid category 3.

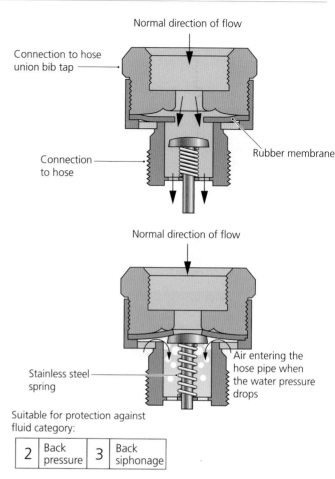

Suitable for protection against fluid category:

2	Back pressure	3	Back siphonage

▲ Figure 1.98 Type HA hose union backflow preventer

Type HUK1 – hose union bib tap incorporating a double check valve arrangement

This hose union bib tap incorporates two single check valves, one situated at the inlet to the tap and one at the outlet. A screw-type test point is also included on the tap body. They are fitted in the same way as a normal HU bib tap. However, they are not suitable for new installations and can only be used as replacements where a hose union bib tap already exists. This is simply because the Water Supply (Water Fittings) Regulations state that any mechanical backflow prevention device should be fitted INSIDE the building to prevent damage by freezing. They are suitable as protection against fluid category 3 for both back pressure and back siphonage.

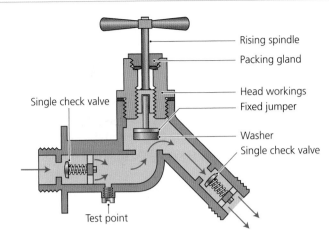

Suitable for protection against fluid category:

3	Back pressure	3	Back siphonage

▲ Figure 1.99 Type HUK1 hose union bib tap with double check valve arrangement

Type HC – diverter with automatic return

This is a mechanical backflow prevention device used in bath/shower combination tap assemblies which automatically returns the bath outlet open to atmosphere if a vacuum occurs at the inlet to the device.

The type HC diverter with automatic return is usually incorporated into the design of an appliance or fitting. It is not a 'standalone' fitting that can be added to the installation. A good example of a type HC diverter would be a bath/shower mixing valve with a diverter valve to operate the shower. Whilst pressure is maintained, the diverter valve remains open and the water is fed to the shower hose. Should loss of pressure occur, the diverter valve closes and any excess water in the shower hose returns to the bath through the open tap, thus preventing the water from backflowing down the cold supply pipe. They are suitable for fluid category 3 to prevent back siphonage only.

Regulations and guidance for backflow prevention

The following clauses are taken from the Water Supply (Water Fittings) Regulations 1999 Guidance Document (published by Defra – Department for Environment Food & Rural Affairs) Section 6.4 Guidance clauses relating to Paragraph 15 of Schedule 2 Backflow prevention.

General

G15.1 Except where expanded water from hot water systems or instantaneous water heaters is permitted to flow back into a supply or distributing pipe, every water fitting through which water is supplied for domestic purposes should be installed in such a manner that no backflow of fluid from any appliance, fitting or process can take place.

G15.2 Avoidance of backflow should be achieved by good system design and the provision of suitable backflow prevention arrangements and devices, the type of which depends on the fluid category to which the wholesome water is discharged. A description of fluid risk categories is shown in Schedule 1 of the Regulations and some suggested examples relating to the fluid categories are shown in Tables 1.5 and 1.7.

G15.3 The type of backflow protection for a given situation is related to the fluid risk categories downstream of the backflow prevention device.

G15.4 Schedules of backflow prevention arrangements and backflow prevention devices, and the maximum permissible fluid risk category for which they are acceptable, are shown in Tables 1.5 and 1.7.

G15.5 Wherever practicable, systems should be protected against backflow without the necessity to rely on mechanical backflow protection devices; this can often be achieved by point of use protection such as a 'tap gap' above the spillover level of an appliance. Minimum air gaps for different sizes of taps and applications are shown in Table 1.5.

G15.6 In cistern-fed systems secondary backflow prevention can often be achieved for appliances by the use of permanently vented distributing pipes.

G15.7 Mechanical backflow protection devices which, depending on the type of device, may be suitable for protection against back pressure or back siphonage, or both, should be installed so that:
- They are readily accessible for inspection, operational maintenance and renewal; and,

- Except for types HA and HUK1, backflow prevention devices for protection against fluid categories 2 and 3, they should not be located outside premises; and,
- They are not buried in the ground; and,
- Vented or verifiable devices, or devices with relief outlets, are not installed in chambers below ground level or where liable to flooding; and,
- Line strainers are provided immediately upstream of all backflow prevention devices required for fluid category 4. Where strainers are provided, servicing valves are to be fitted upstream of the line strainer and downstream of the backflow prevention device; and,
- The lowest point of the relief outlet from any reduced pressure zone valve assembly or similar device should terminate with a type AA air gap located not less than 300 mm above the ground or floor level.

KEY POINT

For information on the installation and maintenance of reduced pressure zone devices (RPZ valve assemblies) see Installation and Guidance Note No. 9-03-02 published by the Water Regulations Advisory Scheme (WRAS).

Appliances incorporating or supplied with water through pumps

G15.8 Where pumped showers, or other appliances supplied through or incorporating pumps, are installed, care should be taken in positioning branches from distributing pipes.

Bidets (including WCs adapted as bidets) with flexible hose and spray handset fittings and with submerged water inlets

G15.9 Bidets with flexible hose and spray handset fittings and/or water inlets below the spillover level of the appliance, are a fluid category 5 risk and should not be supplied with water directly from a supply pipe.

G15.10 Bidets of this type may:

- Be supplied with cold and/or hot water through type AA, AB, or AD backflow prevention arrangements serving the bidet only; or,
- Be supplied with cold water from an independent distributing pipe serving the bidet only or a common distributing pipe serving the bidet and which may also serve a WC or urinal flushing cistern only; or,
- Be supplied with hot water from a water heater, which is supplied from an independent distributing pipe, that serves the bidet only, see Figure 1.100; or,
- Where the bidet is at a lower elevation than any other outlets or appliances, be supplied with water from a common cold and/or hot water vented distributing pipe providing that:
  - The elevation of the spillover level of the bidet, if there is no flexible hose; or,
  - The elevation of the spray outlet, with the hose extended vertically above the spillover level of the bidet,
- Whichever is the highest, is not less than 300 mm below the point of connection of the branch pipe serving the bidet to the main distributing pipe serving other appliances.

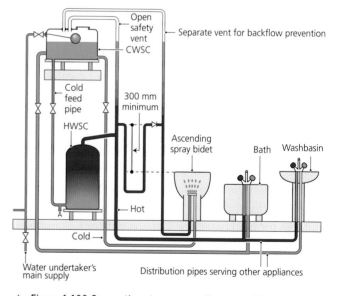

▲ Figure 1.100 Connections to an ascending spray bidet

Bidets with water inlets above spillover level only

G15.11 Bidets in domestic locations with taps or mixers located above the spillover level of the appliance, and not incorporating an ascending spray inlet below spillover level or spray and flexible hose, may be served from either a supply pipe or a distributing pipe provided that the water outlets discharge with a type AUK2 air gap above the spillover level of the appliance. See Table 1.5.

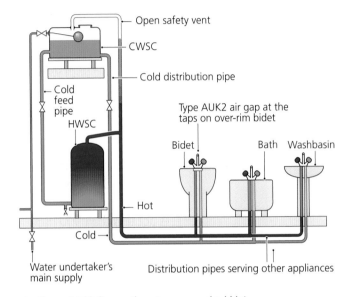

▲ Figure 1.101 Connections to an over-rim bidet

WCs and urinals

G15.12 The water supply to a manually operated WC or urinal flushing valve may be derived either from a supply pipe or a distributing pipe. The flushing valve should be located above the WC pan or urinal and must incorporate, or discharge through, a pipe interrupter with a permanent atmospheric vent; see type DC in Tables 1.5 and 1.7. The lowest part of the vent opening of the pipe interrupter should be located not less than 300 mm above the spillover level of the WC pan or not less than 150 mm above the sparge outlet of a urinal.

> ### KEY POINT
> Flushing valves cannot be used in domestic WCs or urinals.

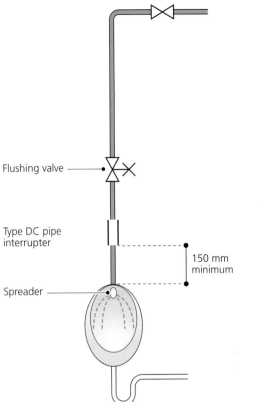

▲ Figure 1.102 Use of flushing valves and DC pipe interrupters

Shower heads or tap inlets to baths, washbasins, sinks and bidets

G15.13 Except where suitable additional backflow protection is provided, all single tap outlets, combination tap assembly outlets, or fixed shower heads terminating over washbasins, baths or bidets in domestic situations should discharge above the spillover level of the appliance with a tap gap (type AUK2) as scheduled in Table 1.5. For a sink in a domestic or non-domestic location, and for any appliances in premises where a higher level of protection is required, such as some appliances in hospitals or other health care premises, a tap gap (type AUK3) is required, see Table 1.5.

Submerged inlets to baths and washbasins

G15.14 Submerged inlets to baths or washbasins in any house or domestic situation are considered to be a fluid category 3 risk and should be supplied with water from a supply or distributing pipe through a double check valve. Submerged inlets to baths or washbasins in other than a house or domestic situation, and sinks in any location, are considered to be a fluid category 5 risk and appropriate backflow protection will be required.

Drinking water fountains

G15.15 Drinking water fountains should be designed so that the outlet of the water delivery jet nozzle is at least 25 mm above the spillover level of the bowl. The nozzle should be provided with a screen or hood to protect it from contamination.

Washing machines, washer-dryers and dishwashers

G15.16 Household washing machines, washer-dryers and dishwashers are manufactured to satisfy a fluid category 3 risk. Where they are likely to be used in a non-domestic situation, appropriate backflow protection for a higher fluid risk category should be provided.

Hose pipes for house garden and other applications

G15.17 Handheld hoses should be fitted with a self-closing mechanism at the outlet of the hose.

▲ Figure 1.103 Typical self-closing handheld garden spray

Commercial and other installations excluding house gardens

G15.18 Any taps and fittings used for supplying water for non-domestic applications, such as commercial, horticultural, agricultural or industrial purposes should be provided with:

- Backflow protection devices appropriate to the downstream fluid category; and,
- Where appropriate, a zone protection system.

G15.19 Soil watering systems installed in close proximity to the soil surface (that is, where the

watered surface is less than 150 mm below the water outlet discharge point), for example irrigation systems and permeable hoses, are considered to be a fluid category 5 risk and should only be supplied with water through a type AA, AB, AD or AUK1 air gap arrangement.

House garden installations

G15.20 Taps to which hoses are, or may be connected and located in house garden locations are to be protected against backflow by means of a double check valve. The double check valve should be located inside a building and protected from freezing.

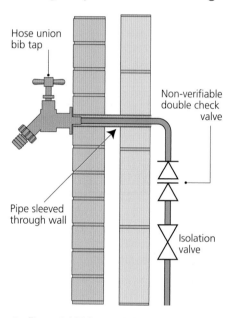

Hose union bib tap

Non-verifiable double check valve

Pipe sleeved through wall

Isolation valve

▲ Figure 1.104 Domestic hose union bib tap installation

G15.21 Where, in existing house installations, a hose pipe is to be used from an existing hose union tap located outside a house and which is not provided with backflow protection, either:

- The existing hose union tap should be provided with a double check valve located inside the building; or,
- The tap should be replaced with a hose union tap that incorporates a double check valve (type HUK1); or,
- A hose union backflow preventer (type HA) or a double check valve should be continuously fitted to the outlet of the tap.

G15.22 Where fixed or handheld devices are used with hose pipes for the application of fertilisers or domestic detergents the minimum backflow protection provided should be suitable for protection against a fluid category 3 risk. Backflow protection against a fluid category 5 risk should be provided where these devices are used for the application of insecticides.

G15.23 Where mini-irrigation systems, such as porous hoses, are installed in house garden situations only, a hose union tap with backflow protection in accordance with clauses G15.20 or G15.21 combined with a pipe interrupter with atmospheric vent and moving element device (type DB) at the connection of the hose to the hose union tap, or not less than 300 mm above the highest point of the delivery point of the spray outlet or the perforated surface of the porous hose, whichever is the highest, is acceptable.

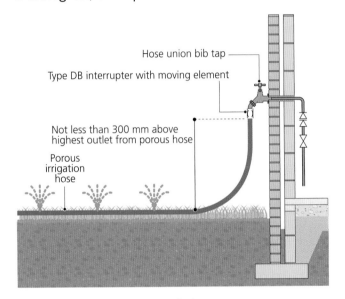

▲ Figure 1.105 Porous hose installation

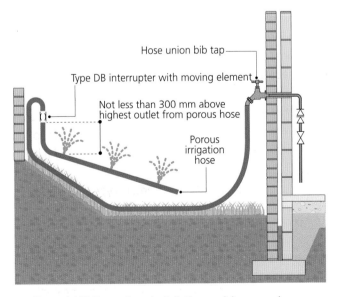

▲ Figure 1.106 Porous hose installation on rising ground

Whole-site and zone protection

G15.24 A whole-site or zone backflow prevention device should be provided on the supply or distributing pipe, such as a single check valve or double check valve, or other no less effective backflow prevention device, according to the level of risk as judged by the water undertaker where:

- A supply or distributing pipe conveys water to two or more separately occupied premises (whether or not they are separately chargeable by the water supplier for a supply of water); or,
- A supply pipe conveys water to premises, which under any enactment are required to provide a storage cistern capable of holding sufficient water for not less than 24 hours ordinary use.

G15.25 The provision of zone or whole-site backflow protection should be in addition to individual requirements at points of use and within the system.

G15.26 Zone protection may be required in other than domestic premises where particular industrial, chemical or medical processes are undertaken.

Fire protection systems

G15.27 Wet sprinkler systems (without additives), first-aid fire hose reels and hydrant landing valves are considered a fluid category 2 backflow risk. Wet sprinkler systems with additives to prevent freezing are considered a fluid category 4 risk.

G15.28 Fluids contained within large cylindrical hydro-pneumatic pressurised vessels are considered to be fluid category 4 risk.

G15.29 Where fire protection systems and drinking water systems are served from a common domestic supply pipe, the connection to the fire systems should be taken from the supply pipe immediately on entry to the building and appropriate backflow protection devices should be installed.

Methods of preventing cross connection in systems that contain non-wholesome water sources

A cross connection is a direct, physical connection between wholesome, potable water and water that is considered non-potable, such as recycled water or harvested rainwater. In extreme circumstances, this can result in serious illness and even death. Cross connections occur during correct plumbing design and installation, such as the hot and cold connections to a shower valve or a mixer tap (cross connection between fluid category 1 and fluid category 2) and these, for the most part, are protected by the correct use of mechanical backflow prevention devices. However, some modern plumbing systems require much more thought and planning, rather than simply the installation of a check valve. The Water Supply (Water Fittings) Regulations 1999 demand that cross connections from a water undertaker's mains to recycled and rainwater harvesting systems, and even connections to private water supplies are eliminated completely in order to safeguard the wholesome water supply. There are several ways in which we can do this:

- correct design of systems taking into account the requirements of the regulations in place
- careful planning and routing of pipework and fittings
- careful use of mechanical backflow prevention devices and air gaps
- using the correct methods of marking and colour coding pipework and systems.

Of these, identification of pipework is most important, especially when additions to the system are required or during routine and emergency maintenance operations.

4 DESIGN TECHNIQUES FOR COLD WATER SYSTEMS

The design of cold water installations involves the calculation of flow rates, capacities of stored water required and pipe sizes. On large contracts it may also involve the planning of pipe routes based upon the architect's drawings. Design processes can be divided into three distinct groups:

- large industrial/commercial contracts – on large contracts a building services engineer (BSE) will design the cold water installation
- new build domestic installations – with these types of installations, the cold water design will

be completed by the plumbing company who may employ their own designer/estimator
- existing dwellings – here, the design of systems is usually completed by an experienced operative who has knowledge of the procedures for completing simple flow rate and pipe sizing calculations.

The design of a system is the first important step towards a successful installation as the calculations completed must allow the finished installation to deliver the specified flow rates and component performances based upon the manufacturer's literature. It is at this stage that the type of materials to be used for the installation will be chosen. This will be based upon the type of building, its uses, the type of system being installed and cost.

Information sources

The installation of cold water systems needs to comply with the Water Supply (Water Fittings) Regulations and we must always consider the recommendations of **BS EN 806** and **BS 8558**. Manufacturer's instructions have to be followed with regard to the appliances installed and materials used. Manufacturer's design flow rates and operating pressures will need to be considered at the system design stage for any installation to operate effectively.

Statutory regulations

Plumbing is one of the most regulated trades within the building services engineering banner. We are governed by sets of regulations, which tell us what we can and what we cannot do and what we must and what we must not do; failure to comply often results in prosecution. Regulations for cold water include:

- Water Act 2003
- Water Supply (Water Fittings) Regulations 1999
- Private Water Supplies Regulations 2016
- Building Regulations.

Aspects of these regulations were discussed at the beginning of this chapter.

To give us a better understanding of the Water Supply (Water Fittings) Regulations, the Water Regulations Advisory Scheme (WRAS) have written the definitive guide to the water regulations.

British Standards and Approved Codes of Practice (ACoPs)

The British Standards provide guidance on interpreting and following the regulations. They are not enforceable, but set out as a series of recommendations so that the minimum standard to comply with the regulations can be achieved. By adhering to the recommendations within the British Standards, the regulations will be seen to be satisfied.

IMPROVE YOUR ENGLISH

The language used in the British Standards can be difficult to interpret; the Advisory Scheme's (WRAS) guidance notes are available to help you ensure you can follow and satisfy the regulations correctly.

KEY POINT

However important the regulations and the British Standards are, they are not our primary source of information when installing equipment and appliances. It must not be forgotten that the manufacturer's literature overrides both of these where a conflict arises.

Manufacturer's instructions

Manufacturer's installation, servicing/maintenance and user instructions are the most important of all documents you will have access to when installing, servicing and maintaining equipment and appliances.

They tell us in installation (technical) language what we can and must do for correct and safe operation of their equipment and they must be followed, otherwise:

- the terms of the warranty will be void
- the installation may be dangerous
- we may inadvertently be breaking the regulations.

KEY POINT

In some instances, it may seem that the instructions contradict the regulations or the British Standards. This is because regulations are only reviewed periodically, whereas manufacturers are moving forward all the time with new, more efficient products, so their information may be more up to date. In these cases, follow a simple but effective rule:

The manufacturer's literature MUST be followed at all times, even if it contradicts the regulations and British Standards.

Verbal and written communication with the customer

There are a number of ways that companies communicate with customers, such as:

- written communication:
  - letters
  - e-mails
  - faxes
  - text messages
- **verbal** communication:
  - face-to-face
  - via the telephone.

KEY TERM

Verbal: the spoken word. Any verbal communication should always be backed up with written confirmation to verify any agreements and clarify any details to prevent confusion.

Written communication

- **Letters** are an official method of communication and are usually easier to understand than verbal communication. Good written communication can help towards the success of any company by portraying and building a professional reputation. Official company business should always:
  - be in written form
  - be on company headed paper
  - have a clear layout (for example, divided into logical paragraphs)

- communicate content clearly using **standard English**.
- **Emails** have emerged as a hugely popular form of communication because of the speed that the information they contain is transferred to the recipient. As with letters, they should be well written and laid out, using correct grammar and spelling to convey professionalism, whether the recipient is a client, customer or colleague.
- **Faxes** are used mainly for conveying documents such as orders, invoices, statements and contracts where the recipient may wish to see an authorising signature. Again, the basic rules apply with regard to layout, standard English and content. Remember to always use a cover page that is appropriate for your company. This is an external communication that reflects the business and company image.
- **Messaging** is used a lot these days as an easy, convenient and cheap way to pass on information. This is an informal method of communication that may use simple text messaging or a smartphone app such as WhatsApp. This method of communication can be used to update a customer about your potential arrival time if you are delayed, for example, but should not be used for any formal information.

KEY TERM

Standard English: use of English following correct spelling and grammar.

INDUSTRY TIP

Examples of business letters are sales letters, information letters, general enquiry or problem-solving letters, etc.

Communication between the company and the customer takes place at every stage of the contract from the initial contact to customer care at the contract completion. Written communication can take the form of:

- **Quotations and estimates** – both of these are written prices as to how much the work will cost to complete. A quotation is a fixed price and cannot vary. An estimate, by comparison, is not a fixed price but can go up or down if the estimate was not accurate or the work was completed ahead of schedule. Most contractors opt for estimates because of this flexibility. Wherever possible,

the estimate/quotation should be accompanied by detailed scale drawings to help the customer understand the work that is to be completed.

- **Invoices/statements** – documents that are issued at the end of any contract as a demand for final payment for services rendered. Usually a period of time is allowed for the payment to be made.
- **Statutory cancellation rights** – a number of laws give the customer the legal right to cancel contracts after signing, providing work has not already started. There is usually no penalty for cancellation, providing the cancellation is confirmed in writing within a specific timeframe. Most cancellation periods start when the customer receives notification of their right to cancel, which should be at least seven days before work commences.
- **Handover information** – at the end of any contract, the customer must be given certain information. For large contracts, this includes the health and safety file. For small domestic contracts, a file should be made which contains any manufacturer's information, installation, servicing and user instructions, the appliance warranty information, contact numbers of key personnel within the company and a letter of thanks for their custom.

Verbal communication

The spoken word is, more often than not, our main method of communication, especially in a work context. In order to present a professional image and communicate effectively, you must:

- consider what you are saying before saying it
- evaluate the response of your listener
- maintain an appropriate **tone of voice**
- use **body language** effectively.

KEY TERMS

Tone of voice: a way of sounding to express meaning or emotion. For example, your tone of voice can communicate confidence and conviction, assuring customers that you are knowledgeable and capable.

Body language: movements and postures which communicate attitudes and feelings.

IMPROVE YOUR ENGLISH

Good verbal communication involves listening carefully as well as speaking clearly.

Verbal communication between the company and the customer is usually:

- **during the handover process** – explain and show the customer where all control valves are and how to use any appliances and controls that have been installed
- **verbal feedback** – most often, discussions with a customer help us to understand the following points:
  - details of ways in which the service to the customer could be improved
  - details of any faults that have developed in relation to the work completed; discussions with the customer can help us to diagnose and identify these faults and, therefore, complete any rectification work quickly and efficiently.

Taking measurements

Before design calculations can be made of an installation, measurements from the building must first be taken. This can be done in two ways:

- direct from the architect's scale drawings
- by visiting the site and taking measurements in situ.

Architect's scale drawings

Architect's working drawings are drawn to scale. The scale is necessary because it would be totally impractical to draw at full size (1:1) for the entire building. The scale resembles a ruler with one exception; the markings represent proportionally smaller or larger distances with the millimetre as its base measurement.

Typically, architect's scale drawings use a variety of scales. The scale can be determined by looking at the drawing legend, which is usually situated to the right of the drawing. The legend is the information that the drawing contains (such as the architect who drew it, the scale, the name of the drawing, what the drawing shows and any notes that should be considered by the person using the drawing).

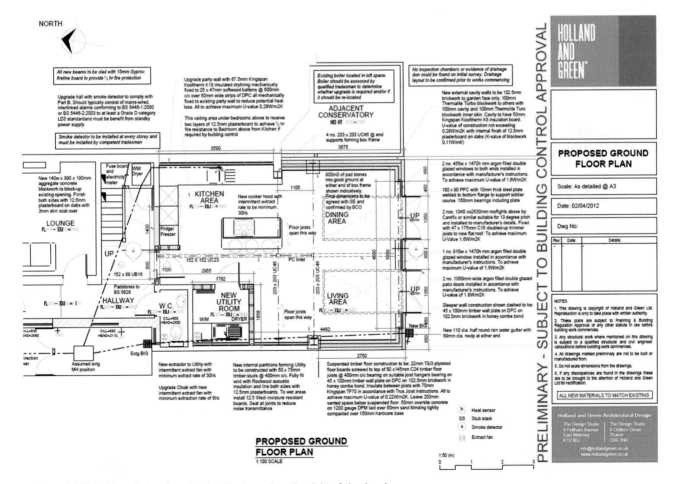

▲ Figure 1.107 Architect's drawing showing the legend on the right of the drawing

Typical drawing scales are:
- 1:1000 for site plans (10 mm = 10 m)
- 1:500 for site plans (10 mm = 5 m)
- 1:100 for plans and elevations (10 mm = 1 m)
- 1:50 for plans, sections and elevations (10 mm = 500 mm)
- 1:20 for part plans, sections and internal elevations (10 mm = 200 mm)
- 1:10 for details and joinery (10 mm = 100 mm)
- 1:5 for details (10 mm = 50 mm).

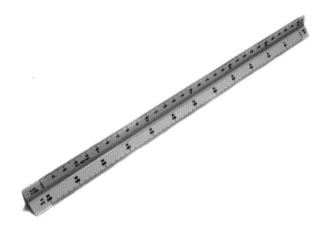

▲ Figure 1.108 Architect's scale rules

Taking measurements using a scale rule, step-by-step

1 Ensure that the drawing is lying flat so that there is no distortion of the drawing. Mistakes in measuring are easy to make when the drawing is not on a flat surface.
2 Identify the scale of the drawing from the legend.
3 Identify the part of the drawing that you wish to measure. This may be a wall, room or a pipe run.
4 Identify the correct scale on the rule that corresponds with the scale on the drawing.
5 Place the zero mark of the rule against the start of the line you wish to measure and read the length. Care must be taken here as it is very easy to misinterpret the length by reading along the wrong scale. Most scale rules have two scales along each edge and it is important that the correct scale is used.
6 Note the length you have measured.

Taking measurements on site

Quite often, especially when dealing with existing properties, it may be necessary to visit the site and take measurements directly from the building. This can be done in several ways:
- Using an architect's tape measure, which is an extra-long tape measure often used by site engineers to plot out a building. Requires two people for accurate measuring of buildings and can be inaccurate if care is not taken to prevent the tape bending and distorting.
- Using a standard tape measure, which requires two people for accurate measuring of buildings and can be inaccurate if care is not taken to prevent the tape bending and distorting.
- Using a laser measure, which is by far the most accurate method as this will not bend like a flexible tape or be affected by obstacles in the way. It uses a pulse of laser light to measure the distance and is accurate to ± 3mm. A single person can operate a laser measure quickly and effectively without distortion.
- Using an ultrasonic measure, which uses an ultrasonic wave to judge the distance, but can be affected by obstacles and obstructions that interrupt the wave.

Calculating component size

When designing cold water systems, there are certain factors that need to be taken into account before progressing to the calculation stage:
- assessment of likely demand; this will include:
  - the number and position of the appliances
  - flow rates required by individual appliances
  - the number of people living in the dwelling
  - frequency of use
- the available incoming pressure
- the length and routes of the pipework
- the type of pipework material to be used
- the requirements of any Regulations and British Standards
- the requirements of the manufacturer's installation and performance data.

Once these areas have been assessed, then the size of the components that the system will contain can be calculated. The components that are likely to be required for domestic dwellings are:
- the capacity of any cisterns that may be included in the design
- the pipe size

- the pump duty, if the system is a boosted cold water installation
- the size and charge of the pressure vessel, if the system is a boosted cold water installation.

Calculation of pipe size

When considering pipe size, there are a number of factors that must be taken into account if the volume flow rate required is to be delivered to the outlets without over sizing the pipework. The factors that need to be considered are:

- the volume flow rate required
- the pressure of the incoming supply
- the length of the pipework run
- the number of changes of direction and valves to be included in the design.

> **KEY POINT**
>
> Pipework sizing is the most important part of any good plumbing design.

The flow rate

The flow velocity is the speed at which the water is moving through the pipework. This needs to be kept to a minimum to prohibit excessive noise within the system.

BS EN 806 Part 3 recommends flow velocities between 0.5 and 2 m/s, with a maximum of 3.5 m/s only in exceptional circumstances. (**BS EN 806** superseded **BS EN 6700**, which recommended higher flow velocities. However, you may still come across designs that reference **BS EN 6700**.)

> **KEY POINT**
>
> Remember – excessive velocity + excessive pressure = excessive noise!

The pipework of any plumbing system must be designed so that the flow rates of the individual appliances and draw-offs are at least equal to those shown in the British Standards or the manufacturer's literature.

▼ Table 1.8 Draw-off flow rates (Qa), minimum flow rates at draw-off points (Q_{min}) and loading units for draw-off points – **BS EN 806**

Draw-off point	Qa (l/s)	Q_{min} (l/s)	Loading units
Washbasin, handbasin, bidet, WC cistern	0.1	0.1	1
Domestic kitchen sink, washing machine*, dish washing machine, sink, shower head	0.2	0.15	2
Urinal flush valve	0.3	0.15	3
Bath domestic	0.4	0.3	4
Taps (garden/garage)	0.5	0.4	5
Non-domestic kitchen sink DN20, bath non-domestic	0.8	0.8	8
Flush valve DN20	1.5	1.0	15

*For non-domestic appliances, check with manufacturer.

Calculating the correct pipe size will ensure that the flow rate at the appliances is adequate and meets the design specification.

Assessing the likely demand

The more outlets and appliances there are on an installation, the number that will be used at any one time generally reduces. The system in use to estimate the likely demand and, therefore, pipe size is based on **loading units**.

> **KEY TERM**
>
> **Loading unit:** a number or a factor, which is allocated to an appliance. It relates to the flow rate at the terminal fitting, the length of time in use and the frequency of use.

Table 1.8 shows the anticipated flow rates and the loading units for the common appliances that are fitted to domestic installations. Let's see how this is used.

IMPROVE YOUR MATHS

Anticipated flow rates and the loading units for the common appliance

Use **BS EN 806 Part 3** and read the tables correctly to add up the loading units and give the required flow rates for a cold water pipework system.

Let's assume that an installation is fed via the main cold water supply and contains the following appliances:

- 1 × sink
- 2 × basins
- 1 × WC
- 1 × bath
- 1 × washing machine

Use Table 2 from **BS EN 806 Part 3** (Table 1.9).

Once we have listed the appliances, we can apply the loading units by looking at Table 1.9 and multiplying by the number of appliances. This is easier in a simple table. Remember – this is just the cold water supply.

▼ Table 1.9 Loading units (**BS EN 806 Part 3**) for a simple domestic cold water system

Appliance	Number	Loading units	Total
Domestic kitchen sink	1	2	2
Washbasin	1	1	1
WC	1	1	1
Bath	1	4	4
Washing machine	1	2	2
Total loading units			**10**

Before we can begin pipe-sizing activities, the need for hot water should also be considered:

- If the system is to be a mains-fed, unvented hot water system, then the manufacturer's data should be consulted with regards to flow rate required.

- If the system is to be a combination boiler or an instantaneous multipoint hot water heater, then the loading units can be added to the table shown because the cold water supply would also have to serve these outlets.

- If, however, the hot water is to be supplied via a vented hot water cylinder, then the capacity of the cistern feeding it will have to be taken into account.

From this, we can calculate the required flow for the cistern.

For this example, we should consider that British Standard **BS EN 806** no longer states a minimum capacity for cisterns. However, it is accepted that there is a minimum of 230 litres storage where the cistern is supplying both cold water outlets via a cold distribution pipe and cold water to a hot water storage vessel (not including a gravity or pumped shower). In the past, it has always been the case that where the cistern is supplying cold water to a hot water storage vessel only, that the capacity of any cistern should be at least equal to the hot water vessel/cylinder it is supplying.

The average hot water cylinder contains around 140 litres of hot water. We will assume therefore that the cold water storage cistern will have the same capacity. The flow rate for the float-operated valve serving the cold water storage cistern can be calculated by dividing the cistern capacity by the required filling time in seconds. (The filling time should not exceed one hour.)

Assuming a filling time of 15 minutes:

140 litres ÷ 900 seconds (15 mins) = 0.15 l/s flow rate

Now work out the flow rate for a 230 litre cistern that needs to fill in 30 minutes.

230 ÷ 1800 seconds (30 mins) = 0.127 or 0.13 l/s flow rate

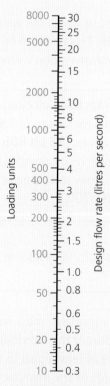

▲ Figure 1.109 Conversion of loading units to design flow rate taken from **BS6700** (Annex D). This is updated in **BS EN 806 Part 3**.

Because the flow rate is small, no loading unit exists for it so once we have worked out the flow rate of the loading units, the flow rate for the cistern can be added on.

The loading unit total is 10. By looking at Figure 1.109 we can determine the flow rate.

From the chart we can see that the flow rate for 10 loading units is 0.3 l/s. Add to this the 0.15 l/s for the storage cistern gives a total flow rate of 0.45 l/s. Now compare the total flow rate by using **BS EN 806** Figure B1 which gives a Q_D flow rate of 0.4 l/s for 10 loading units.

The available pressure (head)

Systems that use the water undertaker's mains pressure must be checked to ascertain the minimum available pressure during peak demand. This can be done on site using a pressure gauge and the property's outside tap or kitchen sink cold supply. The incoming mains pressure to a property is governed by OFWAT (Guaranteed Standards Scheme) and the Water Supply and Sewerage Services (Customer Services Standard) Regulations 2008. They both state that a minimum of 0.7 bar (7 m **head**) should be supplied to a property, so normally water authorities try to achieve 1 bar pressure (10 m head). This will allow most modern appliances, like combination boilers, to work. This minimum standard is backed up by the individual water authority's code of practice.

OFWAT Guaranteed Standards Scheme Section 5 says:

> 'Low pressure: A company must maintain a minimum pressure in the communication pipe of seven metres static head (0.7 bar).'

The same sentence is found in the Water Supply and Sewerage Services (Customer Services Standard) Regulations 2008.

KEY TERM

Head: the pressure exerted by a column of water under gravity.

INDUSTRY TIP

You can read the OFWAT standards at www.ofwat.gov.uk/wp-content/uploads/2017/03/The-guaranteed-standards-scheme-GSS-summary-of-standards-and-conditions.pdf

With cistern-fed supplies, the head pressure is determined by measuring the distance from the base of the cistern to the outlet. The base of the cistern is used because it is a constant (it never moves), whereas the head pressure will be slightly higher than that measurement as the water level in the cistern will give a little more head pressure, but will vary due to water usage. Low pressure supplies generally require an increase in pipe size to compensate for the lack of pressure. This is why the rule of thumb says the minimum size for rising main (the incoming main rising through the building) is 15 mm, but the minimum size for cold distribution is 22 mm (the cold water pipe from the cold water cistern in the loft to the appliances in the property).

INDUSTRY TIP

Flow can be expressed in two ways:
- **Laminar flow** – this is where a fluid, such as water, travels in regular paths. Often called streamline flow, the velocity and pressures at each point remain fairly constant. Laminar flow over a parallel surface such as the internal bore of a pipe consists of layers that are all parallel to each other. The fluid that is in contact with the pipe surface moves only very slowly because of the resistance offered by the pipe material and all other layers slide over it with varying degrees of velocity. The pipe moves at it fastest in the centre of the pipe because there is little or no resistance to flow.
- **Turbulent flow** – this is flow which undergoes irregular fluctuations. The fluid continuously changes direction and velocity. The water swirls and creates eddies while the bulk of the water generally flows in one direction. In a pipe, turbulent flow can be caused by many factors including the internal roughness of the pipe bore or sudden changes in direction, such as an elbow or a tee piece.

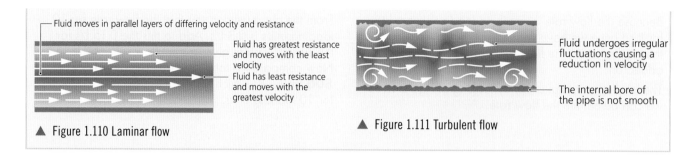

Fluid moves in parallel layers of differing velocity and resistance

Fluid has greatest resistance and moves with the least velocity

Fluid has least resistance and moves with the greatest velocity

Fluid undergoes irregular fluctuations causing a reduction in velocity

The internal bore of the pipe is not smooth

▲ Figure 1.110 Laminar flow

▲ Figure 1.111 Turbulent flow

Pipe sizing

The tabulated method of pipe sizing used to be commonly used when **BS 6700** was active. The tabulated method is now only used for complex domestic and commercial installation, but you still are required to have a working knowledge of the tabulated method. **BS 6700** has now been replaced with **BS EN 806** and **BS 8558**, which offer a simplified method for pipe sizing. This method is found in **BS EN 806 Part 3** and is now the recognised method for pipe sizing on standard domestic installations.

We will work through a domestic installation using both methods, the first method being the tabulated method.

Tabulated pipe sizing method

Firstly, you will need a blank table to fill out with all the figures. The layout of the table should be similar to Table 1.10 below.

The number of lines may vary according to the number of pipework sections, and the mains head will vary with the location supply.

Stage 1 – Make a simple sketch of the system

Make a sketch of the system installation numbering each section of pipework, starting with the incoming mains and then going up through the system. Put any dimensions on the sketch and label the appliance outlets with their LOADING UNITS and MINIMUM FLOW RATES. The loading units and flow rates can be found in **BS EN 806 Part 3** Table 2. Transfer the data to a simple table.

▼ Table 1.10 Blank table for tabulated method

Initial mains head = 30 m (3 bar)

1	2	3	4	5	6	7	8	9	10	11	12	13	14
PIPE REFERENCE	LOADING UNITS	MINIMUM FLOW RATE (l/s)	ASSUMED PIPE SIZE (mm)	HEAD LOSS (m/m RUN)	VELOCITY OF FLOW RATE (m/s)	ACTUAL PIPE LENGTH (m)	EQUIVALENT PIPE LENGTH (m)	EFFECTIVE PIPE LENGTH (m) (column 7 + column 8)	VERTICAL RISE OR DROP	HEAD USED (m) (column 5 × column 9 – head loss of outlets)	RESIDUAL HEAD (m) (column 13 – column 11)	HEAD AVAILABLE (m) (column 12 ± column 10)	FINAL PIPE SIZE (mm)

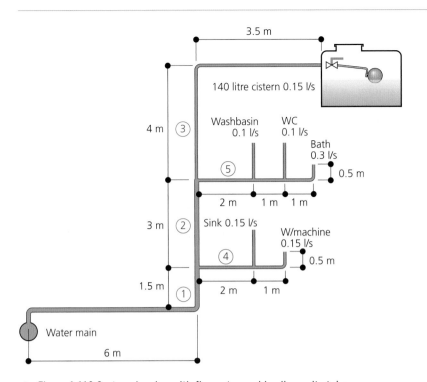

Section	Appliance	Loading units LU	Flow rate l/s
3	Cistern	–	**0.15**
5	Basin	1	0.1
	WC	1	0.1
	Bath	4	0.3
		6	**0.5**
2	Appliances from sections 3 and 5	– 6	0.15 0.5
		6	**0.65**
4	Sink	2	0.15
	Washing m/c	2	0.15
		4	**0.3**
1	Appliances from sections 2, 3, 4 and 5	4 6	0.3 0.65
		10	**0.95**

▲ Figure 1.112 System drawing with flow rates and loading units taken from **BS 6700** (old)

Now add the loading units and flow rates to the calculation table in column 2 and 3.

▼ Table 1.11 The tabular method (stage 1)

Initial mains head = 30 m (3 bar)

1	2	3	4	5	6	7	8	9	10	11	12	13	14
PIPE REFERENCE	LOADING UNITS	MINIMUM FLOW RATE (l/s)	ASSUMED PIPE SIZE (mm)	HEAD LOSS (m/m RUN)	VELOCITY OF FLOW RATE (m/s)	ACTUAL PIPE LENGTH (m)	EQUIVALENT PIPE LENGTH (m)	EFFECTIVE PIPE LENGTH (m) (column 7 + column 8)	VERTICAL RISE OR DROP	HEAD USED (m) (column 5 × column 9 – head loss of outlets)	RESIDUAL HEAD (m) (column 13 – column 11)	HEAD AVAILABLE (m) (column 12 ± column 10)	FINAL PIPE SIZE (mm)
1	10	0.95											
2	6	0.65											
3	–	0.15											
4	4	0.3											
5	6	0.5											

Stage 2 – Assumed pipe diameter

Using your experience and skill, you must assume a pipe size that will cope with the required demand. The basis of your assumption is that the larger the diameter of pipe, the greater the flow rate. Looking at column 3, the flow rates can be identified. Now add your pipe diameters to column 4. At this stage it is only an assumption and the rest of the table will prove whether the choice is correct.

▼ Table 1.12 The tabular method (stage 2)

Initial mains head = 30 m (3 bar)

1 PIPE REFERENCE	2 LOADING UNITS	3 MINIMUM FLOW RATE (l/s)	4 ASSUMED PIPE SIZE (mm)	5 HEAD LOSS (m/m RUN)	6 VELOCITY OF FLOW RATE (m/s)	7 ACTUAL PIPE LENGTH (m)	8 EQUIVALENT PIPE LENGTH (m)	9 EFFECTIVE PIPE LENGTH (m) (column 7 + column 8)	10 VERTICAL RISE OR DROP	11 HEAD USED (m) (column 5 × column 9 – head loss of outlets)	12 RESIDUAL HEAD (m)	13 HEAD AVAILABLE (m) (column 13 – column 11)	14 FINAL PIPE SIZE (mm) (column 12 ± column 10)
1	10	0.95	**28**										
2	6	0.65	**22**										
3	–	0.15	**22**										
4	4	0.3	**15**										
5	6	0.5	**15**										

Stage 3 – Actual and effective pipe lengths

Looking at the sketch made at Stage 1, add up the actual pipe lengths identified and add those lengths to column 7. Using Table 1.13, work out the equivalent pipe lengths which identify the impact that fittings have on the installation, then add those lengths to column 8.

▼ Table 1.13 Equivalent lengths of pipe for fittings (copper, stainless steel and plastics only)

Pipe size (mm)	Elbow (m)	Tee (m)	Stop valve (m)	Check valve (m)
15	0.5	0.6	4.0	2.5
22	0.8	1.0	7.0	4.3
28	1.0	1.5	10.0	5.6
35	1.4	2.0	13.0	6.0
42	1.7	2.5	16.0	7.9
54	2.3	3.5	22.0	11.5

For tees, consider the branch only.
Gate valve resistance is insignificant and will not affect flow rate.
For fittings not shown, consult manufacturer's literature.

▼ Table 1.14 Calculations for column 8

Section	Fitting	Number	Size (mm)	Equivalent length (m)	Total (m)
1	Elbow	2	28	1.0	2.0
	Tee	1	28	1.5	1.5
	Stop valve	1	28	10.0	10.0
					13.5
2	Tee	1	22	1.0	1.0
					1.0
3	Elbow	1	15	0.5	0.5
	Service valve (fullway)	1	15	0	0.0
					0.5
4	Elbow	1	15	0.5	0.5
	Tee	1	15	0.6	0.6
					1.1
5	Elbow	1	22	0.8	0.8
	Tee	2	22	1.0	2.0
					2.8

Once you have put the figures into columns 7 and 8, add them up and put the effective pipe length in column 9.

▼ Table 1.15 The tabular method (stage 3)

Initial mains head = 30 m (3 bar)

1	2	3	4	5	6	7	8	9	10	11	12	13	14
PIPE REFERENCE	LOADING UNITS	MINIMUM FLOW RATE (l/s)	ASSUMED PIPE SIZE (mm)	HEAD LOSS (m/m RUN)	VELOCITY OF FLOW RATE (m/s)	ACTUAL PIPE LENGTH (m)	EQUIVALENT PIPE LENGTH (m)	EFFECTIVE PIPE LENGTH (m) (column 7 + column 8)	VERTICAL RISE OR DROP	HEAD USED (m) (column 5 × column 9 – head loss of outlets)	RESIDUAL HEAD (m) (column 13 – column 11)	HEAD AVAILABLE (m) (column 12 ± column 10)	FINAL PIPE SIZE (mm)
1	10	0.95	28			7.5	13.5	21.0					
2	6	0.65	22			3.0	1.0	4.0					
3	–	0.15	22			7.5	0.5	8.0					
4	4	0.3	15			3.5	1.1	4.6					
5	6	0.5	15			4.5	2.8	7.3					

Stage 4 – Head loss and velocity

The graph shown in Figures 1.113 and 1.114 allows you to determine the head loss for a given flow rate. Using the flow rates on the left-hand side of the graph, move across until you reach the pipe size being used. Then move vertically down to read the head loss figure. Put this into column 5. These figures need to be read as accurately as possible.

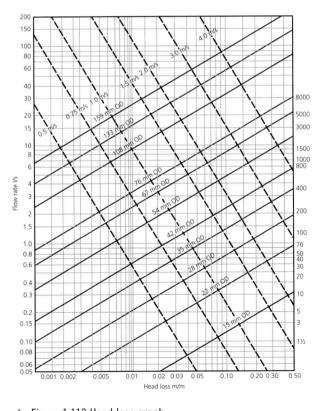

▲ Figure 1.113 Head loss graph

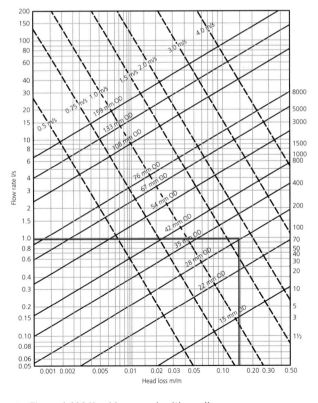

▲ Figure 1.114 Head loss graph with reading

▼ Table 1.16 The tabular method (stage 4 after determining head loss)

Initial mains head = 30 m (3 bar)

1	2	3	4	5	6	7	8	9	10	11	12	13	14
PIPE REFERENCE	LOADING UNITS	MINIMUM FLOW RATE (l/s)	ASSUMED PIPE SIZE (mm)	HEAD LOSS (m/m RUN)	VELOCITY OF FLOW RATE (m/s)	ACTUAL PIPE LENGTH (m)	EQUIVALENT PIPE LENGTH (m)	EFFECTIVE PIPE LENGTH (m) (column 7 + column 8)	VERTICAL RISE OR DROP	HEAD USED (m) (column 5 × column 9 – head loss of outlets)	RESIDUAL HEAD (m)	HEAD AVAILABLE (m) (column 13 – column 11)	FINAL PIPE SIZE (mm) (column 12 ± column 10)
1	10	0.95	28	**0.14**		7.5	13.5	21.0					
2	6	0.65	22	**0.26**		3.0	1.0	4.0					
3	–	0.15	22	**0.14**		7.5	0.5	8.0					
4	4	0.3	15	**0.45**		3.5	1.1	4.6					
5	6	0.5	15	**0.16**		4.5	2.8	7.3					

Using Table D2 from **BS 6700** (Figure 1.115), add the velocity of flow to column 6. This table is read using the edge of a ruler. Place the ruler edge on the outside diameter of the chosen pipe (right-hand line on Figure 1.115). Use that point as a hinge point. Keeping the edge of the ruler on the diameter, align the same edge of the ruler with the flow rate for that section of the pipe (inside right line on Figure 1.115). Keeping the ruler at this point, read off the velocity from the inside left line. You can now add this figure to column 6.

Use the assumed pipe diameter column 4 on the table to line 5 on Table D2 from **BS 6700** (see Figures 1.115 and 1.116).

Use a ruler edge to line up pipe diameter (line 5) to flow rate (column 3) to line 4 on Table D2.

For example:

Section 1 = 28 mm pipe with flow rate of 0.95 l/s

Read off line 1 for velocity.

Now do the same for sections 2, 3, 4 and 5.

▼ Table 1.17 The tabular method (stage 4 after determining head loss and velocity of flow)

Initial mains head = 30 m (3 bar)

1	2	3	4	5	6	7	8	9	10	11	12	13	14
PIPE REFERENCE	LOADING UNITS	MINIMUM FLOW RATE (l/s)	ASSUMED PIPE SIZE (mm)	HEAD LOSS (m/m RUN)	VELOCITY OF FLOW RATE (m/s)	ACTUAL PIPE LENGTH (m)	EQUIVALENT PIPE LENGTH (m)	EFFECTIVE PIPE LENGTH (m) (column 7 + column 8)	VERTICAL RISE OR DROP	HEAD USED (m) (column 5 × column 9 – head loss of outlets)	RESIDUAL HEAD (m)	HEAD AVAILABLE (m) (column 13 – column 11)	FINAL PIPE SIZE (mm) (column 12 ± column 10)
1	10	0.95	28	0.14	**1.80**	7.5	13.5	21.0					
2	6	0.65	22	0.26	**0.26**	3.0	1.0	4.0					
3	–	0.15	22	0.14	**0.14**	7.5	0.5	8.0					
4	4	0.3	15	0.45	**0.45**	3.5	1.1	4.6					
5	6	0.5	15	0.16	**0.16**	4.5	2.8	7.3					

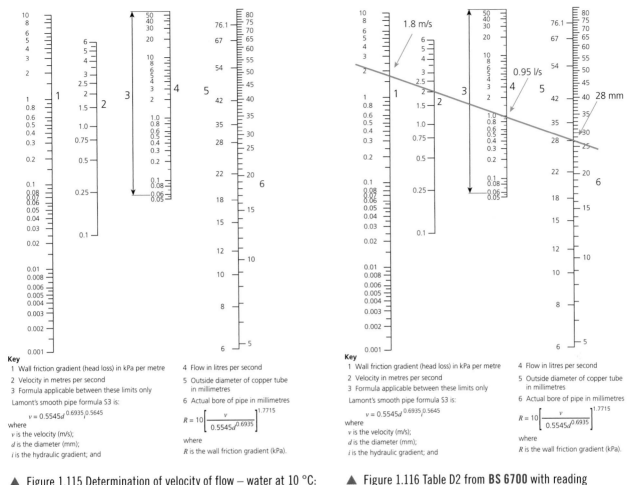

Key
1 Wall friction gradient (head loss) in kPa per metre
2 Velocity in metres per second
3 Formula applicable between these limits only

Lamont's smooth pipe formula S3 is:

$$v = 0.5545 d^{0.6935} i^{0.5645}$$

where
v is the velocity (m/s);
d is the diameter (mm);
i is the hydraulic gradient; and

4 Flow in litres per second
5 Outside diameter of copper tube in millimetres
6 Actual bore of pipe in millimetres

$$R = 10 \left[\frac{v}{0.5545 d^{0.6935}} \right]^{1.7715}$$

where
R is the wall friction gradient (kPa).

▲ Figure 1.115 Determination of velocity of flow – water at 10 °C: Table D2 from **BS 6700**

Key
1 Wall friction gradient (head loss) in kPa per metre
2 Velocity in metres per second
3 Formula applicable between these limits only

Lamont's smooth pipe formula S3 is:

$$v = 0.5545 d^{0.6935} i^{0.5645}$$

where
v is the velocity (m/s);
d is the diameter (mm);
i is the hydraulic gradient; and

4 Flow in litres per second
5 Outside diameter of copper tube in millimetres
6 Actual bore of pipe in millimetres

$$R = 10 \left[\frac{v}{0.5545 d^{0.6935}} \right]^{1.7715}$$

where
R is the wall friction gradient (kPa).

▲ Figure 1.116 Table D2 from **BS 6700** with reading

Stage 5 – Head used

We can now calculate the head used in metres. This is calculated by multiplying column 5 by column 9 and adding the figure to column 11.

▼ Table 1.18 The tabular method (stage 5)

Initial mains head = 30 m (3 bar)

1 PIPE REFERENCE	2 LOADING UNITS	3 MINIMUM FLOW RATE (l/s)	4 ASSUMED PIPE SIZE (mm)	5 HEAD LOSS (m/m RUN)	6 VELOCITY OF FLOW RATE (m/s)	7 ACTUAL PIPE LENGTH (m)	8 EQUIVALENT PIPE LENGTH (m)	9 EFFECTIVE PIPE LENGTH (m) (column 7 + column 8)	10 VERTICAL RISE OR DROP	11 HEAD USED (m) (column 5 × column 9 – head loss of outlets)	12 RESIDUAL HEAD (m)	13 HEAD AVAILABLE (m) (column 13 − column 11)	14 FINAL PIPE SIZE (mm) (column 12 ± column 10)
1	10	0.95	28	0.14	1.80	7.5	13.5	21.0		2.94			
2	6	0.65	22	0.26	0.26	3.0	1.0	4.0		1.04			
3	–	0.15	22	0.14	0.14	7.5	0.5	8.0		1.12			
4	4	0.3	15	0.45	0.45	3.5	1.1	4.6		2.57			
5	6	0.5	15	0.16	0.16	4.5	2.8	7.3		2.47			

Stage 6 – Available head pressure

The head (or pressure) at the outlet can be measured vertically in metre head. The table states that there is 30 m (3 bar) head pressure available. As the water rises through the system, some head pressure is lost due to the pipework rising vertically. Looking at the sketch, section 1 pipework rises 1.5 m (column 10), so the available head is 30 m – 1.5m = 28.5 m (column 13). Now put all the figures in columns 10 and 13.

▼ Table 1.19 The tabular method (stage 6)

Initial mains head = 30 m (3 bar)

1	2	3	4	5	6	7	8	9	10	11	12	13	14
PIPE REFERENCE	LOADING UNITS	MINIMUM FLOW RATE (l/s)	ASSUMED PIPE SIZE (mm)	HEAD LOSS (m/m RUN)	VELOCITY OF FLOW RATE (m/s)	ACTUAL PIPE LENGTH (m)	EQUIVALENT PIPE LENGTH (m)	EFFECTIVE PIPE LENGTH (m) (column 7 + column 8)	VERTICAL RISE OR DROP	HEAD USED (m) (column 5 × column 9 – head loss of outlets)	RESIDUAL HEAD (m) (column 13 – column 11)	HEAD AVAILABLE (m) (column 12 ± column 10)	FINAL PIPE SIZE (mm)
1	10	0.95	28	0.14	1.80	7.5	13.5	21.0	**−1.5**	2.94	**25.56**	**28.5**	
2	6	0.65	22	0.26	0.26	3.0	1.0	4.0	**−4.5**	1.04	**24.46**	**25.5**	
3	–	0.15	22	0.14	0.14	7.5	0.5	8.0	**−8.5**	1.12	**20.38**	**21.5**	
4	4	0.3	15	0.45	0.45	3.5	1.1	4.6	**−2.0**	2.57	**25.43**	**28.0**	
5	6	0.5	15	0.16	0.16	4.5	2.8	7.3	**−5.0**	2.47	**22.53**	**25.0**	

Stage 7 – Final pipe size

To check to see if the assumed pipe diameters are correct, we need to inspect the table for two figures.

Firstly, the velocity in any of the pipework must not be above 3.0 m/s, otherwise there will be system noise (column 6).

Secondly, the available head pressure should be above 10 m (1 bar) at each section of pipework (column 12).

If the velocity is below 3.0 m/s and the residual head is above 10 m, the pipe choice is good and can be installed.

This tabulated method of pipe sizing is now only used for more complex or commercial pipework installations. Today, **BS EN 806 Part 3** allows us to use the simplified pipe sizing method for basic domestic installation.

▼ Table 1.20 The tabular method (stage 7)

Initial mains head = 30 m (3 bar)

1	2	3	4	5	6	7	8	9	10	11	12	13	14
PIPE REFERENCE	LOADING UNITS	MINIMUM FLOW RATE (l/s)	ASSUMED PIPE SIZE (mm)	HEAD LOSS (m/m RUN)	VELOCITY OF FLOW RATE (m/s)	ACTUAL PIPE LENGTH (m)	EQUIVALENT PIPE LENGTH (m)	EFFECTIVE PIPE LENGTH (m) (column 7 + column 8)	VERTICAL RISE OR DROP	HEAD USED (m) (column 5 × column 9 – head loss of outlets)	RESIDUAL HEAD (m) (column 13 – column 11)	HEAD AVAILABLE (m) (column 12 ± column 10)	FINAL PIPE SIZE (mm)
1	10	0.95	28	0.14	**1.80**	7.5	13.5	21.0	−1.5	2.94	**25.56**	28.5	**28**
2	6	0.65	22	0.26	**0.26**	3.0	1.0	4.0	−4.5	1.04	**24.46**	25.5	**22**
3	–	0.15	22	0.14	**0.14**	7.5	0.5	8.0	−8.5	1.12	**20.38**	21.5	**22**
4	4	0.3	15	0.45	**0.45**	3.5	1.1	4.6	−2.0	2.57	**25.43**	28.0	**15**
5	6	0.5	15	0.16	**0.16**	4.5	2.8	7.3	−5.0	2.47	**22.53**	25.0	**15**

The table indicates that 28 mm pipe is suitable for section 1.

IMPROVE YOUR MATHS

Do the same exercise to see if 22 mm pipe could be used in that section.

Simplified pipe sizing method

The simplified pipe sizing method is based on **BS EN 806 Part 3** Section 5 (pages 6–13).

Stage 1 – Make a simple sketch of the system

Make a sketch of the system installation, numbering each section of pipework starting from the incoming mains and going up through the system. Put any dimensions on the sketch and label the appliance

outlets with their **loading units** and **minimum flow rates**. The loading units (LU) and the flow rates can be found on **BS EN 806 Part 3** Section 5 page 6 in Table 2. For this exercise we use the minimum flow rates (Q_{min}) in litres per second.

Add up both the loading units and flow rate for each section of pipe.

You will notice from **BS EN 806** that the table does not list a cold water storage cistern.

The flow rate is calculated by dividing the capacity by the required filling time (in seconds), which should not exceed 1 hour.

For example:

140 litres ÷ 900 seconds (15 mins) = 0.15 l/s flow rate

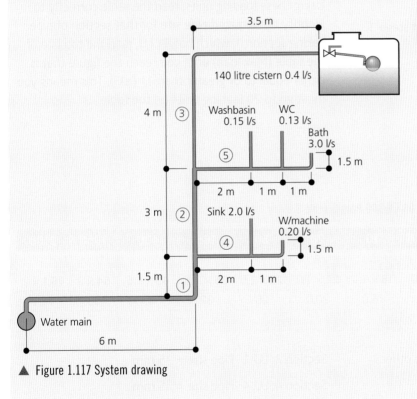

▲ Figure 1.117 System drawing

Section	Appliance	Loading units (LU)	Flow rate (l/s)
1	Sink	2	0.2
	Washing machine	2	0.2
	Basin	1	0.1
	WC	1	0.1
	Bath	4	0.4
	Cistern	1	0.15
		11	**1.15**
2	Basin	1	0.1
	WC	1	0.1
	Bath	4	0.4
	Cistern	1	0.1
		7	**0.7**
3	Cistern	1	1.15
4	Sink	2	0.2
	Washing machine	2	0.2
		4	**0.4**
5	Basin	1	0.1
	WC	1	0.1
		6	**0.6**

→

Stage 2 – Pipework material

Now you have the total loading units for each section to the system.

BS EN 806 contains formulated tables that use the total loading units to give the recommended pipe size.

BS EN 806 Part 3 page 7 Table 3 has a choice of eight different pipework materials including galvanised steel, copper, stainless and various plastics. You need to identify what material(s) is being used for the installation.

We will assume that the system is using **copper** pipework, which means we will be using Table 3.2.

▼ Table 1.21 Recommended pipe sizes for copper (**BS EN 806 Part 3** Table 3.2)

Max load	LU	1	2	3	3	4	6	10	20	50	165	430	1050	2100
Highest value	LU		2			4		5	8					
$d_a \times s$	mm	$12 \times 1{,}0$		$15 \times 1{,}0$				$18 \times 1{,}0$	$22 \times 1{,}0$	$28 \times 1{,}5$	$35 \times 1{,}5$	$42 \times 1{,}5$	54×2	$76{,}1 \times 2$
d_i	mm	10,0		13,0				16,0	20,0	25	32	39	50	72,1
Max length of pipe	m	20	7	5	15	9	7							

Stage 3 – Pipe size

Using the table of loading units completed in stage 1, identify the total LU for each section:

Section 1: 11 LU

Section 2: 7 LU

Section 3: 1 LU

Section 4: 4 LU

Section 5: 6 LU

Using these loading units, read the table correctly to identify the required pipe size for that section of pipe. Taking section 1, which has 11 LU, read the top line of the table (Max load) until you reach the figure that is either equal to or greater than 11 (≥11). This means you must go to 20 loading units on the table.

▼ Table 1.22 Finding recommended pipe sizes using **BS EN 806 Part 3** Table 3.2

Max load	LU	1	2	3	3	4	6	10	20	50	165	430	1050	2100
Highest value	LU		2			4		5	8					
$d_a \times s$	mm	$12 \times 1{,}0$		$15 \times 1{,}0$				$18 \times 1{,}0$	$22 \times 1{,}0$	$28 \times 1{,}5$	$35 \times 1{,}5$	$42 \times 1{,}5$	54×2	$76{,}1 \times 2$
d_i	mm	10,0		13,0				16,0	20,0	25	32	39	50	72,1
Max length of pipe	m	20	7	5	15	9	7							

Now find the recommended pipe size for the other sections of the installation.

Section 1: LU 9 Pipe size = 22 mm

Section 2: LU 7 Pipe size = 22 mm

Section 3: LU 1 Pipe size = 15 mm

Section 4: LU 4 Pipe size = 15 mm

Section 5: LU 6 Pipe size = 15 mm

→

Stage 4 – Design flow rate

In **BS EN 806 Part 3**, Table 2 gives the draw off flow rate (Qa) and the minimum draw off rate (Q_{min}), but there are times where you may be asked about the design flow rate (Qd). If the design flow rate is required, you need to be able to read the graph in **BS EN 806 Part 3** B1 (shown in Figures 1.118 and 1.119).

If we relate this graph to the system we have been using in the above pipe sizing exercise, the highest LU identified is in section 1, which is supplying the whole house. The LU for section 1 is 11 LU: this is read on the bottom x-axis on the graph.

Next, identify the appliance on the system with the largest LU, which in this installation is the bath with 4 LU. This identifies the line on the graph that you need to read from. When you look at the graph, you can see different lines are identified with 2, 3, 4, 5, 8 and 15. As the bath has an LU of 4, you are going to read off the '4' line.

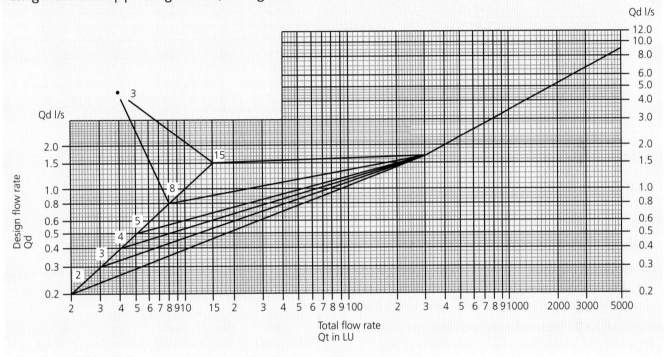

▲ Figure 1.118 Design flow rate Qd in l/s for standard installations in relation to LUs

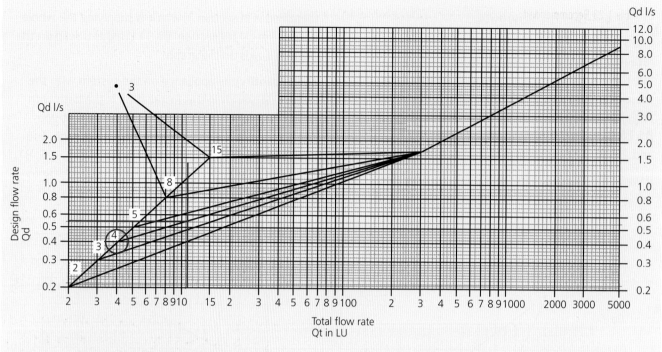

▲ Figure 1.119 Reading the design flow rate

From this you can see that Qd is 0.55 l/s.

Calculation of cistern capacity

Modern plumbing systems have seen an increase in the use of direct cold water systems whereby all the outlets within the building are supplied with water direct from a water undertaker's cold water supply. However, there are situations where the use of cold water storage cisterns is unavoidable, especially in areas where the mains pressure is low. Storing cold water also has the advantage of a reserve supply of water being available should the mains be isolated for any period of time.

British Standard BS 6700 no longer gives storage capacities for domestic dwellings but, in clause 5.3.9.4, recommends a minimum of 230 litres in systems that are supplying cold water to both cold and hot water systems.

For larger buildings, the capacity of any cistern supplying cold water depends upon:

- the type and use of the building
- the number of occupants
- the type and number of the appliances
- the frequency of use
- the likelihood of a breakdown of the supply.

Most capacities are calculated to provide a 24-hour supply should mains failure occur. Table 1.23 provides a reference for cistern capacities based on the type of building and number of occupants. The table can be found in **BS EN 806 Part 2** Clause 19.1.4 Table 6. The standard also states: 'For most dwellings where a constant supply at adequate pressure is a statutory requirement, a maximum capacity of 80 litres per person normally resident should prove satisfactory' – this includes hot water supply storage requirements (see Chapter 2, Hot water systems, planning and design).

▼ Table 1.23 Recommended minimum storage of cold water for domestic installations (**BS EN 806 Part 2** Clause 19.1.4 Table 6)

Type of building	Minimum cold water storage (litres)
Hostel	90 per bed space
Hotel	200 per bed space
Office premises with canteen facilities	45 per employee
Office premises without canteen facilities	40 per employee
Restaurant	7 per meal
Nursery/primary day school	15 per pupil
Secondary/technical day school	20 per pupil
Boarding school	90 per pupil
Children's home/residential nursery	135 per bed space
Nurses' accommodation	120 per bed space
Nursing/convalescent home	135 per bed space

This table can be used in the calculation of cistern capacity.

IMPROVE YOUR MATHS

Calculation of cistern capacities: example

Calculating cistern capacity is quite straightforward provided the correct information is to hand.
To determine cistern capacities, the following calculation should be used:

> Litres of storage required (from Table 1.23)
> × number of people/guests

Consider the following example.

A hotel has 25 beds and 50 restaurant guests. Calculate the amount of cold water storage required.

> $200 \times 25 = 5000$ litres for hotel guests
>
> $7 \times 50 = 350$ litres for restaurant guests
>
> $5000 + 350 = 5350$ litres total cold water storage

Assuming the cistern will have a 6-hour fill time, the design flow rate required to fill the cistern will be as follows:

> Design flow rate = litres required ÷ time in seconds
>
> $= 5350 \div (6 \times 3600) = 0.247$ litres/second (l/s)

This shows that it will require a flow rate of 0.25 l/s (answer rounded up) to fill a cistern with a capacity of 5350 litres in 6 hours.

IMPROVE YOUR MATHS

Cistern sizing

A small primary school is to be built accommodating 350 pupils.
Calculate the cold water storage required and the design flow rate to fill the cistern within a 4-hour fill time.

Calculation of pump power

Calculation of pump power is usually performed to ascertain the power of the pump needed to lift a certain quantity of water at a certain pressure. It is based on the physics of 'work done' relative to 'time'. 'Work done' is the applied force through distance moved and the unit of measurement is the joule. It is thus explained as the work done when a 1-newton force acts through a 1-metre distance or:

> $1 \text{ joule} = 1 \text{ N} \times 1 \text{ m}$

'Time' must be expressed as a period of seconds, which can be combined with work done to become work done over a period of time. It is expressed in the following way:

> Power = work done ÷ time
>
> $\quad$ = (force × distance) ÷ seconds
>
> $\quad$ = (newtons × metres) ÷ seconds (J/s)

where: $1 \text{ J/s} = 1 \text{ watt}$

Force in newtons = kg mass × acceleration due to gravity (9.81 m/s^2)

Power (watts) = mass × 9.81 × distance ÷ time

IMPROVE YOUR MATHS
Calculation of pump power: example

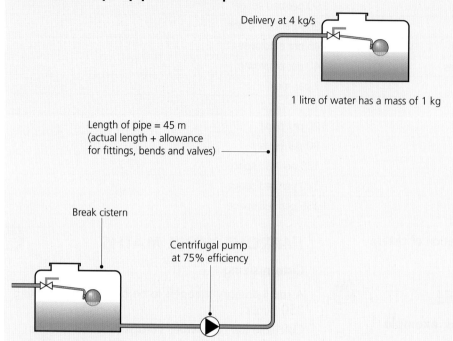

Delivery at 4 kg/s

1 litre of water has a mass of 1 kg

Length of pipe = 45 m
(actual length + allowance
for fittings, bends and valves)

Break cistern

Centrifugal pump
at 75% efficiency

▲ Figure 1.120 System drawing

As can be seen from the drawing, a delivery rate of 4 kg/s is required to fill the cistern. 1 kg = 1 litre, therefore 4 kg/s = 4 l/s. The total length of pipework with all bends, valves and fittings is 45 m.

$$\text{Power} = (\text{mass} \times 9.81 \times \text{distance}) \div \text{time}$$

$$= (4 \times 9.81 \times 45) \div 1$$

$$= 1765.8 \text{ watts}$$

Allowance for pump efficiency:

$$= 1765.8 \times (100 \div 75)$$

$$= 2354.4 \text{ watts}$$

Therefore:

Pump rating (including 75% efficiency allowance) = 2500 watts or 2.5 kW (rating rounded up to the nearest ½ kW)

Pump laws

A pump is manufactured with an impeller of constant diameter. This will have the following characteristics:

- Quantity of water delivered (Q) will vary according to the rotational speed of the impeller (N). This is expressed as:

$$(Q_2 \div Q_1) = (N_2 \div N_1)$$

- Pressure produced (P) will vary with the square of the rotational speed of the impeller (N). This is expressed as:

$$(P_2 \div P_1) = (N_2)^2 \div (N_1)^2$$

- Power (W) required will vary with the cube of the rotational speed of the impeller (N). This is expressed as:

$$(W_2 \div W_1) = (N_2)^3 \div (N_1)^3$$

Where:

Q_2 and Q_1 = discharge of water delivered (l/s)

N_2 and N_1 = rotational speed of the impeller (rpm or rps)

P_2 and P_1 = pressure produced (kPa or kN/m²)

W_2 and W_1 = power required (watts)

IMPROVE YOUR MATHS

Pump laws: example

A 25 kW pump discharges 4 kg/s when the pump impeller rotational speed is 1000 rpm. If the impeller speed is increased to 1200 rpm, what effect will this have on the power required, pressure produced and water delivered?

To calculate these changes, a transposition of the known formula is required.

Quantity of water delivered $= (Q_2 \div Q_1) = (N_2 \div N_1)$

When **transposed**, this becomes:

$Q_2 = (N_2 \times Q_1) \div N_1$

$Q_2 = (1200 \times 4) \div 1000$

$Q_2 = 4.8$ kg/s $= 4.8$ l/s

Power required $= (W_2 \div W_1) = (N_2)^3 \div (N_1)^3$

When transposed, this becomes:

$W_2 = (N_2)^3 \times W_1 \div (N_1)^3$

$W_2 = (1200)^3 \times 2500 \div (1000)^3$

$W_2 = 4320$ watts or 4.32 kW

Pressure produced $= (P_2 \div P_1) = (N_2)^2 \div (N_1)^2$

When transposed, this becomes:

$P_2 = (N_2)^2 \times P_1 \div (N_1)^2$

45 kPa pressure is produced by 1000 rpm. When increased to 1200 rpm, this will produce:

$P_2 = (1200)^2 \times 45 \div (1000)^2$

$P_2 = 64.8$ kPa

If the water pump will accept component change or upgrade and the impeller can be changed with an impeller of a different diameter, then the following formulae will apply.

At a constant rotation speed (N), the quantity of water delivered (Q) will vary according to the cube of impeller diameter (D). This is expressed as:

$(Q_2 \div Q_1) = (D_2)^3 \div (D_1)^3$

Pressure produced (P) will vary with the square of the diameter (D) of the impeller (N). This is expressed as:

$(P_2 \div P_1) = (D_2)^2 \div (D_1)^2$

Power (W) required will vary as the fifth of the diameter of the impeller (D). This is expressed as:

$(W_2 \div W_1) = (D_2)^5 \div (D_1)^5$

Assessment of accumulator capacity

Most accumulators are installed in domestic situations because the water supply entering the property suffers from either low pressure, poor flow rate or a combination of both. Accumulators, when fitted correctly, will boost flow rate and pressure to an acceptable level. It must be remembered, however, that to avoid problems with over-pressurisation during off-peak periods, a pressure reducing valve should be fitted on the mains supply. This will ensure a constant pressure entering the accumulator and prevent bursting of the internal bladder.

Although calculations exist to determine the size of accumulator required, most manufacturers of accumulators work on a fairly accurate rule of thumb. The calculation is based around the incoming pressure and flow rate of the water main. There are several factors, which will need to be taken into account:

- the pressure of the water supply
- the flow rate into the property
- the number of occupants and their water usage requirements
- the location where the accumulator is to be fitted.

Accumulators work on the principle of **Boyle's law** (see Book 1, Chapter 3, Scientific principles), which states that the absolute pressure and volume of a given mass of confined gas (in this case air) are inversely proportional, if the temperature remains unchanged within a closed system. If the space inside the closed system reduces, then the pressure within the system will rise accordingly.

A balloon when inflated contains 4 litres of air at 0.5 bar pressure. If the balloon is squeezed to half its original size, the pressure inside will double to 1 bar pressure because the same amount of air is being forced to occupy only 50 per cent of the space. In other words, it is being compressed.

KEY TERMS

Transpose: to rearrange the information to determine a different part of the formula.

Boyle's law: one of two gas laws that determine the characteristics of a gas.

IMPROVE YOUR MATHS

Assessment of accumulator capacity: example

With this theory in mind, look at the following example.

A property, occupied by two people, has an incoming mains supply pressure of 2 bar. The flow rate of the incoming supply is low at 10 litres per minute (0.1666 litres per second). The property also has a shower fitted delivering a flow rate of approximately 15 litres per minute. Calculate:

- the amount of storage for a given accumulator size
- the air charge required
- the length of time of storage usage
- the length of time required to replenish the accumulator.

The pressure charge in the accumulator needs to be half that of the incoming water supply if it is to fill sufficiently. In this case, if a 2 bar charge was used inside the accumulator, then, because the incoming supply is also 2 bar, the accumulator would not fill at all. Therefore, a lower accumulator charge of 1 bar is required to allow it to fill satisfactorily. This means that if the accumulator has a capacity of 300 litres with a charge of 1 bar pressure of air and an incoming supply pressure of 2 bar, then the accumulator would store 150 litres of water, half of the actual capacity.

If the shower flow rate is 15 litres per minute, then:

Litres ÷ flow rate of the shower = length of time of storage usage

$150 \div 15 = 10$ minutes

The accumulator has a storage capacity that will allow 10 minutes of shower use.

If the incoming supply flow rate is 10 litres per minute, then:

$10 \div 60$ (seconds in 1 minute) = 0.1666 litres/ second.

150 litres ÷ 0.1666 (l/s) ÷ 60 (seconds in 1 minute) = 15 minutes accumulator replenishing time

Accumulators can be situated anywhere within a property but it must be remembered that the higher the accumulator is positioned above the incoming supply, the pressure will drop by 0.1 bar for every metre the accumulator is raised. Delivery pressure to the outlets, however, will increase by 0.1 bar.

IMPROVE YOUR MATHS

Accumulator sizing

A house with a low flow rate of 7 litres per minute and an incoming water supply pressure of 2 bar is occupied by three people. There are two showers in the property, each with a maximum flow rate of 10 litres per minute each. The accumulator size is to be 400 litres. Calculate:

1 the charge in bar pressure required at the accumulator
2 the storage capacity of the accumulator
3 the amount of showering time available when both showers are running
4 the time needed to replenish the accumulator to full capacity.

Presentation of the design

The calculation process for cold water supply takes time to complete and unless the calculations are set out correctly, mistakes are often made. The use of spreadsheets and tables when completing design calculations, especially for pipe sizing, is commonplace amongst most professional building services engineers. These are excellent for including with any quotation or design specification that the company wishes to present to the customer.

Scale drawings and schematic drawings help to show the customer what you are proposing to install to fulfil their requirement. This is especially important when a large installation is to be completed as it helps the customer to keep a track of what is being installed and where. Many companies now also provide three-dimensional drawings and artistic impressions of what the installation will look like when completed.

⑤ DIAGNOSING AND RECTIFYING FAULTS IN COLD WATER SYSTEMS AND COMPONENTS

Whenever mechanical or electrical components are installed on a cold water system, the risk of breakdown and failure is ever present. Periodic preventative maintenance can alleviate some of these problems, but occasionally components simply wear out and must be replaced.

In this part of the chapter, we will investigate some of the methods that help us identify system faults.

Identifying system faults through consultation with the customer

When identifying faults that have occurred on cold water systems, the customer can prove an invaluable source of information as they can often describe when and how the fault first manifested itself and any characteristics that the fault has shown. Verbal discussion with the customer often results in a successful repair without the need for extensive diagnostic tests. The customer should be asked:

- The immediate history of the fault:
  - When did it first occur?
  - How did they notice it?
  - What characteristics did it show?
- Did they notice any unusual noises? This may well indicate the type of component failure that has taken place.
- Did they attempt any repairs themselves? If so, what did they do? This is important because if repairs have been attempted, they may well have to be undone to successfully diagnose the problem.
- What was the result of the fault? Again, this is an important aspect because it can often indicate where the fault lies. For instance, if the customer has noticed a drop in flow rate or pressure, this might indicate a blockage, a blocked filter, scale growth or a mechanical fault, such as a loss of charge from an accumulator or a pump fault.

IMPROVE YOUR ENGLISH

Remember, communicate with your customer first; you can obtain useful information to help you identify systems faults.

Identifying system faults through consultation with the manufacturer's instructions and the British Standards

When attempting to identify faults with cold water systems, the most important document to consult is the manufacturer's instructions. In most cases, these will contain a section on fault finding that will prove an invaluable source of information. Fault finding using manufacturer's instructions usually takes three forms:

- known problems that can occur and the symptoms associated with them
- methods by which to identify the problem in the form of a flow chart. These usually follow a logical, step-by-step approach, especially if the equipment has many parts that could malfunction, such as a cold water boosting set
- the techniques required to replace the malfunctioning component.

A replacement parts list will also be present for those components that can be replaced. When ordering parts, it is advisable to use the model number of the equipment and the parts number from the replacement parts list. This will ensure that the correct part is purchased.

INDUSTRY TIP

Consult the replacement parts list of the manufacturer's instructions when selecting new parts.

British Standard **BS EN 806** may also be consulted as it contains important information regarding minimum flow rates required by certain appliances. Again, this should be used in conjunction with the manufacturer's instructions. Remember – manufacturer's instructions always take precedence over the British Standards and Regulations.

Methods of repairing faults

Routine checks on components and systems can help to identify any potential problems that may be developing within the system as well as keeping the system operating to its maximum performance and within the system design specification.

Checks performed can include:

- **Checking components for correct operating pressures and flow rates** – this can be performed using a pressure and flow rate meter or a weir jug. However, caution must be exercised when using a weir jug as the flow rate of some systems can exceed the flow rate capacity of the weir jug making an accurate reading impossible.
- **Cleaning system components (including dismantling and reassembly)** – again, a very important part of the diagnostic process, especially with certain components such as filters and valves. Sediment and scale can quickly build up in certain water areas and it is important that these components are cleaned on a regular basis to prevent loss of flow rate and pressure. The use of manufacturer's literature is recommended, especially with complex components such as pumps.
- **Checking for correct component operation:**
  - **Pumps** – these should be checked using the manufacturer's commissioning procedures to ascertain whether the pump is performing as the data dictates. A slight fall in performance is to be expected with age. Check to ensure:
    - there are no signs of damage or wear and tear on the pump
    - there are no signs of leakage from the pump
    - that the pump switches on and off at the correct pressure
    - that there are no unusual noises or vibrations when the pump is operating.
  - **Pressure switches (transducers)** – transducers are often checked using a calibrated pressure gauge on the system pipework to ensure that the system is operating within acceptable limits. A transducer that is malfunctioning will often read over 12 bar pressure when tested irrespective of the actual pressure within the system.
  - **Float switches** – when fitted to a cold water storage cistern (CWSC), these can usually be

operated by hand, by moving the float arm, to ensure that the pump starts when the float switch is moved to the 'on' position.
- **Accumulators (pressure vessels)** – accumulators are checked at the Schrader valve with a pressure gauge to ensure that the vessel contains the correct pressure.
- **Gauges and controls** – gauges should correspond to the known pressures within the system. Any deficiency could indicate that the gauge is faulty.
- **Checking for correct operation of water treatment devices** – this must be done by following the recommendations of the manufacturer's servicing and maintenance instructions.
- **Water filters** – these must be regularly cleaned and cleared of all scale and sediment build-up to ensure that flow rates within the system are maintained.
- **Water softeners** – again, these should be checked using the manufacturer's set of specific maintenance instructions, especially where the softener uses external additives. Water softeners can be checked for correct operation by the use of a pH indicator.
- **Cold water storage cisterns (CWSC) and break cisterns** – when used to supply wholesome water supply, they should be checked and cleaned at regular intervals, and at least every six months for large cisterns supplying wholesome water to many properties.

▲ Figure 1.121 A pressure gauge

Why a water treatment device may not be operating as required

Base exchange (ion exchange) water softeners operate by passing water containing calcium and magnesium (hard water) through a sealed vessel that contains resin beads. The beads attract and absorb the calcium and magnesium, replacing the hardness with sodium, effectively softening the water. After time, the beads become saturated with the hardness salts, reducing the effectiveness of the water softener. When this happens, the beads need to be regenerated using a brine or salt solution. Backwashing removes the hardness from the beads before the sodium is replenished by the salt in the brine solution. Most modern softeners complete this procedure automatically either by a timed backwashing programme or when the beads become saturated with the hardness salts.

Base exchange water softeners often begin to lose their effectiveness when:
- limescale builds up within the softener
- the salt crystals require replenishment
- the water softener fails to backwash.

It is therefore important that water softeners are maintained at regular intervals to ensure that the softening process remains effective.

Isolating cold water systems

Isolating cold water systems or sections of a cold water system is required during:
- maintaining and servicing components and equipment
- repairing and replacing pipework and components

- decommissioning procedures (permanent and temporary)
- commissioning new pipework and systems
- an emergency (burst pipes or failed components).

For this reason, it is important that stop valves, isolation valves and gate valves are kept in full working order.

> **KEY POINT**
>
> Always put a notice by the stop valve/isolation valve informing that the system is isolated and must not be turned on.

VALUES AND BEHAVIOURS

Do not forget to keep the householder/responsible person informed of the areas that are going to be isolated during repair and maintenance tasks and operations, and always ask the customer if they need to 'draw off' a temporary supply of water (kettle, saucepans, bucket, etc.) for the short period of system isolation.

Carrying out diagnostic checks and repairs

In this part of the chapter, we will investigate diagnostic checks that we can perform on boosting pumps and backflow prevention devices.

Booster pumps

Diagnostic tests on booster pumps are fairly straightforward to undertake. Table 1.24 illustrates some of the problems associated with boosting pumps and their diagnostic approaches.

▼ Table 1.24 Booster pump troubleshooting

Fault	Diagnostic test
No power at the motor	Check for voltage at the motor terminal box. Check the feeder panel for tripped circuit breakers and reset as necessary. If the circuit breakers continue to trip out, then initiate tests on the wiring and the pump motor as detailed below.
Defective capacitor (single-phase pumps only)	Isolate the power supply and discharge the capacitor by touching the leads together. Check the capacitor with an analogue ohmmeter as detailed on the next page. If the capacitor is working correctly, the needle should jump to almost zero (0) ohms and slowly return to infinity (∞). If this does not happen, replace the capacitor.

→

▼ Table 1.24 Booster pump troubleshooting (continued)

Fault	Diagnostic test
Loose electrical connections Motor winding has shorted out Damage to the pump, causing overload Burned contacts on the starter motor Motor winding defect (this must be verified by conducting a motor winding resistance test)	**Checked using an ammeter** These defects will cause the amp draw of the pump to exceed that stated in the manufacturer's data. They can be diagnosed using an ammeter, which is used to check the current of electricity as it flows through the wiring: 1 Make sure the pump is set to run. 2 Set the scale on the front of the ammeter to 100 amps. 3 Place the tongs of the ammeter around the cable. 4 Slowly rotate the scale on the ammeter back towards 0 (zero) until an exact reading is shown. 5 Record the measurement. 6 Check against the manufacturer's data.
Motor winding defect	**Checked using an ohmmeter** This is to check the condition of the motor winding and is known as a motor winding resistance test. It is conducted using an ohmmeter: 1 Turn OFF the POWER and check for safe isolation. 2 Disconnect the electrical leads to the motor. 3 Set the ohmmeter scale selector to R × 1 (if you anticipate ohm values under 10) or R × 10 (for ohm values over 10). 4 Touch the leads of the ohmmeter to two motor leads. Touching the leads of the ohmmeter to the live and neutral motor lead will measure the main winding's resistance. 5 Watch the ohmmeter scale and compare the figure with the appropriate chart in the pump installation and maintenance instructions. If all ohm values are normal, the motor windings are neither shorted nor open. If any one ohm value is less than normal, the motor winding may be starting to short out. If any one ohm value is higher than normal, the winding may be starting to open. If some ohm values are lower than normal (25% and above) and some are less than normal (25% and below), the leads may be crossed.
The pump runs too often	There are several possibilities: ● The pressure switch is not properly adjusted or is defective. Check the pressure setting on the switch and the voltage across the closed contacts. If it is defective, then it must be replaced. ● The accumulator is too small. It must be replaced with a larger accumulator. As a rule of thumb, the accumulator should have at least 45 litres for every litre per minute flow rate of the pump. ● The air charge in the accumulator is insufficient. Check the air charge in the accumulator and repair or replace as necessary as detailed earlier in this chapter.
Pump runs but does not produce enough flow (l/m)	Again, there are a number of possibilities: ● The shaft is rotating the wrong way. The pump usually rotates anti-clockwise when viewed from the top. Check the wiring against the manufacturer's instructions. Re-wire as necessary. ● The pump is airlocked. Close the isolation valves, open the priming plug and prime the pump by slowly opening the isolation valves. The suction lines from the water source should be opened first. ● The strainers and check valves are clogged. Remove and inspect the filters and valves. Clean as necessary.
Verifying the boosting pump performance	When any faults have been repaired, then the pump must be checked to ensure that it is working correctly. This can be achieved installing pressure gauges on both upstream and downstream pipework, close to the pump. Correct operation shows a drop in pressure on the upstream pipework as the suction of the pump acts on the upstream gauge and an increase in pressure to that shown on the manufacturer's instructions or the pump data plate. The flow rate should also be checked.

Backflow prevention devices

Most backflow prevention devices contain spring-loaded valves and diaphragms internally and these, occasionally, can fail. When this happens, the risk of contamination from back siphonage and back pressure becomes greater. Therefore, periodic testing is vital to maintain the protection offered by these components and in the case of RPZ valves, it is a **mandatory** requirement.

KEY TERM

Mandatory: required by law or regulation.

If a fault in a backflow prevention device is suspected, the best way to test them is to simulate a reduced pressure upstream of the suspect device. This can be quite a precarious operation. The idea is just to test the valve and not to create a potential contamination situation during the course of the diagnostic process and so any such simulation must be approached with extreme care. For those devices that have a test point on the valve body, such as a verifiable single check valve or a verifiable double check valve, testing becomes easier as the valve can be tested by simulating an upstream pressure loss and removing the test plug. If water is seen at the test plug, this may be an indication that the valve has failed and the valve should be replaced as a precaution.

HEALTH AND SAFETY

Extreme care must be taken not to create a potential contamination situation during the course of the diagnostic process of potential faults in backflow prevention devices.

RPZ valves are quite complex in their design. These too contain test points and can be tested in a similar manner to verifiable check valves. However, if water is seen to be dripping from the relief port, then this could indicate that the internal diaphragm has ruptured and is allowing water past the seal. In this case, only a registered operative can undertake repairs.

Diagnosing and preventing corrosion within cold water system pipework

Corrosion within plumbing systems occurs constantly. It is almost impossible to prevent. In many circumstances, it is the result of poor choice when mixing different metals in the same installation.

In this part of the chapter, we will investigate common types of corrosion, their causes and ways in which we can prevent it to ensure the long life of the systems we install.

Electrolytic corrosion

Where cold water systems are concerned, the electrolytic environment is the presence of the water itself. Water is classified as an electrolyte simply because it allows the passage of an electric current. When the two metals are placed together in the presence of water, a minute electric current passes between them due to the reaction between the two metals as one metal destroys the other. The most noble of the metals is classed as the cathode and suffers no adverse effects. The least noble metal is known as the anode and will suffer a complete molecular degradation with the passage of time. A by-product of this is the formation of hydrogen gas.

The amount of **electrolytic corrosion** that takes place will depend upon where the metals appear on the electrochemical series of metals, which is shown in Table 1.25. The closer they are on the table then the corrosion is slow and controllable. The further they are apart the corrosion becomes rapid and eventually a failure will occur. An example of this is where ordinary duplex brass fittings are used within plumbing systems. Brass is a mixture of copper (cathodic) and zinc (anodic). Eventually, the zinc will be eaten away by the copper, leaving the fitting brittle and susceptible to sudden fracture. A tell-tale sign that corrosion is taking place is the presence of a white powdery substance (zinc oxide) on the fitting body.

KEY TERM

Electrolytic corrosion: a process of accelerated corrosion between two or more differing metals when placed in an electrolytic environment.

▼ Table 1.25 The electrochemical series of metals

Material		Potential (volts)
Gold (Au)	Cathodic	+1.50
Platinum (Pt)		+1.20
Silver (Ag)		+0.80
Copper (Cu)		+0.15
Lead (Pb)		−0.13
Tin (Sn)		−0.14
Nickel (Ni)		−0.23
Cobalt (Co)		−0.28
Thallium (Tl)		−0.34
Cadmium (Cd)		−0.40
Iron (Fe)		−0.41 to 0.44
Chromium (Cr)		−0.74
Zinc (Zn)		−0.76
Manganese (Mn)		−1.18 to 1.19
Aluminium (Al)		−1.68
Magnesium (Mg)	Anodic	−2.36 to 2.37

Preventing electrolytic corrosion

Electrolytic corrosion is notoriously difficult to prevent because many of the brass fittings available have a very high zinc content, which assists in corrosion taking place. The Water Regulations are very specific here as they stipulate that fittings used for hot and cold water supply that are to be used above and below ground must be made from a material that will not corrode in normal use. The solution is to use fittings that are made from gunmetal (a mixture of copper, zinc and tin) as this is highly resistant to electrolytic corrosion. The fittings are known as DZR fittings (de-zincification resistant) and are marked with **CR** to show that they are corrosion resistant.

Many plumbing systems use sacrificial anodes to assist in preventing electrolytic corrosion. By using a metal with high anodic properties, the corrosion is prevented from attacking the pipework and fittings as any electrolytic reaction will corrode the anode instead. Although usually used in hot water systems, sacrificial anode rods manufactured from magnesium significantly reduce the effects of electrolytic corrosion.

Blue water corrosion

Blue water corrosion occurs in copper plumbing systems and is identified by a blue-green colouration to the water when water is first drawn from the tap. It is often (but not always) associated with pitting corrosion. The colour of the water is the result of very fine copper corrosion products suspended in the water and is more common in copper pipes that are used infrequently. Occasionally, the signs of blue water corrosion are apparent even after a very short period of stagnation.

KEY TERM

Blue water corrosion: occurs in copper plumbing systems causing a blue-green colouration to the water.

The colour of the water can vary greatly from pale blue to dark blue/green. The particles become very visible if the water is left to stand in a container overnight as the particles will sink to the bottom of the container leaving a blue/green copper corrosion residue. This can sometimes be evident as blue/green staining on sanitary ware.

Blue water corrosion is more likely to occur in soft water areas where the pH value of the water is high or areas where the residual chlorine content of the water is low. However, such is the nature of blue water corrosion, that copper pipe can develop the symptoms even in other, low pH water areas.

Treating and minimising blue water corrosion

There is some evidence to show that blue water corrosion could be linked to microbial activity interacting with the copper pipe and that, because the signs of corrosion appear in a non-uniform manner, the internal surface of the pipe may also play a part in its formation. It has been shown, that in nearly all cases of blue water corrosion, flushing the system with water in excess of 70 °C followed by disinfection with water containing a chlorine based solution will prevent its appearance. However, if the residual chlorine content is not maintained, then it is likely that the corrosion will re-appear.

⑥ COMMISSIONING COLD WATER SYSTEMS AND COMPONENTS

Inadequate commissioning, system flushing and maintenance operations can affect the quality of drinking water, irrespective of the materials that have been used in the system installation. Building debris and swarf (pipe filings) can easily block pipes and these can also promote bacteriological growth. In addition, excess flux used during the installation can cause corrosion and may lead to the amount of copper that the water contains exceeding the permitted amount for drinking water. This could have serious health implications and, in severe cases, may cause blue water corrosion.

It is obvious, then, that correct commissioning procedures must be adopted if the problems stated are to be avoided. There are four documents that must be consulted:

- The Water Supply (Water Fittings) Regulations 1999
- British Standard **BS 6700** and **BS EN 806** (in conjunction with **BS 8558**)
- The Building Regulations Approved Document G1 and G2
- the manufacturer's instructions of any equipment and appliances.

The Water Supply (Water Fittings) Regulations

These are the national requirements for the design, installation, testing and maintenance of cold and hot water systems in England and Wales. Scotland follows the almost identical Scottish Water Byelaws 2014. Their purpose is to prevent contamination, wastage, misuse, undue consumption and erroneous metering of the water supply used for domestic purposes. Schedule 2 of the Regulations states that:

> The whole installation should be appropriately pressure tested, details of which can be found in the Water Regulations Guide (Section 4: Guidance clauses G12.1 to G12.3). This requires that a pressure test of 1.5 times the maximum operating pressure for the installation or any relevant part.

> Every new water service, cistern, distribution pipe, hot water cylinder or other appliance and any extension or modification shall be thoroughly flushed with drinking water prior to being taken into service.

> Under certain circumstances, the whole system should be disinfected before being put into use. This will be discussed later in this section.

This schedule should be read in conjunction with the British Standards.

The British Standards (**BS EN 806.4** and **BS 8558**)

The main British Standard for commissioning, testing, flushing and disinfection of systems is **BS EN 806.4:2010** Specifications for installations inside buildings conveying water for human consumption. Installation (in conjunction with guidance document **BS 8558:2011** Guide to the design, installation, testing and maintenance of services supplying water for domestic use within buildings and their curtilages).

The Building Regulations

The Building Regulations make reference to cold water services and systems. These are mentioned briefly in Approved Document G1 Cold water supply and Approved Document G2 Water efficiency. Additional recommendations can be found in Annex 1 Wholesome water and Annex 2 Competent persons self-certification schemes.

Manufacturer's instructions

Where appliances and equipment are installed on a system, the manufacturer's instructions are a key document when undertaking testing and commissioning procedures and it is important that these are used correctly at both installation and commissioning operations. Only the manufacturers will know the correct procedures that should be used to safely put the equipment into operation so that it performs to its maximum specification.

KEY POINT

Remember:
- Always read the instructions before operations begin.
- Always follow the procedures in the correct order.
- Always hand the instructions over to the customer upon completion.
- Failure to follow the instructions may invalidate the manufacturer warranty.

Visual inspections

Before soundness testing a cold water system, visual inspections of the installation should take place. This should include:

- walking around the installation. Check that you are happy that the installation is correct and meets installations standards
- checking that all open ends are capped off and all valves are isolated
- checking that all capillary joints are soldered and that all compression joints are fully tightened
- checking that enough pipe clips, supports and brackets are installed and that all pipework is secure
- checking that the equipment (such as boosting pumps, float switches, accumulators, etc.) are installed correctly and that all joints and unions on and around the equipment are tight
- checking that cisterns and tanks are supported correctly and that float operated valves are provisionally set to the correct water level
- checking that all appliances' isolation valves and taps are off. These can be turned on and tested when the system is filled with water
- where underground services have been installed, checking that any pipework is at the minimum depth required by the Water Supply (Water Fittings) Regulations.

The water undertaker must be given the opportunity to view and inspect the installation, preferably before it is tested and commissioned to ensure that the Water Supply (Water Fittings) Regulations have been complied with. Any remedial work pointed out by the water inspector can then be completed without the need to drain the system.

Filling and venting systems – the initial system fill

The initial system fill is always conducted at the normal operating pressure of the system. The system must be filled with fluid category 1 water direct from the water undertaker's mains cold water supply. It is usual to conduct the fill in stages so that the filling process can be managed comfortably. There are several reasons for this:

- Filling the system in a series of stages allows the operatives time to check for leaks stage by stage.

Only when the stage being filled is leak free should the next stage be filled.

- Air locks from cistern-fed supplies are less likely to occur as each stage is filled slowly and methodically. Any problems can be assessed and rectified as the filling progresses without the need to isolate the whole system and initiate a full drain down. Allowing cisterns to fill to capacity and then opening any gate valves is the best way to avoid air locks. This ensures that the full pressure of the water is available and the pipes are running at full bore. Trickle filling can encourage air locks to form causing problems later during the fill stage.
- It is possible that less manpower will be used by staged filling procedures. On very large and multi-storey systems, the use of two-way radios greatly helps the operatives during the filling process and isolation of a potential problem becomes quicker. Operatives should be stationed at the main isolation points to initiate a rapid turn off should a problem occur.

When the system has been filled with water, the system should be allowed to stabilise and any float operated valves should be allowed to shut off. The system will then be deemed to be at normal operating pressure.

Once the filling process is complete, another thorough visual inspection should take place to check for any possible leakage. The system is then ready for pressure testing.

Soundness testing

Pressure testing can commence when the initial fill to test the pipework integrity has been completed. Again, on large systems, this is best done in stages to avoid any possible problems.

The requirements of the Water Regulations

Regulation 12 of the Water Regulations requires that:

> The water system shall be capable of withstanding an internal water pressure not less than 1½ times the maximum pressure to which the installation or relevant part is designed to be subjected in operation.

In practice, this means that a system that has an operating pressure of 2 bar:

$$2 \text{ bar} \times 1.5 = 3 \text{ bar}$$

Regulation 12 also states that the regulation shall be deemed satisfied in the case of water systems that do not contain elastomeric (plastic) pipe where the whole system is subjected to the test pressure by pumping, after which the test should continue for 1 hour without further pumping and without any visible leakage and any pressure loss.

Where the system does contain elastomeric (plastic) pipework, there are two acceptable tests that can be conducted. These are classified as test type A and test type B.

Cold water systems testing is detailed in **BS 6700** and **BS EN 806.4**:

- **Copper tubes and low carbon steel pipes** – systems installed in copper tube and low carbon steel pipes should be tested to 1 1/2 times normal operating pressure (1.5 times normal operating pressure or 50 per cent above normal operating pressure). It should be left for a period of 30 minutes to allow for temperature stabilisation and then left for a period of 1 hour with no visible pressure loss. **BS EN 806.4** states a test pressure of 1.1 times the maximum design pressure (MDP) for a minimum period of 10 minutes.

- **Elastomeric (plastic polybutylene) pressure pipe systems** – these are tested differently to rigid pipes. There are two tests that can be carried out. These are known as test type A and test type B and are detailed in **BS 6700**. It is important that the correct tests are used for elastomeric pipes as these can become stressed very easily at high pressures.

  - **Test type A** – Slowly fill the system with water and raise the pressure to 1 bar (100 kPa). Check and repump the pressure to 1 bar if the pressure drops during this period provided there are no leaks. Check for leaks. After 45 minutes, increase the pressure to 1.5 times normal operating pressure and let the system stand for 15 minutes. Now release the pressure in the system to 1/3 of the previous pressure and let it stand for a further 45 minutes. The test is successful if there are no leaks. **BS EN 806.4** states that test type A can be carried out at 1.1 times the MDP for a period of 10 minutes.

  - **Test type B** – Slowly fill the system with water and pump the system up to the required pressure and maintain the pressure for a period of

30 minutes and note the pressure after this time. The test must continue with no further testing. Check the pressure after a further 30 minutes. If the pressure loss is less than 60 kPa (or 0.6 bar), the system has no visible leakage.

Visually check for leakage for a further 120 minutes. The test is successful if the pressure loss is less than 20 kPa (0.2 bar). **BS EN 806.4** states Test B can be carried out at 1.1 times MDP over a period of 30 minutes, reducing the pressure in accordance with Figure 3 in **BS EN 806.4**.

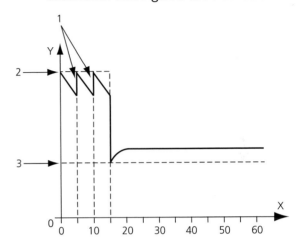

Key
1 Pumping
2 Test pressure 1.5 times maximum working pressure
3 0.5 times maximum working pressure
X Time (minutes)
Y Pressure

▲ Figure 1.122 Test type A

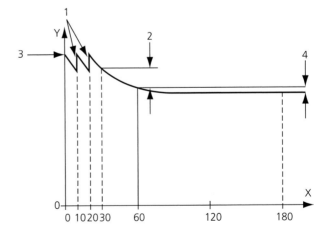

Key
X Time (minutes)
Y Pressure
1 Pumping
2 Pressure drop < 60 kPa (0.6 bar)
3 Test pressure
4 Pressure drop < 20 kPa (0.2 bar)

▲ Figure 1.123 Test type B

Planning the test

Before the test is conducted, a risk assessment should be carried out.

Personal protective equipment should also be used.

The following factors should be carefully considered:

- Is the test being used appropriate for the service and the building environment?
- Will it be necessary to divide the vertical pipework into sections to limit the pressures in multi-storey buildings?
- Will the test leave pockets of water that might cause frost damage or corrosion later?
- Can all valves and equipment withstand the test pressure? If not, these will need to be removed and temporary pipework installed.
- Are there enough operatives available to conduct the test safely?
- Can different services be interconnected as a temporary measure to enable simultaneous testing?
- How long will it take to fill the system using the available water supply?
- When should the test be started when the size of the system is considered? Preparation should also be taken into account.

HEALTH AND SAFETY

Pressure testing involves stored energy, the possibility of blast and the potential hazards of high velocity missile formation due to pipe fracture and fitting failure. A safe system of work should be adopted and a permit to work sought where necessary.

Preparing for the test

- Check that the high points of the system have an air vent or a tap to help with the removal of air from the system during the test. These should be closed to prevent accidental leakage.
- Blank or plug any open ends and isolate any valves at the limit of the test where the test is being conducted in stages.
- Remove any vulnerable equipment and components and install temporary piping.
- Open the valves within the section to be tested.

- Attach the test pump to the pipework and install extra pressure gauges if necessary.
- Check that a suitable hose is available for draining down purposes.

KEY POINT

IMPORTANT! Check that the test pump is working correctly and that the pressure gauge is calibrated and functioning correctly.

▲ Figure 1.124 Hydraulic pressure test pump

The hydraulic test procedure, step-by-step

1 Using the test pump, begin to fill the system. When the pressure shows signs of rising, stop and walk the route of the section under test. Listen for any sounds of escaping air and visually check for any signs of leakage.
2 Release air from the high points of the system or section and completely fill the system with water.
3 When the system is full and free of air, pump the system up to the required test pressure.
4 If the pressure falls, check that any isolated valves are not passing water and visually check for leaks.
5 Once the test has been proven sound, the test should be witnessed and a signature obtained on the test certificate.
6 When the pressure is released, open any air vents and taps to atmosphere before draining down the system.
7 Refit any vulnerable pieces of equipment, components and appliances.

Disinfection of large cold water systems

The British Standard requires that water piping systems be disinfected in the following circumstances:

- in new installations (except private dwellings occupied by a single family)
- where large extensions or alterations to systems have been undertaken
- where it is suspected that the system has been compromised by contamination (such as fouling by drainage, sewage or animals or the physical entry by operatives during maintenance operations or repair)
- where the system has not been in regular use nor regularly flushed
- where underground pipework has been installed. This does not include localised repairs or where a junction has been installed after the fittings have been disinfected by immersion in a solution of sodium hypochlorite to dilution of 200 parts per million (ppm). 1 ppm is equivalent to 1 mg/l.

The requirements of the disinfection process

Disinfection requires that the system is first thoroughly flushed with water and then refilled and a disinfection agent added as the system refills. The disinfection agent should have an initial concentration of 50 ppm, which should be calculated according to the total water capacity of the pipework and any storage vessels. The chlorinated water should then be allowed to stand within the pipework for a contact period of no less than 1 hour. The process will have been successfully completed if, after one hour, the free residual chlorine content is at least 30 ppm at the furthest outlet on the system.

During the disinfection process, the use of household chemicals, bleaches and toilet cleaners must be avoided as these can react with the disinfectant to produce highly toxic fumes. All personnel and residents within the building, including those that are not normally present during working hours, should be informed of the disinfection operation before it commences.

The disinfectant types

The disinfectants used for water supplies must be approved for use by the Drinking Water Inspectorate (1990) and must conform to the following standards:

- **EN 900** for calcium hypochlorite, or
- **EN 901** for sodium hypochlorite.

Since most disinfection processes use sodium hypochlorite, we will concentrate on this disinfectant for the purpose of this chapter.

Sodium hypochlorite solution contains around 5 per cent available chlorine content, which is equivalent to 50,000 ppm of chlorine. Obviously, for disinfection of water supply systems this is far too much. To create an initial content of 50 ppm for the disinfection process:

$$50 \div 1,000,000 \times 100 \div 5 = 0.001 \text{ part of sodium hypochlorite solution to 1 part water}$$

This equates to a ratio of 1 litre of sodium hypochlorite containing 5 per cent chlorine to every 1000 litres of system water capacity.

For commercial strength sodium hypochlorite solutions that contain 10 per cent available chlorine, the ratio is much less:

$50 \div 1,000,000 \times 100 \div 10 = 0.0005$ part of sodium hypochlorite solution to 1 part water

This equates to a ratio of 0.5 litres of sodium hypochlorite containing 10 per cent chlorine to every 1000 litres of system water capacity.

INDUSTRY TIP

For more information from the Drinking Water Inspectorate (1990), visit: www.dwi.gov.uk

Calculating system capacity

To administer the correct concentration of disinfectant, the total system volume in litres must be calculated. This is completed by:

- calculating the capacity in litres up to the water line of any storage cisterns, and
- calculating the capacity in litres of the pipework.

Calculating the capacity of a cistern

Calculation of cistern capacity is a simple process:

- measure the length of the cistern in metres (m)
- measure the width of the cistern in metres (m)
- measure the height to the water line in metres (m).

$L \times W \times H$ (to water line) = volume (m^3)

volume $\times 1000$ = capacity in litres (l)

IMPROVE YOUR MATHS
Cistern capacity calculations

A cistern measures 2.5 m × 1.75 m × 1.5 m to the water line. Calculate the cistern capacity in litres. You will use this figure later in the chapter.

IMPROVE YOUR MATHS
Calculating the capacity: example

A cold water storage cistern measures 2 m × 1.5 m × 1 m to the water line. What is the capacity in litres of the cistern?

$2 \times 1.5 \times 1 = 3\, m^3$

$3 \times 1000 = 3000$ litres

The capacity of the cistern is therefore 3000 litres.

Calculating the capacity of the system pipework

Calculating the capacity of the pipework involves measuring the length of the pipework by pipe size. The bigger the pipe, the more water it will hold. The calculation to determine the capacity of a pipe is as follows:

$\pi r^2 \times$ length $\times 1000$

Where:

$\pi = 3.142$

r = radius of the pipe in metres (m)

IMPROVE YOUR MATHS
Calculating the capacity of the system pipework: example 1

What is the capacity of 70 mm diameter pipe measuring 1 m in length? The radius of the pipe is half the diameter, therefore:

$70 \div 2 = 35$ mm or 0.035 m

So, the calculation looks like this:

$3.142 \times 0.035^2 \times 1 \times 1000$

$= 3.142 \times (0.035 \times 0.035) \times 1 \times 1000$

$= 3.142 \times 0.001225 \times 1 \times 1000$

$= 3.84$ litres

Fortunately, the copper tube manufacturers provide the capacities of the common tube sizes per metre within their data sheets and this is provided in Table 1.26.

▼ Table 1.26 The capacity of copper tube per metre

Standard/spec	Tube dimension (OD)	Carrying capacity (l/m)
BS EN 1057 R250 half-hard straight lengths	6	0.018
	8	0.036
	10	0.061
	12	0.092
	15	0.145
	22	0.320
	28	0.539
	35	0.835
	42	1.232
	54	2.091
	66.7	3.247
	76.1	4.197
	108	8.659

The calculation of the capacity of the pipework can be carried out using a table. The section of pipe, its size and length must first be identified and numbered and then added to the table. Consider the drawing in Figure 1.125.

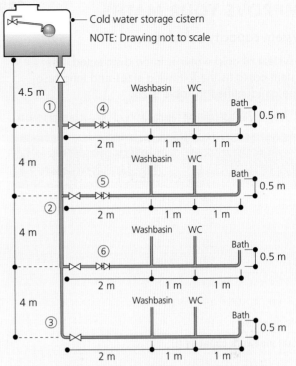

Cold water storage cistern

NOTE: Drawing not to scale

Pipework to the washbasin and WC have been assumed at 1 m for each appliance per floor

Pipework sizes have not been calculated and are assumed for the purpose of system capacity calculations

▲ Figure 1.125 System drawing

IMPROVE YOUR MATHS

Calculating the capacity of the system pipework: example 2

A system of cold water is to be disinfected. The system contains the following sizes and lengths of pipe as identified in Table 1.27 below.

▼ Table 1.27 Determining the capacity of the system pipework

Pipe	Size (mm)	Length (m)	Capacity per metre (l)	Capacity of the pipe (l)
1	42	4.5	1.232	5.544
2	35	8	0.835	6.68
3	22	8.5	0.320	2.72
4	22	4.5	0.320	1.44
5	22	4.5	0.320	1.44
6	22	4.5	0.320	1.44
Appliance legs	15	8	0.145	1.16
Total pipework capacity (l)				20.424

Therefore, the capacity of the pipework shown in the example drawing is:

20.424 litres

If the capacity of the cistern from the previous example is taken into account, the total system capacity is:

3000 + 20.424 = 3020.424 litres

If we assume that sodium hypochlorite solution with 5 per cent chlorine content is to be used, then to calculate the amount of sodium hypochlorite solution required:

3020.424 × 0.001 = 3.020424 litres

Therefore, to successfully disinfect the example system, 3.020 litres of sodium hypochlorite solution will need to be added during the system fill.

IMPROVE YOUR MATHS

System capacity calculations

A system of cold water is to be disinfected. The system contains the following sizes and lengths of pipe as identified in Table 1.28.

▼ Table 1.28 Determining the capacity of the system pipework

Pipe	Size (mm)	Length (m)	Capacity per metre (l)	Capacity of the pipe (l)
1	54	10		
2	42	22.5		
3	35	15.5		
4	28	24.5		
5	22	14.5		
6	15	10		
Appliance legs	15	12		
Total pipework capacity (l)				

Now add to the system capacity total and determine the amount of sodium hypochlorite solution (5 per cent chlorine content) that must be used to disinfect the system.

The method of disinfection

The method of disinfection is as follows:

1 Thoroughly flush the system with water from the undertaker's main to remove any swarf, flux residues or other contaminants.

2 Calculate the cistern capacity and the amount of disinfectant required to give the required 50 ppm.

3 Add this to the cistern as it is filling to give a good mix with the cistern water.

4 Working away from the cistern, open each draw-off until a smell of disinfectant is detected. This should be done at every draw-off to ensure that the system is full of disinfectant.

5 As the water from the cistern is drawn off, it will be necessary to add more disinfectant solution to ensure that the required 50 ppm throughout the installation is achieved.

6 Once the system is full and the float operated valves on the system have shut, the 1-hour test period can begin.

Testing for residual chlorine

At the end of the 1-hour period, the free residual chlorine content should be 30 ppm at the furthest draw-off point. If this is not achieved, then the system must be drained and the test started afresh.

To test for residual chlorine content, a special chemical analysis test is used, which indicates the strength of the chlorine by the colour it turns when the chemical is added to the water.

The test consists of a clear plastic tube with a sample of water from the chlorinated system in it. A tablet is then added to the water and the tube is shaken. As the tablet dissolves, the water will change colour, the colour depends upon the chlorine strength. These are indicated in Table 1.29.

▼ Table 1.29 Determining the residual chlorine after disinfection

Water colour	Chlorine level (ppm or mg/l)
Clear	None
Faint pink/pink	0.2–1
Pink/red	1–5
Red/purple	5–10
Purple/blue	10–20
Blue/grey-green	20–30
Grey-green/yellow	30–50
Muddy brown	Over 50
Colour develops but then disappears	Excessive

Flushing the system after disinfection

Heavily chlorinated water is dangerous to wildlife and fish. Prolonged contact with copper pipe can also cause cuprous chloride corrosion, which will continue to attack the tube, even after flushing.

VALUES AND BEHAVIOURS

It is your duty as a plumber to ensure your work does not have a negative impact on the environment. Disinfection of systems is one process where improper handling of dangerous chemicals can cause pollution to water supplies and threaten wildlife.

Before draining the system, the advice of the local water undertaker should be sought. Alternatively, a neutralising chemical (sodium thiosulphate) can be administered to the system to clean the water before draining takes place. The chemical should be added at the rate of:

System volume × ppm of chlorine × 2 = grams of chemical required

Once the system has been drained, it should be flushed with clean water until the free residual chlorine content is no more than that in the water undertaker's mains cold water supply.

KEY POINT

- **Do** take care to warn people before disinfection begins. The chemicals are dangerous.
- **Do** handle chemicals with care and always use PPE.
- **Do** calculate the capacity of the system. Using too much chemical disinfectant will not produce better results and may result in the test having to be completed again.
- **Do** use chlorinated water when topping up cisterns and pipework during the test. This will help keep the correct disinfectant level.
- **Do** check the chlorine level at the end of the test period.
- **Do not** leave the water in the system for more than 1 hour and **never** overnight.
- **Do not** discharge the test water direct into a water course or drain without first contacting the local water undertaker or the Environment Agency.
- **Always** complete the disinfection record paperwork correctly.

The flushing procedure for cold water systems and components not requiring disinfection

The flushing of cold water systems is a requirement of the British Standards. All systems, irrespective of their size, must be thoroughly flushed with clean water direct from the water undertaker's main supply before being taken into service. This should be completed as soon as possible after the installation has been completed to remove potential contaminants, such as flux residues, PTFE, excess jointing compounds and swarf.

Simply filling a system and draining down again does not constitute a thorough flushing. In most cases, this will only move any debris from one point in the system to another. In practice, the system should be filled and the water run at every outlet until the water runs completely clear and free of any discoloration. It is extremely important to ensure that all equipment and appliances and every water fitting is flushed completely. When flushing boosted cold water supplies, water from the undertaker's water main must first be introduced into the break cistern and then boosted, using the boosting pumps, to all points on the system.

It is generally accepted that systems should not be left charged with water once the flushing process has been completed, especially if the system is not going to be used immediately, as there is a very real risk that the water within the system could become stagnant. In practice, it is almost impossible to effect a complete drain down of a system, particularly large systems, where long horizontal pipe runs may hold water. This in itself is very detrimental as corrosion can often set in and this can also cause problems with water contamination. It is recommended, therefore, that to minimise the risk of corrosion and water quality problems to leave systems completely full and flush through at regular intervals of no less than twice weekly, by opening all terminal fittings until the system has been taken permanently into operation. If this is the case, then provision for frost protection must be made.

Operational checks
Taking flow rate and pressure readings

When the system has been commissioned and put into operation, the flow rates and pressures should be checked against the specification and the manufacturer's instructions. This can be completed in several ways:

- Flow rates can be checked using a weir gauge. This is sometimes known as a weir cup or a weir jug. The method of use is simple. The gauge has a slot running vertically down the side of the vessel, which is marked with various flow rates. When the gauge is held under running water, the water escapes out of the slot. The height that the water achieves before escaping from the slot determines the flow rate. Although the gauge is accurate, excessive flow rates will cause a false reading because the water will evacuate out of the top of the gauge rather than the side slot.

- System pressures (static) can be checked using a Bourdon pressure gauge at each outlet or terminal fitting. Bourdon pressure gauges can also be permanently installed either side of a boosting pump to indicate both inlet and outlet pressures.
- Both pressure (static and running) and flow rate can be checked at outlets and terminal fittings using a combined pressure and flow rate meter.

▲ Figure 1.126 A Bourdon pressure gauge

Dealing with defects with cold water systems that are discovered during commissioning

Commissioning is the part of the installation where the system is filled and run for the first time. It is now that we see if it works as designed. Occasionally problems will be discovered when the system is fully up and running, such as:

- **Systems that do not meet correct installation requirements.** This can take two forms:
  - Systems that do not meet the design specification – problems such as incorrect flow rates and pressures are quite difficult to deal with.

If the system has been calculated correctly and the correct equipment has been specified and installed to the manufacturer's instructions, then problems of this nature should not occur. However, if the pipe sizes are too small in any part of the system, then flow rate and pressure problems will develop almost immediately downstream of where the mistake has been made. In this instance, the drawings should be checked and confirmation with the design engineer that the pipe sizes that have been used are correct before any action is taken. It may also be the case that too many fittings or incorrect valves have been used causing pipework restrictions.

- Another cause of flow rate and pressure deficiency is the incorrect set-up of equipment such as boosting pumps and accumulators. In this instance, the manufacturer's data should be consulted and set-up procedures followed in the installation instructions. It is here that mistakes are often made. If problems still continue, then the manufacturer's technical support should be contacted for advice. In a very few cases, the equipment specified is at fault and will not meet the design specification. If this is the case then the equipment must be replaced.

- **Poor installation techniques.** Installation is the point where the design is transferred from the drawing to the building. Poor installation techniques account for:
  - Noise – incorrectly clipped pipework can often be a source of frustration within systems running at high pressures because of the noise that it can generate. Incorrect clipping distances and, often, lack of clips and supports can put strain on the fittings and cause the pipework to reverberate throughout the installation, even causing fitting failure and leakage. To prevent these occurrences, the installation should be checked as it progresses and any deficiencies brought to the attention of the installing engineer. Upon completion, the system should be visually checked before flushing and commissioning begins.
  - Undue warming of cold water systems – this generally occurs if the cold water pipework has been installed too close to either hot water pipework or heating system pipework.

The Health and Safety Executive state that to prevent micro-bacterial growth due to temperature the cold water must not exceed 20 °C. To prevent this, the cold water pipework should be insulated across the entire system including cisterns and storage vessels.

- Leakage – water causes a huge amount of damage to a building and can even compromise the building structure. Leakage from pipework if left undetected causes damp, mould growth and an unhealthy atmosphere. It is, therefore, important that leakage is detected and cured at a very early stage in the system's life.
- It is almost impossible to ensure that every joint on every system installed is leak free. Manufacturing defects on fittings and equipment and damage sometimes cause leaks. Leakage due to badly jointed fittings and poor installation practice are much more common, especially on large systems where literally thousands of joints have to be made until the system is complete. These can often be avoided by taking care when jointing tubes and fittings, using recognised jointing materials and compounds and using manufacturers' recommended jointing techniques.
- **Microbiological contamination within cold water systems.** Contamination of cold water systems can occur for a variety of reasons:
    - the ingress of insects and vermin into stored water
    - the ingress of debris during installation
    - poor installation practice
    - undue warming of the cold water.

    There are many forms of microbiological contamination, including *Legionella pneumophila*, *E. coli* and *Pseudomonas*, and most are centred on an increase in water temperature where the bacteria can reproduce. In such instances, the system must be disinfected using the recognised techniques discussed previously. In all cases of microbiological contamination, advice should be sought from a recognised company dealing with such outbreaks and consultation with the HSE.
- **Defective components and equipment.** Defective components cause frustration and cost valuable installation time. If a component or piece of equipment is found to be defective, do not attempt a repair as this may invalidate any manufacturer's warranty. The manufacturer should first be contacted as they may wish to send a representative to inspect the component prior to replacement. The supplier should also be contacted to inform them of the faulty component. In some instances where it is proven that the component is defective and was not a result of poor installation, the manufacturer may reimburse the installation company for the time taken to replace the component.

Commissioning

Notification of work carried out

At all stages of the installation from design to commissioning, notification of the installation will need to be given so that the relevant authorities can check that the installation complies with the regulations and to ensure that the installation does not constitute a danger to health. Notification must be given to:

- **The water undertaker** – the Water Supply (Water Fittings) Regulations dictate that the local water authority must be informed before work starts, at various stages during its installation and on completion. It also states that where backflow prevention devices are installed, such as RPZ valves, notification must be given and they must be inspected prior to being commissioned. A list of notifiable works is provided on page 5. Notification should be completed by the installing plumber and involves completing a water undertaker's notification form.
- **The local building control office** – under Building Regulations Approved Document G, cold water installations and water conservation are notifiable to the local authority building control office. Building Regulations approval can be sought from the local authority by submitting a building notice. Plans are not required with this process so it is quicker and less detailed than the full plan's application. It is designed to enable small building works to get under way quickly. Once a 'building notice' has been submitted and the local authority has been informed that work is about to start, the work will be inspected as it progresses. The authority will notify if the work does not comply with the Building Regulations.

Notice should be given to building control not later than 5 days after work completion and until this is received no completion certificates can be issued. A list of notifiable works is given in section 1 and includes:

- the installation of a new bathroom where alterations have to be made to the drainage system or soil stack or where no bathroom existed before
- the installation of a new kitchen where alterations have to be made to the drainage system or soil stack or where no kitchen existed before
- the installation of sanitary appliances, such as new washroom facilities
- the installation of water conservation components, such as rainwater harvesting or grey water recycling systems.

Building Regulations Compliance certificates

From 1 April 2005 the Building Regulations demanded that all installations must be issued with a Building Regulations Compliance certificate. This is to ensure that all Building Regulations relevant to the installation have been followed and complied with.

Commissioning records

Commissioning records for large cold water systems should be kept for reference during maintenance and repair and to ensure that the system meets the design specification. Typical information that should be included on the record is as follows:

- the date, time and the name(s) of the commissioning engineer(s)
- the location of the installation
- the amount of cold water storage (if any)
- the types and manufacturer of equipment and components installed
- the type of pressure test carried out and its duration
- disinfection processes, the disinfection chemicals used and the disinfection readings
- the flow rates and pressures at the outlets
- the pressures on both the suction side and the discharge side of any boosting pumps
- the accumulator pressure.

Handover procedures

When the system has been tested and commissioned, it can then be handed over to the customer. The customer will require all documentation regarding the installation and this should be presented to the customer in a file, which should contain:

- all manufacturers' installation, operation and servicing manuals for the taps, sanitary ware and any other equipment fitted to the installation such as boosting pumps and accumulators
- the commissioning records and certificates
- the water undertaker's acceptance of the system
- the Building Regulations Compliance certificate
- an 'as fitted' drawing showing the position of all isolation valves, backflow prevention devices, etc.

The customer must be shown around the system and shown the operating principles of any controls. Emergency isolation points on the system should be pointed out and a demonstration of the correct isolation procedure in the event of an emergency. Explain to the customer how the systems work and ask if they have any questions. Finally, point out the need for regular servicing of the appliances and leave emergency contact numbers.

7 SERVICING AND MAINTENANCE OF COLD WATER SYSTEMS

Maintenance tasks on cold water services, appliances and valves are essential to ensure the continuing correct operation of the system. The term used when isolating a water supply during maintenance operations is 'temporary decommissioning'.

Before undertaking the repair or replacement of components, we must first ascertain what the problem is. The customer will be able to tell you what is happening with the component.

IMPROVE YOUR ENGLISH

Remember, the customer may not know the technical language, but they will be able to explain the problem well enough for you to understand.

If a component requires replacing, we must ensure that we get as near to a like-for-like replacement as possible, that we have the correct tools available and that the customers' property is either removed or protected with dust sheets and other coverings before we begin.

Manufacturer's instructions

When repairing or replacing components, the manufacturer's instructions give step-by-step methods of how this should be done. These should be followed wherever possible. In some instances where the component is old or the customer has lost the original instructions, a copy may be available on the manufacturer's web site.

If the component being serviced develops a fault, a fault-finding flow chart is often included in the servicing and maintenance instructions. These are generally easy to follow and often include replacement parts and their catalogue numbers.

Maintenance

There are basically two types of maintenance:
- planned preventative maintenance
- unplanned/emergency maintenance.

Planned preventative maintenance

Planned preventative maintenance is usually performed on larger systems and commercial/industrial installations. It is performed to a pre-arranged maintenance schedule, which may mean out-of-hours working if the supply of water cannot be disrupted during normal working hours. It is designed to stop problems from occurring by catching faults at their early stages. Planned preventative maintenance could include:
- periodic system inspection – checking for leaks
- re-washering of float operated valves
- re-washering and re-seating of terminal fittings and taps
- inspection and cleaning of cisterns
- re-adjustment of water levels in cisterns
- re-washering of drain valves
- cleaning of filters and strainers
- maintenance of water softeners
- check correct operation of stop valves
- checking flow rates at all outlets.

When a maintenance task involves isolating the cold water supply, a notice will need to placed at the point of isolation stating 'system off – do not turn on' to prevent accidental turn on of the system. In most systems, it will be possible to isolate specific parts of the installation without the need to have the whole supply turned off. Where no such isolation exists, it may be of benefit to use a pipe freezing kit so that total system isolation is not undertaken.

Unplanned and emergency maintenance

Unplanned maintenance and emergency maintenance occurs when a fault suddenly develops, such as a burst pipe, or a small problem suddenly becomes a larger issue, such as a dripping tap or sudden loss of water. Unplanned and emergency maintenance can include:
- burst pipes and leaks
- running overflows
- dripping taps
- loss of low pressure, cistern-fed cold water supply due to faulty FOVs
- poor past installation practices, such as incorrectly positioned overflow pipes
- complete component breakdown necessitating the replacement of the component.

Many of the maintenance practices we use involve the decommissioning of systems so that parts and pipes can be replaced.

The risk of *Legionella*

Legionella (or *Legionella pneumophila* to give it its correct name), is a water-borne bacterium that develops and multiplies in stagnant water at temperatures between 20 and 45 °C. The bacteria are dormant at temperatures below 20 °C and will not survive temperatures above 60 °C. It can develop in any hot or cold water system where the water has been left to stand for a long period of time, in, say, cold water cisterns or hot water storage cylinders. It can even develop in unused branches (known as dead legs) of hot and cold water systems.

Legionnaires' disease develops from the *Legionella* bacteria and this is potentially fatal if contracted. Legionnaires' disease develops by breathing water into the lungs that is contaminated with *Legionella*, either in aerosol form (mist) or by minute water droplets from air conditioning coolers and whirlpool/hydro-therapy baths.

Anyone can develop the disease. However, the elderly, smokers, people with respiratory and kidney disease and the young are particularly at risk.

> **HEALTH AND SAFETY**
>
> Any hot or cold water system left to stand for a long period of time is at risk of containing *Legionella pneumophila*.

Dealing with *Legionella*

Any system that is installed in healthcare, social care or public buildings must be tested for the bacteria. A full risk assessment should be undertaken of the hot and cold water services and measures put in place to control the risk.

When testing for *Legionella*, temperature readings must be made at the furthest and the closest outlets to the cisterns and hot water cylinders on a monthly basis. Hot water storage cylinders should be checked monthly and cold water cisterns on a six-monthly cycle.

To reduce the *Legionella* risk, remove all non-used outlets and corresponding pipework. Rarely used outlets and shower heads should be flushed and cleaned weekly, and clean and de-scale shower heads every three months. Cold water cisterns and tanks should be cleaned periodically, and the hot water drained from hot water storage cylinders and a check made for debris and signs of corrosion.

Water samples should be analysed for *Legionella* periodically to demonstrate that bacteria counts are acceptable. The frequency of such tests will be determined by the level of risk, in accordance with the risk assessment. Water treatments and disinfection by the use of chlorine dioxide can also be undertaken by specialist companies.

> **INDUSTRY TIP**
>
> The Health and Safety Executive produce a guide called 'Legionnaires' Disease: The control of *Legionella* bacteria in water systems' and a free copy can be downloaded from: www.hse.gov.uk/pUbns/priced/l8.pdf

Keeping maintenance records

A record of all repairs and maintenance tasks completed will need to be recorded on the maintenance schedule at the time of completion, including their location, the date when they were carried out and the type of tests performed. This will ensure that a record of past problems is kept for future reference.

Where appliance servicing is carried out, the manufacturer's installation and servicing instructions should be consulted. Any replacement parts may be obtained from the manufacturers.

SUMMARY

As we have seen as we have worked through this chapter, there is so much more we need to learn about cold water systems other than the simple systems we became accustomed to in Book 1. Cold water systems and their regulations are often complex in their design and require great skill in their installation and commissioning. Add to this the difficulties surrounding private water supplies and it is obvious that the modern plumber must be competent in many different areas of work. This chapter gives an excellent insight into the complexities of modern cold water systems that will build on the knowledge from Book 1 and enable you to make a worthwhile contribution within the work environment.

Test your knowledge

1 In which of the following situations are the Water Supply (Water Fittings) Regulations not enforceable?

   a Water distribution pipework within an agricultural building fed from the mains

   b Domestic heating systems connected to the water supply by a quick fill loop

   c Water feeding terminal fittings fed via a private borehole

   d Cold water service pipework under ground between properties fed at mains pressure

2 Which of the following works need to be notified and consent given by the local water undertaker before work commences?

   a Installation of a pump or booster drawing more than 12 litres per minute

   b Replacement of a combination boiler above 30 kW heat input

   c The installation of an outside tap

   d The installation of a bath with a volume of 200 litres

3 Which Building Regulations Approved Document deals with cold water supply?

   a G1

   b G2

   c G3

   d G4

4 To prevent pumps running dry within a boosted cold water system, the break cistern should be sized to allow the pumps to run at maximum output for:

   a Not less than 5 minutes

   b Not less than 10 minutes

   c Not less than 15 minutes

   d Not less than 20 minutes

5 The item below is suitable for protecting against backflow up to which category?

   a Category 1

   b Category 2

   c Category 3

   d Category 4

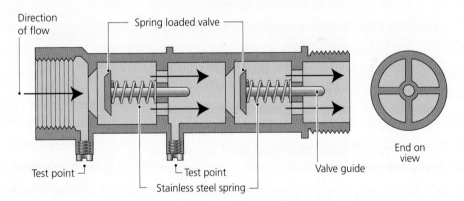

Direction of flow

Spring loaded valve

Test point

Test point

Stainless steel spring

Valve guide

End on view

6 What is the purpose of the item labelled X in the image below?

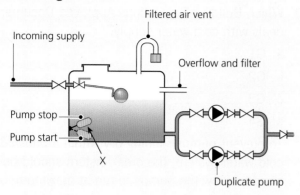

a To start the pump when the water level falls to within 225 mm of the cistern outlet

b To stop the pump when the water level falls to within 225 mm of the cistern outlet

c To automatically isolate the cold water supply if a leak is detected downstream of the pump

d To automatically isolate the cold water supply if a leak is detected upstream of the pump

7 Identify the correct meaning of the term 'actual capacity' when referring to storage cisterns:

a The maximum volume which it could hold when filled to its overflowing level

b The total volume it could hold when filled to the top of the cistern

c The total volume it can hold when shut off by the float operated valve

d The total volume it contains including any expansion through heating

8 What type of water treatment plant is shown in the image below?

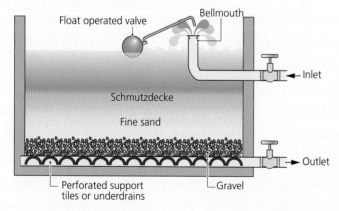

a Rapid sand filter   c Pressure filter

b Slow sand filter   d Absorption filter

9 Which type of non-mechanical backflow prevention is most suitable to protect against contamination at a kitchen sink?

a AD

b AG

c AUK2

d AUK3

10 A HUK1 type mechanical backflow prevention device is suitable for protection up to which fluid category?

a 1

b 2

c 3

d 4

11 Which British Standard is used alongside BS EN 806?

a BS 12058

b BS 8558

c BS EN 1212

d BS 1984

12 In a boosted cold water system, what item activates the pump?

a Float switch

b Solenoid

c Transducer

d Pressure switch

13 What must be avoided when cold water cisterns are joined together in a customer's loft area?

a Back flow

b Cross flow

c Stagnation

d Pressurisation

14 What does the infrared sensor eye operate to release water out of the tap?

a Transformer

b Transducer

c Flow switch

d Solenoid

15 What is the purpose of the retaining ring on a shower hose connected to a combination bath tap and shower head?

a Prevent scalding when in use

b Increase pressure on a low pressure system

c Keep the shower head above the bath rim

d Restrict flow to the shower head

16 What would this item be used to connect?

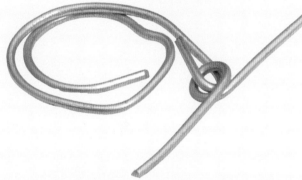

a Shower hose to bath tap

b Hot feed from cylinder to shower pump

c Rising main to cold water storage cistern

d Expansion vessel to cold water mains

17 Which of the following does NOT have category 2 water flowing through it?

a Hot water cylinder

b Rising main

c Mixer shower

d Combination tap and shower

18 Which of the following appliances does NOT have a risk of category 5 water being present?

a Bath

b WC

c Kitchen sink

d Bidet

19 What backflow prevention device MUST be installed on an outside tap?

a Single check valve

b RPZ valve

c Double check valve

d DC pipe interrupter

20 Which of the following is NOT a statutory regulation?

a Water Act 2003

b Water Supply (Water Fittings) Regulations 1999

c BS EN 806

d Building Regulations

21 According to BS EN 806, what is the loading unit for a bath?

a 2

b 3

c 4

d 5

22 A cold water storage cistern in a loft area is sized to provide cold water to the property over what period?

a 24 hours

b 12 hours

c 8 hours

d 48 hours

23 What is a fitting marked with DZR resistant to?

a Electrolytic corrosion

b De-zincification

c Blue water corrosion

d Contamination

24 Which metal is used to prevent electrolytic corrosion taking place in plumbing systems?

 a Copper

 b Zinc

 c Magnesium

 d Nickel

25 When disinfecting a system, how many parts per million must the chlorine mixture be diluted to?

 a 200 ppm

 b 500 ppm

 c 175 ppm

 d 50 ppm

26 For a single occupancy dwelling, what is the water extraction rate of a private water supply?

27 Which system of cold water uses a pneumatic pressure vessel?

28 Describe a break cistern.

29 Describe the purpose of a transducer within a boosted cold water system.

30 Which backflow prevention device is notifiable under the regulations?

31 List the Regulative and Guidance documentation that will be required when designing and installing a new domestic cold water system.

32 Plumbers who install systems and fitting have a legal responsibility to ensure that the materials and fitting comply with the regulations.
What does the Water Supply (Water Fittings) Regulations state in Regulation 3 about waste and contamination of water?

33 Explain the function of a float switch in a boosted cold water system.

34 List the documentation that the customer will require at the handover stage after a system has been tested and commissioned.

Answers can be found online at www.hoddereducation.co.uk/construction.

CHAPTER 2

HOT WATER SYSTEMS, PLANNING AND DESIGN

This chapter provides learning in application of design techniques, installation and use of specialist components, maintenance, diagnostics and rectification of faults and commissioning procedures, along with the backflow protection in plumbing systems to comply with current legislation and regulations. The chapter covers open vented and unvented systems in multi-storey dwellings.

By the end of this chapter, you will have knowledge and understanding of the following learning outcomes in this chapter:

- different fuels in domestic hot water systems
- the types of hot water system and their layout requirements
- specialist components in hot water systems
- design techniques for hot water systems
- the installation requirements of hot water systems and components
- testing and commissioning requirements of hot water systems and components
- diagnosing and rectifying faults in hot water systems and components
- servicing and maintenance of hot water systems.

Return to Book 1 and remind yourself of the topics covered in Chapter 6, Hot water systems, which included:
- sources of information relating to work on hot water systems
- hot water systems and components

- system safety and efficiency
- prepare for the installation of systems and components
- install and test systems and components
- decommission systems and components
- replace defective components.

1 DIFFERENT FUELS IN DOMESTIC HOT WATER SYSTEMS

There are several different fuel sources available for heating the supply of hot water in a dwelling. They can be divided into two categories: fossil fuels and renewable energy.

Fossil fuels

Gas (natural gas) – probably the most popular way of heating the hot water supply in the UK. It is also the cheapest and cleanest of all fossil fuel types, with CO_2 emissions much lower than with solid fuel, for example. However, it is not available to all buildings, as some remote parts of the UK are not connected to the natural gas network.

Gas (liquid petroleum gas, or LPG) – an alternative gas source for properties not connected to the gas network. It is supplied from a pressurised storage tank positioned some distance from the building. Supplies are replenished by tanker delivery. Liquid propane is preferred to liquid butane, as it boils at a much lower temperature (–45 °C) than butane (–4 °C). LPG tends to be very expensive.

Electricity – this fuel can be used to heat water in a variety of ways, such as via immersion heaters, instantaneous water heaters or showers from a 230 V single-phase supply.

Oil – this is another alternative gas source for properties not connected to the gas network. It is supplied from a storage tank positioned some distance from the building. Supplies are replenished by tanker delivery. There are many types of heating oil, with the most common being C2 grade, 28-second viscosity oil or kerosene. They are very similar to diesel fuel. Appliances tend to be big and quite noisy and they require regular servicing.

Solid fuel – there are many types of solid fuel used to produce energy and provide heating and hot water, although the use of some solid fuels (e.g. coal) is restricted or prohibited in some urban areas, due to unsafe levels of toxic emissions. In some areas, smokeless coal and coke are the only solid fuels used.

Renewable energy

Solar thermal – solar thermal technology utilises heat from the Sun to generate domestic hot water supply to offset the water heating demand from other sources, such as electricity or gas.

Geothermal – geothermal energy is heat directly from the Earth. It is a clean, renewable resource. Geothermal heat can be used directly, without involving a power plant or heat pump, for a variety of applications such as space heating and cooling, hot water supply and industrial processes. Its uses for bathing can be traced back to the ancient Romans.

Biomass – the term 'biomass' can be used to describe many different types of solid and liquid fuels. It is defined as any plant matter used directly as a fuel or that has been converted into other fuel types before combustion. Generally, solid biomass is used as heating fuel, including wood pellets, vegetable waste (such as wood waste and crops used for energy production), animal materials/wastes and other solid biomass.

2 THE TYPES OF HOT WATER SYSTEM AND THEIR LAYOUT REQUIREMENTS

In this section, we will take a look at the component layout features and functions for hot water system and the methods of providing water supplies.

Hot water system types

Hot water systems can be divided into two categories. These are:

- centralised systems where hot water is delivered from a central point to all hot water outlets in the dwelling. The water may be heated by a boiler or an immersion heater
- localised systems, often called single point or point of use systems. With these systems, the hot water is delivered by a small water heater at the point where it is needed.

Centralised systems

Centralised systems are those where the source of hot water supply is sited centrally in the property for distribution to all of the hot water outlets. They are usually installed in medium to large domestic dwellings, such as a three-bedroomed house. These can be further divided into:

- centralised hot water storage systems
- centralised instantaneous hot water systems.

Centralised hot water storage systems are divided into:

- open vented systems – those hot water storage systems that are fed from a cistern in the roof space and contain a vent pipe that is open to the atmosphere
- unvented systems – those hot water storage systems that are fed directly from the cold water main and utilise an expansion vessel or an internal air bubble to allow for expansion.

Centralised instantaneous hot water systems can be divided into:

- gas fired instantaneous multi-point hot water heaters – those heaters that heat the water instantaneously
- gas or oil fired combination boilers – operate in a similar fashion to instantaneous hot water heaters but also have a central heating capability
- thermal stores – sometimes referred to as water jacketed tube heaters
- gas or oil fired combined primary storage units – again these are very similar in operation to the thermal store.

Open vented hot water storage systems

In an open vented storage hot water system, water is heated, generally by a boiler or an immersion heater, and stored in a hot water storage vessel sited in a central location in the property usually in the airing cupboard. Open vented systems contain a vent pipe, which remains open to the atmosphere ensuring that the hot water cannot exceed 100°C. The vent pipe acts as a safety relief should the system become overheated. It must be sited over the cold feed cistern in the roof space.

The cylinder is fed with water from the cold feed cistern. The capacity of the cistern will depend upon the capacity of the hot water storage vessel. The cistern feeding cold water to a hot water storage vessel must be at least equal to that of the hot water storage vessel. Here are some important points to note about open vented hot water systems:

- The open vent pipe must not be smaller than 22 mm pipe and must terminate over the cold feed cistern.
- The open vent pipe must not be taken directly from the top of the hot water storage vessel.
- The hot water draw-off pipe should rise slowly from the top of the cylinder to the open vent pipe and incorporate at least 450 mm of pipe between the storage cylinder and its connection point to the open vent. This is to prevent parasitic circulation (also known as one pipe circulation) from occurring.
- The cold feed pipe should be sized correctly. The cold feed is the main path for expansion of water to take place within the cylinder when the water is heated. The heated water from the cylinder expands up the cold feed pipe raising the water level in the cold feed cistern.

- The cistern should be placed as high as possible to ensure good supply pressure. The higher the cistern then the greater the pressure at the taps. Poor pressure can be increased by raising the height of the cistern.
- All pipes should be laid with a slight fall (except the hot water draw-off) to prevent air locks within the system.
- The cold feed pipe from the storage cistern must only feed the hot water storage cylinder.
- A drain off valve should be fitted at the lowest point of the cold feed pipe.

As we discovered in Book 1, there are two types of open vented hot water storage systems. These are:
- the direct system
- the indirect system.

Open vented direct hot water storage system

The direct open vented hot water storage system uses a direct type hot water storage cylinder. The direct cylinder contains no form of heat exchanger and so is not suitable for use with central heating systems. The connections for the cold feed and draw-off are usually male thread connections with the primary flow and return connections are female thread. They are usually heated by either one or two immersion heaters, depending on the cylinder type or they may be heated by a gas fired hot water circulator. Existing installations may also use a back boiler placed behind a solid fuel fire. Because the water in the boiler comes directly from the hot water storage cylinder, the boiler must be made of a material that does not rust. This is to prevent rusty water being drawn off at the taps. Suitable boiler materials are:
- copper
- stainless steel
- bronze.

A typical direct system using immersion heaters is shown in Figure 2.1.

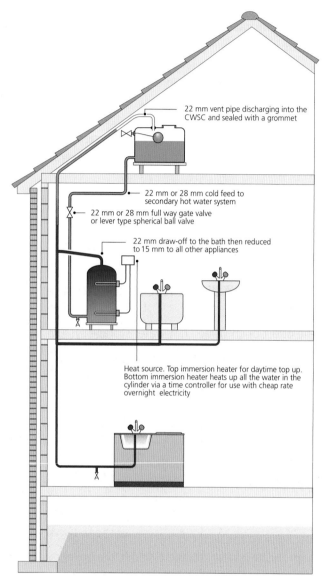

22 mm vent pipe discharging into the CWSC and sealed with a grommet

22 mm or 28 mm cold feed to secondary hot water system

22 mm or 28 mm full way gate valve or lever type spherical ball valve

22 mm draw-off to the bath then reduced to 15 mm to all other appliances

Heat source. Top immersion heater for daytime top up. Bottom immersion heater heats up all the water in the cylinder via a time controller for use with cheap rate overnight electricity

▲ Figure 2.1 Direct open vented hot water storage system

The indirect system

An indirect system uses an indirect-type hot water storage cylinder, which contains some form of heat exchanger to heat the secondary water. There are two distinct types:

- the double feed indirect hot water storage cylinder
- the single feed, self-venting indirect hot water storage cylinder.

The heat exchanger contains primary water and so is classified as part of the central heating system to the dwelling.

Open vented indirect (double feed type) hot water storage system

This is probably the most common of all hot water delivery systems installed in domestic properties. It uses a double feed indirect hot water storage cylinder, which contains a heat exchanger, at the heart of the system. The heat exchanger within the cylinder is usually a copper coil but, in older type cylinders, it can also take the form of a smaller cylinder called an annular. It is called indirect simply because the secondary water in the cylinder is heated indirectly by the primary water via the heat exchanger.

In a double feed indirect system two cisterns are used – a large cistern for the domestic hot water and a smaller one for the heating. It is now general practice to install indirect cylinders in preference to direct types, even if the indirect flow and return are capped off.

The double feed indirect hot water storage cylinder allows the use of boilers and central heating systems that contain a variety of metals, such as steel and aluminium because the water in the cylinder is totally separate from the water in the heat exchanger. This means that there is no risk of dirty or rusty water being drawn off at the taps. The system is designed in such a way that the water in the boiler and primary pipework is hardly ever changed, the only loss of water being in the feed and expansion cistern through evaporation.

The secondary water is that which is drawn from the hot water storage cylinder to supply the hot taps. It is heated by conduction as the water in the cylinder is in contact with the heat exchanger.

A feed and expansion cistern feeds the primary part of the system, and this must be large enough to accommodate the expansion of the water in the system when it is heated. The vent pipe from the primary system must terminate over the feed and expansion cistern. An alternative method would be to use a sealed heating system, which is fed with water from the cold water main via a filling loop. Expansion of water is accommodated in an expansion vessel.

Hot water storage cylinders must conform to **BS 1566**, which specifies the minimum heating surface area of the heat exchanger.

A typical open vented indirect (double feed type) hot water storage system utilising fully pumped primary circulation is shown in Figure 2.2.

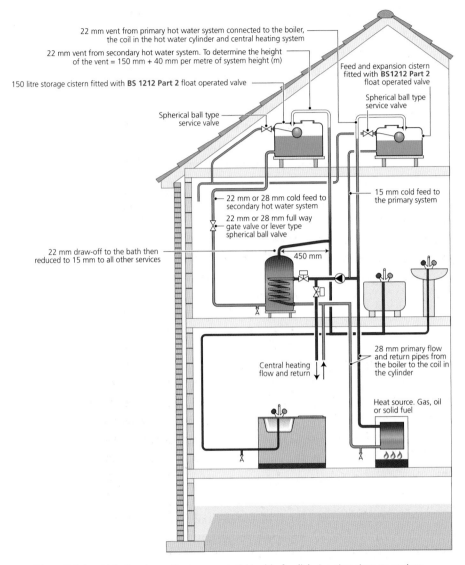

22 mm vent from primary hot water system connected to the boiler, the coil in the hot water cylinder and central heating system

22 mm vent from secondary hot water system. To determine the height of the vent = 150 mm + 40 mm per metre of system height (m)

150 litre storage cistern fitted with **BS 1212 Part 2** float operated valve

Feed and expansion cistern fitted with **BS1212 Part 2** float operated valve

Spherical ball type service valve

Spherical ball type service valve

22 mm or 28 mm cold feed to secondary hot water system

22 mm or 28 mm full way gate valve or lever type spherical ball valve

15 mm cold feed to the primary system

22 mm draw-off to the bath then reduced to 15 mm to all other services

450 mm

Central heating flow and return

28 mm primary flow and return pipes from the boiler to the coil in the cylinder

Heat source. Gas, oil or solid fuel

▲ Figure 2.2 An old indirect gravity open vented (double feed) hot water storage system

Open vented indirect (single feed, self-venting type) hot water storage system

This system uses a single feed, self-venting indirect cylinder, often referred to by its trade name – the 'primatic' cylinder. It contains a special heat exchanger, which uses air entrapment to separate the primary water from the secondary water.

It is fitted in the same way as a direct system, with only one cold feed cistern in the roof space but, unlike the direct system, it allows a boiler and central heating to be installed. It does not require a separate feed and expansion cistern. The heat exchanger works in such a way that the primary and secondary water are separated by a bubble of air that collects in the heat exchanger, preventing the waters from mixing. According to the Domestic Building Services Compliance Guide, these cylinders are no longer allowed for new or replacement cylinders. A 'double feed' type cylinder must be used on all replacement installations.

A typical open vented indirect (single feed, self-venting type) hot water storage system utilising gravity circulation is shown in Figure 2.3.

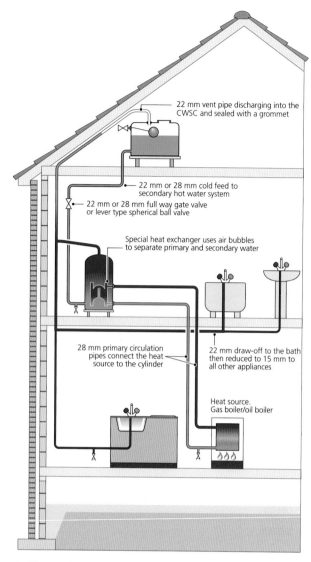

22 mm vent pipe discharging into the CWSC and sealed with a grommet

22 mm or 28 mm cold feed to secondary hot water system

22 mm or 28 mm full way gate valve or lever type spherical ball valve

Special heat exchanger uses air bubbles to separate primary and secondary water

28 mm primary circulation pipes connect the heat source to the cylinder

22 mm draw-off to the bath then reduced to 15 mm to all other appliances

Heat source. Gas boiler/oil boiler

▲ Figure 2.3 Indirect open vented single feed self-venting hot water storage system

INDUSTRY TIP

On no account must central heating inhibitors be used in the primary water if a single feed cylinder is installed, as this would cause contamination of the water if the air bubbles were to rupture.

Unvented hot water storage systems

An unvented hot water storage system is simply a sealed system of pipework and components that is supplied with water above atmospheric pressure. The system does not require the use of a feed cistern. Instead, it is fed with water directly from a water undertaker's mains supply or with water supplied by

a booster pump and a cold water accumulator if the mains pressure is low.

An unvented hot water system differs from open vented types because there is no vent pipe. Expansion of water due to the water being heated is accommodated either in an external expansion vessel or an expansion bubble within the storage cylinder. The system also requires other mechanical safety devices for the safe control of the expansion of water and to ensure that the water within the storage cylinder does not exceed 100 °C. There are two categories of centralised unvented hot water storage systems:

- directly fired/heated storage systems
- indirectly fired/heated storage systems.

Unvented hot water storage systems and pipework arrangements

An unvented hot water storage system is not always the best type of system for any domestic situation. There are many factors that must be considered before this arrangement is installed into a property, such as:

- available pressure and flow rate – this is probably the most important factor simply because poor pressure and flow rate will affect the operating performance of the installation. Pressure and flow rate readings should be taken at peak times to ensure adequate water supply before recommending this type of system
- the route of the discharge pipework, termination and discharge pipework size
- the type of terminal fittings to be used – this is especially important when retro-fitting unvented installations on to existing hot water systems, as the existing taps, etc. may not be suitable
- cost – unvented systems tend to be very expensive.

The types of unvented hot water storage cylinders

There are two types of unvented hot water storage cylinders, both are manufactured to **BS EN 12897:2016+A1:2020** Specification for indirectly heated unvented (closed) storage water heaters, and available as direct fired/heated or indirectly heated vessels:

- unvented hot water storage cylinders using an external expansion vessel
- unvented hot water storage cylinders incorporating an internal expansion air gap.

Most unvented cylinders are manufactured from high-grade duplex stainless steel for strength and corrosion resistance. Some older cylinders may be manufactured from copper or steel with a polyethylene or cementitious lining.

Unvented hot water storage cylinders can be purchased as 'units' or 'packages':

- Units are delivered with all the components already factory fitted and require less installation time.
- Packages are delivered with all components separately packaged (except those required for safety, such as temperature relief valves). These have to be fitted by the installer in line with the manufacturer's instructions.

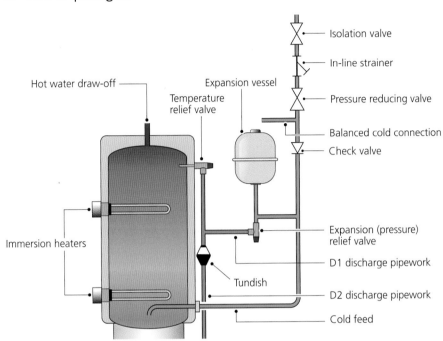

▲ Figure 2.4 A typical unvented cylinder with external expansion vessel

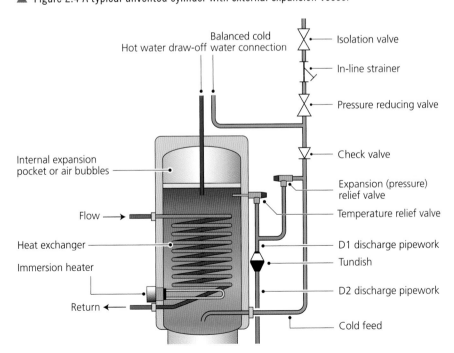

▲ Figure 2.5 A typical unvented cylinder with internal expansion

The various types of unvented hot water system

There are three basic types of unvented hot water system. They are defined by how the water is heated. These are:

- indirect storage systems
- direct storage systems:
  - electrically heated
  - gas or oil fired
- small point of use (under sink).

Indirect storage systems

Indirect unvented hot water storage systems utilise an indirect unvented hot water storage cylinder at the heart of the system. As with open vented systems, the cylinder contains a coiled heat exchanger to transfer the heat indirectly from the primary system to the secondary system. This can be done in one of two ways:

- by the use of a gas fired condensing boiler
- by the use of an oil fired condensing boiler.

Older, non-condensing boilers may be used if the boiler is an existing appliance, providing the boiler contains both a control thermostat and a high energy cut-out (high limit) thermostat to limit the water temperature at the coil should the control thermostat fail. On no account must solid fuel appliances and boilers be used to provide heat to the coil. The primary hot water system may either be open vented or sealed system.

An immersion heater provides back-up hot water heating for use during the summer or for when the boiler malfunctions.

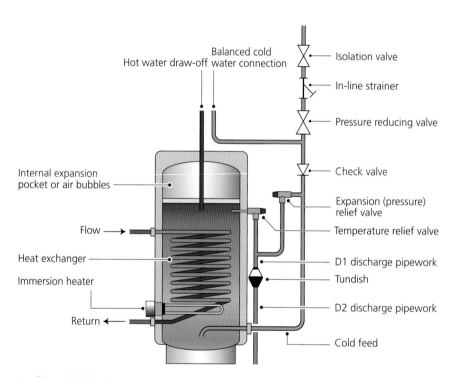

Hot water draw-off — Balanced cold water connection

Isolation valve

In-line strainer

Pressure reducing valve

Check valve

Internal expansion pocket or air bubbles

Expansion (pressure) relief valve

Temperature relief valve

Flow →

Heat exchanger

D1 discharge pipework

Immersion heater

Tundish

D2 discharge pipework

Return ←

Cold feed

▲ Figure 2.6 The indirect type unvented hot water storage cylinder

Direct storage systems

The direct system uses a direct type unvented hot water cylinder that does not contain any form of heat exchanger. There are two very different types:

- **Electrically heated** – this type of cylinder does not contain a heat exchanger. Instead, the water is heated directly by two immersion heaters controlled by a time switch. One immersion heater is located close to the bottom of the cylinder to heat all of the contents of the cylinder at night and another located in the top third to top up the hot water during the day if required via a one-hour boost button on the time switch. Both immersion heaters are independently controlled and cannot be used simultaneously.

The immersion heaters are manufactured to **BS EN 60335.2.73** and **must** contain a user thermostat, usually set to 60 °C, and a non-resetting thermal cut out (high limit stat).

- **Gas or oil fired** – the design of these water heaters originated in North America. They consist of a hot water storage vessel with a flue pipe that passes through the centre. Expansion of the water is catered for by the use of an external expansion vessel. Below the storage vessel is a burner to heat the water and this can be fuelled by either gas or oil depending on the type. The burner is controlled by a thermostat and a gas/oil valve. An energy cut-out prevents the water exceeding the maximum of 90 °C. The safety and functional controls and components layout is almost identical to other unvented hot water storage systems.

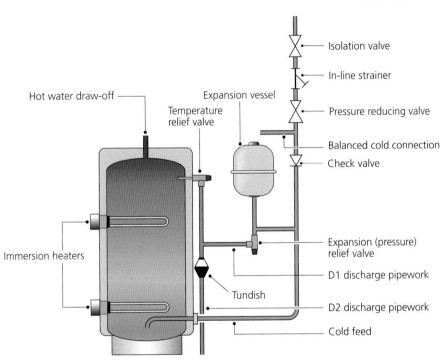

▲ Figure 2.7 The electrically heated direct type unvented hot water storage cylinder

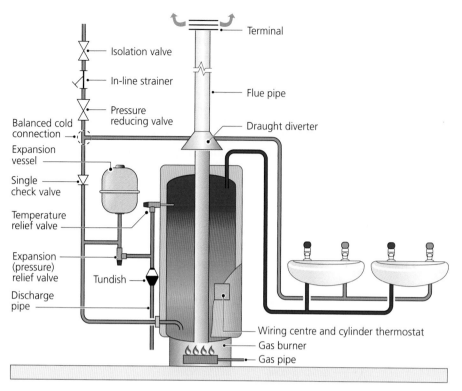

Terminal

Isolation valve

In-line strainer

Pressure reducing valve

Flue pipe

Balanced cold connection

Draught diverter

Expansion vessel

Single check valve

Temperature relief valve

Expansion (pressure) relief valve

Tundish

Discharge pipe

Wiring centre and cylinder thermostat

Gas burner

Gas pipe

▲ Figure 2.8 The gas fired direct type unvented hot water storage cylinder

Direct unvented under sink storage heaters

Unvented under sink hot water storage heaters are connected direct to the mains cold water supply and deliver hot water at near mains cold water pressure. Because they have less than 15 litres of storage, they are not subject to the stringent regulations that surround the installation of larger unvented hot water storage units.

The expansion of water may be taken up within the pipework, provided the pipework is of sufficient size to cope with the water expansion. If not, then an external expansion vessel will be required.

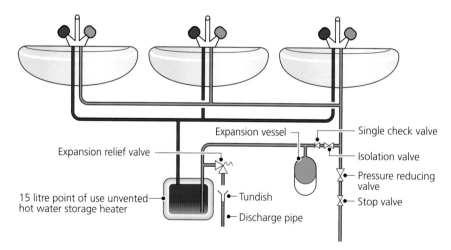

Expansion vessel

Single check valve

Expansion relief valve

Isolation valve

Pressure reducing valve

15 litre point of use unvented hot water storage heater

Tundish

Stop valve

Discharge pipe

▲ Figure 2.9 The unvented type under sink storage water heater pipework layout

▼ Table 2.1 Comparisons between vented and unvented storage hot water systems

Advantages	Disadvantages
Open vented systems	
• Storage is available to meet the demands at peak times • Low noise levels • Always open to the atmosphere • Water temperature can never exceed 100 °C • Reserve of water available if the mains supply is interrupted • Low maintenance • Low installation costs	• Space needed for both the hot water storage vessel and the cold water storage • Risk of freezing • Increased risk of contamination • Low pressure and often, poor flow rate • Outlet fittings can be limited because of the low pressure
Unvented systems	
• Higher pressure and flow rates at all outlets giving a larger choice of outlet fittings • Balanced pressures at both hot and cold taps • Low risk of contamination • The hot water storage vessel can be sited almost anywhere in the property making it a suitable choice for houses and flats alike • The risk from frost damage is reduced • Less space required because cold water storage is not needed • Installation is quicker as less pipework is required • Smaller diameter pipework may be used in some circumstances	• No back up of water should the water supply be isolated • If the cold water supply suffers from low pressure or flow rate, the system will not operate satisfactorily • There is the need for discharge pipes that will be able to accept very hot water and there will be restrictions on their length • A high level of maintenance is required • Higher risk of noise in the system pipework • Initial cost of the unvented hot water storage vessel is high

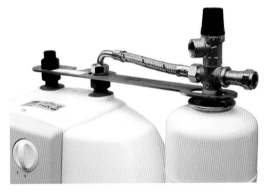

▲ Figure 2.10 The unvented type under sink storage water heater with expansion vessel

Comparisons between open vented and unvented hot water storage systems

There are important differences between these two types of systems. Table 2.1 compares open vented and unvented hot water storage systems.

Gas fired instantaneous multi-point hot water heaters

With this type of hot water heater, cold water is taken from the water undertaker's main and heated in a heat exchanger as demand requires before being distributed to the outlets. As long as the tap is running, hot water will be delivered to the taps. There is no limit to the amount of hot water that can be delivered. There is no storage capacity.

Expansion of water due to being heated is accommodated by back pressure within the cold water main. However, if this is not adequate or the cold water system contains pressure reducing valves or check valves, then an expansion vessel must be fitted.

The heater works on **Bernoulli's principle** by using a venturi tube to create a pressure differential across the gas valve when the cold water is flowing into the heater.

KEY TERM

Bernoulli's principle: states that when a pipe is suddenly reduced in size, the velocity of the water increases, but the pressure decreases. The principle can also work in reverse. If a pipe suddenly increases in size, then the velocity will decrease but the pressure will increase slightly.

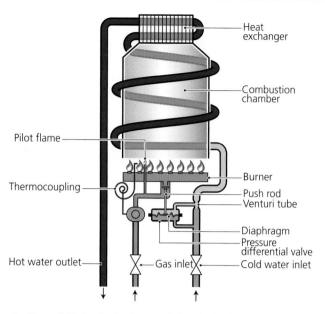

▲ Figure 2.11 Gas instantaneous hot water heater

Gas or oil fired combination boilers

Combi boilers are dual function appliances. They provide instantaneous hot water and central heating within the same appliance. In normal working mode, combination boilers are central heating appliances, supplying a proportion of their available heat capacity to heat the central heating water. When a hot tap is opened, a diverter valve diverts the boiler water around a second heat exchanger, which heats cold water from the water undertaker's cold water mains to supply instantaneous hot water at the hot taps. In this mode, the entire heat output is used to heat the water. Temperature control is electronic and this automatically adjusts the burner to suit the output required. Typical flow rates are around 9 litres per minute (35 °C temperature rise). Some combination boilers incorporate a small amount of storage and this can double the flow rate to around 18 litres per minute.

Thermal stores

Sometimes called water jacketed tube heaters, thermal stores work by passing mains cold water through two heat exchangers which are encased in a large storage vessel of primary hot water fed from a boiler. They are very similar to an indirect system but they work in reverse.

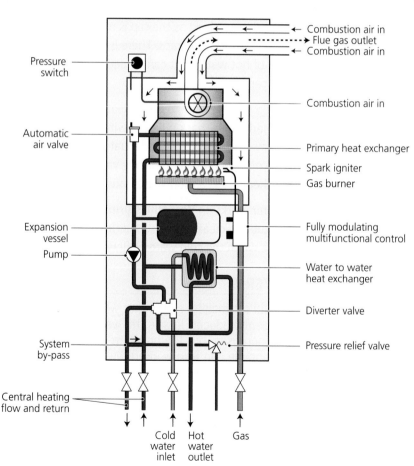

▲ Figure 2.12 Combination boiler

Inside the unit are two heat exchangers which the mains cold water passes through and a small expansion chamber. The expansion chamber allows for the small amount of expansion of the secondary water. The primary water can reach temperatures of up to 82 °C, which can, potentially, be transferred into the secondary water. Because of this, an adjustable thermostatic mixing valve blends the secondary hot water with mains cold water so that the water does not exceed 60 °C.

Gas or oil fired combined primary storage units

These are very similar in design to thermal stores and work in exactly the same way, in that cold water from the mains supply is passed through a heat exchanger. The difference here is that the unit has its own heat source, in the form of a gas burner, to heat the primary water, eliminating the need for a separate boiler.

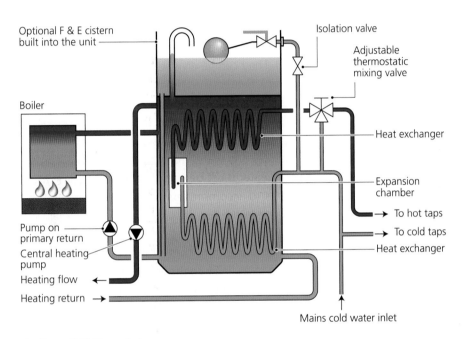

▲ Figure 2.13 Thermal store

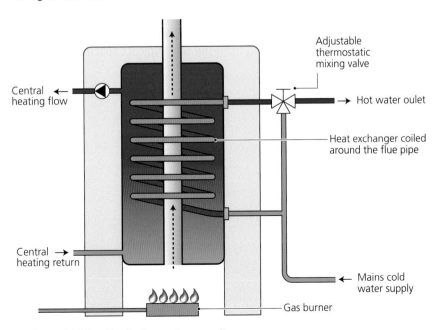

▲ Figure 2.14 Combined primary storage unit

Localised systems

Localised systems are often called single point or point of use systems. They are designed to serve one outlet at the position where it is needed and are usually installed where the appliance is some distance away from the centralised hot water supply.

Again, these can be divided into two categories:
- instantaneous type heaters
- storage type heaters.

Instantaneous type localised water heaters

These can either be fuelled by gas or electricity and are generally described as inlet controlled. This simply means that the water supply is controlled at the inlet to the heater. The water is heated as it flows through the heater and will continue to do so as long as the water is flowing. When the control valve is closed, the water flow stops and the heat source shuts down.

This type of heater is generally used to supply small quantities of hot water, such as washbasins and showers. Typical minimum water pressure is 1 bar.

▲ Figure 2.15 An instantaneous localised water heater

Storage type localised water heaters

This type of heater is often referred to as the displacement type heater, as the hot water is displaced from the heater by cold water entering the unit. Typical storage capacities are between 7 and 10 litres (for over sink type). They can be divided into:

- Over sink heaters – as the name suggests, these are fitted over an appliance such as a sink. The water is delivered from a spout on the heater. A common complaint with this type of heater is that they constantly drip water from the spout. This is normal as the heater must be open to the atmosphere at all times to accommodate the expansion when the water is heated. The dripping water is the expansion taking place and will stop once the heater has reached its operating temperature. They are often referred to as 'inlet controlled' water heaters.
- Under sink heaters – the under sink heater works in exactly the same way as the over sink heater. The main difference is that these heaters usually require a special tap or mixer tap that allows the outlet to be open to the atmosphere at all times to allow for expansion. The inlet of water to the heater is still controlled from the tap. Typical capacities are up to 15 litres.

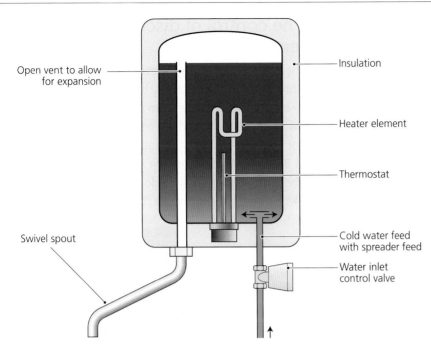

Open vent to allow for expansion

Insulation

Heater element

Thermostat

Swivel spout

Cold water feed with spreader feed

Water inlet control valve

▲ Figure 2.16 A typical over sink storage water heater

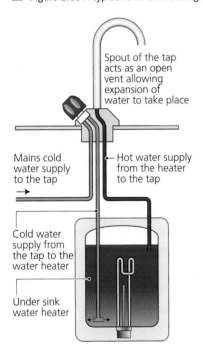

Spout of the tap acts as an open vent allowing expansion of water to take place

Mains cold water supply to the tap

Hot water supply from the heater to the tap

Cold water supply from the tap to the water heater

Under sink water heater

▲ Figure 2.17 A typical under sink storage water heater

Design temperatures for hot water systems

There is some form of hot water delivery system in almost all domestic properties in the UK, whether this is from a centralised or localised hot water system. The overriding concerns with hot water are:

- hot water temperatures that kill the *Legionella* bacteria can also cause scalding

- hot water temperatures that do not cause scalding are ideal for the *Legionella* bacteria to multiply.

These may seem, at first glance, to be contradictory, but in fact they are equally as important. Let's look at the facts.

Legionella pneumophila, better known as Legionnaires' disease, presents a very real risk, especially in any hot water system that contains a storage vessel. Between 20 and 45 °C, *Legionella* bacteria multiply roughly every two minutes. At 60 °C *Legionella* bacteria is dead within 32 minutes. However, water at this temperature is likely to cause a partial thickness burn in about five seconds.

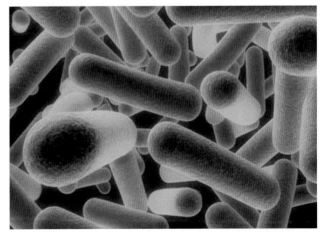

▲ Figure 2.18 *Legionella pneumophila* bacteria

129

These two contradictory subjects – *Legionella* and scalding – are of major importance to designers as they both present different restrictions on the temperatures that can safely be used in the domestic hot water systems we install. Because of this, temperature control must be exercised both in the storage vessels/heaters and at the point of use.

The recommended temperatures are:

- Hot water storage vessel – these should store water at a temperature of **60–65°C** as this temperature offers protection against the reproduction of *Legionella* bacteria.
- Hot water outflow – the Water Supply (Water Fittings) Regulations state that hot water should be distributed at a temperature not less than **55°C** and should reach the hot water outlets at **50°C** within **30 seconds** of the tap being turned on. Whilst this is possible with most hot water storage systems, it is not, realistically, possible with instantaneous hot water delivery types.
- Secondary return – the Health and Safety Executive (HSE) publication 'The control of *Legionella* bacteria in water systems: Approved Code of Practice and Guidance' states that:

  The circulation pump is sized to compensate for the heat losses from the distribution circuit such that the return temperature to the calorifier is not less than 50°C.

- At point of use:
  - Instantaneous heaters – most instantaneous hot water heaters have a varying temperature range between **35** and **55°C**, but this depends on the flow rate through the heater, the flow rate being higher with the lower temperature. Typically power outputs of between 3 and 12 kW are available.
  - Storage heaters – hot water temperatures typically range up to **75°C** for vented type storage water heaters with a 5–10 litre capacity and up to **80°C** with an energy cut-out set at **85°C** for unvented types up to 15 litres capacity.
  - Thermostatic blending valve installations – the subject of thermostatic blending valves and the control of discharge temperatures requires careful consideration and so will be discussed separately.

The control of discharge temperatures

The objective of any hot water storage system is to store water at the relatively high temperature of **60°C** to ensure that it is free from any bacteria, to distribute the water at **55°C** and yet to deliver the water at the hot water outlets at the relatively low temperature of **35** to **46°C** to ensure the safety of the end user. The most efficient way to do this is by the use of thermostatic mixing valves (TMVs).

▲ Figure 2.19 A typical thermostatic mixing valve

Thermostatic mixing valves (sometimes known as a thermostatic blending valve) are designed to mix hot and cold water to a predetermined temperature to ensure that the water is delivered to the outlet at a temperature that will not cause injury but is hot enough to facilitate good personal hygiene. There are three methods of installing TMVs.

Single valve installations

This is probably the most common of all TMV installations. The maximum pipe length to a single appliance is 2 m from the TMV to the outlet. Back-to-back installations are acceptable from a single valve, provided that the use of one appliance does not affect the other and that both appliances have a similar flow rate requirement (such as two washbasins). Typical installations are:

- Baths – it is now a requirement of Building Regulations Approved Document G3 that all bath installations in new and refurbished properties incorporate the use of a TMV. This would normally be set to a temperature of between **41°C** and **44°C**, depending on personal comfort levels.

Temperatures above this can only be used in exceptional circumstances.

- Showers – these installations usually require a temperature of not more than **43 °C**. In residential care homes and other medical facilities, a temperature of not more than **41 °C** should be used according to NHS guidelines.
- Washbasins – careful consideration must be made with washbasin installations because this is probably the only appliance used in domestic dwellings where the user puts their hands directly in the running water without waiting for the water to get hot. When the water reaches maximum temperature, scalding can occur. Therefore, typical temperatures between **38 °C** and **41 °C** can be used depending upon the application. Again, NHS guidelines recommend a temperature of no more than **41 °C**.
- Bidets – a maximum of **38 °C** should be used with bidet installations.
- Kitchen sinks – this is probably the area where the user is most at risk. The need to ensure that bacteria and germs are killed and that grease is thoroughly removed dictates that a water temperature of between **46 °C** and **48 °C** is used. However, as the kitchen is an area with no published recommendations on hot water temperature, a safe temperature similar to that of washbasins should be considered to lessen the risk of scalding unless notices warning of very hot water are used.

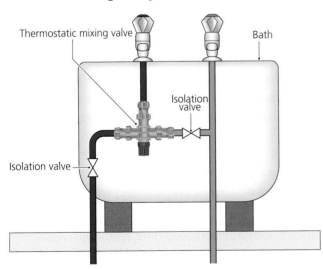

▲ Figure 2.20 A single thermostatic mixing valve installation

Digital showers use state-of-the-art technology to give very accurate control of the showering temperature and flow rate. The shower control and the water mixer have digital intelligence built in.

Digital mixer showers take water from both the hot and cold water supplies and mix them in an electronically controlled mixing valve to accurately reach the desired temperature. An electronic control panel mounted in the showering area provides separate control of the water flow rate and the temperature. A processor box sited remotely adjusts the flow and temperature to the settings selected by the user on the control panel. These are then controlled electronically to provide the desired temperature and flow at the shower.

Accurate control is maintained by the adjustment of separate proportioning valves and pumps or by motorised control of a mixing valve.

Group mixing

Installations where a number of appliances of a similar type are fed from a single TMV are allowed in certain installations. However, installations of this type are not recommended where the occupants are deemed to be high risk, such as nursing homes. If a group installation is to be considered, then the following points should be followed:

- The operation of any one appliance should not affect others on the run.
- When one TMV is used with a number of similar outlets, the length of the pipework from the valve to the outlets should be kept as short as possible so that the mixed water reaches the furthest tap within 30 seconds.
- With group shower installations, it is not unusual to see pipe runs in excess of 10 m. Pipework runs of this length carry an unacceptable *Legionella* risk. These situations can be dealt with by:
- careful monitoring of the water at the showerheads and appropriate treatment should *Legionella* be detected
- regular very hot water disinfection when the system is not in use.

Typical group installations are:

- Group showers – with the correct sized TMV a number of shower outlets may be served at a temperature of between **38** and **40 °C**. For safety reasons, the temperature must not exceed **43 °C**.
- Washbasins – rows of washbasins may be served from a single TMV. Temperatures of between **38** and **40 °C** are typical, but it should not exceed **43 °C** for safety reasons.

131

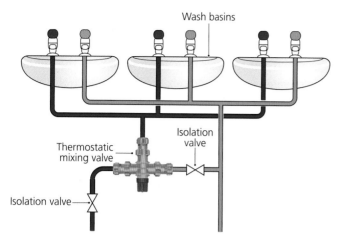

Wash basins

Thermostatic mixing valve

Isolation valve

Isolation valve

▲ Figure 2.21 A group thermostatic mixing valve installation

Centralised mixing

Centralised mixing is very similar to group mixing but occurs when there are groups of different hot water appliances to be served from a single TMV. The following recommendations should be followed:

- If the mixed water is recirculated within the *Legionella* growth temperature range, then anti-*Legionella* precautions similar to those recommended for group mixing will need to be implemented.
- If the mixed water is to be recirculated at about *Legionella* growth temperature regimes, then the recommendations for single TMV installations are appropriate.
- The operation of any one outlet should not affect other outlets.

The types of thermostatic mixing valves

Thermostatic mixing valves are certificated under a third-party certification scheme set up and administrated by NSF International. Under this scheme, thermostatic mixing valves are certificated and approved for use, depending on their application. They are divided into two groups:

- TMV2 – Approved Document G Sanitation, hot water safety and water efficiency of the Building Regulations in England and Wales requires that the hot water outlet to a bath should not exceed 48 °C. It also states that valves conforming to **BS EN 1111** or **BS EN 1287** are suitable for this purpose. Similar requirements exist in Scotland.

TMV2 approval is for domestic thermostatic installations and uses **BS EN 1111** and **BS EN 1287** as a basis for the thermostatic valves performance testing.

- TMV3 – these valves are manufactured and tested for healthcare and commercial thermostatic installations and use the NHS specification D08 as a basis for the thermostatic valves performance testing.

Table 2.2 is a guide for the selection of TMVs for a given application.

INDUSTRY TIP

Look up the products and become more familiar with the styles and temperatures. For examples, see www.rwc.co.uk/product-category/temperature-control/

Balanced and unbalanced supply pressures in unvented hot water storage systems

Balanced pressure means that both the hot and the cold water are supplied at the same pressure. Most modern mixer taps, shower mixer valves and thermostatic blending valves require a balanced pressure to operate correctly and to ensure that the correct mixing of hot and cold water takes place without causing backflow problems or presenting the danger of scalding.

With unvented hot water systems, balancing the hot and cold water supplies is a relatively easy task to perform. The balanced cold supply must be taken from the cold supply to the unvented storage vessel. It must be connected after the pressure reducing valve and before the single check valve as shown in Figure 2.22. The pressure reducing valve ensures that the same pressure is supplied to both hot and cold outlets, and the non-return valve ensures that the balanced cold water supply is not affected by backflow from the pressurised hot water storage vessel in the event of a sudden loss of pressure from the cold water supply.

On systems that use a composite valve (a valve which combines the strainer, PRV, non-return valve and pressure relief/expansion valve), a balanced cold connection is usually designed into the valve in the form of a compression fitting connection.

▼ Table 2.2 TMV selector chart

Environment	Appliance	Is a TMV: required by legislation or authoritative guidance?	Is a TMV: recommended by legislation or authoritative guidance?	Is a TMV: suggested best practice?	Valve type?
Private dwelling	Bath	Yes		Yes	TMV2
	Basin	Yes		Yes	
	Shower				
	Bidet				
Housing Association dwelling	Bath	Yes		Yes	TMV2
	Basin	Yes		Yes	
	Shower				
	Bidet				
Housing Association dwelling for the elderly	Bath	Yes			TMV2
	Basin	Yes			
	Shower	Yes			
	Bidet	Yes			
Hotel	Bath			Yes	TMV2
	Basin			Yes	
	Shower			Yes	
NHS nursing home	Bath		Yes		TMV3
	Basin		Yes		
	Shower		Yes		
Private nursing home	Bath		Yes		TMV3
	Basin		Yes		
	Shower		Yes		
Young persons' care home	Bath	Yes			TMV3
	Basin	Yes			
	Shower	Yes			
Schools, including nursery	Bath	Yes, but 43°C max	Yes		TMV2
	Basin	Yes			
	Shower				
Schools for the severely disabled, including nursery	Bath	Yes, but 43°C max	Yes		TMV3
	Basin	Yes			
	Shower				
NHS hospital	Bath	Yes			TMV3
	Basin	Yes			
	Shower	Yes			
Private hospital	Bath		Yes		TMV3
	Basin		Yes		
	Shower		Yes		

Source: Caretaking Support Services

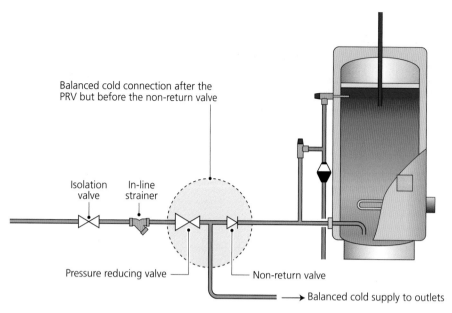

Balanced cold connection after the
PRV but before the non-return valve

Isolation
valve

In-line
strainer

Pressure reducing valve

Non-return valve

Balanced cold supply to outlets

▲ Figure 2.22 Balanced cold supply connection position

Secondary circulation in domestic dwellings

Secondary circulation is required where the length of any draw-off pipework is excessive. British Standard **BS EN 806** and the Water Supply (Water Fittings) Regulations give the maximum length a hot water draw-off pipe may travel without a secondary circulation system being installed. These lengths are reproduced in Table 2.3 below.

▼ Table 2.3 Maximum recommended lengths of uninsulated hot water pipes

Outside diameter of pipe (mm)	Maximum length (m)
12	20
Over 12 up to and including 22	12
Over 22 up to and including 28	8
Over 28	3

It should be remembered that no DHWS dead leg should exceed a volume of 0.5 litres. The table is designed around this figure with pipe volumes /m giving a maximum length per diameter.

Secondary circulation is a method of returning the hot water draw-off back to the storage cylinder in a continuous loop to eliminate cold water **dead legs** by reducing the distance the hot water must travel before it arrives at the taps.

In all installations, secondary circulation must use forced circulation via a bronze or stainless steel-bodied circulating pump to circulate the water to and from the storage cylinder. The position of the pump will depend upon the type of hot water system installed.

▲ Figure 2.23 A bronze-bodied secondary circulation pump

Secondary circulation installations on unvented hot water storage systems

In most cases, a secondary circulation connection is not fitted on an unvented hot water storage vessel and, unlike open vented hot water storage vessels, it is not possible to install a connection of the vessel itself. Where secondary circulation is required, this must be taken to the cold water feed connection using a swept tee just before the cold feed enters the unit. To safeguard against reverse circulation, a non-return valve or single check valve must be fitted after the circulating pump and just before the swept tee branch. The pump should be fitted on the secondary return, close to the hot water storage vessel.

Secondary circulation installations on open vented hot water storage systems

With secondary circulation on open vented systems, the return pipe runs from the furthest hot tap back to the cylinder where it enters at about a quarter of the way down. A circulating pump is placed on the return, close to the hot water cylinder, pumping into the vessel. As with all secondary circulation systems, the pump must be made from bronze or stainless steel to ensure that corrosion does not pose a problem. Isolation valves must be installed either side of the pump so that the pump may be replaced or repaired. The system is shown in Figure 2.25.

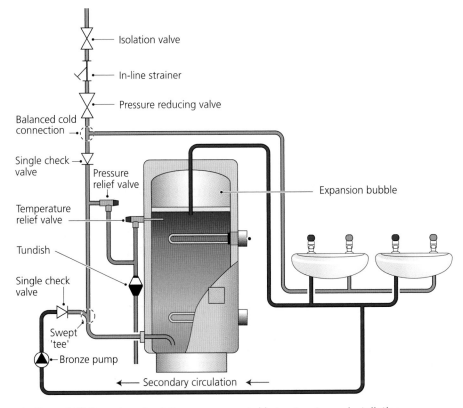

▲ Figure 2.24 Secondary circulation on an unvented hot water storage installation

Some open vented cylinders can be purchased with a secondary return connection already installed on the cylinder. Alternatively, an Essex flange can be used on cylinders where no connection exists.

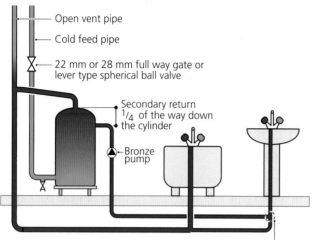

Open vent pipe

Cold feed pipe

22 mm or 28 mm full way gate or lever type spherical ball valve

Secondary return ¹/₄ of the way down the cylinder

Bronze pump

Secondary return connection at the furthest appliance

▲ Figure 2.25 Secondary circulation on an open vented hot water storage installation

▲ Figure 2.26 An Essex flange

Positioning secondary circulation components

The secondary flow (the hot water draw-off), as we have already seen, should have a temperature of at least **60°C+.** The secondary return of the secondary circulation circuit should have a return temperature of **50°C** when it reaches the cylinder at the end of the circuit. In this way, the hottest part of the cylinder will always be the top where the hot water is drawn off. If reversed circulation were to occur, the water in the

cylinder would never reach the disinfecting temperature of **60°C** and so would always be at risk of a *Legionella* outbreak, however remote.

By installing a single check valve on the return and positioning it between the pump and the cylinder, reverse circulation is prevented.

Time clocks for secondary circulation

If secondary circulation is used on hot water systems, it should be controlled by a time clock so that the circulating pump is not running 24 hours a day. The time clock should be set to operate only during periods of demand and should be wired in conjunction with pipe thermostats (also known as aquastats) to switch off the pump when the system is up to the correct temperature and circulation is not required and to activate the pump when the water temperature drops.

Insulating secondary circulation pipework

If secondary circulation systems are installed, they should be insulated for the entire length of the system. This is to prevent excessive heat loss through the extended pipework due to the water being circulated by a circulating pump. The insulation should be thick enough so as to maintain the heat loss below the values in Table 2.4.

▼ Table 2.4 Insulation thickness for secondary circulation pipework

Tube/pipe size (mm)	Maximum heat loss per metre (w/m)
15	7.89
22	9.12
28	10.07

Secondary circulation on large open vented hot water storage systems

Figure 2.27 shows a large domestic hot water system with secondary circulation. As can be seen, there are some significant differences from other secondary circulation systems:

- The hot water vessel includes a shunt pump (see Figure 2.27). This is to circulate the water within the cylinder to ensure that the varying temperature (stratification) of the water inside is kept to a minimum and to ensure an even heat distribution throughout, thereby preventing the growth of the

Legionella bacteria. Stratification is desirable during the day so that the draw-off water is maintained at its hottest for the longest period of time. Because of this, the shunt pump should only operate during periods of low demand (such as at night).

- The secondary circulation pump (component **5** in Figure 2.27) is installed on the secondary flow and not the secondary return as with other, smaller systems.
- A non-return valve (component **6** in Figure 2.27) is installed on the secondary flow to ensure that reverse circulation does not occur.
- A cylinder thermostat (component **3** in Figure 2.27) is provided to maintain the temperature within the cylinder at a maximum of 60 °C.
- A pipe stat (component **2** in Figure 2.27) installed on the secondary flow maintains the temperature at a minimum of at least 50 °C.
- A motorised valve (component **4** in Figure 2.27) is installed on the secondary return close to the hot water storage vessel to prevent water being drawn from the secondary return when the pump is not operating.
- Lockshield gate valves (components **7** and **8** in Figure 2.27) are provided to balance the system to ensure even circulation throughout the secondary water system.
- The secondary circulation system, shunt pumps and thermostats are controlled through a control box (component **1** in Figure 2.27).

Here are some points to remember regarding large centralised hot water systems:

- The pipework should be carefully designed to prevent dead legs as this is a major concern with regard to *Legionella pneumophila*.
- The hot water storage vessel should be capable of being heated to 70 °C, again, to kill any *Legionella* that may be present.
- There should be easy access for draining, cleaning, inspection and maintenance.
- If a shunt pump is installed, the storage vessel should be insulated on its underside to prevent excessive heat loss.

The use of trace heating instead of secondary circulation

Electric trace heating uses an electric cable that forms a heating element. It is positioned directly in contact with the pipe along the whole length of the pipe length. The pipe is then covered in thermal insulation. The heat generated by the element keeps the pipe at a specific temperature.

The operation of the trace heating element should be timed to a period when the hot water system is in most use (such as early in the morning and evening). If the pipe is well insulated and installed with a timer, the amount of energy usage will be minimal.

By using trace heating, the additional cost of the extra pipework for the secondary return and its associated pump and running costs is removed.

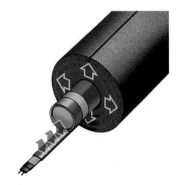

▲ Figure 2.28 Trace heating

INDUSTRY TIP

Trace heating can also be used as frost protection on cold water systems.

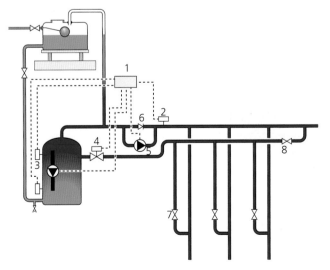

▲ Figure 2.27 Secondary circulation on a large domestic open vented hot water storage installation

3 SPECIALIST COMPONENTS IN HOT WATER SYSTEMS

In this section, we will look at some of the specialised components that are used in unvented hot water storage systems. These are important because of the operating characteristics of mains-fed hot water storage systems.

The function and position of components in unvented hot water systems controls

The controls for unvented hot water storage systems (UHWSS) fall under two categories. These are:
- safety
- functional.

In this part of the chapter, we will look at the various controls and components for unvented hot water storage systems, their function and the position that they occupy within the system.

Safety controls – control thermostat, high limit thermostat and temperature relief valve

With the water inside the storage vessel at a pressure above atmospheric pressure, the control of the water temperature becomes vitally important. This is because as the pressure of the water rises, so the boiling point of the water rises. In simple terms, if total temperature control failure were to occur, the water inside the vessel would eventually exceed 100 °C, with disastrous consequences. The line graph in Figure 2.30 demonstrates the pressure/temperature relationship.

On the graph it can be seen that at the relatively small pressure of 1 bar, the boiling point of the water has risen to 120.2 °C! If a sudden loss of pressure at the hot water storage vessel were to occur due to vessel fracture, at 120.2 °C the entire contents of the cylinder would instantly flash to steam with explosive results causing structural damage to the property.

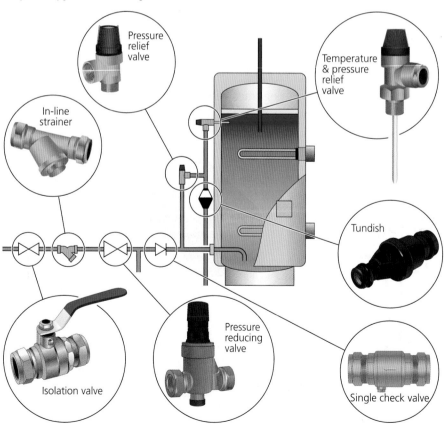

Pressure relief valve

In-line strainer

Temperature & pressure relief valve

Tundish

Isolation valve

Pressure reducing valve

Single check valve

▲ Figure 2.29 The controls on a modern UHWSS

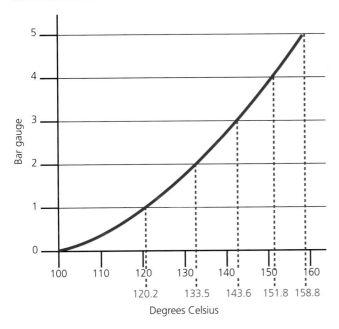

▲ Figure 2.30 Boiling point/pressure relationship

IMPROVE YOUR MATHS

Calculating how much steam would be produced, illustrates the point further:

1 cm³ of water creates 1600 cm³ of steam. If the storage vessel contains 200 litres of water and each litre of water contains 1000 cm³, then the amount of steam produced would be:

$$200 \times 1000 \times 1600 = 320{,}000{,}000 \text{ cm}^3 \text{ of steam!}$$

The Building Regulations Approved Document G3 states that unvented hot water storage systems must have a three-tier level of safety built into the system. This takes the form of three components that are fitted to the storage vessel. The aim of these components is to ensure that the water within the system never exceeds 100°C. These components are:

- Control thermostat (set to 60 to 65 °C) – this can take two forms depending on the type of storage vessel:
  - with direct heated vessels this is the immersion heater user thermostat
  - with indirectly heated vessels it is the cylinder thermostat wired to the central heating wiring centre.

Indirectly fired systems are also controlled, in part, by the boiler thermostat (82 °C maximum setting) and the boiler high limit stat, designed to operate at typically 90 °C).

- Overheat thermostat (thermal cut-out 90 °C maximum but more usually factory set at between 85 and 89 °C) – again, this can take two forms:
  - with direct heated systems, it is incorporated into the immersion heater thermostat
  - with indirectly heated systems, it is a separate component factory wired into the vessel and designed to operate the motorised valve at the primary hot water coil.
- Temperature/pressure relief valve (95 °C) – a standard component used on most vessels that is designed to discharge water when the temperature exceeds 95 °C. Most types have a secondary pressure relief function.

▲ Figure 2.31 A temperature/pressure relief valve

Functional controls

The functional controls of an unvented hot water storage system are designed to protect the water supply:

- To avoid contamination, the storage cylinder or vessel must be of an approved material, such as copper or duplex stainless steel, or have an appropriate lining that will not cause corrosion or contamination of the water contained within it. Where necessary it must be protected by a sacrificial anode.
- A single check valve (often referred to as a non-return valve) must be fitted to the cold water inlet to prevent hot or warm water from entering the water undertaker's mains supply.
- A means of accommodating and containing the increase in volume of water due to expansion must be installed. This can either be by the use of an externally fitted expansion vessel or via an integral air bubble.

- An expansion valve (also known as the pressure relief valve) must be installed and should be designed to operate should a malfunction occur with either the pressure reducing valve or the means of accommodating the expanded water. The expansion valve must be manufactured to **BS EN 1491:2000**, Building valves. Expansion valves. Tests and requirements.

The Water Supply (Water Fittings) Regulations also state that:

> Water supply systems shall be capable of being drained down and fitted with an adequate number of servicing valves and drain taps so as to minimise the discharge of water when water fittings are maintained or replaced.

To comply with this requirement, a servicing valve should be fitted on the cold supply close to the storage vessel but before any other control. The valve may be a full-bore spherical plug, lever action type isolation valve or a screw-down stop valve. Any drain valves fitted should be manufactured to **BS 2879** and be 'type A' drain valves with a locking nut and an 'O' ring seal on the spindle.

▲ Figure 2.32 A BS 1010 screw-down stop valve

▲ Figure 2.33 A lever action spherical plug isolation valve

▲ Figure 2.34 A 'type A' drain off valve

The functional controls of an unvented hot water storage system are listed below. We will look at each one in turn and identify its position within the system

In-line strainer

The in-line strainer is basically a filter designed to prevent any solid matter within the water from entering and fouling the pressure reducing valve and any other mechanical components sited downstream. In modern storage systems, this is incorporated into the composite valve, which will be discussed later in this section.

▲ Figure 2.35 In-line strainer

Pressure reducing valve

Pressure reducing valves (PRVs) were looked at in detail in Chapter 1.

The pressure reducing valve of an unvented hot water storage system reduces the pressure of the incoming water supply to the operating pressure of the system. In all cases this will be set by the manufacturer and sealed at the factory. The outlet pressure will remain constant even during periods of fluctuating pressures. Should the pressure of the water supply drop below that of the operating pressure of the PRV, it will remain fully open to allow the available pressure to be used.

Replacement internal cartridges are available and easily fitted without changing the valve body should a malfunction occur.

Modern PRVs for unvented hot water storage systems are supplied with a balanced cold connection already fitted.

▲ Figure 2.36 Pressure reducing valve

Single check valve

The single check valve (also known as a non-return valve) is fitted to prevent hot water from backflowing from the hot water storage vessel causing possible fluid category 2 contamination of the cold water supply. The single check valve also ensures that the expansion of water when it is heated is taken up within the systems expansion components or expansion bubble. Single check valves are classified as either type EA or EB backflow prevention devices.

> ### KEY POINT
> Backflow prevention devices were discussed in detail in Chapter 1, Cold water systems, planning and design.

▲ Figure 2.37 Single check (non-return) valve

In most cases, the check valve will be part of the composite valve, to be discussed later in this section.

Expansion device (vessel or integral to cylinder)

Water expands when heated. Between 4 and 100 °C it will expand by approximately 4 per cent. Therefore 100 litres of water at 4 °C becomes 104 litres at 100 °C. It is this expansion of water that must be accommodated in an unvented hot water storage system. This can be achieved in one of two ways:
- by the use of an externally fitted expansion vessel
- by the use of a purpose designed internal expansion space or 'expansion bubble'.

Expansion vessels

An expansion vessel is a cylindrical shaped vessel that is used to accommodate the thermal expansion of water to protect the system from excessive pressures. It is installed as close to the storage vessel as possible and preferably higher. There are two basic types: the bladder (bag) type expansion vessel and the diaphragm type expansion vessel.

The bladder (bag) type expansion vessel

Also known as the bag type expansion vessel, it is usually made from steel and contains a neoprene rubber bladder to accept the expanded water. At no time does the water come into contact with the steel vessel as it is contained at all times within the bladder.

The inside of the steel vessel is filled with either air or nitrogen to a predetermined pressure. The initial pressure charge from the manufacturer is usually made with nitrogen to negate the corrosive effects on the steel vessel's interior. A Schrader valve is fitted to allow the pressures to be checked and to allow an air 'top-up' if this becomes necessary. Figure 2.38 shows the workings of a bladder type expansion vessel.
- Diagram A shows the bladder in its collapsed state. This is because the only pressure is the air/nitrogen charge compressing the empty bladder. There is no water in the bladder.
- Diagram B shows that water under pressure has entered the bladder during the initial cold fill of the storage cylinder, causing the bladder to expand pressurising the air in relation to the water pressure. The bladder has expanded because the water pressure is greater than the pressure of the air.
- Diagram C shows the bladder fully expanded due to the hot water expansion when the system is heated.

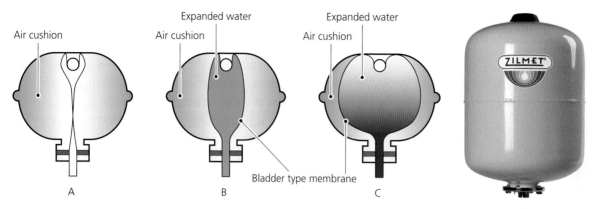

▲ Figure 2.38 Workings of a bladder (bag) type expansion vessel

▲ Figure 2.39 Bladder (bag) type expansion vessel

With some bladder expansion vessels, the bladder is replaceable in the event of bladder failure. A flange at the base of the vessel holds the bladder in place. By releasing the air and removing the bolts, the bladder can be withdrawn and replaced.

The diaphragm type expansion vessel

Diaphragm expansion vessels are used where the water has been de-oxygenated by the use of inhibitors or because the water has been repeatedly heated, such as in a sealed central heating system. They must not be used with unvented hot water storage systems because the water is always oxygenated and comes into direct contact with the steel of the vessel.

They are made in two parts with a neoprene rubber diaphragm separating the water from the air charge.

Again, like the bladder type expansion vessel, a Schrader valve is fitted to allow top-up and testing of the air pressure. Figure 2.40 shows the workings of a diaphragm type expansion vessel.

Internal expansion

With internal expansion, an air pocket is formed as the hot water storage vessel is filled. A floating baffle plate provides a barrier between the air and the water so that there is minimum contact between the air and the water in the cylinder. When the water is heated, the expansion pushes the baffle plate upwards in a similar manner to an expansion vessel.

Over a period of time, the air within the air bubble will dissipate as it is leeched into the water. When this happens, expansion cannot take place and the pressure relief valve will start to discharge water. However,

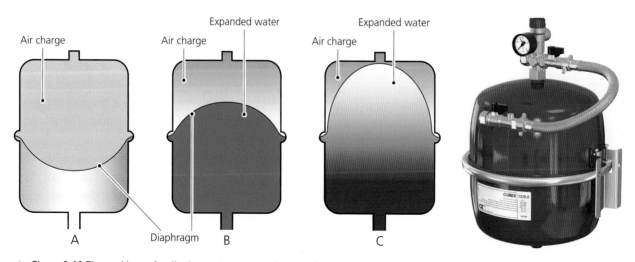

▲ Figure 2.40 The workings of a diaphragm type expansion vessel

▲ Figure 2.41 Diaphragm type expansion vessel

this will only occur as the water heats up. Once the cylinder is at its full temperature, the pressure relief valve will close and will only begin to discharge water again when expansion is taking place. Because of this, manufacturers of bubble top units and packages recommend that the cylinder is drained down completely and refilled to recharge the air bubble.

> ### INDUSTRY TIP
>
> The cylinder should be drained on an annual basis, or as and when required.

The scientific principles of expansion vessels

The principle of an expansion vessel is that a gas is compressible but liquids are not. That principle is based upon Boyle's law. In this case the gas is air or nitrogen and the liquid is water.

Boyle's law states:

The volume of a gas is inversely proportional to its absolute pressure provided that the temperature remains constant.

In other words, if the volume is halved, the pressure is doubled.

Mathematically, Boyle's law is expressed as:

$$P_1 V_1 = P_2 V_2$$

Where:

P_1 = initial pressure = 1 bar
V_1 = initial volume = 20 litres
P_2 = final pressure = to be found
V_2 = final volume = 20 litres – 10 litres of expanded water

So, to find the pressure in the vessel, the formula must be transposed:

$$P_2 = \frac{P_1 \times V_1}{V_2}$$

Therefore:

$$P_2 = \frac{1 \text{ bar} \times 20 \text{ litres}}{10 \text{ litres}}$$

$$= 2 \text{ bar final cold pressure}$$

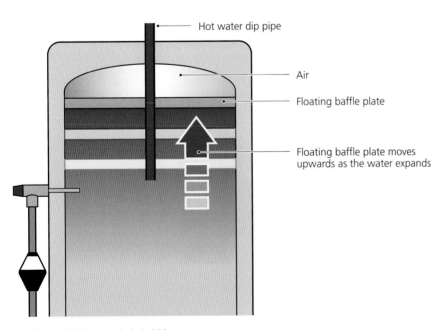

Hot water dip pipe

Air

Floating baffle plate

Floating baffle plate moves upwards as the water expands

▲ Figure 2.42 Integral air bubble

IMPROVE YOUR MATHS

If, on the initial cold fill of the system, the vessel required, say, 5 litres of water to be taken in, the air pressure to apply to the vessel can be calculated. We can assume a water pressure of 1 bar.

P_1 = 1 bar

V_1 = 20 litres

V_2 = 20 litres – 5 litres = 15 litres

P_2 = pressure to be calculated

$$P_2 = \frac{P_1 \times V_1}{V_2}$$

Therefore:

$$P_2 = \frac{1 \text{ bar} \times 20 \text{ litres}}{15 \text{ litres}} = 1.33 \text{ bar}$$

The capacity left in the vessel after the initial fill is 15 litres with a cold fill pressure of 1 bar and that 10 litres of water are to expand inside the vessel, the final pressure of the system will be:

$$P_2 = \frac{P_1 \times V_1}{V_2}$$

Therefore:

$$P_2 = \frac{1 \times 15}{15 - 10} = \frac{15}{5} = 3 \text{ bar}$$

The initial pressure of the empty 20 litre vessel was 1.33 bar. On initial cold fill 5 litres of water entered the vessel reducing the capacity to 15 litres. As a result, the air was compressed even more when the expansion of water takes place and instead of 2 bar final pressure, the pressure when the water is heated will be 3 bar.

IMPROVE YOUR MATHS

Transposing the formula $P_1V_1 = P_2V_2$ as shown in the example above, find the final hot operating pressure of the storage cylinder.

Where:

P_1 = initial pressure = 1.5 bar

V_1 = initial volume = 18 litres

P_2 = final pressure = to be found

V_2 = final volume = 18 litres – 9 litres of expanded water

$P_1V_1 = P_2V_2$

This equals:

1.5 bar × 18 litres = P_2 × 9 litres of water

Now transpose the equation to find P_2.

Pressure relief valve

Often referred to as the expansion relief valve, the pressure relief valve is designed to automatically discharge water in the event of excessive mains pressure or malfunction of the expansion device (expansion vessel or air bubble). It is important that no valve is positioned between the pressure relief valve and the storage cylinder.

The pressure at which the pressure relief valve operates is determined by the operating pressure of the storage vessel and the working pressure of the pressure reducing valve. The valve is pre-set by the manufacturer and must not be altered.

The pressure relief valve will not prevent the storage vessel from exploding should a temperature fault occur and, as such, is not regarded as a safety control.

▲ Figure 2.43 Pressure relief valve

Tundish arrangements

The tundish is part of the discharge pipework and is supplied with every unvented hot water storage system. It is the link between the D1 and D2 pipework arrangements. It has three main functions:

- to provide a visual indication that either the pressure relief or temperature relief valves are discharging water due to a malfunction
- to provide a physical, type A air gap between the discharge pipework and the pressure relief/temperature relief valves
- to give a means of releasing water through the opening in the tundish in the event of a blockage in the discharge pipework.

The tundish must always be fitted in the upright position in a visible place close to the storage vessel. The tundish will be looked at in more detail when discharge pipework arrangements are discussed later in this section.

▲ Figure 2.44 Tundish

Composite valves

These days, it is very rare to see individual controls fitted on an unvented hot water storage system unless it is an early type manufactured in the 1990s. Most manufacturers now prefer to supply composite valves which incorporate many components into one 'multi-valve'. A typical composite valve will contain:

- a strainer
- a pressure reducing or pressure limiting valve, followed immediately by
- a balanced cold take off, and finally
- a pressure relief valve.

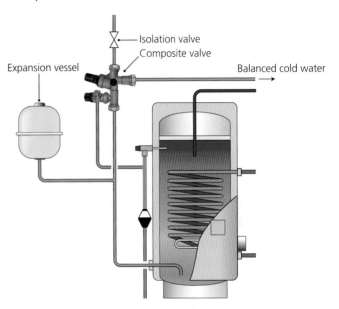

▲ Figure 2.46 Position of a composite valve

Some composite valves may also contain an isolation valve. With all controls contained in a single valve, making the connection to an unvented hot water storage vessel is a simple matter of just connecting the cold supply, without the need to ensure that the controls have been fitted in the correct order.

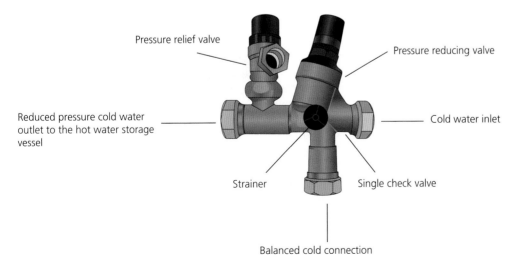

▲ Figure 2.45 A typical composite valve

The discharge pipework

The layout features for temperature and expansion relief (discharge) pipework

With unvented hot water systems, there is always the possibility, however undesirable, that the pressure relief and temperature relief valves may discharge water. The discharge pipework is designed specifically to remove the discharged water away from the building safely. It is, therefore, very important that it is installed correctly with correct size of pipe and that the pipework is made from the correct material, especially since the water discharged may be at near boiling point. There are three sections to the discharge pipework:

- D1 pipework arrangement
- the tundish
- D2 pipework arrangement.

As we have already established the role of the tundish earlier in the chapter, we will concentrate specifically on the D1 and D2 sections of the discharge pipework.

To ensure that there is no damage to the property, the discharge pipework should be positioned in a safe but visible position and should conform to the following:

- The discharge must be via an air break (tundish) positioned within 600 mm of the temperature relief valve.
- The tundish must be located within the same space as the hot water storage vessel.
- It should be made of metal or other material capable of withstanding the temperature of the discharged water. The pipe should be clearly and permanently marked to identify the type of product and its performance standards (see Key point).
- The discharge pipe must not exceed the hydraulic resistance of a 9 m straight length of pipe without increasing the pipe size.
- It must fall continuously throughout its entire length with a minimum fall of 1 in 200.
- The D2 pipework from the tundish must be at least one pipe size larger than the D1 pipework.

- The discharge pipe should not connect to a soil discharge pipe unless the pipe material can withstand the high temperatures of discharge water, in which case it should:
  - contain a mechanical seal (such as a Hepworth Hep$_v$O valve), not incorporating a water trap to prevent foul air from venting through the tundish in the event of trap evaporation
  - be a separate branch pipe with no sanitary appliances connected to it
  - where branch pipes are to be installed in plastic pipe they should be either polybutalene (PB) to class S of **BS 7291.2:2010** or cross linked polyethylene (PE-X) to Class S of **BS 7291.3:2010**
  - be marked along the entire length with a warning that no sanitary appliances can be connected to the pipe.
- The D1 pipework must not be smaller than the outlet of the temperature relief valve.
- The D1 discharge from both the pressure relief and temperature relief valves may be joined by a tee piece provided that all of the points above have been complied with.
- There must be at least 300 mm of vertical pipe from the tundish to any bend in the D2 pipework.

> **KEY POINT**
> Paragraph 3.9 of Approved Document G3 Guidance notes specifies metal pipe for the discharge pipework. However, G3 itself states only that hot water discharged from a safety device should be safely conveyed to where it is visible but will not cause a danger to persons in or about the building. Since many types of plastic pipe are now able to withstand the heat of the discharge water, the responsibility for the choice of material rests with the installer, the commissioning engineer and the local building control officer to ensure that G3 is complied with. It is also important that if plastic pipes are used, the type of plastic is clearly indicated for future reference when inspections and servicing are carried out.

Figure 2.47 illustrates some of the requirements mentioned above.

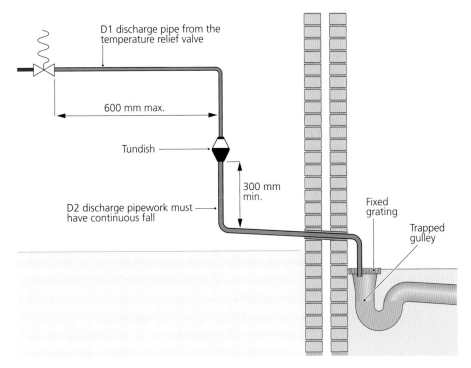

▲ Figure 2.47 The layout of the discharge pipework

The pipe size and positioning methods for safety relief (discharge) pipework connected to unvented hot water cylinder safety valves

As we have already seen, the discharge pipework must not exceed the hydraulic resistance of a 9 m straight length of pipe without increasing the pipe size. Where the discharge pipework exceeds 9 m, the size of the discharge pipe will require calculating including the resistance of any bends and elbows. You can use Table 2.5 to do this.

▼ Table 2.5 Sizing discharge pipework

Valve outlet size	Maximum size of discharge to tundish (D1) (mm)	Maximum size of discharge pipe from tundish (D2) (mm)	Maximum resistance allowed, expressed as a length of straight pipe without bends or elbow (m)	Resistance created by each bend or elbow (m)
G ½	15	22	Up to 9	0.8
		28	Up to 18	1.0
		35	Up to 27	1.4
G ¾	22	28	Up to 9	1.0
		35	Up to 18	1.4
		42	Up to 27	1.7
G 1	28	35	Up to 9	1.4
		42	Up to 18	1.7
		54	Up to 27	2.3

IMPROVE YOUR MATHS

Let's look at how Table 2.5 works.

The temperature and pressure relief valves both have ½" BSP outlets. Therefore, the D1 pipework, as can be seen from the table, can be installed in 15 mm tube. The discharge pipe run is 6 m long to the final termination and there are 6 elbows installed in the run of pipe.

Using the first row in the table, the first option has to be 22 mm because the D2 pipework must be at least one pipe size larger than the D1 pipework. The maximum length of 22 mm pipe is 9 m, but there are six elbows in the run and each of these has a resistance of 0.8 m.

$$6 \times 0.8 = 4.8 \, \text{m}$$

If we add the original length of 6 m, we get:

$$4.8 + 6 = 10.8 \, \text{m}$$

The length of 22 mm discharge pipe, as we have already seen, is 9 m so, at 10.8 m, 22 mm pipe is not large enough for the discharge pipe run. Another pipe size will have to be chosen.

Looking at 28 mm, we see that the maximum run of pipe is 18 m, but the 28 mm elbows now have a resistance of 1 m and there are six of them. Therefore:

$$6 \times 1 = 6 \, \text{m}$$

Add this to the original length of 6 m:

$$6 + 6 = 12 \, \text{m}$$

In this case the discharge pipework is well within the 18 m limit and so 28 mm discharge pipework can be installed.

IMPROVE YOUR MATHS

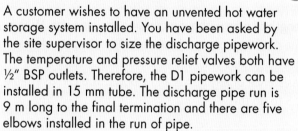

A customer wishes to have an unvented hot water storage system installed. You have been asked by the site supervisor to size the discharge pipework. The temperature and pressure relief valves both have ½" BSP outlets. Therefore, the D1 pipework can be installed in 15 mm tube. The discharge pipe run is 9 m long to the final termination and there are five elbows installed in the run of pipe.

What size of discharge pipework should be installed?

Correct termination of the discharge pipework

A risk assessment is likely to be needed where any termination point for the discharge pipework is to be considered. This will determine whether any special requirements are needed in relation to the termination point and its access. Points to be considered here are:

- areas where the public may be close by or have access to
- areas where children are likely to play or have access to
- areas where the discharge may cause a nuisance or a danger
- termination at height
- the provision for warning notices in vulnerable areas.

The Building Regulations Approved Document G3 states that the discharge pipe (D2) from the tundish must terminate in a safe place with no risk to persons in the discharge vicinity. Acceptable discharge arrangements are:

- To trapped gully with pipe below gully grate but above the water seal.
- Downward discharges at low level up to a maximum 100 mm above external surfaces, such as car parks, hard standings and grassed areas, are acceptable, provided a wire cage or similar guard is provided to prevent contact, whilst maintaining visibility.
- Discharges at high level, onto a flat metal roof or other material capable of withstanding the temperature of the water may be used provided that any plastic guttering system is at least 3 m away from the point of discharge to prevent damage to the guttering.
- Discharges at high level, into a metal hopper and metal downpipe may be used provided that the end of the discharge pipe is clearly visible. The number of discharge pipes terminating in a single metal hopper should be limited to six to ensure that the faulty system is traceable.
- Discharge pipes that turn back on themselves and terminate against a wall or other vertical surface should have a gap of at least one pipe diameter between the discharge pipe and the wall surface.

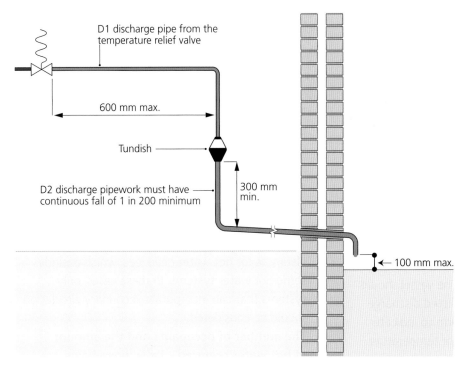

▲ Figure 2.48 The low-level termination of discharge pipework 1

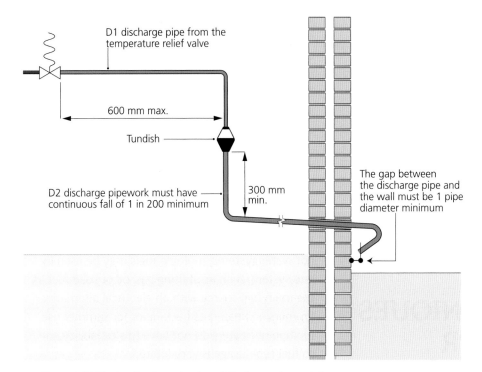

▲ Figure 2.49 The low-level termination of discharge pipework 2

KEY POINT

The discharge may consist of high temperature water and steam. Asphalt, roofing felt and other non-metallic rainwater goods may be damaged by very high temperature hot water discharges.

Termination of the discharge pipework where the storage vessel is sited below ground level

When storage vessels are sited below ground, such as in a cellar, the removal of the discharge becomes a problem because it cannot be discharged safely away from the building. However, with the approval of the local authority and the vessel manufacturer it may be possible to pump the discharge to a suitable external point. A constant temperature of 95 °C should be allowed for when designing a suitable pumping arrangement. The pump should include a suitable switching arrangement installed in conjunction with a discharge collection vessel made from a material resistant to high temperature water. The vessel should be carefully sized in line with the predicted discharge rate and should include an audible alarm to indicate discharge from either of the pressure or temperature relief valves is taking place.

Specialist components

Our work as plumbers covers a multitude of various installations, systems and components. Occasionally, we may be asked to install specialist components that we may only come into contact with on a limited number of occasions. Even so, it is important that we become familiar with these 'specialist' components to ensure that we position and install them correctly and according to the manufacturer's instructions and in line with any regulations or recommendations.

Specialist components used on hot water systems are much the same as those used on cold water systems, and these were covered in much detail in Chapter 1, Cold water systems, planning and design.

4 DESIGN TECHNIQUES FOR HOT WATER SYSTEMS

This section will cover the design techniques used for hot water systems, including calculating system requirements and installation requirements.

Choosing the right hot water system

The type of system we choose will depend on the following points:

- **The size of the property and the distance from the outlets** – the Water Supply (Water Fittings) Regulation stipulates the maximum distance that a hot water supply pipe may run without constituting wastage of water. This is because of the amount of cold water that is drawn off before hot water arrives at the taps. This 'dead', cold water must be limited. Large properties may exceed the maximum distances for hot water dead legs, which excludes some hot water systems. In these cases, only systems that can incorporate secondary circulation should be considered.

- **The number of occupants and the amount of hot water required** – larger households will, obviously, require more hot water, which can be supplied in a number of ways (such as an instantaneous water heater giving unlimited hot water amounts or a large hot water storage cylinder) but other factors must also be considered before a decision is made.

- **The number of hot water outlets** – again, an important point because this may automatically exclude such appliances as combi boilers and instantaneous heaters because, although classed as multipoint heaters, only one outlet at a time may be opened satisfactorily whereas other types of hot water system may allow multiple open taps with good flow rate.

- **The type(s) of fuel to be used** – with most storage hot water systems, multiple fuels may be used in one system, such as utilising gas, oil or solid fuel as the main fuel source with an electrical alternative (immersion heater) as back-up or for summer use. Multipoint heaters do not have this capability and so fuel type usage is very limited.

- **Installation and maintenance costs** – again, a very important point because of the size of the system, initial cost of the appliance and materials. Add to this the installation costs and any maintenance costs over the lifetime of the system.

- **Running costs and energy efficiency** – new, more efficient methods of heating water are being developed constantly. Perhaps the most important recent development is that of solar hot water heating, which can, theoretically, offer a 60 per cent saving on domestic hot water heating costs, despite its initial costly installation. The development of fuel-efficient condensing oil and gas boilers and storage cylinders with fast heat recovery times have also helped in terms of energy efficiency.

When these points are considered, the choice of hot water system should be quite a straightforward affair. Certain dwellings almost dictate the system that should be fitted. For example, it would be unwise to install a combi boiler in a dwelling with three bathrooms, a kitchen, utility room and downstairs washroom: the hot water demand would be more than the boiler could cope with.

> **KEY POINT**
>
> By far, the main considerations that must be considered when selecting the right hot water system are the type and number of appliances and their pattern and frequency of use. Knowing this will indicate the correct choice of system to install and the customer can then be advised accordingly.

Statutory regulations and sources of information

Statutory regulations

The installation of hot water systems is governed strictly by various regulations:

- Building Regulations Approved Document G3 2010
- Building Regulations Approved Document L1A/B 2013 (with 2016 amendments)
- Water Supply (Water Fittings) Regulations 1999
- Gas Safety (Installation and Use) Regulations
- **BS 7671** The IET Wiring Regulations.

The Building Regulations Approved Document G3 2010

In the past, Building Regulations Approved Document G3 only related to unvented hot water supply systems. In 2010 it was updated to encompass all hot water delivery systems in domestic dwellings. It is divided into four parts:

- Part 1 of G3 is a new requirement. It states that heated wholesome water must be supplied to any washbasin or bidet that is situated in or adjacent to a room containing a sanitary convenience, to any washbasins, bidets, fixed baths or showers installed in a bathroom and any sink in an area where food is prepared.

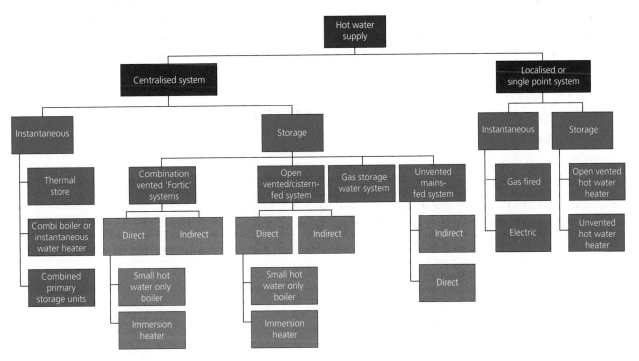

▲ Figure 2.50 Choosing the right hot water system

- Part 2 is an expanded requirement. It states that any hot water system, including associated storage (including any cold water storage cistern) or expansion vessel must resist the effects of any temperature or pressure that may occur during normal use as a consequence of any reasonably anticipated fault or malfunction. This amendment was enforced after the failure of an immersion heater thermostat that caused the collapse of a storage cistern containing water almost at boiling point.

- Part 3, again, is an amended requirement. It states that any part of a hot water system that incorporates a hot water storage vessel must include precautions to ensure that the temperature of the stored water does not exceed 100 °C and that any discharge from such safety devices is safely conveyed to a point where it is visible without constituting a danger to persons in or about the building.

- Part 4 states that any hot water supply to a fixed bath must include provision to limit the temperature of the discharged water from any bath tap so that it cannot exceed 48 °C. This requirement applies to any new-build or property conversions. It is a new requirement that is intended to prevent scalding.

It is interesting to note that Regulation G3 applies to all domestic dwellings, including greenhouses, small detached buildings, extensions and conservatories, but only if they are served with hot water supplied from a dwelling.

It should be noted that the local building control officer should be informed before commencing any installation of a hot water system.

The Building Regulations Approved Document L1A/B 2010

This document promotes the conservation of fuel and power. The basic outline to this document is that the building and services contained within a dwelling must be designed and installed to actively reduce the amount of CO_2 produced. The building fabric must contain insulation to limit heat loss, and heating appliances, associated controls and equipment and lighting systems must all reduce the energy wasted. Pipes and storage vessels must also be insulated to reduce the waste of energy.

This document should be read in conjunction with the Domestic Building Compliance Guide.

INDUSTRY TIP

Copies of the Building Regulations Approved Documents G3 2010 and L 2013 can be downloaded free from these links:

www.gov.uk/government/publications/sanitation-hot-water-safety-and-water-efficiency-approved-document-g

www.gov.uk/government/publications/conservation-of-fuel-and-power-approved-document-l

The Water Supply (Water Fittings) Regulations 1999

In many respects, the Water Regulations mirror the Building Regulations and these two documents should be consulted before undertaking any design or installation of hot water systems.

Hot water supply is covered in Section 8 of Schedule 2 of the Water Supply (Water Fittings) Regulations. It is reproduced here complete with the guidance notes attached to the regulations (published by Defra – Department for Environment Food & Rural Affairs).

KEY POINT

Remember! It is an offence to contaminate, misuse, waste, unduly consume or erroneously meter water from a water undertaker's water main. The Water Supply Regulations are enforceable in a court of law.

SECTION 8
Schedule 2: Paragraphs 17, 18, 19, 20, 21, 22, 23 and 24: Hot water services

17 (1) Every unvented water heater, not being an instantaneous water heater with a capacity not greater than 15 litres, and every secondary coil contained in a primary system shall:

a Be fitted with a temperature control device and either a temperature relief valve or a combined pressure and temperature relief valve; or

b Be capable of accommodating expansion within the secondary hot water system.

(2) An expansion valve shall be fitted with provision to ensure that water is discharged in a correct manner in the event of a malfunction of the expansion vessel or system.

18 Appropriate vent pipes, temperature control devices and combined temperature pressure and relief valves shall be provided to prevent the temperature of the water within a secondary hot water system from exceeding 100 °C.

▲ Figure 2.51 Pipe insulation

19 Discharges from temperature relief valves, combined temperature pressure and relief valves and expansion valves shall be made in a safe and conspicuous manner.

20 (1) No vent pipe from a primary circuit shall terminate over a storage cistern containing wholesome water for domestic supply or for supplying water to a secondary system.

(2) No vent pipe from a secondary circuit shall terminate over any combined feed and expansion cistern connected to a primary circuit.

21 Every expansion cistern or expansion vessel, and every cold water combined feed and expansion cistern connected to a primary circuit, shall be such as to accommodate any expansion water from that circuit during normal operation.

22 (1) Every expansion valve, temperature relief valve or combined temperature and pressure relief valve connected to any fitting or appliance shall close automatically after a discharge of water.

(2) Every expansion valve shall:

a Be fitted on the supply pipe close to the hot water vessel and without any intervening valves; and

b Only discharge water when subjected to a water pressure of not less than 0.5 bar (50 kPa) above the pressure to which the hot water vessel is, or is likely to be, subjected in normal operation.

Guidance

Unvented hot water systems

G17.1 Every unvented water heater or storage vessel, and every secondary coil contained in a heater and not being an instantaneous water heater or a thermal storage unit of 15 litres or less capacity, should be fitted with:

a A temperature control device; and either a temperature relief valve or combined temperature and pressure relief valve; and

b An expansion valve; and

c Unless the expanded water is returned to the supply pipe in accordance with Regulation 15(2)(a), either;

   i An expansion vessel; or

   ii Contain an integral expansion system, such that the expansion water is contained within the secondary system to prevent waste of water.

G17.2 An expansion valve should be fitted to all unvented hot water storage systems, with a capacity in excess of 15 litres, to ensure that expansion water is discharged in a correct manner in the event of a malfunction of the expansion vessel or system.

G17.3 Where expansion water is accommodated separately the expansion vessel should preferably be of an approved 'flow through type' and should comply with the requirements of **BS 6144** and **BS 6920**.

Temperature of hot water within a storage system

G18.1 Irrespective of the type of fuel used for heating, the temperature of the water at any point within a hot water storage system should not exceed 100 °C and appropriate vent pipes, temperature control devices and other safety devices should be provided to prevent this occurring.

Hot water distribution temperatures

G18.2 Hot water should be stored at a temperature of not less than 60 °C and distributed at a temperature of not less than 55 °C. This water distribution temperature may not be achievable where hot water is provided by instantaneous or combination boilers.

G18.3 The maintenance of acceptable water temperatures may be achieved by efficient routing of pipes, reducing the lengths of pipes serving individual appliances and the application of good insulation practices to minimise freezing of cold water pipes and to promote energy conservation for hot water pipes. For references, see Comments and Recommendations of Clause 4.3.32.2 of **BS 8558**.

Temperature of hot water supplies at terminal fittings and on surfaces of hot water pipes

G18.4 Where practicable the hot water distribution system should be designed and installed to provide the required flow of water at terminal fittings to sanitary and other appliances at a water temperature of not less than 50 °C and within 30 seconds after fully opening the tap. This criteria may not be achievable where hot water is provided by instantaneous or combination boilers.

→

SECTION 8 Schedule 2: Paragraphs 17, 18, 19, 20, 21, 22, 23 and 24: Hot water services	Guidance

23 (1) A temperature relief valve or combined temperature and pressure relief valve shall be provided on every unvented hot water storage vessel with a capacity greater than 15 litres.
 	
(2) The valve shall:
a Be located directly on the vessel in an appropriate location, and have a sufficient discharge capacity, to ensure that the temperature of the stored water does not exceed 100 °C; and
b Only discharge water at below its operating temperature when subjected to a pressure of not less than 0.5 bar (50 kPa) in excess of the greater of the following:
   i The maximum working pressure in the vessel in which it is fitted, or
   ii The operating pressure of the expansion valve.
 	
(3) In this paragraph "unvented hot water storage vessel" means a hot water storage vessel that does not have a vent pipe to the atmosphere.
 	
24 No supply pipe or secondary circuit shall be permanently connected to a closed circuit for filling a heating system unless it incorporates a backflow prevention device in accordance with a specification approved by the regulator for the purposes of this Schedule.
	G18.5 Terminal fittings or communal showers in schools or public buildings, and in other facilities used by the public, should be supplied with water through thermostatic mixing valves so that the temperature of the water discharged at the outlets does not exceed 43 °C.
	
	G18.6 The temperature of water discharged from terminal fittings and the surface temperature of any fittings in health care premises should not exceed the temperatures recommended in HS(G)104 – Safe hot water and surface temperatures.
	
	Energy conservation
	
	G18.7 All water fittings forming part of a primary or secondary hot water circulation system and all pipes carrying hot water to a tap that are longer than the maximum length given in the below table should be thermally insulated in accordance with **BS 5422**.
	
	▼ Maximum recommended lengths of uninsulated hot water pipes
	
	{{TABLE}}

Outside diameter (mm)	Maximum length (m)
12	20
Over 12 and up to 22	12
Over 22 and up to 28	8
Over 28	3

G19.1 Discharge pipes from expansion valves, temperature relief valves and combined temperature and pressure relief valves should be installed in accordance with the guidance given in this document and should also comply with the requirements of Building Regulation G3.

Discharge pipes from safety devices

G19.2 Where discharge pipes pass through environments outside the thermal envelope of the building they should be thermally insulated against the effects of frost.

G19.3 The discharge pipe from a temperature relief valve or combined temperature and pressure relief valve should:

- be through a readily visible air gap discharging over a tundish located in the same room or internal space and vertically as near as possible, and in any case within 600 mm, of the point of outlet of the valve
- be of non-ferrous material, such as copper or stainless steel, capable of withstanding any temperatures arising from a malfunction of the system

- have a vertical drop of 300 mm below the tundish outlet, and thereafter be laid to a self-draining gradient
- be at least one size larger than the nominal outlet size of the valve, unless its total equivalent hydraulic resistance exceeds that of a straight pipe 9 metres long. Where the total length of the pipe exceeds 9 metres equivalent resistance, the pipe shall be increased in size by one nominal diameter for each additional, or part of, equivalent 9 metres resistance length. The flow resistance of bends in the pipe should be taken into consideration when determining the equivalent length of pipe
- terminate in a safe place where there is no risk to persons in the vicinity of the point of discharge. See Building Regulation G3.

> **KEY POINT**
>
> Alternatively, the size of the discharge pipe may be determined in accordance with Annex D of **BS EN 806.3** (which refers back to **BS 6700:1997** Annex C).

Discharge pipes from expansion valves

G19.4 The discharge pipe from an expansion valve may discharge into the tundish used for the discharge from a temperature relief valve or from a combined temperature and pressure relief valve as described in G19.1, or:

- discharge through a readily visible air gap over a tundish located in the same room or internal space and vertically as near as possible and in any case within 600 mm of the point of outlet of the valve; and,
- be of non-ferrous material, such as copper or stainless steel; and,
- discharge from the tundish through a vertical drop outlet and thereafter be laid to a self-draining gradient; and,
- not be less than the nominal outlet size of the expansion valve and discharge external to the building at a safe and visible location.

Vent pipes

G20.1 Vent pipes from primary water systems should be of adequate size but not less than 19 mm internal diameter. They may terminate over their respective cold water feed and expansion cisterns, or elsewhere providing there is a physical air gap, at least equivalent to the size of the vent pipe, above the top of the warning pipe, or overflow if there is one, at the point of termination.

G20.2 Vent pipes from hot water secondary storage systems should be of adequate size but not less than 19 mm internal diameter and be insulated against freezing.

G20.3 Where vent pipes, from either a primary or secondary system, terminate over their respective cold water feed cisterns, they should rise to a height above the top water level in the cistern sufficient to prevent any discharge occurring under normal operating conditions.

Hot water systems supplied with water from storage cisterns

G20.4 In any cistern-fed vented or unvented hot water storage system the storage vessel should:

- be capable of accommodating any expansion water; or
- be connected to a separate expansion cistern or vessel; or

- be so arranged that expansion water can pass back through a feed pipe to the cold water storage cistern from which the apparatus or cylinder is supplied with water.

G20.5 Where the cold water storage cistern supplying water to the hot water storage vessel is also used to supply wholesome water to sanitary or other appliances, any expansion water entering the cistern through the feed pipe should preferably not raise the temperature of the wholesome water in the cistern to more than 20 °C.

Vented systems requiring dedicated storage cisterns or mechanical safety devices

G20.6 Every vented and directly heated hot water storage vessel, single feed indirectly heated hot water storage vessel, or any directly or indirectly heated storage vessel where an electrical immersion heater is installed, should be supplied with water from a dedicated storage cistern unless:

- where the energy source is gas, oil or electricity, a non-self-setting thermal energy cut-out device is provided in addition to the normal temperature-operated automatic-reset cut-out; or,
- where the energy source is solid fuel, a temperature relief valve complying with **BS EN 1490:2000**, or a combined temperature and pressure relief valve complying with **BS EN 1490:2000**, is provided complete with a readily visible air-break to drain device and discharge pipe as described in G19.3.

G20.7 Every double feed indirectly heated hot water storage system which is heated by a sealed (unvented) primary circuit, or the primary circuit heating medium is steam or high temperature hot water, or where an electric immersion heater is installed, should:

- be supplied with water for the secondary circuit from a dedicated cold water storage cistern; or,
- be provided with a non-self-setting thermal energy cut-out device to control the primary circuit, and any electric immersion heaters, in addition to any temperature-operated automatic-reset cut-out.

G20.8 No water in the primary circuit of a double feed indirect hot water storage vessel should connect hydraulically to any part of a hot water secondary storage system.

G20.9 Vent pipes from primary circuits should not terminate over cold water storage cisterns containing wholesome water for supply to sanitary appliances or secondary hot water systems.

G20.10 Vent pipes from secondary hot water systems should not terminate over feed and expansion cisterns supplying water to primary circuits.

G20.11 No water in the primary circuit of a single feed indirect hot water storage vessel, under normal operating conditions, should mix with water in the secondary circuit. Single feed indirect hot water storage vessels should be installed with a permanent vent to the atmosphere.

Primary feed and expansion cisterns

G21.1 Every expansion cistern, and every cold water combined feed and expansion cistern connected to a primary or heating circuit should be capable of accommodating any expansion water from the circuit and installed so that the water level is not less than 25 mm below the overflowing level of the warning pipe when the primary or heating circuit is in use.

Expansion and safety devices

G22.1 Expansion valves, temperature relief valves or combined temperature and pressure relief valves connected to any fitting or appliance should close automatically after an operational discharge of water and be watertight when closed.

G22.2 Expansion valves should comply with **BS EN 1490:2000**. They should be fitted on the supply pipe close to the hot water vessel and without any intervening valves, and only discharge water when subjected to a water pressure of not less than 0.5 bar (50 kPa) above the pressure to which the hot water vessel is, or is likely to be, subjected to in normal operation.

Temperature and combined temperature relief valves

G23.1 Except for unvented hot water storage vessels of a capacity of 15 litres or less, a temperature relief valve complying with **BS EN 1490:2000**, or a combined temperature and pressure relief valve complying with

BS EN 1490:2000, should be provided on every unvented hot water storage vessel. The valve should:

- be located directly on the storage vessel, such that the temperature of the stored water does not exceed 100°C; and,
- only discharge water at below its operating temperature when subjected to a pressure not less than 0.5 bar (50 kPa) greater than the maximum working pressure in the vessel to which it is fitted, or 0.5 bar (50 kPa) greater than the operating pressure of the expansion valve, whichever is the greater.

Non-mechanical safety devices

G23.2 If a non-mechanical safety device such as a fusible plug is fitted to any hot water storage vessel, that vessel requires a temperature relief valve or combined temperature and pressure relief valve designed to operate at a temperature not less than 5°C below that at which the non-mechanical device operates or is designed to operate.

Filling of closed circuits

G24.1 No primary or other closed circuit should be directly and permanently connected to a supply pipe unless it incorporates an approved backflow prevention arrangement.

G24.2 A connection may be made to a supply pipe for filling or replenishing a closed circuit by providing a servicing valve and an appropriate backflow prevention device, the type of which will depend on the degree of risk arising from the category of fluid contained within the closed circuit, providing that the connection between the backflow prevention device and the closed circuit is made by:

- a temporary connecting pipe which must be completely disconnected from the outlet of the backflow prevention device and the connection to the primary circuit after completion of the filling or replenishing procedure; or
- a device which in addition to the backflow prevention device incorporates an air gap or break in the pipeline which cannot be physically closed while the primary circuit is functioning; or
- an approved backflow prevention arrangement.

The Gas Safety (Installation and Use) Regulations 1998

Many hot water supply appliances utilise gas as their main fuel source for both direct and indirect domestic hot water heating. This, obviously, means that the Gas Regulations play an important part in any hot water installation.

The Gas Safety (Installation and Use) Regulations deal with the safe installation, maintenance and use of these appliances and any gas pipework and fittings connected to them in both domestic and industrial/commercial premises. The main requirement of the regulations is that only a competent person, deemed by the Health and Safety Executive to be any person that is a member of an approved body, must carry out work on any gas fitting. In this case, installers of gas appliances, pipework and fittings must by registered with Gas Safe.

BS 7671 **The IET Wiring Regulations**

As with the Gas Regulations, heating hot water often uses electricity either as a direct or indirect fuel source.

All domestic and industrial electrical installations must conform to the IET Wiring Regulations. In England and Wales, the Building Regulations Approved Document P 2010 requires that domestic installations are designed and installed according to British Standard **BS 7671**, Chapter 13. This document was written to standardise electrical installations in line with international document IEC60364-1 and equivalent standards from other countries. Guidance is given in installations manuals such as the IET on-site guide and IET Guidance notes 1 to 7.

Installations in industrial and commercial premises must also satisfy various other legislative documents such as the Electricity at Work Regulations 1989. Again, the recognised standards and practices contained in **BS 7671** will help meet these requirements.

Industry standards

There are a number of industry standards that we can reference to ensure that we conform to the regulations when installing hot water systems.

- British Standard **BS EN 806 Parts 1 to 5** – this standard contains extensive information regarding the design and installation hot water supply systems.
- British Standard **BS 8558:2011** – this provides complementary guidance to **BS EN 806**. It is a guide to the design, installation, testing, operation and maintenance of services supplying water for domestic use.
- The Domestic Building Services Compliance Guide – this guide provides guidance to the Building Regulations Approved Documents L1 and L2 when installing fixed building services within new and existing dwellings to help them comply with the Building Regulations. The guide specifically targets space heating, domestic hot water services, mechanical ventilation, comfort cooling and interior lighting. New technologies, such as heat pumps, solar thermal panels and micro-combined heat and power systems are also discussed. The guide also refers to other publications that refer to techniques to assist in the design and installation of systems that are over and above the standard that is required by the Building Regulations.

Manufacturer technical installation and maintenance instructions

Unvented hot water storage systems must be fitted, commissioned and maintained strictly in accordance with the manufacturer's instructions. These contain vital information for the correct and safe installation, operation and maintenance of the system and its components, such as:

- the minimum required pressure and flow rate of the incoming supply for satisfactory operation of the system
- the minimum size of the incoming cold water supply
- the minimum size of any hot water distribution pipework
- the required heat input and heat recovery time
- any electrical installation requirements
- the operation of any controls
- the calculation required to ascertain the correct size of the discharge pipework
- fault-finding techniques.

Design and calculations for hot water systems

Designing hot water systems follows the same basic principles of designing cold water systems. The main difference being if the system requires storage, then this will have to be calculated using the information contained in **BS EN 806** for the amount of water required for hot water purposes. Calculating the size of the pipework follows the same calculation techniques that were discussed in Chapter 1, Cold water systems, planning and design.

Taking measurements of building features in order to carry out design calculations

The subject of taking measurements from site and from plans and drawings was discussed in detail in Chapter 1, Cold water systems, planning and design.

Calculating the size of hot water system components

Here we will look at the methods of calculating the size of the components used in hot water systems. The components are:

- cistern – only required for open vented hot water storage systems
- hot water storage vessel
- pipework
- secondary circulation pump
- booster pump (shower and full system).

Cistern

The subject of cold water storage has been covered in detail elsewhere in this book. However, elements of this are worth re-iterating here.

The storage capacities for dwellings should be calculated based upon usage and the number of occupants. Storage for supplying cold water to a hot water storage vessel – the amount of cold water storage should equal the storage capacity of the hot water storage vessel.

For the purposes of this chapter, 230 litres should be the figure to remember.

When calculating the amount of cold water storage for hot water supply, the figures in Table 2.6 should be considered.

▼ Table 2.6 Recommended minimum storage of hot water for domestic installations (**BS EN 806** Table 6)

Type of building	Minimum hot water storage (litres)
Hostel	90 per bed space
Hotel	200 per bed space
Office premises with canteen facilities	45 per employee
Office premises without canteen facilities	40 per employee
Restaurant	7 per meal
Nursery/primary day school	15 per pupil
Secondary/technical day school	20 per pupil
Boarding school	90 per pupil
Children's home/residential nursery	135 per bed space
Nurses' accommodation	120 per bed space
Nursing/convalescent home	135 per bed space

The calculation for determining the size of a cistern was covered in Chapter 1, Cold water systems, planning and design.

Hot water storage vessels

The minimum storage requirements of a hot water storage vessel are given as:

- 35 litres to 45 litres per occupant, unless the hot water storage vessel provides a quick reheat of stored water
- 100 litres for systems that use solid fuel as heat source
- 200 litres for systems that use off-peak electricity as the main heat source.

There are two methods for calculating the capacity of a hot water storage vessel. The first method is based, in part, on information contained within the Chartered Institute of Plumbing and Heating Engineers (CIPHE) design guide. The second method takes a slightly different approach and is shown in detail in **BS 8558**. This is the method we will look at here.

The **BS 8558** method of hot water storage capacity calculation considers the following factors:

- the pattern of use
- the rate of heat input to the stored water
- the recovery period for the hot water storage vessel
- the stratification (if any) of the vessel.

Annex B of **BS 8558** gives guidance on the calculation of hot water storage capacity. It states 'The storage capacity required to achieve an acceptable quality of service depends on the rate of heat input to the stored water as well as the pattern of use.'

Table 2.7 shows typical heat inputs.

▼ Table 2.7 Typical heat inputs

Appliance	Heat input (kW)
Electric immersion heater	3
Gas fired circulator	3
Small boiler with direct cylinder	6
Medium boiler and indirect cylinder	10
Large domestic boiler with indirect cylinder	15
Directly fired gas hot water storage heater	10

KEY POINT

A brief overview of stratification

Stratification was covered in Book 1, but here is a reminder.

Stratification is where the hot water 'floats' on the layer of colder water entering the storage vessel. The hot water sits in temperature layers with the hottest water at the top of the storage cylinder, gradually cooling towards the bottom.

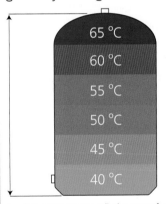

In a hot water storage cylinder, water forms in layers of temperature from the top of the cylinder, where the water is at its hottest, to the base where it is at its coolest

▲ Figure 2.52 Stratification of hot water storage vessels

Stratification is necessary if the cylinder is to perform to its maximum efficiency, and manufacturers will purposely design storage vessels and cylinders with stratification in mind. Designers will generally design:

- a vessel that is cylindrical in shape
- a vessel that is designed to be installed upright rather than horizontal
- with the cold feed entering the cylinder horizontally.

It is generally accepted that stratification occurs more readily in vertical rather than horizontal cylinders.

Calculation of heat input

In Table 2.7, we saw some of the more common heat inputs. However, heat input can be calculated based upon the efficiency of the heat source. If a boiler is the main source of heat for the generation of hot water, then the efficiency of the boiler is required. This is one of the most important factors for improving energy efficiency for the selection of a hot water storage vessel. The formula for calculating heat input is as follows:

$$\frac{SHC \times litres \times temperature\ difference\ (\Delta t) \times boiler\ efficiency}{Time\ in\ seconds \times 100}$$
$$= kW$$

Where:

SHC = specific heat capacity of water. This is taken as being 4.19 kJ/kg/°C

Litres = the storage of water in the hot water storage vessel

Temperature difference = the difference in temperature between the incoming cold supply and the required temperature of the stored hot water

Boiler efficiency = usually taken as 93% for a condensing boiler

Time in seconds = the time limit for the water to get hot in seconds

IMPROVE YOUR MATHS

Calculation of heat input: example

A hot water storage cylinder has a capacity of 210 litres. The cylinder is required to be at 65 °C within two hours. The temperature of the incoming water is 4 °C. What is the kW required?

$$\frac{SHC \times litres \times temperature\ difference\ (\Delta t) \times boiler\ efficiency}{Time\ in\ seconds \times 100} = kW$$

SHC = 4.19

Litres of water = 210 litres

$\Delta t = 61\ °C$

Time in seconds = 7200 seconds (2 hours)

$$\frac{SHC \times litres \times temperature\ difference\ (\Delta t) \times boiler\ efficiency}{Time\ in\ seconds \times 100} = kW$$

$$\frac{4.19 \times 210 \times 61 \times 93}{7200 \times 100} = \frac{4991672.7}{720000} = 6.93\ kW$$

IMPROVE YOUR MATHS

Calculation of heat input

A hot water storage cylinder has a capacity of 140 litres. The cylinder is required to be at 60 °C in two hours. The temperature of the incoming water is 4 °C. What is the kW required?

Calculating capacity based on recovery time

The capacity of the hot water storage vessel depends upon the rate of heat input to the stored water and the pattern of use. The calculation used for this considers the time M (in minutes) taken to heat a specific quantity of water through a specific temperature rise. The formula is as follows:

$$M = VT \div (14.3P)$$

Where:

V = volume of water heated (l)

T = temperature of water (°C)

P = rate of heat input into the water (kW)

The above formula ignores any heat loss from the cylinder as this is likely to be negligible over a short period of time and the formula can be applied to any hot water storage situation whether stratification occurs or not.

So, how does the formula work?

The following examples, taken from **BS 8558** Annex C, assume that a small domestic dwelling has one bath installed, and assumptions on pattern of usage have been made. The maximum requirements are as follows.

IMPROVE YOUR MATHS

Calculating capacity based on recovery time: example 1 – assuming good stratification

If the water in the cylinder is heated via a top entry 3 kW immersion heater, then good stratification is likely to occur. The time needed to heat 60 litres of water from 10 °C to 60 °C for the second bath is, therefore:

$$M = VT \div (14.3P)$$

$$60 \times 50 \div (14.3 \times 3) = 3000 \div 42.9$$

$$= 69.93 \text{ minutes (rounded up to 70 minutes)}$$

However, the second bath is required within 25 minutes of the first, and so this water must come from the storage cylinder. To calculate this, the original formula must be transposed to:

$$V = M(14.3P) \div T$$

One bath using 60 litres of hot water @ 60 °C + 40 litres of cold water + 10 litres of hot water @ 60 °C for kitchen use followed by a second bath fill 25 minutes later.

Totals are 70 litres of hot water @ 60 °C followed 25 minutes later by 100 litres for a second bath fill, which may be achieved by mixing hot water @ 60 °C with cold water @ 10 °C.

The volume of water heated in 25 minutes is, then, as follows:

$$V = M(14.3P) \div T$$

$$V = (25 \times 14.3 \times 3) \div 50$$

$$V = 21.45 \text{ rounded to 21 litres}$$

Total minimum storage requirement is calculated as:

$$70 \text{ litres} + 60 \text{ litres} - 21 \text{ litres} = 109 \text{ litres}$$

The total of 109 litres is shown in the Table 2.8, from **BS 8558**, as the minimum storage capacity of a hot water storage vessel with stratification for a small domestic dwelling.

▼ Table 2.8 Minimum hot water storage capacity

Heat input to the water	Small dwelling with one bath		Large dwelling with two baths	
	With stratification (litres)	With mixing (indirect type) (litres)	With stratification (litres)	With mixing (indirect type) (litres)
3	109	122	165	250
6	88	88	140	200
10	70	70	130	130
15	70	70	120	130

IMPROVE YOUR MATHS

Calculating capacity assuming good stratification

A small dwelling has the following hot water requirements.

One bath using 55 litres of hot water @ 65 °C + 45 litres of cold water + 10 litres of hot water @ 65 °C for kitchen use followed by a second bath fill 30 minutes later.

Totals are 65 litres of hot water @ 65 °C followed 30 minutes later by 100 litres for a second bath fill, which may be achieved by mixing hot water @ 65 °C with cold water @ 10 °C.

Calculate the minimum hot water capacity required assuming good stratification using a 3 kW top entry immersion heater.

IMPROVE YOUR MATHS

Calculating capacity based on recovery time: example 2 – assuming good mixing

Good water mixing occurs when the hot water storage vessel is heated by a primary heat exchanger coil of the type found in a double feed indirect open vented hot water storage cylinder (or the equivalent unvented type storage cylinder).

Because mixing is occurring, as soon as hot water is drawn off, to be replaced by colder water, mixing takes place and the whole of the cylinder becomes cooler. If 70 litres of hot water is used (60 litres for the first bath and 10 litres for kitchen use), then the remaining hot water in the cylinder and the 70 litres of cold water @ 10°C replacing the used hot water will equal the heat energy of the entire contents simply because mixing has taken place.

The heat energy in the cylinder is the product of the volume of the hot water storage vessel and its resultant temperature.

A further transposition of the original formula is necessary.

Therefore, if V is the storage vessel volume and T is its temperature after being refilled with 70 litres of water @ 10°C:

$$(\text{Storage vessel volume} - 70 \text{ litres})$$
$$\times 60\,°C + (70 \text{ litres replacement} \times 10\,°C) = VT$$

Therefore:

$$T = (60V - 4200 + 700) \div V$$
$$T = (60V - 3500) \div V$$
$$T = 60 - 3500 \div V$$

From the original requirement, a second bath is required after 25 minutes. Therefore, with a heat input of 3 kW:

$$25 = VT \div (14.3 \times 3)$$
$$T = (25 \times 14.3 \times 3) \div V$$
$$T = 1072.5 \div V$$

The temperature required for the second bath is 40°C. Therefore, after the first draw-off of 70 litres and its subsequent replenishment with 70 litres of water @ 10°C, the temperature must be at least 40°C (or above) after 25 minutes. To achieve this, the minimum storage capacity must be:

$$(60 - 3500 \div V) + (1072.5 \div V)$$
$$60 - 2427.5 \div V = 40\,°C$$

When the calculation is transposed:

$$2427.5 \div (60 - 40) = 121.375 \text{ rounded up to 122 litres}$$

IMPROVE YOUR MATHS

Calculating capacity assuming good mixing

If 65 litres of hot water at 65°C is used (55 litres for the first bath and 10 litres for kitchen use), then the remaining hot water in the cylinder and the 65 litres of cold water @ 10°C replacing the used hot water will equal the heat energy of the entire contents simply because mixing has taken place.

From the original requirement, a second bath is required after 25 minutes with a heat input of 3 kW.

The temperature required for the second bath is 40°C. Therefore, after the first draw-off of 65 litres and its subsequent replenishment with 65 litres of water @ 10°C, the temperature must be at least 40°C (or above) after 25 minutes.

Using the above example as a guide, calculate the capacity of the hot water storage cylinder.

Hot water pipework sizing

The calculation for determining the size of the pipework for hot water systems is identical to that used for cold water systems. This was discussed in Chapter 1, Cold water systems, planning and design.

Sizing a secondary circulation circulating pump

Sizing a hot water secondary circulation circulating pump involves the calculation of the **mass flow rate** of the circuit based upon the heat loss from the circuit and the temperature difference between the flow and the return. Look at Figure 2.53.

KEY TERM

Mass flow rate: the mass of a substance (in kilograms) which passes in a unit of time. In this case, it is kilograms per second.

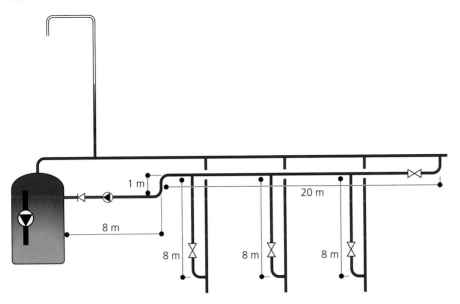

▲ Figure 2.53 Secondary circulation sizing

IMPROVE YOUR MATHS

From the drawing in Figure 2.53, we can see that the length of the return pipe is 53 m. From this, we can calculate the size of the return pipe and the size of the pump. The data we need for this is as follows:

Length of pipe = 53 m

Temperature of the secondary flow (t_f) = 65 °C

Temperature of the secondary return (t_r) = 55 °C

Specific heat capacity of water = 4.19 kJ/kg

Heat loss from the insulated pipe = 8 W/m²/C

By calculating the mass flow rate of the return pipe in kg/s, the size of the return pipe and the circulating pump size (in pascals pressure drop) can be calculated by using the reference data for copper tubes from CIBSE guide C. First, however, the mass flow rate must be calculated by using the following formula:

$$Ma = \frac{kW}{SHC \times \Delta t\left(t_f - t_r\right)} = kg/s$$

To find the kW:

$$kW = length \times W/m^2/C$$

$$kW = 53 \times 8 = 425\,W = 0.425\,kW$$

Therefore:

$$Ma = \frac{kW}{SHC \times \Delta t\left(t_f - t_r\right)} = kg/s$$

$$Ma = \frac{0.425}{4.19 \times (65-55)} = kg/s$$

$$Ma = \frac{0.425}{4.19 \times 10} = kg/s$$

$$Ma = \frac{0.425}{41.90} = kg/s$$

$$Ma = 0.010\,kg/s$$

Looking at CIBSE guide C for copper tube (Figure 2.54), it can be seen that the nearest mass flow rate to the calculated figure is boxed in red.

Pipe sizing tables   1

q_m	= mass flow rate — kg/s
c	= velocity — m/s
$\Delta p/l$	= pressure loss per unit length — pa/m

Copper tubes BS2871 Table X
Water at 75 °C

Δp/l	c	10 mm qm	10 mm c	12 mm qm	12 mm c	15 mm qm	15 mm c	22 mm qm	22 mm c	28 mm qm	28 mm c	35 mm qm	35 mm c	42 mm qm	42 mm c	c	Δp/l
0.1								0.001	0.00	0.00	0.00	0.007	0.01	0.015	0.01		0.1
0.2								0.002	0.01	0.005	0.01	0.014	0.02	0.023	0.02		0.2
0.3								0.003	0.01	0.008	0.02	0.019	0.02	0.026	0.02		0.3
0.4								0.004	0.01	0.011	0.02	0.019	0.02	0.032	0.03		0.4
0.5						0.001	0.01	0.005	0.02	0.014	0.03	0.021	0.03	0.036	0.03		0.5
0.6						0.001	0.01	0.006	0.02	0.015	0.03	0.023	0.03	0.040	0.03		0.6
0.7						0.001	0.01	0.007	0.02	0.015	0.03	0.026	0.03	0.044	0.04		0.7
0.8						0.001	0.01	0.008	0.03	0.015	0.03	0.028	0.03	0.048	0.04		0.8
0.9						0.001	0.01	0.009	0.03	0.016	0.03	0.030	0.04	0.051	0.04		0.9
1.0						0.002	0.01	0.010	0.03	0.017	0.03	0.032	0.04	0.055	0.05	0.05	1.0
1.5				0.001	0.01	0.003	0.02	0.012	0.04	0.022	0.04	0.040	0.05	0.070	0.06		1.5
2.0				0.001	0.01	0.004	0.03	0.012	0.04	0.026	0.05	0.048	0.06	0.083	0.07		2.0
2.5				0.002	0.02	0.005	0.04	0.014	0.04	0.030	0.06	0.055	0.07	0.094	0.08		2.5
3.0		0.001	0.02	0.002	0.02	0.006	0.04	0.016	0.05	0.033	0.06	0.061	0.07	0.105	0.09		3.0
3.5		0.001	0.02	0.003	0.03	0.007	0.05	0.017	0.05	0.036	0.07	0.067	0.08	0.114	0.09		3.5
4.0		0.001	0.02	0.003	0.03	0.008	0.06	0.019	0.06	0.039	0.07	0.072	0.09	0.124	0.10		4.0
4.5		0.001	0.02	0.003	0.03	0.008	0.06	0.020	0.06	0.042	0.08	0.078	0.10	0.132	0.11		4.5
5.0		0.001	0.02	0.004	0.04	0.008	0.06	0.022	0.07	0.045	0.09	0.083	0.10	0.141	0.12		5.0
5.5		0.002	0.03	0.004	0.04	0.008	0.06	0.023	0.07	0.048	0.09	0.087	0.11	0.149	0.12		5.5
6.0		0.002	0.03	0.005	0.06	0.008	0.06	0.024	0.08	0.050	0.09	0.092	0.11	0.156	0.13		6.0
6.5		0.002	0.03	0.005	0.06	0.008	0.06	0.025	0.08	0.053	0.10	0.096	0.12	0.164	0.14		6.5
7.0		0.002	0.03	0.006	0.07	0.008	0.06	0.027	0.09	0.055	0.10	0.100	0.12	0.171	0.14		7.0
7.5		0.002	0.03	0.006	0.07	0.009	0.06	0.028	0.09	0.057	0.11	0.105	0.13	0.178	0.15	0.15	7.5
8.0		0.003	0.05	0.006	0.07	0.009	0.06	0.029	0.09	0.059	0.11	0.108	0.13	0.185	0.15		8.0
8.5		0.003	0.05	0.006	0.07	0.010	0.07	0.030	0.10	0.061	0.12	0.112	0.14	0.191	0.16		8.5
9.0		0.003	0.05	0.006	0.07	0.010	0.07	0.031	0.10	0.064	0.12	0.116	0.14	0.198	0.16		9.0
9.5		0.003	0.05	0.006	0.07	0.010	0.07	0.032	0.10	0.066	0.13	0.120	0.15	0.204	0.17		9.5
10.0	0.05	0.003	0.05	0.006	0.07	0.011	0.08	0.033	0.11	0.068	0.13	0.123	0.15	0.210	0.17		10.0
12.5		0.004	0.07	0.006	0.07	0.012	0.08	0.037	0.12	0.077	0.15	0.140	0.17	0.239	0.20		12.5
15.0		0.005	0.08	0.007	0.08	0.014	0.10	0.042	0.13	0.086	0.16	0.156	0.19	0.265	0.22		15.0
17.5		0.005	0.08	0.008	0.09	0.015	0.11	0.046	0.15	0.094	0.18	0.170	0.21	0.289	0.24		17.5
20.0		0.005	0.08	0.008	0.09	0.016	0.11	0.049	0.16	0.101	0.19	0.184	0.23	0.312	0.26		20.0
22.5		0.005	0.08	0.009	0.10	0.017	0.12	0.053	0.17	0.108	0.21	0.197	0.24	0.334	0.28		22.5
25.0		0.005	0.08	0.010	0.11	0.019	0.13	0.056	0.18	0.115	0.22	0.209	0.26	0.354	0.29	0.30	25.0
27.5		0.005	0.08	0.010	0.11	0.020	0.14	0.060	0.19	0.122	0.23	0.221	0.27	0.374	0.31		27.5

▲ Figure 2.54 CIBSE guide C for copper tube (The Chartered Institution of Building Services Engineers (CIBSE), www.cibse.org)

This shows that a 15 mm copper pipe will deliver 0.010 kilograms per second (kg/s) at a velocity of 0.07 litres per second (l/s) with a pressure loss of 9.5 pascals per metre (Pa/m). In this case, the mass flow rate matches perfectly. In most instances, the nearest figure up would be chosen. However, there may be instances where there is more than one alternative. Look at the table again, you will see that 12 mm pipe will also deliver 0.010 kg/s but the pascals per metre is 27.5 with a velocity of 0.11 litres per second. This means that the pump power will have to be over twice the size than for 15 mm to deliver the same flow rate, so in this case 15 mm would be chosen.

IMPROVE YOUR MATHS

Calculating pump size

Now the pump size can be calculated. To do this, the frictional resistance of the return pipe must be found considering the fittings and valves.

Each change of direction and valve resist flow. This is measured in length of pipe. In other words, the resistance of an elbow or bend will have the same resistance as a specific amount of pipe. Consider Table 2.9.

▼ Table 2.9 Typical equivalent lengths for copper tube

Bore of pipe	Equivalent length			
	Elbow (m)	Tee (m)	Stop valve (m)	Check valve (m)
15	0.5	0.6	4.0	2.5
22	0.8	1.0	7.0	4.3
28	1.0	1.5	10.0	5.6
35	1.4	2.0	13.0	6.0
42	1.7	2.5	16.0	7.9
54	2.3	3.5	22.0	11.5

The return pipe has been calculated to 15 mm pipe. Looking at the system drawing in Figure 2.53, you will see that there are:

$$6 \text{ elbows @ } 0.5 \text{ m} = 3 \text{ m}$$
$$3 \text{ tees @ } 0.6 \text{ m} = 1.8 \text{ m}$$
$$1 \text{ check valve @ } 2.5 \text{ m} = 2.5 \text{ m}$$
$$= 7.3 \text{ m}$$

So, the resistance in the fittings totals another 7.3 m of pipe. When this is added to the actual length of pipe, it totals:

$$53 + 7.3 = 60.3 \text{ m}$$

Looking at the CIBSE table again (Figure 2.54), the pascals pressure drop was 9.5. This is now multiplied by the length of pipe:

$$9.5 \times 60.3 = 572.85 \text{ Pa}$$

So, the circulating pump must have at least 572.85 pascals of pressure to circulate the return water through 15 mm copper pipe at 0.07 l/s to guarantee the temperature of the return water is 55 °C when it reaches the hot water storage cylinder.

IMPROVE YOUR MATHS

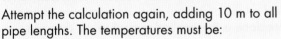

Attempt the calculation again, adding 10 m to all pipe lengths. The temperatures must be:

$$t_f = 60 \text{ °C}$$
$$t_r = 50 \text{ °C}$$

Assume the following fittings:

- 10 elbows
- 6 tees
- 1 check valve.

Sizing a hot water shower boosting pump

A shower boosting pump is designed to boost water at low pressure and flow rate to a higher pressure and flow rate to give a better showering experience. Calculation of the size of a shower boosting pump is based upon the pressure and flow rate that the pump is designed to deliver at the showerhead and the height of the column of water that the pump has to move.

Pump duty is calculated in kilopascals, the pascal being the unit of pressure:

$$1 \text{ bar pressure} = 10 \text{ m head} = 100 \text{ kPa}$$

Flow rate is calculated in litres per second. Shower pumps can deliver between 11 and 25 litres per minute (l/m) or 0.41 litres per second (l/s), depending on the type and their application.

To correctly size a shower pump, you must find the pump duty. The pump duty is the ability of the pump to overcome frictional resistances and the additional static head to displace water from one point (such as the cistern in an open vented hot water storage system) to another, often higher, point (such as the shower head). When calculating the pump duty, both the desired flow rate and head (pressure) must be taken into account. Once this has been calculated then the correct pump can be chosen from the duty point on the manufacturer's design charts.

The duty point is defined as that point on the H–Q system curve where the actual pump performance is in line with the calculated design criteria. Figure 2.55 illustrates a typical duty point where the design criteria and the pump performance coincide.

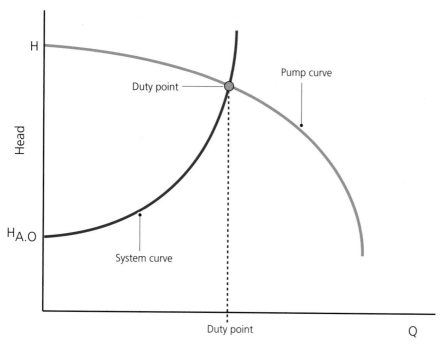

▲ Figure 2.55 Pump duty point

The actual duty point is always at the point where the pump H–Q and the system H–Q points intersect.

IMPROVE YOUR MATHS

Calculating pump duty

A shower boosting pump is required to deliver 15 litres per minute at a pressure of 2 bar to a shower head 3 m higher than the pump itself. Frictional losses in the system due to pipework, valves and fittings can be calculated as static head × 0.05.

Static head from the pump to the shower head = 3 m or 0.3 bar or 30 kPa

Frictional losses = 30 × 0.05 = 1.5 kPa

Design head = 2 bar or 200 kPa

Therefore, the pump duty is:

30 + 200 + 1.5 = 231.5 kPa

Now, a pump can be chosen from the manufacturer's literature with a duty of 231.5 kPa that will deliver 15 l/m or 0.25 l/s flow rate.

IMPROVE YOUR MATHS

Calculation of pump power

Calculation of pump power is usually performed to ascertain the power of the pump needed to lift a certain quantity of water at a certain pressure. It is based on the physics of work done relative to time. Work done is the applied force through distance moved and the unit of measurement is the joule. It is thus explained as the work done when a 1 newton force acts through 1 metre distance or:

1 joule = $1 \, \text{N} \times 1 \, \text{m}$

The time must be expressed as a period of seconds, which can be combined with work done to become work done over a period of time. It is expressed in the following way:

Power = work done ÷ time

= (force × distance) ÷ seconds

= (newtons × metres) ÷ seconds (J/s)

where: 1 J/s = 1 watt

Force (newtons) = kg mass × acceleration due to gravity $(9.81 \, \text{m/s}^2)$

Power (watts) = mass × 9.81 × distance ÷ time

IMPROVE YOUR MATHS

Calculate the pump duty

A shower boosting pump is required to deliver 11 litres per minute at a pressure of 1.5 bar to a shower head 2.5 m higher than the pump itself. Frictional losses in the system due to pipework, valves and fittings can be calculated as static head × 0.05.

Calculate the pump duty in kPa. Consult the manufacturer's literature to source a suitable shower pump (either inlet or outlet type).

IMPROVE YOUR MATHS

Consider the example below.

A delivery rate of 4 kg/s is required to fill the cistern. 1 kg = 1 litre, therefore 4 kg/s = 4 l/s. The total length of pipework with all bends, valves and fittings is 45 m.

Power = (mass × 9.81 × distance) ÷ time

= (4 × 9.81 × 45) ÷ 1

= 1765.8 watts

Allowance for pump efficiency:

= 1765.8 × (100 ÷ 75)

= 2354.4 watts

Therefore:

Pump rating (including 75% efficiency allowance) = 2500 watts or 2.5 kW (rating rounded up to the nearest ½ kW).

Sizing expansion vessels for sealed heating systems and feed and expansion cisterns for open vented systems

Water, when it is heated, expands. The amount of expansion will depend on the temperature of the water. At atmospheric pressure, water is at its greatest density at 4°C. At this temperature, 1 m³ of water has a mass of 1000 kg. From this point forwards as the water temperature increases, 1 m³ will lose density. At 100°C, 1 m³ of water has a mass of 958 kg or an expansion rate of 4 per cent. The densities of water at various temperatures and pressures are shown in Table 2.10.

▼ Table 2.10 Density of water by temperature

Temperature (°C)	Density (kg/m³)	Specific volume 10⁻³ (m³/kg)
0 (ice)	916.8.8	
0.01	999.8	1.00
4 (maximum density)	1000.0	
5	999.9	1.00
10	999.8	1.00
15	999.2	1.00
20	998.3	1.00
25	997.1	1.00
30	995.7	1.00
35	994.1	1.01
40	992.3	1.01
45	990.2	1.01
50	988	1.01
55	986	1.01
60	983	1.02
65	980	1.02
70	978	1.02
75	975	1.03
80	972	1.03
85	968	1.03
90	965	1.04
95	962	1.04
100	958	1.04

The expansion of the water in a central heating system, if not accommodated, will lead to an increase in system pressure and possibly component or appliance failure as a result. In a sealed heating system, the expansion of water is accommodated in an expansion vessel. In an open vented system, the feed and expansion cistern accommodates the expansion of water. Both of these vital parts of the installation will need to be sized correctly.

Sizing an expansion vessel

There are several methods for sizing expansion vessels. All methods must consider the volume of cold water in the system and the amount by which it will expand in order to reach its design temperature. The CIBSE method is shown below.

If the system volume is known, expansion vessels can be sized with the following formula:

$$V = \frac{eC}{1 - \frac{P_1}{P_2}}$$

Where:

V = the total volume of the expansion vessel

C = the total volume of water in the system in litres

P_1 = the fill pressure in bars absolute (gauge pressure + 1 bar)

P_2 = the setting of the pressure relief valve + 1 bar

e = the expansion factor that relates to the maximum system requirements

▼ Table 2.11 Expansion factors

Expansion factor 'e'	Temperature (°C)
0.0324	85
0.0359	90
0.0396	95
0.0434	100

'e' can be found from the formula:

$$e = \frac{d_1 - d_2}{d_2}$$

Where:

d_1 = density of water at filling temperature kg/m³

d_2 = density of water at operating temperature kg/m³

IMPROVE YOUR MATHS

Sizing an expansion vessel: example 1

An unvented hot water system has a total water volume of 300 litres. The pressure of the water main is 1.5 bar and the pressure relief maximum pressure is 6 bar. The system is designed to operate at a maximum temperature of 60°C, which means the expansion factor will have to be calculated. The fill temperature of the water is 10°C.

Calculate the expansion factor using:

$$e = \frac{d_1 - d_2}{d_2}$$

Calculate the expansion vessel volume using:

$$V = \frac{eC}{1 - \frac{P_1}{P_2}}$$

Step 1: Calculate the expansion factor 'e'

The temperature of the fill water is 10°C with a density of 999.8 kg/m³. The maximum operating temperature is 60°C with a density of 983 kg/m³. Therefore the 'e' factor is:

$$e = \frac{999.8 - 983}{983} = 0.0170$$

Step 2: Calculate the expansion vessel volume

V = the total volume of the expansion vessel

C = 300 litres

$P_1 = 1.5 + 1$

$P_2 = 6 + 1$

e = 0.0170

$$V = \frac{eC}{1 - \frac{P_1}{P_2}}$$

Therefore, the expansion vessel volume is:

$$V = \frac{0.0170 \times 300}{1 - \frac{2.5}{7}}$$

$$= \frac{5.1}{1 - 0.357}$$

$$= \frac{5.1}{0.643}$$

$$= 7.931 \text{ litres or } 2.64\% \text{ expansion}$$

So, the expansion vessel volume is: 7.931 litres or 2.64% of total system volume.

IMPROVE YOUR MATHS

A sealed central heating system has a total water volume of 400 litres. The pressure of the water main is 1 bar and the pressure relief maximum pressure is 4 bar. The system is designed to operate at a maximum temperature of 85 °C, which means the expansion factor will have to be calculated. The fill temperature of the water is 4 °C.

Calculate the expansion factor using:

$$e = \frac{d_1 - d_2}{d_2}$$

Calculate the expansion vessel volume using:

$$V = \frac{eC}{1 - \dfrac{P_1}{P_2}}$$

Design calculations in an acceptable format

The methods that can be used to present the design of a hot water system and the results of calculations performed were covered in Chapter 1. However, it is worth revisiting the basic principle of design and calculation presentation.

Scale and not-to-scale drawings

Scale drawings and schematic drawings produced using computer aided design (CAD) packages help to show the customer what you are proposing to install to fulfil their requirement. This is especially important when a large installation is to be completed as it helps the customer to keep a track of what is being installed and where. Many companies now also provide 3D drawings and artistic impressions of what the installation will look like when completed.

Presentation of designs and calculations

The calculation process for hot water supply takes time to complete and unless the calculations are set out correctly, mistakes are often made. The use of spreadsheets and tables when completing design calculations, especially for pipe sizing, is commonplace amongst most professional building services engineers. These are excellent for including with any quotation or design specification that the company wishes to present to the customer and can be saved as a hard copy in portable document format (PDF) or as either Word or Excel documents.

⑤ THE INSTALLATION REQUIREMENTS OF HOT WATER SYSTEMS AND COMPONENTS

This is an important section. Here, you will learn how to use 'best practice' when installing unvented hot water storage systems (UHWSS) that comply with both the Water Supply (Water Fittings) Regulations and Approved Document G of the Building Regulations.

The installation of unvented hot water storage cylinders

The installation of unvented hot water storage systems is subject to strict requirements of Building Regulations Approved Documents G3 and L and the Water Supply (Water Fittings) Regulations. Typical pipework layouts are shown in Figures 2.56 and 2.57.

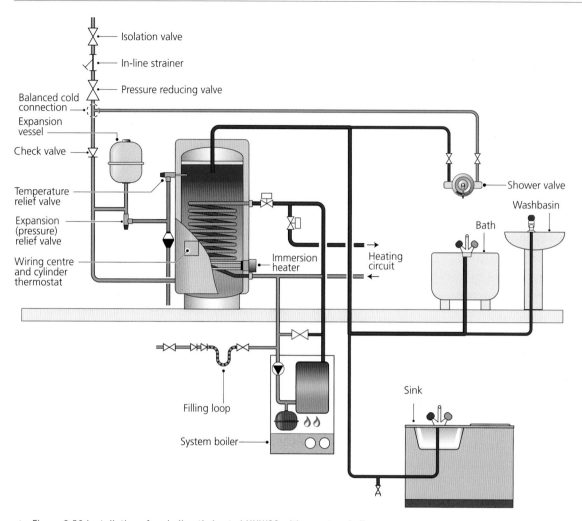

Isolation valve

In-line strainer

Pressure reducing valve

Balanced cold connection

Expansion vessel

Check valve

Temperature relief valve

Expansion (pressure) relief valve

Wiring centre and cylinder thermostat

Immersion heater

Heating circuit

Shower valve

Washbasin

Bath

Filling loop

System boiler

Sink

▲ Figure 2.56 Installation of an indirectly heated UHWSS with a system boiler

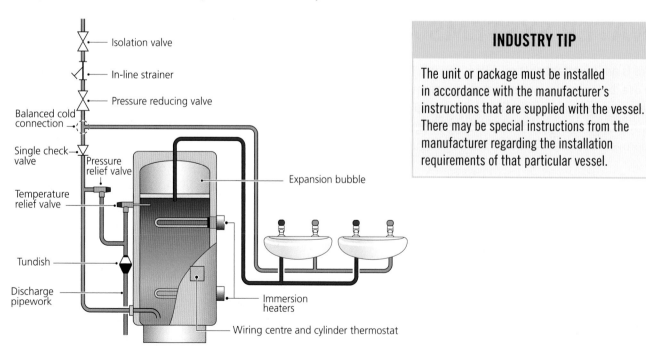

Isolation valve

In-line strainer

Pressure reducing valve

Balanced cold connection

Single check valve

Pressure relief valve

Temperature relief valve

Expansion bubble

Tundish

Discharge pipework

Immersion heaters

Wiring centre and cylinder thermostat

INDUSTRY TIP

The unit or package must be installed in accordance with the manufacturer's instructions that are supplied with the vessel. There may be special instructions from the manufacturer regarding the installation requirements of that particular vessel.

▲ Figure 2.57 Installation of a directly fired UHWSS with immersion heaters as the primary heat source

The floor on which the vessel is to be sited must be substantial enough to accommodate the weight of the vessel and its water contents.

The pipework must be fitted in accordance with **BS EN 806** and **BS 8558**. Unvented hot water storage systems (UHWSS) require at least a 22 mm cold water feed supplied by a water undertaker because of the high flow rate and pressure that the vessels operate at. Water can be supplied through a boosting pump and cold water accumulator if necessary (this will be discussed later in the chapter). A 22 mm hot water draw-off is required in all installations but this may be reduced for particular appliances such as washbasins, sinks and bidets. Isolation valves should be fitted at all appliances in line with good practice.

> **KEY POINT**
>
> It doesn't actually state that isolation valves are needed at every appliance in the Water Regulations but it is considered good practice to install them. The Water Regulations say that every float operated valve must have a service valve fitted as near to the float operated valve as possible. Other appliances aren't mentioned.

The order that the functional and safety components are installed is of paramount importance if the system is to operate safely and efficiently and this can be studied from Figure 2.56 for indirectly heated vessels and Figure 2.57 for directly heated vessels.

Unvented hot water storage systems require the installation of discharge pipework to safely convey any water that may be discharged as the result of a defect or malfunction. Discharge pipework will be discussed later in this section.

The use of cold water accumulators in unvented hot water systems

The use of cold water accumulators is becoming increasingly popular, especially in areas where the water pressure is exceptionally low. Accumulators and boosting pumps, as we saw in Chapter 1, offer a positive solution to the problem of low water pressure and poor low flow rate by storing water at night for use during the day. Both flow rate and pressure are

critical factors when fitting unvented hot water storage systems as these rely on a good flow rate and pressure to provide a satisfactory operation. It should be borne in mind, however, that boosting pumps that deliver more than 12 litres per minute are not allowed under the Water Supply (Water Fittings) Regulations when the cold water supply is being taken directly from a water undertaker's mains supply.

The issue of poor mains supply

Water supply pressures have consistently diminished over the last 30 years. As more and more homes, factories, offices and shops are built, the loading on the UK water system has increased with little or no upgrading of the water mains supply network. The pipework that serves our towns and cities is now supplying more properties than ever before and that has resulted in a gradual degradation of both pressure and flow rate. In some areas of the UK, the supply pressure can be as little as 1 bar, which is unsatisfactory for an unvented hot water storage system.

Pressure of water takes two forms: **static pressure** and **dynamic pressure**.

> **KEY TERMS**
>
> **Static pressure:** this is the water pressure when no flow is occurring. This is always greater than the dynamic pressure.
>
> **Dynamic pressure:** also known as 'running pressure', this is the water pressure when outlets are open and water is flowing.

During periods of peak use, both static and dynamic pressures will decrease. If, during this time, a property has a static pressure of, say 2 bar, then the dynamic pressure could drop to below 1 bar. At off-peak times, say during the night, this could rise significantly to 3 bar static and 2 bar dynamic simply because less water is being used in the surrounding area. An accumulator would take advantage of the night-time rise in pressures to replenish its storage capacity while the mains pressure is at its highest. With the accumulator fully replenished, a good pressure and flow rate would be available throughout the day provided that the accumulator has been sized correctly.

Figure 2.58 shows a typical unvented hot water storage system with an accumulator installed to increase both the pressure and the flow rate. An important factor here is the use of two pressure reducing valves. The first PRV regulates the pressure entering the property so that any pressure fluctuations can be controlled to a predetermined pressure at night when the accumulator is replenishing. The second PRV reduces the pressure to that of the UHWSS manufacturer's recommendations.

Accumulators require a minimum incoming supply pressure to replenish successfully, usually around 2 bar. If the incoming supply cannot deliver this even at off peak periods, then a booster pump should also be installed as shown in Figure 2.59.

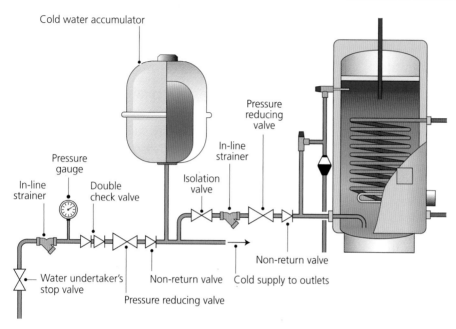

▲ Figure 2.58 An accumulator installed on an unvented system

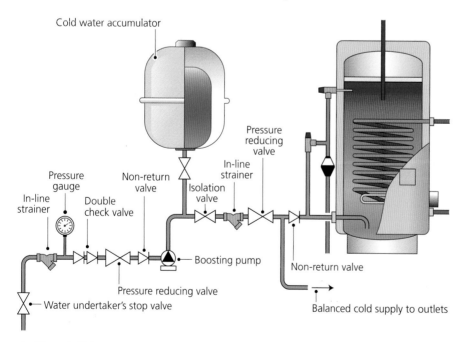

▲ Figure 2.59 An accumulator with a boosting pump installed on an unvented system

The installation of hot water pipework – general

The materials used for hot water installations are copper tubes to **BS EN 1057** and polybutylene pipes and fittings, as these are the only materials that do not cause contamination of the water and can withstand the temperatures associated with hot water distribution pipework. The pipework should be capable of withstanding at least 1½ times the normal operating pressure of the system and sustained temperatures of 95 °C with occasional temperature increases up to 100 °C to allow for any malfunctions of any hot water heating appliances that may occur. All systems must be capable of accommodating thermal expansion and movement within the pipework. Care should be taken when pressure testing open vented cylinders to ensure that the maximum pressure that the cylinder can withstand is not exceeded. If necessary, the cylinder should be disconnected and the pipework capped before testing commences.

The installation methods for hot water systems are very similar to those for cold water installations. Care should be taken when installing hot and cold water pipework side-by-side so that any cold water installation is not unduly warmed by the hot water pipework.

Temperature control of hot water systems

Hot water systems must not be allowed to exceed 100 °C at any time. A maximum normal operating temperature of 60 °C is required to kill off *Legionella* bacteria. There are several methods by which we can maintain and control the temperature of hot water systems and prevent it from exceeding the maximum temperature specified.

A thermostat should be installed and set to the temperature required. A second, high-limit thermostat operates should the maximum temperature be exceeded. This is known as a second-tier level of temperature control:

- Immersion heaters that have a re-settable double thermostat – one thermostat can be set between 50 and 70 °C, the other is a re-settable high limit thermostat designed to switch off the power to the unit when the maximum temperature is exceeded. It must be manually reset.
- Immersion heaters with a non-resettable double thermostat – one thermostat can be set between 50 and 70 °C, the other is a high-limit thermostat designed to permanently switch off the power to the unit until the immersion heater is replaced and the fault rectified.
- Open vented double feed indirect cylinders with gravity or pumped primary circulation must be fitted with a minimum of a cylinder thermostat and a motorised zone valve which closes when the water in the cylinder reaches a pre-set level.
- Open vented cylinders with no high-limit thermostat can be fitted with a temperature relief valve which opens automatically at a specified temperature to discharge water via a tundish and discharge pipework safely to outside the property.

The insulation of hot water pipework

When installing new hot water installations in domestic properties, pipes should be wrapped with thermal insulation that complies with the Domestic Heating Compliance Guide. There are four main considerations:

- Primary circulation pipes for heating and hot water circuits should be insulated wherever they pass outside the heated living space such as below ventilated suspended timber floors and unheated roof spaces. This is for protection against freezing.
- Primary circulation pipes for domestic hot water circuits should be insulated throughout their entire length except where they pass through floorboards, joists and other structural obstructions.
- All pipes connected to hot water vessels, including the vent pipe, should be insulated for at least one metre (1 m) from their points of connection to the cylinder or at least up to the point where they become concealed.
- If secondary circulation, such as a pumped circuit feeding bath and basin taps in a large property, is installed, all pipes fed with hot water should be insulated to prevent excessive heat loss through the secondary circulation circuit.

Expansion of hot water pipework

When the pipework of the hot water system is filled with hot water, the heated pipework will expand. As the pipework cools down, it will contract. This expansion and contraction must be accommodated for during the installation process or noise within the installation will result. Pipes that pass through walls and floors, and not enough room has been left for expansion, will 'tick' and 'creak' as the expansion and contraction take place.

The rate of expansion will depend upon the material the pipe is made from. Generally, pipework made from plastic materials tends to expand more than that made from copper. The coefficients of linear expansion for polybutylene and copper are as follows:

- The coefficient of linear expansion of plastic pipe is 0.00018 m per metre per °C.
- The coefficient of linear expansion of copper pipe is 0.000016 m per metre per °C.

This means that for every degree rise in temperature, Polybutylene pipe will expand 0.00018 m in every metre and copper will expand 0.000016 m in every metre.

Protection against backflow and back siphonage

Hot water is categorised as fluid category 2 because heat has been added to the cold, wholesome water. Other considerations here are that many of the bathroom appliances that are connected to the hot and cold supply are also at risk from fluid categories 3 and 5. Appliances that may be at risk from backflow are listed in Table 2.12.

▼ Table 2.12 Backflow risks

Wash basins **Fluid cat. 2 and 3 risk**	Taps for use with wash basins should discharge at least 20 mm above the spillover level of the appliance (AUK2 air gap). Mixer taps should be protected by the use of single check valves on the hot and cold supplies. Twin-flow mixer taps do not require any backflow protection as the water mixes on exit of the tap.
Kitchen sinks **Fluid cat. 5 risk**	No backflow protection is required as the height of the outlet is well above the spillover level of the appliance. This is classified as an AUK3 air gap. If a mixer tap, where both hot and cold water mix in the tap body is installed then single check valves must fitted on both hot and cold supplies. Twin-flow mixer taps do not require any backflow protection as the water mixes on exit of the tap.
Baths **Fluid cat. 2, 3 and 5 risk**	As for wash basins except that the air gap should be 25 mm. Bath/shower mixer taps, where the water is fed from the mains cold water supply and there is a risk of the shower head being below the water level in the bath, should be protected by double check valves, or a shower hose retaining ring which maintains an AUK2 air gap above the spillover level of the bath.
Bidets **Fluid cat. 2, 3 and 5 risk**	There are two types of bidet that are at risk from backflow. These are: • The ascending spray type – special consideration must be made when fitting this type of bidet (see Figure 1.100). These cannot be used with mains-fed hot and cold water systems. Fluid cat 5 risk. • The over rim with shower hose connection – with this installation, there is fluid cat. 5 risk as well as a fluid cat. 2 risk.
Shower valves **Fluid cat. 2 and 3 risk**	When both hot and cold supplies are fed from a cistern, no backflow protection is required. However, when both are fed from mains-fed supplies then single check valves are required with a hose retaining ring to prevent the hose entering the water. If no retaining ring is fitted then both hot and cold supplies should have a double check valve installed.
Electric shower units **Fluid cat. 2 and 3 risk**	A double check valve is required where a hose retaining ring is not fitted.

Backflow and methods of prevention were discussed in detail in Chapter 1, Cold water systems, planning and design.

⑥ TESTING AND COMMISSIONING REQUIREMENTS OF HOT WATER SYSTEMS AND COMPONENTS

Testing and commissioning of hot water systems is probably the most important part of any installation, as it is here that the system design is finally put into operation. For an installation to be successful, it has to comply with both the manufacturer's installation instructions and the regulations in force. It also has to satisfy the design criteria and flow rates that have been calculated and the customer's specific requirements.

Testing and commissioning performs a vital role and its importance cannot be overstated. Correct commissioning procedures and system set-up often make the difference between a system working to the specification and failing to meet the required demands.

In this part of the chapter, we will look at the correct methods of testing and system commissioning.

Information sources required to complete commissioning work on hot water systems

Inadequate commissioning, system set-up, system flushing and maintenance operations can affect the performance of any hot water system, irrespective of the materials that have been used in the system installation. Building debris and swarf (pipe filings) can easily block pipes and these can also promote bacteriological growth. In addition, excess flux used during the installation can cause corrosion and may lead to the amount of copper that the water contains exceeding the permitted amount for drinking water. This could have serious health implications and, in severe cases, may cause corrosion of the pipework, fittings and any storage vessel installed.

It is obvious, then, that correct commissioning procedures must be adopted if the problems stated are to be avoided. There are four documents that must be consulted:
- Water Supply (Water Fittings) Regulations 1999
- British Standard **BS EN 806** (in conjunction with **BS 8558**)
- Building Regulations Approved Document G3
- manufacturer's instructions of any equipment and appliances.

The documents required for correct testing and commissioning were investigated in Chapter 1.

The checks to be carried out during a visual inspection of an unvented hot water storage system to confirm that it is ready to be filled with water

Before soundness testing a hot water system, visual inspections of the installation should take place. This should include:
- Walking around the installation. Check that you are happy that the installation is correct and meets installations standards.
- Check that all open ends are capped off and all valves are isolated.
- Check that all capillary joints are soldered and that all compression joints are fully tightened.
- Check that enough pipe clips, supports and brackets are installed and that all pipework is secure.
- Check that the equipment (such as unvented hot water storage cylinder, shower boosting pumps, expansion vessels and subsequent safety and functional controls) are installed correctly and that all joints and unions on and around the equipment are tight.
- Check that the pre-charge pressure in the expansion vessel is correct and in accordance with the manufacturer's data.
- Check that any cisterns installed on open vented hot water storage systems are supported correctly and that float operated valves are provisionally set to the correct water level.
- Check that all appliances' isolation valves and taps are off. These can be turned on and tested when the system is filled with water.
- Check that the D1 and D2 discharge pipework complies with the Building Regulations and that it terminates in a safe but visible position.

The initial system fill

The initial system fill is always conducted at the normal operating pressure of the system. The system must be filled with fluid category 1 water direct from the water undertaker's mains cold water supply. It is usual to conduct the fill in stages so that the filling process can be managed comfortably. There are several reasons for this:

- Filling the system in a series of stages allows the operatives time to check for leaks stage by stage. Only when the stage being filled is leak free should the next stage be filled.
- **Open vented systems** – air locks from cistern-fed open vented systems are less likely to occur as each stage is filled slowly and methodically. Any problems can be assessed and rectified as the filling progresses without the need to isolate the whole system and initiate a full drain down. Allowing cisterns to fill to capacity and then opening any gate valves is the best way to avoid air locks. This ensures that the full pressure of the water is available and the pipes are running at full bore. Trickle filling can encourage air locks to form causing problems later during the fill stage.
- **Unvented systems** – before an unvented hot water storage system is filled, the pressure at the expansion vessel (if fitted) should be checked with a Bourdon pressure gauge to check the pre-charge pressure. Unvented hot water storage systems should be filled with all hot taps open. This is to ensure that pockets of air at high pressure are not trapped within the storage vessel as this can cause the system to splutter water, even after the system has filled. Water should be drawn from every hot water outlet to evacuate any air pockets from the system. The taps can be closed when the water runs freely without spluttering. The temperature and pressure relief valves should be opened briefly to ensure their correct operation and to test the discharge pipework arrangement.
- When the system has been filled with water, the system should be allowed to stabilise to full operating pressure. Any float operated valves should be allowed to shut off. The system will then be deemed to be at normal operating pressure.

Once the filling process is complete, another thorough visual inspection should take place to check for any possible leakage. The system is then ready for pressure testing.

Soundness testing hot water systems

The procedure for soundness testing hot water systems is described in Water Supply (Water Fittings) Regulations 1999, as well as **BS EN 8064**. There are two types of tests:

- testing metallic pipework installations
- testing plastic pipework systems.

Both of these test procedures are covered in detail in Chapter 1.

Flushing procedures for hot water systems and components

Again, this subject was covered in detail in Chapter 1, but differs slightly because of the appliances and equipment installed on hot water systems.

Like cold water installations, the flushing of hot water systems is a requirement of the British Standards. All systems, irrespective of their size, must be thoroughly flushed with clean water direct from the water undertaker's main supply before being taken into service. This should be completed as soon as possible after the installation has been completed to remove potential contaminants, such as flux residues, PTFE, excess jointing compounds and swarf. Simply filling a system and draining down again does not constitute a thorough flushing. In most cases, this will only move any debris from one point in the system to another. In practice, the system should be filled and the water run at every outlet until the water runs completely clear and free of any discoloration. It is extremely important that any hot water storage vessels and cold water storage cisterns should be drained down completely.

It is generally accepted that systems should not be left charged with water once the flushing process has been completed, especially if the system is not going to be used immediately, as there is a very real risk that the water within the system could become stagnant. In practice, it is almost impossible to effect a complete drain down of a system, particularly large systems, where long horizontal pipe runs may hold water. This in itself is very detrimental as corrosion can often set in and this can also cause problems with water contamination. It is recommended, therefore, that to minimise the risk of corrosion and water quality problems to leave systems completely full and flush through at regular intervals of no less than twice weekly, by opening all terminal fittings until the system has been taken permanently into operation. If this is the case, then provision for frost protection must be made.

Taking flow rate and pressure readings

Once the hot water system has been filled and flushed, the heat source should be put into operation and the system run to its operating temperature. Thermostats and high-limit thermostats should be checked to ensure that they are operating at their correct temperatures. When the system has reached full operating temperature and the thermostats have switched off, the flow rates, pressures and water temperatures can then be checked against the specification and the manufacturer's instructions. This can be completed in several ways:

- Flow rates can be checked using a weir gauge. This is sometimes known as a weir cup or a weir jug. The method of use is simple. The gauge has a slot running vertically down the side of the vessel, which is marked with various flow rates. When the gauge is held under running water, the water escapes out of the slot. The height that the water achieves before escaping from the slot determines the flow rate. Although the gauge is accurate, excessive flow rates will cause a false reading because the water will evacuate out of the top of the gauge rather than the side slot.
- System pressures (static) can be checked using a Bourdon pressure gauge at each outlet or terminal fitting. Bourdon pressure gauges can also be permanently installed either side of a boosting pump to indicate both inlet and outlet pressures.
- Both pressure (static and running) and flow rate can be checked at outlets and terminal fittings using a combined pressure and flow rate meter.
- The temperature should be checked using a thermometer at the hot water draw-off to ensure that it is at least 60 °C but does not exceed 65 °C. Each successive hot water outlet, moving away from the storage vessel should be temperature checked

to ensure that any thermostatic mixing valves are operating at the correct temperature and that the hot water reaches the outlet within the 30-second limit. If a secondary return system is installed, then the circulating pump should be running when the tests are conducted and the temperature of the return checked just before it re-enters the cylinder to ensure that the temperature is no less than 10 °C lower than the draw-off, 50 °C minimum.

▲ Figure 2.60 Checking hot water flow rates

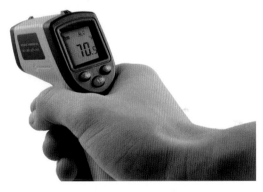

▲ Figure 2.61 Checking the hot water temperature using an infrared thermometer

Balancing a secondary circulation system

Large secondary circulation systems should contain bronze lockshield valves on every return leg of the hot water secondary circuit. These should be fitted as close to the appliances as possible and are used to balance the system so that the flow rates to each leg are such that:

- heat loss through the circuit is kept to a minimum
- the temperature of each leg is constant
- the temperature of the return at the cylinder is not less than 50 °C.

Correct balancing is achieved by opening the valves on the longest circuits and then successively closing the lockshield valves a little at a time working towards the cylinder until the flow rate through each circuit is equal. The flow rate should be balanced so that all of the circuits achieve the same temperature at the same time. This is especially important with those systems that operate through a time clock.

Dealing with defects found during commissioning

Commissioning is the part of the installation where the system is filled and run for the first time. It is now that we see if it works as designed. Occasionally, problems will be discovered when the system is fully up and running, such as systems that do not meet correct installation requirements. This can take several forms:

- **Systems that do not meet the design specification** – problems such as incorrect flow rates and pressures are quite difficult to deal with. If the system has been calculated correctly and the correct equipment has been specified and installed to the manufacturer's instructions, then problems of this nature should not occur. However, if the pipe sizes are too small in any part of the system, then flow rate and pressure problems will develop almost immediately downstream of where the mistake has been made. In this instance, the drawings should be checked and confirmation with the design engineer that the pipe sizes that have been used are correct before any action is taken. It may also be the case that too many fittings or incorrect valves have been used causing pipework restrictions.

  Another cause of flow rate and pressure deficiency is the incorrect set-up of equipment such as boosting pumps and accumulators. In this instance, the manufacturer's data should be consulted and set-up procedures followed in the installation instructions. It is here that mistakes are often made. If problems still continue, then the manufacturer's technical support should be contacted for advice. In a very few cases, the equipment specified is at fault and will not meet the design specification. If this is the case then the equipment must be replaced.

- **Poor installation techniques** – installation is the point where the design is transferred from the drawing to the building. Poor installation techniques account for:

  - **Noise** – incorrectly clipped pipework can often be a source of frustration within systems running at high pressures because of the noise that it can generate. Incorrect clipping distances and, often, lack of clips and supports can put strain on the fittings and cause the pipework to reverberate throughout the installation, even causing fitting failure and leakage. To prevent these occurrences, the installation should be checked as it progresses and any deficiencies brought to the attention of the installing engineer. Upon completion, the system should be visually checked before flushing and commissioning begins.

  - **Leakage** – water causes a huge amount of damage to a building and can even compromise the building structure. Leakage from pipework if left undetected causes damp, mould growth and an unhealthy atmosphere. It is, therefore, important that leakage is detected and cured at a very early stage in the system's life.

It is almost impossible to ensure that every joint on every system installed is leak free. Manufacturing defects on fittings and equipment and damage sometimes cause leaks. Leakage due to badly jointed fittings and poor installation practice are much more common, especially on large systems where literally thousands of joints have to be made until the system is complete. These can often be avoided by taking care when jointing tubes and fittings, using recognised jointing materials and compounds and using manufacturer's recommended jointing techniques.

▲ Figure 2.62 A plumber's nightmare! A badly designed plumbing system makes fault finding almost impossible

The risk from *Legionella pneumophila* in hot water systems

According to the Health and Safety Executive, the instances of Legionnaires' disease derived from hot water supply has diminished over recent years due to better installation techniques and more awareness of sterilisation methods. However, large hot water systems can often be complex in their design and, therefore, still present a significant risk of exposure. The environments where the *Legionella* bacteria proliferates are listed below:

- At the base of the cylinder or storage vessel where the cold feed enters and cold water mixes with the hot water within the vessel. The base of the storage vessel may well eventually contain sediments, which support the bacterial growth of *Legionella*.
- The water held in a secondary circulation system between the outlet and the branch to the secondary circulation system as this may not be subject to the high-temperature sterilisation process.

HEALTH AND SAFETY

In general, hot water systems should be designed to aid safe operation by preventing or controlling conditions which allow the growth of *Legionella*. They should, however, permit easy access for cleaning and disinfection. The following points should be considered:

- Materials such as natural rubber, hemp, linseed oil-based jointing compounds and fibre washers should not be used in domestic water systems. Materials and fittings acceptable for use in water systems are listed in the directory published by the Water Research Centre.
- Low-corrosion materials (copper, plastic, stainless steel, etc.) should be used where possible.

Defective components and equipment

Defective components cause frustration and cost valuable installation time. If a component or piece of equipment is found to be defective, do not attempt a repair as this may invalidate any manufacturer's warranty. The manufacturer should first be contacted as they may wish to send a representative to inspect the component prior to replacement. The supplier should also be contacted to inform them of the faulty component. In some instances where it is proven that the component is defective and was not a result of poor installation, the manufacturer may reimburse the installation company for the time taken to replace the component.

The procedure for notifying works carried out to the relevant authority

At all stages of the installation from design to commissioning, notification of the installation will need to be given so that the relevant authorities can check that the installation complies with the regulations and to ensure that the installation does not constitute a danger to health. It must be remembered that only operatives that are registered to do so can install unvented hot water storage systems. The operative's registration number must be given on any paperwork submitted to the local authority.

Under Building Regulations Approved Document G, hot water installations are notifiable to the local authority building control office. Building Regulations approval can be sought from the local authority by submitting a 'building notice'. Plans are not required with this process so it is quicker and less detailed than the full plans application. It is designed to enable small building works to get under way quickly. Once a building notice has been submitted and the local authority has been informed that work is about to start, the work will be inspected as it progresses. The authority will notify if the work does not comply with the Building Regulations.

Notice should be given to building control not later than five days after work completion and until this is received no completion certificates can be issued.

Building Regulations Compliance certificates

From 1 April 2005, the Building Regulations demanded that all installations must be issued with a Building Regulations Compliance certificate. This is to ensure that all Building Regulations relevant to the installation have been followed and complied with.

Commissioning records for hot water systems

Commissioning records such as benchmark certificates for hot water systems should be kept for reference during maintenance and repair and to ensure that the system meets the design specification. Typical information that should be included on the record is as follows:

- the date, time and the name(s) and I.D. numbers of the commissioning engineer(s)
- the location of the installation
- the amount of hot water storage and cold water storage (if any)
- the types and manufacturer of equipment and components installed
- the type of pressure test carried out and its duration
- the incoming static water pressure
- the flow rates and pressures at the outlets
- the expansion vessel pressure
- whether temperature and pressure relief valves have been fitted
- the results of tests on the discharge pipework.

The benchmark certificate should be signed by the operative and the customer and kept in a file in a secure location.

Hand over to the customer or the end user

When the system has been tested and commissioned, it can then be handed over to the customer. The customer will require all documentation regarding the installation and this should be presented to the customer in a file, which should contain:

- all manufacturers' installation, operation and servicing manuals for the unvented hot water storage vessel and associated controls
- the commissioning records and certificates
- the Building Regulations Compliance certificate
- an 'as fitted' drawing showing the position of all isolation valves, backflow prevention devices, etc.

VALUES AND BEHAVIOURS

Remember, it leaves a positive impression to take time at the end of a job to ensure the customer fully understands how to operate their new system. Ask if they have any questions and try to answer any questions asked with a plain, non-technical explanation.

The customer must be shown around the system and shown the operating principles of any controls. Emergency isolation points on the system should be pointed out and a demonstration of the correct isolation procedure in the event of an emergency. Explain to the customer how the systems work and ask if they have any questions. Finally, point out the need for regular servicing of the appliances and leave emergency contact numbers.

IMPROVE YOUR ENGLISH

Clear communication skills are vital to ensure the customer fully understands how their system works. Give them the opportunity to ask any questions and ensure you answer clearly in terms they are likely to understand.

7 DIAGNOSING AND RECTIFYING FAULTS IN HOT WATER SYSTEMS AND COMPONENTS

As with cold water systems, the risk of breakdown and failure of hot water systems is ever present.

In this part of the chapter, we will investigate some of the methods that help us identify system faults.

Obtaining details of system faults from end users

When identifying faults that have occurred on hot water systems, the customer can prove an invaluable source of information as they can often describe when and how the fault first manifested itself and any characteristics that the fault has shown. Verbal discussion with the customer often results in a successful repair without the need for extensive diagnostic tests.

IMPROVE YOUR ENGLISH

When consulting the client to obtain information about system faults, the customer should be asked:

- The immediate history of the fault:
  - When did it first occur?
  - How did they notice it?
  - What characteristics did it show?
- Did they notice any unusual discharge of water around the storage vessel or an increase or decrease in flow rate or pressure? This may well indicate the type of component failure that has taken place.
- Did they attempt any repairs themselves? If so, what did they do? This is important because if repairs have been attempted, they may well have to be undone to successfully diagnose the problem.
- What was the result of the fault? Again, an important aspect because it can often indicate where the fault lies. For instance, if the customer has noticed a drop in flow rate or pressure, this might indicate a blockage, a blocked strainer or scale growth.

How to use manufacturer instructions and industry standards to establish the diagnostic requirements of hot water system components

When attempting to identify faults with hot water systems, the most important document that can be consulted is the manufacturer's instructions. In most cases these will contain a section on fault finding that will prove an invaluable source of information. Fault-finding using manufacturer's instructions usually takes three forms:

- known problems that can occur and the symptoms associated with them
- methods by which to identify the problem in the form of a flow chart; these usually follow a logical, step-by-step approach, especially if the equipment has many parts that could malfunction, such as a pressure reducing valve or an expansion vessel
- the techniques required for replacement of the malfunctioning component.

A replacement parts list will also be present for those components that can be replaced. When ordering parts, it is advisable to use the model number of the equipment and the parts number from the replacement parts list. This will ensure that the correct part is purchased.

BS 8558 Table 5 may also be consulted as it contains important information regarding minimum flow rates required by certain appliances. Again, this should be used in conjunction with the manufacturer's instructions.

KEY POINT

Remember! Manufacturer's instructions always take precedence over the British Standards and Regulations.

The routine checks and diagnostics performed on hot water system components as part of a fault-finding process

Routine checks on components and systems can help to identify any potential problems that may be developing within the system as well as keeping the system operating to its maximum performance and within the system design specification. Checks performed can include:

- **Checking components for correct operating pressures, temperatures and flow rates** – these are important checks, simply because they can indicate whether a component has started to fail and will require replacement or whether the component will require recalibration. Those components that are pressure and temperature related such as expansion vessels, pressure reducing valves and thermostats are particularly vulnerable and susceptible to failure.
- **Cleaning system components (including dismantling and reassembly)** – components such as in-line strainers should be checked during periodic maintenance or when there is a noticeable drop off on flow rate. A blocked strainer will dramatically reduce the flow of water and may well affect the dynamic pressure of the system also. Pressure reducing valves and composite valves can also be checked and cleaned but usually these contain sealed cartridges where dismantling is not advisable. New cartridges should be installed wherever a blockage in the PRV is suspected.
- **Checking for correct operation of system components:**
  - **Thermostats** – these can be checked using a thermometer in the hot water flow once the thermostat has shut off. This will indicate whether the thermostat is operating at the correct temperature.
  - **Pumps** – these should be checked using the manufacturer's commissioning procedures to ascertain whether the pump is performing as the data dictates. A slight fall in performance is to be expected with age. Check to ensure:
    - there no signs of damage or wear and tear on the pump

- there are no signs of leakage from the pump
- that the pump switches on and off at the correct pressure
- that there are no unusual noises or vibrations when the pump is operating.
- **Timing devices** – time clocks can be checked to see if they activate at the correct time and that any advance timings such as 1-hour boost buttons, work correctly. The time display should be checked against the correct time of the check and any alterations to the time made.
- **Expansion and pressure vessels** – these should be checked for the correct pressure using a portable Bourdon pressure gauge. The type used to check tyre pressures is ideal for this. Any signs of water leakage should be investigated. Always refer to the manufacturer's instructions for the correct charge and pre-charge pressures.
- **Gauges and controls** – gauges are notorious for requiring replacement or recalibration as they often display an incorrect pressure. They should be replaced as necessary.
- **Checking for correct operation of system safety valves:**
  - **Temperature relief and expansion/pressure relief valves** – these can be checked by twisting the top and holding the valve open for 30 seconds. Always ensure that the valve closes completely and that the water stops without any drips.

Should any components require replacing, they should be replaced with like-for-like components or, if this is not possible, check with the manufacturers that the part is approved for use with the storage vessel.

The methods of repairing faults in hot water system components

Repairing system components should be undertaken using the manufacturer's servicing and maintenance instructions, as these will contain the order in which the component should be dismantled and re-assembled. As with all components, there will be occasions when it cannot be repaired and replacement is the only option. Some of the components that may be repaired and/or replaced are listed in Table 2.13.

▼ Table 2.13 Hot water components

Component	Known fault	Symptom	Repair
Pumps	Worn/broken impeller	Motor working but water not being pumped. No water at the outlets	No repair possible. Replace the pump
	Burnt out motor	Voltage detected at the pump terminals but pump not working	No repair possible. Replace the pump
	Cracked casing	Water leaking from the pump body	No repair possible. Replace the pump
	Faulty capacitor	Slow starting pump	Replace the capacitor if possible. Check manufacturer's instructions
Expansion vessels	Pressure loss due to faulty Schrader valve	No pressure in the expansion vessel. Water discharging from the pressure relief valve during water heat up	Pump air into the expansion vessel using a foot pump and check the Schrader valve with leak detector fluid. Check for bubbles. Replace Schrader valve as necessary
	Ruptured bladder/diaphragm	Water discharging from the Schrader valve. Water discharging from the pressure relief valve on water heat up	It is possible to replace the bladder/diaphragm of some accumulators. Check the manufacturer's instructions
Expansion (pressure) relief valve	Water dripping intermittently when the water is being heated	Usually an indication that the expansion vessel has lost its air charge or internal expansion bubble has disappeared	Check and recharge the expansion vessel or internal air bubble.
	Water running constantly	Usually an indication of incorrect pressure due to a malfunction of the pressure reducing valve	Check and replace the pressure reducing valve
Temperature/pressure relief valve	Cold water running constantly	Usually an indication of two potential faults: <ul><li>incorrect pressure due to a malfunction of the pressure reducing valve</li><li>faulty pressure relief valve</li></ul>	Check and replace the pressure reducing valve and the pressure relief valve
	Hot water running constantly	Usually a sign of thermostat and high limit stat malfunction	Isolate the system from the electrical supply and allow to cool before attempting a repair. Check and replace the thermostat and high limit stat as necessary
Thermostats	Hot water too hot	System thermostat is not operating at the correct temperature	Check the temperature of the hot water with a thermometer against the setting on the thermostat. Replace the thermostat as necessary
	No hot water	System thermostat not operating	Check the thermostat with a GS38 electrical voltage indicator for correct on/off functions. Replace as necessary
High-limit thermostat	No hot water	Usually an indication that the system thermostat has malfunctioned, and the high limit thermostat has activated to isolate the heat source	Check the main system thermostat and reset the high limit thermostat
Pressure (Bourdon) gauges	Sticking pressure indicator needle	Gauge not reading the correct pressure and does not move when the pressure is raised or lowered	No repair possible. Replace the gauge

KEY POINT

When replacing or repairing valves and controls, it is important to ensure that the water supply is isolated and the section of pipework is completely drained before beginning to repair or replace the valve.

▼ Table 2.14 Unvented hot water storage system fault finding

Fault	Probable cause	Recommended solution
No hot water flow	Mains cold water off	Check and open isolation valve/stop valve
	Strainer blocked	Turn off water and clean filter
	Cold water connection incorrectly installed	Check and refit mains cold water as necessary
Poor flow rate	Strainer blocked	Turn off water and clean filter
Water from the hot taps is cold	Immersion heater not switched on	Check and switch on as necessary
	Thermal cut-out (high limit thermostat) has operated	Check and reset by pushing the reset button or replace immersion heater as necessary
	Indirect boiler is not working	Check boiler operation. If a fault is detected, repair boiler
	Indirect boiler thermal cut-out 5 (high limit thermostat) has operated	Reset the boiler cut-out and check the operation of the boiler thermostat
	Motorised valve is not working properly	Check wiring of motorised valve and replace/repair as required
Water discharges from the pressure relief valve	**Intermittently** Air bubble has reduced or expansion vessel has lost its air charge	Recharge the air bubble by draining down, or; Check and recharge expansion vessel as necessary
	Continually Pressure reducing valve not working correctly Pressure relief valve seating damaged	1 Check and replace pressure reducing valve as required 2 Replace pressure relief valve as required
Water discharges from the temperature relief valve	**Hot** Thermal control failure	SWITCH OFF ELECTRICAL POWER. DO NOT TURN OFF WATER SUPPLY. When the discharge stops, check all thermal controls. Replace as necessary
	Cold Joint failure of the pressure reducing valve and the pressure relief valve	1 Check and replace pressure reducing valve as required 2 Replace pressure relief valve as required

▼ Table 2.15 Open vented hot water storage system fault finding

Fault	Probable cause	Recommended solution
No hot water flow	Mains cold water off	Check and open isolation valve/stop valve
	Float operated valve stuck in the off position	Check and clean or replace the float operated valve as necessary
	System air locked	Drain down the cylinder and refill the system
Water flowing cold	Immersion heater failure	Check and replace immersion heater
	Heat source not working	Check the boiler for correct operation. Repair as necessary
Water flowing only lukewarm	Cylinder thermostat or immersion heater thermostat set too low	Increase the temperature setting to 60 °C and check the correct operation
Poor flow rate	Cold feed pipe and/or hot water draw-off blocked with scale	Check the cold feed connection and the hot water draw-off and de-scale as required
Poor pressure	Cold water feed/storage cistern too low	Raise the cistern to increase the distance to the hot water outlets
Very hot water discharging into the cold water feed/storage cistern from the open vent	Failure of the immersion heater thermostat and energy cut-out	Check and test the thermostat and energy cut-out and replace

Safe isolation of hot water systems or components during maintenance and repair

Repair and maintenance tasks on hot water services, appliances and valves are essential to ensure the continuing correct operation of the system. The term used when isolating a water supply during maintenance operations is temporary decommissioning. There are basically two types:

- planned preventative maintenance
- unplanned/emergency maintenance.

When a maintenance task involves isolating the hot water supply, a notice will need to be placed at the point of isolation stating 'system off – do not turn on' to prevent accidental turn on of the system. Where key components such as the expansion vessel, pressure relief and temperature relief valves are found to be faulty, then the system should be isolated and temporarily decommissioned until replacement parts are obtained and fitted. In most systems, it will be possible to isolate specific parts of the installation without the need to have the whole supply turned off. It may be a good idea to remove any levers from the isolation valves to

prevent accidental turn on whilst the system is drained down. Where no such isolation exists, it may be of a benefit to use a pipe freezing kit so that total system isolation is not undertaken. If a component requires removal and replacing, it is always a good idea to cap off any open ends until the new component is installed.

Where the equipment also uses an electrical supply, safe isolation of the electricity supply is vital and the safe isolation procedure should be followed and the fuse/supply locked off for safety (see Chapter 5, Electrical principles).

A record of all repairs and maintenance tasks completed will need to be recorded on the maintenance schedule at the time of completion, including their location, the date when they were carried out and the type of tests performed. This will ensure that a record of past problems is kept for future reference.

Where appliance servicing is carried out, the manufacturer's installation and servicing instructions should be consulted. Any replacement parts may be obtained from the manufacturers.

Do not forget to keep the householder/responsible person informed of the areas that are going to be isolated during maintenance tasks and operations.

Procedures for carrying out diagnostic tests to locate faults on shower pumps, expansion vessels and thermostats

In this part of the chapter, we will investigate diagnostic checks that we can perform on shower pumps, expansion vessels and thermostats.

Shower pumps

Diagnostic tests on shower pumps are fairly straightforward to undertake. Table 2.16 illustrates some of the problems associated with shower pumps and their diagnostic approaches.

▼ Table 2.16 Shower pump fault finding

Fault	Probable cause	Recommended solution
Pump will not start	Electrical	Check power supply Check fuse Check circuit breaker Loose wiring connections
	Inlet/outlet connections incorrectly installed or reversed	Check that connections are plumbed in the correct way round and that all valves are open
	Insufficient gravity flow	Check that the installation complies with the instructions Check inlet filters are not blocked Check flow rate is a minimum required by the manufacturer's instructions on both hot and cold
	Float switch sticking in the outlet	Ensure there is no debris in the outlet area
	Float switch malfunction	Replace float switch
Reduced or intermittent flow	Incorrect or no anti-aeration flange/pipework arrangement fitted	Check that the pipework connections comply with the instructions
	Insufficient gravity flow	See above
	Blocked inlet filters	Ensure that the filters and the showerhead are free of debris. Always fit the filters to the pump inlets
	Couplers are restricting flow	Ensure that the flexible anti-vibration couplers at the shower pump are not bent or distorted
	Air in the system	Run the system on full hot with the pump isolated from the electrical supply (gravity only) for several minutes Check that the cold water storage requirements are correct for the installation and that the pump is fitted to the manufacturer's instructions Ensure that the cold water fill rate of the cisterns is adequate and that water starvation is not occurring Ensure that automatic air vents are fitted at all high points where airlocks may occur
	Wrong size pump for the system	Ensure that the pump size is correct for the system
	Hot water temperature too high	Ensure that the temperature of the hot water does not exceed 65°C. Fit a hot water system thermostatic mixing valve at the hot water storage cylinder if required

➡

▼ Table 2.16 Shower pump fault finding (continued)

Fault	Probable cause	Recommended solution
Pump starts with all outlets closed	Leak in the system	Check pipework for leaks
	Outlet open	Ensure all outlets are closed and all open ends are capped
Pump is noisy	Air in the system	See above
	Pump is vibrating on the surface	Ensure rubber shock absorbers are fitted
	Flexible pump connections are causing vibration	Ensure that the flexible anti-vibration couplers at the shower pump are not bent or distorted
Pump is leaking	Pump is subjected to mains pressure	Check that the installation complies with the instructions
	Pump has suffered chemical damage	Ensure that the pump has not come into contact with a chemical substance such as flux
	Pump is exposed to excessive temperature	Ensure that the temperature of the hot water does not exceed 65 °C. Fit a hot water system thermostatic mixing valve at the hot water storage cylinder if required
	Pump appears to have leaked but not sure	Check that the leak is not from pipework directly above the pump
Pump motor appears to be running hot and water flow rate is minimal	Pump has suffered a major terminal internal fault	Pump needs replacing
	Pump is air locked	Bleed the air from the pump

Expansion vessels

Expansion vessels are checked at the Schrader valve with a pressure gauge to ensure that the vessel contains the correct pressure.

▼ Table 2.17 Expansion vessel fault finding

Fault	Probable cause	Recommended solution
No air charge in the expansion vessel	Faulty Schrader valve	Recharge the vessel with air and check the Schrader valve with leak detection fluid. If the valve is leaking, then replace the expansion vessel
Water detected at the Schrader valve	Expansion vessel full of water due to a ruptured membrane/diaphragm in the expansion vessel	Replace the membrane if possible (check the manufacturer's instructions). If not replace the expansion vessel with one of similar capacity

Note: If a faulty expansion vessel is diagnosed, the system should be isolated and temporarily decommissioned until a replacement vessel is obtained and fitted.

Thermostats

Faulty thermostats are usually indicated by one of two symptoms:
- excessive hot water
- no hot water.

The type of thermostat on the system will depend on whether the system is directly heated or indirectly heated. The use of manufacturer's instructions when diagnosing faults with thermostats and high-limit thermostats is recommended.

Figure 2.63 shows a fault-finding chart to determine the cause of excessive hot water in an indirectly heated hot water storage system.

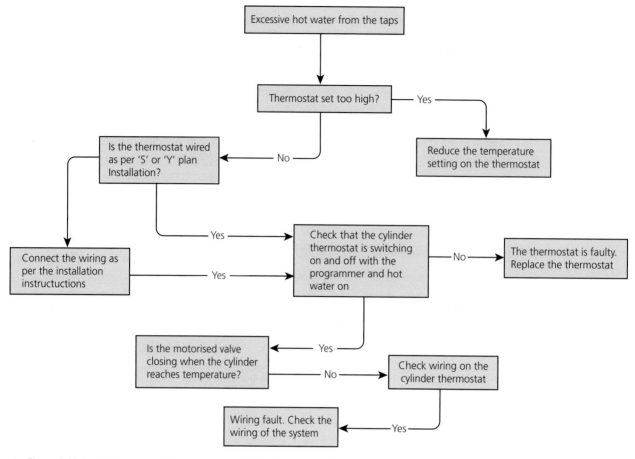

▲ Figure 2.63 Fault-finding chart to assess excessive hot water in an indirectly heated hot water storage system

⑧ SERVICING AND MAINTENANCE OF HOT WATER SYSTEMS

Work related to servicing accounts for the majority of problems found in hot water systems and components, although the lack of periodic maintenance can also cause a significant amount of hot water related problems.

In this part of the chapter, we will investigate some of the more common problems found in both open vented and unvented hot water systems.

The periodic servicing requirements of hot water systems

Hot water systems, like other plumbing systems in the home, require a certain amount of periodic maintenance to ensure a continued and efficient operation. Open vented and unvented systems have different maintenance requirements.

▼ Table 2.18 Maintenance requirements of hot water systems

Open vented systems	Unvented systems
1 Check the installation for signs of leakage around the storage vessel and associated pipework.	1 Unvented hot water storage systems should be serviced every 12 months.
2 Check the cold water storage cistern in the roof space and clean as necessary. Check that the base and the bearers that the cistern is sitting on are in good condition and fit for purpose.	2 Check the installation for signs of leakage around the storage vessel and associated pipework.
3 Adjust the float operated valve to the correct water level. If there have been signs of the water overflowing, then the FOV should be re-washered.	3 Check that the components are approved for use with the storage vessel. Ask the customer if any of the components have been replaced during the lifetime of the vessel.
4 Check the operation of the isolation valves and gate valves to ensure that they operate correctly. Advise the customer if they require replacement.	4 Check the pressure in the expansion vessel and top-up the pressure with a foot pump as necessary.
5 Check and replace as necessary the sacrificial anode inside the cylinder.	5 If the system has an internal expansion bubble, the system should be drained down and refilled to recharge the air.
6 Run the heat source and check the temperature of the hot water.	6 While the system is drained, remove the in-line strainer (filter) and clean of any debris.
7 Run the system to 65 to 70 °C to ensure that the cylinder has been disinfected. Do not forget to reset the thermostat to 60 °C for safety purposes.	7 Check the discharge pipework to ensure it complies with the regulations. Check the termination point.
8 Check the system flow rates using a weir gauge.	8 Operate the pressure relief valve by twisting the top and holding open for 30 seconds.
	9 Operate the temperature/pressure relief valve by twisting the top and holding open for 30 seconds.
	10 Check the tundish to ensure that water is not discharging from the air gap.
	11 Run the heat source(s) and check the temperature of the hot water. Check the operation of the system thermostat(s) to ensure that they shut down at the desired temperature.
	12 Check that the non-self-setting high-limit (energy cut-out) thermostat operates to the manufacturer's specification. This is a requirement of the Building Regulations.
	13 Run the system to 65 to 70 °C to ensure that the cylinder has been disinfected. Do not forget to reset the thermostat to 60 °C for safety purposes.
	14 Check the static and dynamic pressures of the system to determine if the pressure reducing valve is operating within the manufacturer's limits.
	15 Check the temperature of the water at the outlets.
	16 Check the flow rates using a weir gauge.
	17 Check that any information and warning notices required by the unvented cylinders are displayed permanently on the storage vessel. This is a requirement of the Building Regulations.
	18 Complete the servicing log in the benchmark handbook.

SUMMARY

As we have worked through this chapter, we have seen that hot water is a very complex subject. It becomes obvious that careful consideration must be given to the requirements of the customer if the system that we fit is to meet their specific needs. The subsequent system choice is often the result of the calculations we make to determine flow rates, pipe sizes and quantity of hot water required. This chapter gives you the knowledge needed to install good, well thought out, well planned hot water storage systems as well as an insight into the complexities of good hot water system design.

Test your knowledge

1 What is the minimum diameter of the open vent pipe within an open vented hot water system?

   a 15 mm

   b 22 mm

   c 28 mm

   d 35 mm

2 Which of the following materials would be most suitable for the manufacture of a secondary circulator to be installed within a hot water system?

   a Bronze

   b Plastic

   c Steel

   d Aluminium

3 What minimum distance should the draw-off pipework from the top of an open vented hot water cylinder rise before connection to the open vent pipe?

   a 150 mm   c 350 mm

   b 250 mm   d 450 mm

4 Which of the following appliances is most likely to offer the customer low pressure at a hot water outlet?

   a Combination boiler

   b Direct cylinder

   c Combination unit

   d Indirect cylinder

5 An unvented hot water storage system requires safety controls to be fitted to ensure that the water it contains does not exceed what temperature?

   a 60°C   c 85°C

   b 65°C   d 100°C

6 Identify the type of DHW system shown in the image:

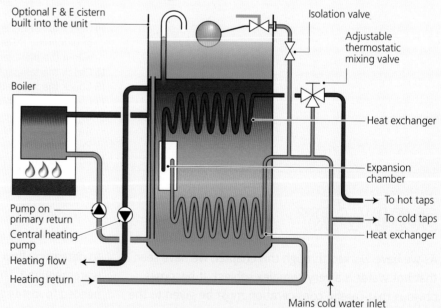

   a Unvented direct storage unit

   b Gas fired hot water storage heater

   c Thermal store

   d Combination boiler

7 Under Building Regulations Part G, what is a new or replacement immersion heater required to have?

a Temperature relief valve

b Non-resetting overheat thermostat

c Restricted thermostat to a maximum temperature of 60 °C

d Protective coating to avoid electrolytic corrosion

8 What is the maximum distance between the outlet of a temperature relief valve and the tundish forming part of the D1 pipework?

a 300 mm

b 550 mm

c 600 mm

d 650 mm

9 Water at the outlet of a shower installed within a school should be fitted with a TMV so that the water does not exceed what temperature?

a 43 °C

b 48 °C

c 50 °C

d 60 °C

10 Which Building Regulation Approved Document relates to unvented hot water storage systems?

a L1A

b L2B

c G3

d G2

11 At what temperature should a TMV that is supplying a domestic bath be set?

a 44 °C

b 38 °C

c 48 °C

d 31 °C

12 When installing a balanced cold supply with an unvented hot water system, what is positioned immediately after the pressure reducing valve?

a Single check valve

b Isolator

c Cold supply

d Strainer

13 What material is used to make the secondary circulation pump that could be fitted to an unvented system?

a Brass

b Copper

c Cast iron

d Bronze

14 What item could have failed if the pressure relief valve is letting by in to the tundish?

a Expansion vessel

b Strainer

c Pressure regulating valve

d Cylinder stat

15 Between what temperatures does *Legionella* multiply?

a 15–25 °C

b 20–30 °C

c 20–45 °C

d 30–45 °C

16 The Water Regulations state that hot water should be distributed at a temperature of not less than 55 °C and should reach the outlets at 50 °C within what time period?

a 30 seconds

b 15 seconds

c 45 seconds

d 60 seconds

17 At what temperature will the temperature and pressure relief valve discharge water?

a 65 °C

b 85 °C

c 75 °C

d 95 °C

18 What is a typical kW rating of an immersion heater in a hot water cylinder?

a 1 kW

b 2 kW

c 3 kW

d 4 kW

19 The minimum storage requirement of a hot water storage vessel allows how many litres of water per person?

a 15–25 litres per person

b 25–35 litres per person

c 35–45 litres per person

d 45–55 litres per person

20 In what units is specific heat capacity measured?

a °C/litre

b kJ/kg/°C

c kJ/litre

d litre/kJ/kg

21 At what temperature is water most dense?

a 0 °C

b 4 °C

c 60 °C

d 100 °C

22 By how much, approximately, does water expand when heated in a hot water system?

a 10%

b 7.5%

c 4%

d 2%

23 If the coefficient of linear expansion of copper pipe is 0.000016 mm/m/°C, by how much would a 5.0 m length of copper pipe expand by if heated from 10 °C to 60 °C?

a 4 mm

b 0.048 mm

c 0.48 mm

d 48 mm

24 If a customer had a problem with low pressure at a shower, which of the following would NOT improve the installation?

a Install an unvented hot water system with balanced cold feed

b Install a twin impellor pump

c Install a larger open vented indirect cylinder

d Install an electric shower

25 If a hot water cylinder contains 150 litres of cold water before being heated, what is the approximate volume of the expansion?

a 1.5 litres

b 15 litres

c 6 litres

d 8 litres

26 What is the requirement for the installation of a TMV if installing a wash hand basin in an infant school?

27 Describe what happens to *Legionella* bacteria when water is held at a temperature of 60 °C.

28 Name the functional controls of an unvented hot water storage system.

29 Using the formula below, calculate the kW input required to raise the temperature of a 300-litre cylinder from 4 to 65 °C in 2 hours:

$$\frac{\text{SHC} \times \text{litres of water} \times \text{temperature difference } (\Delta t) \times \text{boiler efficiency (93)}}{\text{Time in seconds} \times 100} = \text{kW}$$

When:

SHC = 4.19

litres of water = 300 litres

$\Delta t = 61\,°C$

time in seconds = 7,200 seconds (2hrs)

30 A discharge pipe from a ¾" temperature relief valve is to be installed. The route from the tundish will include 7.5 m of pipework and two elbows. What size D1 and D2 pipework will be required and why?

31 Describe a centralised hot water system.

32 Outline the three tiers of safety and their temperatures that are required for an unvented hot water cylinder.

33 When designing a hot water system, state what the Water Supply (Water Fittings) Regulations state about the temperature of the hot water at the outlet point.

34 A customer states that there is a continual drip through the tundish when the hot water system gets up to temperature. Explain what the most likely cause of this is.

Answers can be found online at www.hoddereducation.co.uk/construction.

Central heating is a vast and complex subject. There are now more options with regard to sources of heat, pipe materials and heat emitters than ever before. Environmentally friendly technology and the re-emergence of underfloor heating has meant that the customer can now afford to be selective about the system they have installed into their property. The advent of heat pumps and solar systems, with the savings on fuel and running costs, has dramatically lowered the carbon footprint of domestic properties. No longer does the customer have to rely appliances that burn carbon rich fuels such as gas and oil. Zero carbon and carbon neutral fuels have revolutionised domestic heating whilst advances in technology have lowered the cost of the energy saving appliances that were once only available to a select few.

In this chapter, we will build on the knowledge of central heating systems you learned in Book 1.

This chapter provides learning in installation, maintenance, application of design techniques to include heat and ventilation loss through the building fabric, diagnostics and rectification of faults and commissioning procedures. It provides in-depth learning of system types, components, controls, and servicing requirements in systems up to large domestic dwellings and/or systems of equal size in commercial and industrial premises. We will investigate new and exciting technology that has the potential to dramatically cut the cost of heating our homes whilst, at the same time, reducing our carbon emissions. We will also look at new controls and components that can transform an existing wasteful installation into an energy efficient system.

By the end of this chapter, you will have knowledge of the following six areas:
- types of central heating system and their layout requirements
- design techniques for central heating systems
- positioning central heating system components
- installation, connection and testing requirements of electrically operated central heating components
- commissioning central heating systems and components
- diagnosing and rectifying faults in central heating systems and components.

Return to Book 1, and remind yourself of the topics covered in Chapter 7, Central heating systems, which included:
- understanding central heating systems and their layouts
- installing central heating systems and components
- understanding the decommissioning requirements of central heating systems and their components.

1 TYPES OF CENTRAL HEATING SYSTEM AND THEIR LAYOUT REQUIREMENTS

As highlighted, there are now more options with regard to sources of heat, pipe materials and heat emitters than ever before. In this section we will consider the types and layouts of central heating systems.

The space heating zoning requirements of single occupancy dwellings

Central heating systems in dwellings are subject to the strict requirements of the Building Regulations, which stipulate that every home must be divided into at least two heating zones. Approved Document L is driven by the need for energy efficiency. It lays down specific requirements as to how this must be carried out.

INDUSTRY TIP

You can access the Conservation of fuel and power: Approved Document L at: www.gov.uk/government/publications/conservation-of-fuel-and-power-approved-document-l

---

The current Building Regulations came into force in 2013 (with 2016 amendments) for England and Wales. Part L is currently being updated and should be re-issued in the coming months. This update should include any amendments along with elements from Boiler plus, CHeSS, Domestic Heating Compliance Guide and The Building Services Compliance Guide. The document is divided into four specific sections:

- Part L1A Installations in new domestic dwellings
- Part L1B Installations in existing domestic dwellings
- Part L2A Installations in new industrial/commercial buildings
- Part L2B Installations in existing industrial/commercial buildings.

The requirements of both of these very different aspects of space heating are explained in two accompanying documents:

- the Domestic Building Services Compliance Guide for Parts L1A and L1B (see Table 3.1 on page 195)
- the Non-Domestic Building Services Compliance Guide for Parts L2A and L2B.

Both of these documents were produced by the Government as guidance to help installers comply with Approved Document L. To support these documents, the Energy Saving Trust published a set of standards known as the CHeSS standards (Central Heating Systems Specifications), which lay down both good practice and best practice with regard to controls of central heating systems.

Approved Document L1A and L1B

The main requirement of Approved Document L is for a boiler interlock. A boiler interlock is a series of controls (cylinder thermostats, programmable room thermostats, programmers and time switches) that prevents the boiler from cycling when there is no demand for heat. In addition:

- every home must be divided into at least two heating zones, using a thermostat controlling a motorised valve
- if the house is less than 150 m^2, then these can be controlled by the same time clock or programmer
- if the house is larger than 150 m^2, then each zone must be controlled by its own time clock/programmer
- living and sleeping areas (zones) must be controlled at different temperatures by means of a thermostat
- every radiator should be fitted with a thermostatic radiator valve, unless the radiator is being used as the reference radiator for a thermostat situated elsewhere in the room.

These requirements apply every time a home is built.

Where existing installations are concerned, the requirements of Document L are made retrospectively. In other words, if an existing system does not comply with the regulations, then the system must be updated:

- every time a home has an extension or change of use
- every time more than one individual component, such as a boiler, is replaced in a heating system.

Simple boiler servicing is exempt from this, but the recommendation is made that radiator thermostats should be fitted when the system is drained down.

▼ Table 3.1 Recommended minimum controls for new gas-fired wet central heating systems (BEAMA Domestic Building Services Compliance Guide 2021, Table 3)

Control type	Minimum standard
Boiler interlock	System controls should be wired so that when there is no demand for space heating or hot water, the boiler and pump are switched off.
Zoning	Dwellings with a total floor area > 150 m^2 should have at least two space heating zones, each with an independently controlled heating circuit[1]. Dwellings with a total floor area[2] ≤ 150 m^2 may have a single space heating zone[3].
Control of space heating	Each space heating circuit should be provided with: ● independent time control, and either: ● a room thermostat or programmable room thermostat located in a reference room[4] served by the heating circuit, together with individual radiator controls such as thermostatic radiator valves (TRVs) on all radiators outside the reference rooms, or ● individual networked radiator controls in each room on the circuit.
Control of hot water	Domestic hot water circuits supplied from a hot water store (i.e. not produced instantaneously as by a combination boiler) should be provided with: ● independent time control, and ● electric temperature control using, for example, a cylinder thermostat and a zone valve or three-port valve. (If the use of a zone valve is not appropriate, as with thermal stores, a second pump could be substituted for the zone valve.)

The standards in this table apply to new gas-fired wet central heating systems. In existing dwellings, the standards set out in Table 4 will apply in addition.

Always also follow manufacturers' instructions.

[1] A heating circuit refers to a pipework run serving a number of radiators that is controlled by its own zone valve.

[2] The relevant floor area is the area within the insulated envelope of the dwelling, including internal cupboards and stairwells.

[3] The SAP notional dwelling assumes at least two space heating zones for all floor areas, unless the dwelling is single storey, open plan with a living area > 70% of the total floor area.

[4] A reference room is a room that will act as the main temperature control for the whole circuit and where no other form of system temperature control is present.

▼ Table 3.2 Recommended minimum standards when replacing components of gas-fired wet central heating systems (BEAMA Domestic Building Services Compliance Guide 2021, Table 4)

Component	Reason	Minimum standard	Good practice[1]
1. Hot water cylinder	Emergency	For copper vented cylinders and combination units, the standard losses should not exceed Q = 1.28 × (0.2 + 0.0521V$^{2/3}$) kWh/day, where V is the volume of the cylinder in litres. Install an electric temperature control, such as a cylinder thermostat. Where the cylinder or installation is of a type that precludes the fitting of wired controls, install either a wireless or thermo-mechanical hot water cylinder thermostat or electric temperature control. If separate time control for the heating circuit is not present, use of single time control for space heating and hot water is acceptable.	Upgrade gravity-fed systems to fully pumped. Install a boiler interlock and separate timing for space heating and hot water.
	Planned	Install a boiler interlock and separate timing for space heating and hot water.	Upgrade gravity-fed systems to fully pumped.

→

▼ Table 3.2 Recommended minimum standards when replacing components of gas-fired wet central heating systems (BEAMA Domestic Building Services Compliance Guide 2021, Table 4) (continued)

2. Boiler	Emergency/ planned	**All boiler types except heating boilers that are combined with range cookers** The ErP[2] seasonal efficiency of the boiler should be a minimum of 92% and not significantly less than the efficiency of the appliance being replaced. In the exceptional circumstances defined in the *Guide to the condensing boiler installation assessment procedure for dwellings* (ODPM, 2005), the boiler SEDBUK 2009 efficiency should not be less than 78% if natural gas-fired, or not less than 80% if LPG-fired. In these circumstances the additional requirements for combination boilers would not apply. Install a boiler interlock as defined for new systems. Time and temperature control should be installed for the heating system. **Combination boilers** In addition to the above, at least one of the following energy efficiency measures should be installed. The measure(s) chosen should be appropriate to the system in which it is installed: ● Flue gas heat recovery ● Weather compensation ● Load compensation ● Smart thermostat with automation and optimisation.	Upgrade gravity-fed systems to fully pumped. Fit individual radiator controls such as thermostatic radiator valves (TRVs) on all radiators except those in the reference room.
3. Radiator	Emergency		Fit a TRV to the replacement radiator if in a room without a room thermostat.
	Planned		Fit TRVs to all radiators in rooms without a room thermostat.
4. New heating system – existing pipework retained	Planned	The new boiler and its controls should meet the standards in section 2 of this table. Fit individual radiator controls such as TRVs on all radiators except those in the reference room.	In dwellings with a total floor area > 150 m^2, install at least two heating circuits, each with independent time and temperature control, together with individual radiator controls such as TRVs on all radiators except those in the reference rooms.

Always also follow manufacturers' instructions.

[1] Best practice would be as for a new system.

[2] Refers to the efficiency methodology set out in Directive 2009/125/EC for energy performance related products.

The types of central heating systems and their layout requirements

In this part of the chapter, we will investigate central heating systems and their pipework layouts and controls. These will be restricted to sealed heating systems, as open vented systems were covered in depth in Book 1.

Sealed heating systems

Sealed heating systems are those that do not contain a feed and expansion cistern but are filled with water directly from the mains cold water supply via a temporary filling loop. Large systems would be filled via an automatic pressurisation unit. The expansion of water is taken up by the use of an expansion vessel and the open vent is replaced by a pressure relief valve which is designed to relieve the excess pressure by releasing the system water and discharging safely to a drain point outside of the dwelling. This is vital as the water may be in excess of 80°C. A pressure gauge is also included so that the pressure can be set when the system is filled and periodically checked for rises and falls in the pressure as these could indicate a potential component malfunction. The system is usually pressurised to around 1 bar. There are several types:

- sealed systems with an external pressure vessel
- system boilers that contain all necessary safety controls
- combination boilers.

All fully pumped systems, such as those with two or three 2-port zone valves (known as the S-plan and the S-plan plus) or a 3-port mid-position valve (known as the Y-plan) or a 3-port diverter valve (known as the W-plan), can be installed as sealed systems or they can be purpose designed 'heating only' systems using a combination boiler with instantaneous hot water supply.

It is important to remember that the safety controls are installed to prevent explosions caused by high pressure and also products of steam.

The advantages of sealed central heating systems

There are several advantages to sealed heating systems.

- Less pipework – because the need for a feed and expansion cistern is removed, less pipework is often used.
- Smaller pipework – sealed heating systems operate at a slightly higher temperature than vented systems, which means that the heat delivery for the size of pipe is increased. This means that smaller pipework can often be used.
- Higher heat emitter temperatures – because sealed systems operate at a slightly higher temperature,

this means that slightly smaller heat emitters may be used in some cases.

- Greater range of heat emitters – the slightly higher operating temperature offers a greater range in heat emitters, including fan convectors and skirting heating.
- Quicker installation time – sealed systems are often easier to install and this results in quicker installation time.
- Quicker filling – sealed systems fill much more quickly than vented systems because the filling water is coming straight from the mains cold water supply.
- Fewer airlocks – filling the system from the mains cold water supply eliminates the problems of airlocks.
- Sealed system components inside the boiler – system boilers are supplied with all necessary sealed system components already installed as part of the boiler within the boiler casing. This simplifies the system installation.
- Fewer components – if a system boiler is used, the problems of siting and installing the expansion vessel and its associated components are eliminated.

Fully pumped systems with two or three 2-port zone valves (known as the S-plan and the S-plan plus)

The S-plan has two 2-port motorised zone valves to control the primary and heating circuits separately by the cylinder and room thermostats respectively. This system is recommended for dwellings with a floor area greater than 150 m² because it allows the installation of an additional 2-port zone valve to zone the upstairs heating circuit from the downstairs circuit (the S-plan plus). A separate room thermostat and, possibly, a second time clock/programmer would also be required for upstairs zoning.

A system bypass is required for **overheat protection** of the boiler.

KEY TERM

Overheat protection: when water cannot circulate or the thermostats have been satisfied and the motorised valves have closed, the boiler will continue to heat up for a short period even though the burner has shut down. This is because of the latent heat (see Book 1) in the boiler casing. If the boiler overheats, the high-limit thermostat will activate, and the boiler will fail to operate when it is next required. A pump-overrun circuit, which is fitted to most modern boilers, will ensure that the pump continues to run when the boiler has shut down to dissipate any latent heat. If the motorised valves are closed, the automatic bypass valve opens from the pump pressure to allow water circulation, allowing the excess heat to dissipate, keeping the boiler temperature below high-limit shut-down.

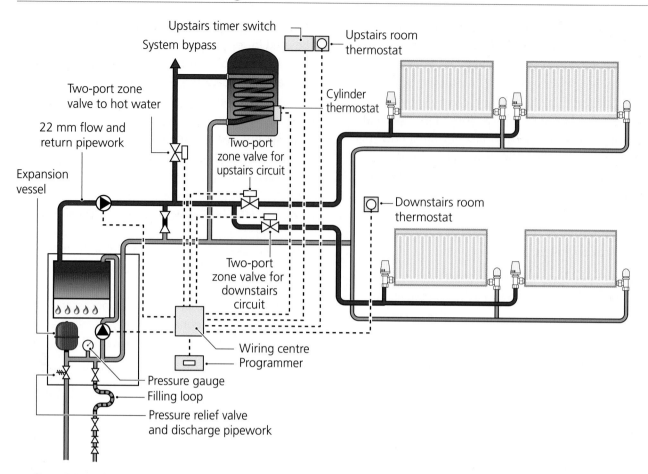

Upstairs timer switch

System bypass

Upstairs room thermostat

Two-port zone valve to hot water

Cylinder thermostat

22 mm flow and return pipework

Expansion vessel

Two-port zone valve for upstairs circuit

Two-port zone valve for downstairs circuit

Downstairs room thermostat

Wiring centre

Programmer

Pressure gauge

Filling loop

Pressure relief valve and discharge pipework

▲ Figure 3.1 The S-plan plus system

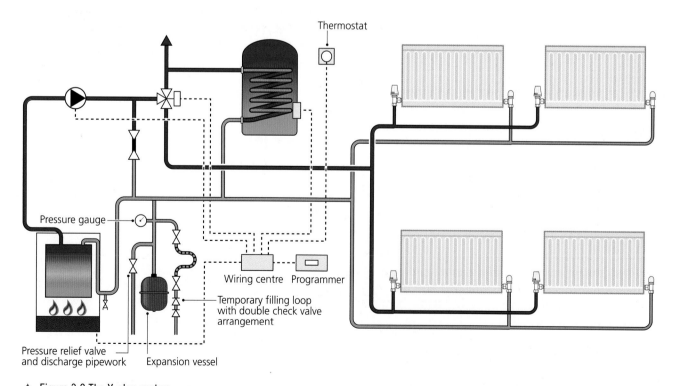

Thermostat

Pressure gauge

Wiring centre   Programmer

Temporary filling loop with double check valve arrangement

Pressure relief valve and discharge pipework

Expansion vessel

▲ Figure 3.2 The Y-plan system

Fully pumped systems with 3-port mid-position valve (known as the Y-plan) or a 3-port diverter valve (known as the W-plan)

The 3-port valve mid-position (Y-plan) or diverter valve (W-plan) controls the flow of water to the primary (cylinder) circuit and the heating circuit. The valve reacts to the demands of the cylinder thermostat or the room thermostat.

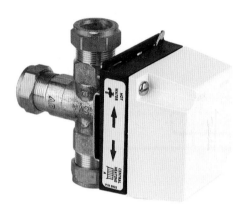

▲ Figure 3.3 The 3-port mid-position valve

The system contains a system bypass fitted with an automatic bypass valve which simply connects the flow pipe to the return pipe. The bypass is required when all circuits are closed either by the motorised valve or the thermostatic radiator valves as the rooms reach their desired temperature. The bypass valve opens automatically as the circuits close to protect the boiler from overheating by allowing water to circulate through the boiler keeping the boiler below its maximum high temperature. This prevents the boiler from 'locking out' on the overheat energy cut-out.

System boilers

A system boiler is an appliance where all necessary safety and operational controls are included and fitted directly to the boiler. There is no need for a separate expansion vessel, pressure relief valve or filling loop and this makes the installation much simpler.

The system boiler has all the components for a sealed system contained within the boiler unit. It is filled directly from the mains cold water via a filling loop which is often fitted by the boiler manufacturer.

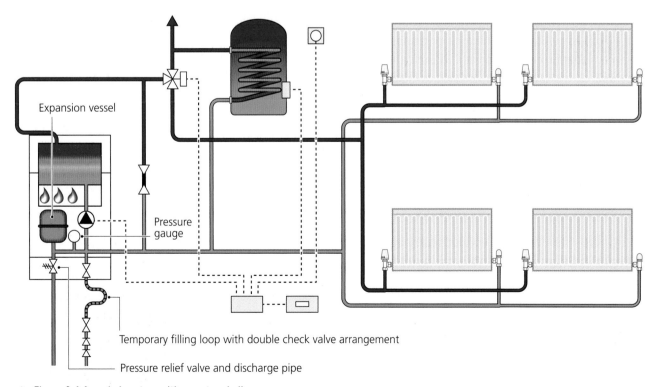

Expansion vessel

Pressure gauge

Temporary filling loop with double check valve arrangement

Pressure relief valve and discharge pipe

▲ Figure 3.4 A sealed system with a system boiler

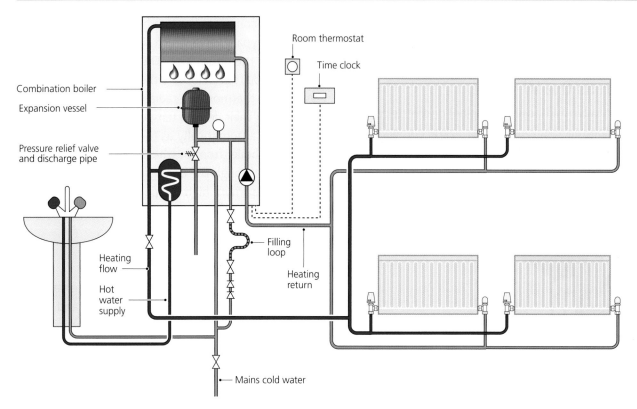

▲ Figure 3.5 A sealed system with a combination boiler

Combination boilers

In recent years, combination boilers have become one of the most popular forms of central heating in the UK. A combination boiler provides central heating and instantaneous hot water supply from a single appliance. Modern combination boilers are very efficient and they contain all the safety controls (such as expansion vessel, pressure relief valve) of a sealed system. Most 'combis' also have an **integral filling loop**.

KEY TERM

Integral filling loop: a filling loop that is designed and installed as part of the boiler by the manufacturer.

Multiple boiler installations and the use of low loss headers

For a boiler to work at its maximum efficiency, the water velocity passing through the heat exchanger needs to be maintained within certain parameters. This is especially important for condensing boilers as they rely on a defined temperature drop across the flow and return

before the condensing mode begins to work effectively. Installation of a low loss header allows the creation of two separate circuits. These are shown in Figure 3.6.

- The primary circuit – the flow rate within the primary circuit can be maintained at the correct flow rate for the boilers so that the maximum efficiency of the boilers is maintained regardless of the demand placed on the secondary circuit. Each boiler has its own shunt pump so that equal velocity through the boilers is maintained.

- The secondary circuits – the secondary circuits allow for varying flow rates demanded by the individual balanced zones or circuits. Each zone would be controlled by a shunt pump set to the flow rate for that particular zone. A 2-port motorised zone valve, time clock and room thermostat control each zone independently and these are often fitted in conjunction with other controls such as outdoor temperature sensors. In some cases, the flow rates through each secondary circuit will exceed that required by the boilers. In other cases, the opposite is true and the boiler flow rate will be greater than the maximum flow rate demanded by the secondary circuits, especially where multiple boiler installations are concerned.

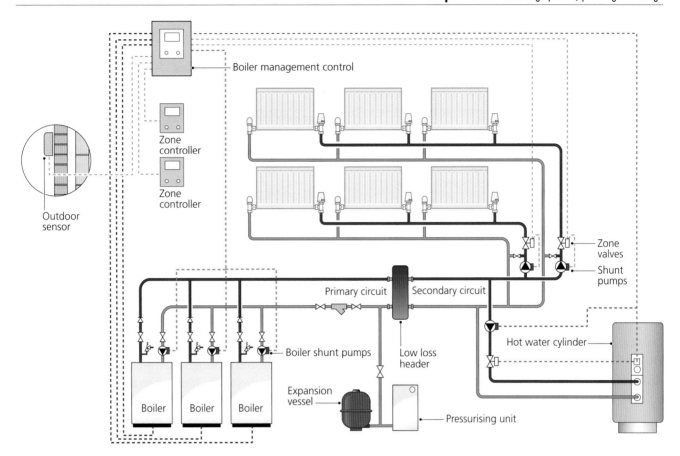

▲ Figure 3.6 A multiple boiler installation with a low loss header

Water velocity is just part of the problem. Water temperature is also important. There are two potential problems here:

- If the difference in temperature between the flow and return is too great, it puts a huge strain on the boiler heat exchangers because of the expansion and contraction. This is known as **thermal shock**.
- For a condensing boiler to go into condensing mode, the return water temperature must be in the region of 55 °C. In some instances, temperature sensors are fitted to the low loss header to allow temperature control over the primary circuit.

KEY TERM

Thermal shock: the rapid cooling or heating of a substance that can lead to failure of the material.

The low loss header is ideal for use with systems that have a variety of different heat emitters. It is the perfect place for installing an automatic air valve for removing unwanted air from the system. Drain points can also be fitted for removing any sediment

that may collect in the header. Both of these features are usually fitted as standard on most low loss headers.

How it works – the low loss header

Low loss headers act as an intermediary between the boiler and the heating and hot water circuits. They are designed to provide a hydraulic separation between the boiler and the heating circuits. This helps to regulate the flow rate through the boiler at optimum velocity for the best efficiency of the appliance(s), ensuring greater performance of the heating system.

Controls for heating systems

No matter how good the central heating design or how accurate the calculations, the system requires proper control to be effective, efficient and economical to run. The types of controls that are added to a system can greatly improve its performance. Even older systems can benefit from the addition of modern and effective controls.

In this part of the chapter, we will look at the various controls for central heating systems, their function and how they 'fit' into modern systems.

The function of components used in central heating systems

In Book 1, we investigated some of the more common controls, such as room thermostats, cylinder thermostats, time clocks and programmers. Here we will investigate some of the more advanced controls that can be installed alongside these as well as recapping the working principles of other common components.

Zone control valves

Zone control of multiple spaces within a dwelling is achieved by the use of motorised valves activated and controlled by a time clock/programmer/room thermostat arrangement or a programmable room thermostat that will do the same function.

The most common types of motorised valves are 2-port zone valves and 3-port mid-position and diverter valves. The method of use of these valves will depend on the pipework layout and installer/end user preference.

- 3-port valves provide separate hot water and heating circuits. Zoning of the living spaces can be achieved by the inclusion of additional 2-port valves on the individual space circuits (such as upstairs and downstairs circuits). 3-port valves include a mid-position which allows shared flow.
- 2-port motorised zone valves are probably the most common of all zone arrangements used. They provide zoning of individual circuits and are used where more than one zone is needed. A separate zone valve is used for each zone.

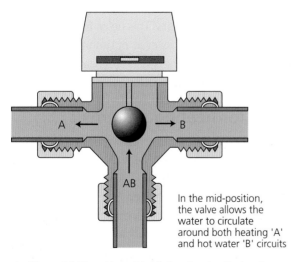

In the mid-position, the valve allows the water to circulate around both heating 'A' and hot water 'B' circuits

▲ Figure 3.7 The mid-position valve showing the heating connection (A), hot water connection (B) and the common or primary connection (AB)

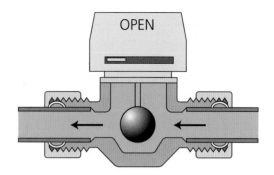

OPEN

▲ Figure 3.8 The 2-port zone valve

Valves of 22 mm can be used on boilers up to around 20 kW. 28 mm or larger should be used where the system is greater than 20 kW.

Advanced controls – weather compensation, optimum start and delayed start

Domestic central heating systems can benefit from more advanced controls, which are often digital, especially when a condensing boiler is fitted. Condensing boilers respond to lower flow and return temperatures better than non-condensing appliances. Advanced controls enhance system efficiency.

- **Weather compensation** – compensates for the external temperature. This control uses an externally fitted temperature sensor fitted on a north or north-east facing wall so as not to be in the direct path of solar radiation. As the external temperature rises, the weather compensator reduces the circulation temperature of the flow from the boiler to compensate for the warmer outside temperature. Similarly, the reverse occurs if the weather gets colder.
- **Delayed start** – here, the end user sets the time to bring on the heating taking into account the time it would normally take to warm the dwelling (for example, people might set the heat to come on at 5 p.m. if they were due to arrive home from work at 6 p.m.). A delayed start unit will, at the time the heat is due to come on, compare the current indoor temperature to that required by the room thermostat. It will then delay the start of the boiler firing if required. The benefits are that during milder weather when the heat requirement is less, energy will be saved. Room thermostats with a delayed start function are now available.

- **Optimum start** – the end user sets the required occupancy times and the required room temperature and the controller calculates the necessary heat up time so that the rooms are at the required temperature irrespective of the outside temperature. The idea is based around comfort rather than energy savings.

> ## INDUSTRY TIP
>
> Central heating control technology now includes smart devices that can communicate remotely via Wi-Fi or Bluetooth. Such devices will increasingly be connected to the internet of things. You should aim to keep up to date with all developments in this fast-moving field.

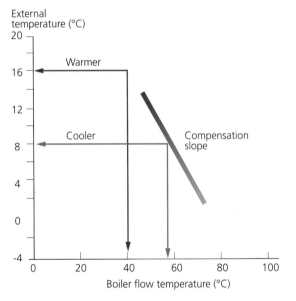

▲ Figure 3.9 A weather compensation graph

▲ Figure 3.10 A weather compensator

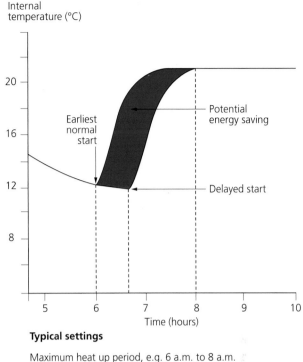

Typical settings

Maximum heat up period, e.g. 6 a.m. to 8 a.m.
Normal occupancy period, e.g. 8 a.m. to 10 a.m.

▲ Figure 3.11 Delayed/optimum start function

Domestic boiler management systems (home automation systems)

A boiler management system (BMS) is an electronic controller that provides bespoke control solutions for domestic central heating systems.

Standard functions of BMS control include real-time temperature and boiler/controls monitoring, room temperatures (known as set points) and time schedule adjustment, optimisation, and night setback control.

The system remembers key points, such as how quickly the building heats up or cools down and makes its own adjustments so that energy savings can be made. If it is very cold outside at, for example, 2 a.m., the BMS will switch the heating on at 4.15 a.m. to allow the building to be at the correct temperature by the time the user has set the heating to come on, say, at 7 a.m., irrespective of the time that the user has set the time for the heating to activate. On milder nights, the heating may not come on until 6.15 a.m. but it will still reach its set point by 7 a.m.

It will also learn how well your house retains heat and may shut down early if it calculates that your set point will still be maintained at your OFF time of, for example, 10 p.m.

These systems provide a cost-effective means of monitoring system efficiency and can reduce heating costs by up to 30 per cent.

Electronic sensors are fitted to the flow and return pipework and an external temperature sensor is fitted for weather compensation. The information is used to accurately vary the system output according to demand. This helps to significantly reduce fuel wastage caused by temperature overshoot, heat saturation of the heat exchanger, unnecessary boiler cycling and flue gas losses whilst maintaining internal comfort levels and reducing CO_2 emissions.

The selection of system and control types for single family dwellings

The installation of an effective system of central heating controls has a major effect on the consumption of energy and the effectiveness of the system. Choosing the right controls will lead to:

- improved energy efficiency
- reduced fuel bills
- lower CO_2 emissions.

The establishment of a minimum standard of heating controls is vital if the heating system is to achieve satisfactory efficiencies when the system is in use. The efficiency of the boiler is only part of the story. For the boiler to achieve these efficiencies, at least a minimum standard of controls MUST be installed.

So, what is a good system of controls?

A good system of controls must:

- ensure that the boiler does not operate unless there is demand. This is known as **boiler interlock**
- only provide heat when it is required to achieve the minimum temperatures.

KEY TERM

Boiler interlock: this is NOT a single control. It is a combination of several controls working in conjunction to ensure that the boiler does not fire unless it is required. It is key in ensuring good system efficiency and saving energy.

There are two levels of controls for domestic properties and these are set out in Central Heating System Specification (CHeSS) CE51 2008:

1 **Good practice:** this set of controls achieves good energy efficiency in line with Approved Document L 2010. This is described in detail in the CHeSS document:

a **HR7:** Good practice for systems with a regular and a separate hot water storage:

   i   full programmer
   ii room thermostat
   iii cylinder thermostat
   iv boiler interlock (see note 1)
   v   TRVs on all radiators, except in rooms with a room thermostat
   vi automatic bypass (see note 2).

b **HC7:** Good practice for systems using a combination boiler or combined primary storage unit:

   i   time switch
   ii room thermostat
   iii boiler interlock (see note 1)
   iv TRVs on all radiators, except in rooms with a room thermostat
   v   automatic bypass valve (see note 2).

2 **Best practice:** this standard uses enhanced controls to further enhance energy efficiency in line with Approved Document L1A/B 2010. This is described in detail in the CHeSS document:

a **HR8:** Best practice for systems with a regular and a separate hot water storage:

   i   programmable room thermostat, with additional timing capability for hot water
   ii cylinder thermostat
   iii boiler interlock (see note 1)
   iv TRVs on all radiators, except in rooms with a room thermostat
   v   automatic bypass valve (see note 2)
   vi more advanced controls, such as weather compensation, may be considered.

b **HC8:** Best practice for systems using a combination boiler or combined primary storage unit:

 i programmable room thermostat

 ii boiler interlock.

 iii TRVs on all radiators, except in rooms with a room thermostat

 iv automatic bypass valve (see note 2)

 v more advanced controls, such as weather compensation, may be considered.

Note 1 (from CHeSS): Boiler interlock is not a physical device but an arrangement of the system controls (room thermostats, programmable room thermostats, cylinder thermostats, programmers and time switches) so as to ensure that the boiler does not fire when there is no demand for heat. In a system with a combi boiler this can be achieved by fitting a room thermostat. In a system with a regular boiler this can be achieved by correct wiring interconnection of the room thermostat, cylinder thermostat, and motorised valve(s). It may also be achieved by more advanced controls, such as a boiler energy manager. TRVs alone are not sufficient for boiler interlock.

Note 2 (from CHeSS): An automatic bypass valve controls water flow in accordance with the water pressure across it, and is used to maintain a minimum flow rate through the boiler and to limit circulation pressure when alternative water paths are closed. A bypass circuit must be installed if the boiler manufacturer requires one, or specifies that a minimum flow rate has to be maintained while the boiler is firing. The installed bypass circuit must then include an automatic bypass valve (not a fixed position valve).

Care must be taken to set up the automatic bypass valve correctly, in order to achieve the minimum flow rate required (but not more) when alternative water paths are closed.

Source: Energy Saving Trust (2008) *Central heating system specifications (CHeSS)*

The application of system controls – time and temperature – to space heating zones

The number of homes that require both time and temperature zone control has increased in recent years. In 2006, a survey showed that the average floor area of a domestic property with four bedrooms was around 157 m² and over 200 m² for a five-bedroom domestic property. With properties of this size, zoning becomes a necessity and in 2006, Document L1A/B of the Building Regulations requested that zoning of the heating system must be installed in all properties of 150 m² or more. This was updated in 2010 to include any property.

In most instances zoning requires the separating of the upstairs circuit from the downstairs or, in the case of single-storey dwellings, separating the living space from the rest of the property. Separate time and temperature control of the individual circuits is a necessity.

Zoning with separate temperature control

Separate temperature-controlled zones provide a much better living environment because different parts of the dwelling can be maintained at different temperatures without relying on a single room to dictate the temperature across the whole system. Lower temperatures can be maintained in those rooms within the dwelling that are not occupied, allowing the dwelling to take full advantage of any solar gains, especially in rooms that face south, south-east or south-west. This can be quite pronounced, even in the winter Sun. Significant savings on both energy usage and fuel costs can be made by simply taking advantage of the free heat that the Sun can provide. Outside sensors linked to weather compensators and delayed start and optimum start controls further help to reduce energy usage and cost.

Zoning with separate time control

Zoning with separate time control offers another dimension to the concept of zoning by allowing the heating to be controlled at different times of the day in different zones. The heat can be focussed in those rooms that are occupied throughout the day with the heating to other parts of the dwelling timed to come on in the early morning and evening. Separate zones reduce energy usage and costs whilst maintaining improved comfort levels throughout the property.

Zoning in practice

The choice of controls for the zones should be decided by the predicted activity in those zones. There are many options that can be used individually or collectively to achieve a good system control:

- Using individual temperature and timing controls in every zone.
- Multi-channel programmers allow the timing of individual rooms or multiple zones to be set from a single point. This is often more desirable than many individual programmers at different locations within the dwelling.
- Thermostatic radiator valves (TRVs) vary the heat output of individual heat emitters. This can be beneficial where solar gain adds to the room temperature as they are very fast reacting in most circumstances. Some TRVs also have electronically timed thermostatic heads which can be linked to a wireless programmer.

VALUES AND BEHAVIOURS

Zoning is required by Approved Document L1A/B of the Building Regulations 2013 (with 2016 amendments) and the installer must make decisions on the best way to arrange those zones to take the best advantage of energy savings whist complying with the wishes of the customer/ end user and complying with the regulations. The only way that this can be achieved is by talking to the customer and finding out their usage patterns. The main aim of zoning is to avoid overheating areas that require less heat to maintain the warmth or because the set point could be lower than in other areas. The point here is that the number of zones laid down by Document L is the minimum and there are real benefits to adding additional zones in key areas of the property.

Zoning can help make significant energy savings. It allows the optimisation of the heating system whilst maintaining the dwelling at a comfortable temperature and saving money at the same time.

Underfloor heating systems

Underfloor heating has been around for many years – the Romans used a warm air system 1500 years ago to good effect. It is only fairly recently that its benefits have been rediscovered. It is very suitable for use with the lower flow temperatures from new environmental technologies such as air and ground source heat pumps and solar heating, and so it has become not only a viable option for the domestic dwelling but one that will also save money, energy, reduce CO_2 emissions and, as a consequence, significantly reduce environmental impact.

The design principles of underfloor central heating systems

An underfloor heating system provides invisible warmth and creates a uniform heat, eliminating cold spots and hot areas. The temperature of the floor needs to be high enough to warm the room without being uncomfortable underfoot. There is no need for unsightly radiator/convectors because the heat literally comes from the ground up. Underfloor heating creates a low temperature heat source that is spread over the entire floor surface area. The key word here is low temperature. Whereas most wet central heating systems containing radiators and convectors operate at around 70 to 80°C, underfloor heating operates at a much lower temperatures, making it an ideal system for air and ground source heat pump fuel sources. Typical temperatures are:

- 40–45°C for concrete (screeded) floors
- 50–60°C for timber floor constructions.

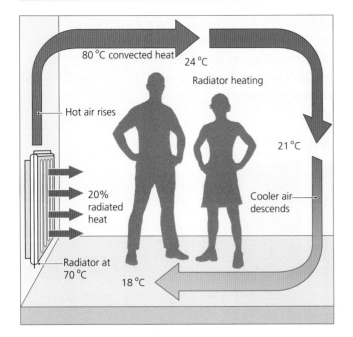

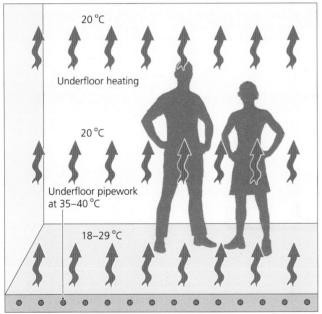

▲ Figure 3.12 The principle of underfloor heating

Traditional wet central heating systems generate convection currents and radiated heat. Around 20 per cent of the heat is radiated from the hot surface of the radiators and if furniture is placed in front of the radiator, the radiation emission is reduced. 80 per cent of the heat is convection currents, which makes the hot air rise. This adds up to a very warm ceiling! Underfloor heating systems, however, rely on both conduction and radiation. The heat from the underfloor heating system conducts through the floor warming the floor structure,

making the floor surface a large storage heater, the heat is then released into the room as radiated heat. Around 50 per cent to 60 per cent of the heat emission is in the form of radiation providing a much more comfortable temperature at low room levels when compared to a traditional wet system with radiators and, with the whole floor being heated, furniture positioning no longer becomes a problem because as the furniture gains heat, it too emits warmth.

During the design stage, the pipe coils are fixed at specific centres depending on the heat requirement of the room and the heat emission (in watts) per metre on pipe. The whole floor is then covered with a screed to a specific depth creating a large thermal storage heat emitter. The water in the pipework circulates from and to a central manifold and this heats the floor. The heat is then released into the room at a steady rate. Once the room has reached the desired temperature, a room thermostat actuates a motorised head on the return manifold and closes the circuit to the room.

Such is the nature of underfloor heating that many fuel types can be used, some utilising environmentally friendly technology. Gas and oil fired boilers are common, but biomass fuels, solar panels and heat pumps are also used.

VALUES AND BEHAVIOURS

More information regarding green energy heating solutions can be found at: www.gov.uk/green-deal-energy-saving-measures

Floor coverings are an important aspect for underfloor heating. Some floor coverings create a high thermal resistivity, making it difficult for the heat to permeate through. Carpet underlays and some carpets have particularly poor thermal transmittance, which means the heat is kept in and not released. Thermal resistivity of carpets and floor coverings is known as TOG rating. The higher the TOG rating, the less heat will get through. Floor coverings used with underfloor heating should have a TOG rating of less than 1 and must never exceed 2.5.

Quite often, underfloor heating is used in conjunction with traditional wet radiators, especially on properties such as barn conversions. The higher temperatures required for radiators do not present a problem because the flow water for the underfloor system is blended with the return water via a thermostatic blending valve to maintain a steady temperature required for the underfloor system. Zoning the upstairs and downstairs circuits with 2-port motorised zone valves and independent time control for the heat emitters also helps in this regard.

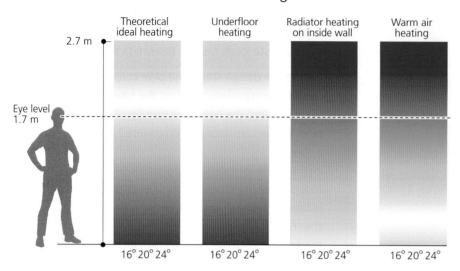

▲ Figure 3.13 Heating theory

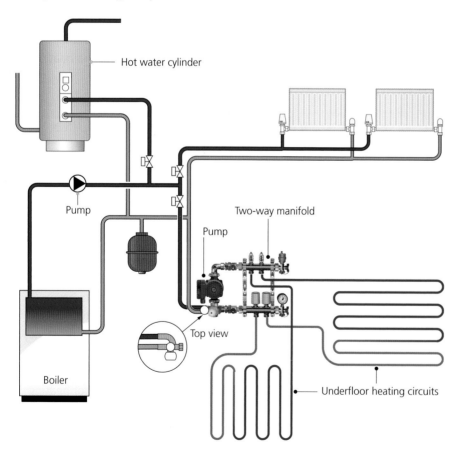

▲ Figure 3.14 Typical underfloor heating system combined with wet radiators

▼ Table 3.3 Advantages and disadvantages of underfloor heating

Advantages	Disadvantages
● The pipework is hidden underfloor. This allows better positioning of furniture and interior design. ● The heat is uniform giving a much better heat distribution than traditional systems. ● These systems are very energy efficient with low running costs. ● Environmentally friendly fuels can be used. ● Underfloor heating is almost silent with low noise levels when compared to other systems. ● Cleaner operating with little dust carried on convector currents. This can help those people who suffer allergies, asthma and other breathing problems. ● System maintenance is low and decorating becomes easier as there are no radiators to drain and remove. ● Individual and accurate room temperatures as every room has its own room thermostat that senses air temperature. ● Lower possibility of leaks. ● Greater safety, as there are no hot surfaces that can burn the elderly, infirm or the very young. ● Better zone control as each room is in effect a separate zone.	● Not very suitable for existing properties unless a full renovation means the removal of floor surfaces. ● Can be expensive to install when compared to more traditional systems. ● Heat up time is longer as the floor will need to get to full temperature before releasing heat. ● Slower cool-down temperatures mean the floors may still be warm when heat is not required. ● Greater installation time. ● More electrical installation of controls is required, as each room will need its own room thermostat and associated wiring.

The layout features of underfloor heating

Underfloor heating uses a system of continuous pipework, laid under a concrete or timber floor in a particular pattern and at set centre-to-centre pipe distances. Each room served by an underfloor heating system is connected at a central location to a flow and return manifold, which regulates the flow through each circuit. The manifold is connected to flow and return pipework from a central heat source, such as a boiler or a heat pump.

The manifold arrangement also contains a thermostatic mixing valve to control the water to the low temperatures required by the system and an independent pump to circulate the water through every circuit.

Each underfloor heating circuit is individually controlled by a room thermostat, which activates a motorised head on the return manifold to precisely control the heat to the room to suit the needs of the individual.

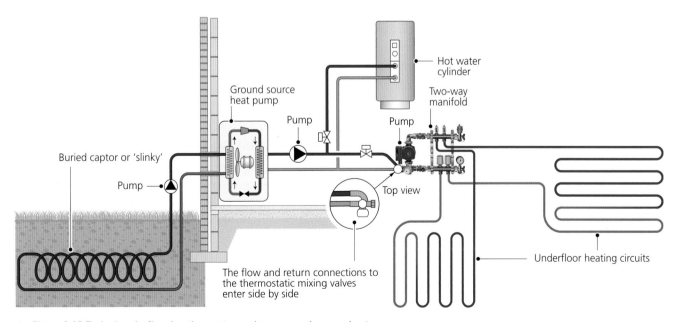

▲ Figure 3.15 Typical underfloor heating system using a ground source heat pump

The working principles of underfloor central heating system pipework and components

As we have already seen, underfloor heating works by distributing heat in a series of pipes laid under the floor of a room. To do this, certain components are required to distribute the flow of heat to ensure that the system warms the room. However, the components must be controlled in such a way so as to maintain a steady flow of heat whilst ensuring that the floor does not become too hot to walk on. This is achieved by the use of:

- manifolds
- a thermostatic blending valve
- a circulating pump
- various pipework arrangements to suit the floor and its coverings
- the application of system controls – time and temperature – to space heating zones.

The use of manifolds

In technical terms, the manifold is designed to minimise the amount of uncontrolled heat energy from the underfloor pipework. The manifold is at the centre of an underfloor heating system. It is the distribution point where water from the heat source is distributed to all of the individual room circuits, and as such, should be positioned as centrally in the property as possible. Room temperature is maintained via thermostatic motorised actuators on the return manifold whilst the correct flow rate through each coil is balanced via the flow meters on the flow manifold. Both the flow and return manifolds contain isolation valves for maintenance activities, an automatic air valve to prevent air locks and a temperature gauge so that the return temperature can be monitored.

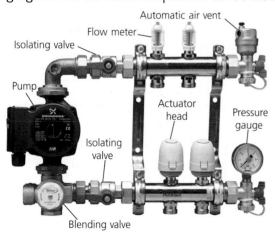

▲ Figure 3.16 Typical underfloor heating manifold

Most manifolds contain a circulating pump and a thermostatic mixing valve, often called a blending valve. These will be discussed below.

The thermostatic mixing (blending) valve

The thermostatic mixing or blending valve is designed to mix the flow and return water from the heat source to the required temperature for the underfloor heating circuits. They are available in many different formats, the most common being as part of the circulating pump module as shown in Figure 3.17. The temperature of the water is variable by the use of an adjustable thermostatic cartridge inside the valve.

▲ Figure 3.17 Underfloor heating circulating pump/blending valve module

The circulating pump

The circulating pump is situated between the thermostatic mixing valve and the flow manifold to circulate the blended water through every circuit. Most models are variable speed.

Underfloor heating pipework arrangements

The success of the underfloor heating system depends upon the installation of the underfloor pipework and the floor pattern installed. There are many variations of pipe patterns based upon two main pattern types. These are:

- the series pattern
- the snail pattern.

In general, underfloor heating pipes should not be laid under kitchen or utility room units.

The series pattern

The series pattern (also known as the meander pattern) is designed to ensure an even temperature across the floor especially in systems incorporating long pipework runs. It is often used in areas of high heat loss.

The flow pipe must be directed towards any windows or the coldest part of the room before returning backwards and forwards across the room at the defined pipe spacing centres.

The snail pattern

The snail pattern (also known as the bifilar pattern) is used where an even uniform temperature is required such as under hardwood floors and vinyl floor tiles.

The flow pipe runs in ever decreasing circles until the centre of the room is reached, it then reverses direction and returns with parallel runs back to the starting point.

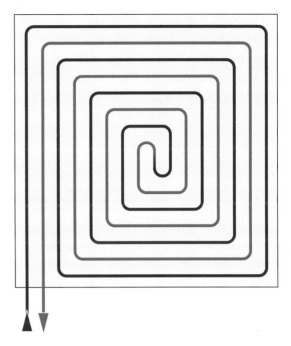

▲ Figure 3.19 The snail pattern

Positioning components in underfloor central heating systems

For an underfloor heating system to work effectively, the components require careful positioning to ensure that the efficiency of the system is maintained. All too often, systems fail to live up to their potential because of poor positioning of key components.

Manifolds

The longer the circuit, the more energy is needed to push the water around it. Water will always take the least line of resistance and shorter circuits will always be served first. In many instances, balancing the system will help even out the circulation times so that all circuits receive the heat at the same time but the system will only be as good as the slowest circuit. If the longest circuit is slow, once the system is balanced, then ALL circuits will be slow. In this regard, the positioning of the manifold is of great importance. By positioning the manifold centrally within the dwelling, the length of each circuit is balanced so that long circuits become shorter. Even if the short circuits become longer, the time for the heating system to reach full temperature will be shortened and balancing the system will become much easier.

▲ Figure 3.18 The series pattern

A potential problem that may occur where the manifold is located is that the area may become a potential 'hot spot' on the system because of the pipework congestion around the manifold. This can be prevented by insulating the pipework around the manifold until the pipework enters the room it is serving.

Pipework arrangements (cabling)

There are many variations to these two basic layouts. The pattern should be set out in accordance with the orientation and the shape of the room. Window areas may be colder and may require the bulk of the heat in that area. Other considerations include the type of floor construction and the floor coverings. The pipework should be laid in one continuous length without joints. In some instances, the pipe is delivered on a continuous drum of up to 100 m to enable large areas to be covered without the need for joints. Large rooms may require more than one zone and the manufacturer's instructions should be checked for maximum floor coverage per zone.

▲ Figure 3.20 The series pattern laid out

▲ Figure 3.21 The snail pattern laid out

Pipework installation techniques

Solid floor

There are many types of underfloor heating installation techniques for a solid floor. The drawing above shows one of the more common types using a plastic grid where the underfloor heating pipe is simply walked into the pre-made castellated grooves for a precise centre-to-centre guide for the pipework using a minimum radius bend.

▲ Figure 3.22 Solid floor underfloor heating installation method

The panels are laid on to pre-installed sheets of insulation to ensure a good performance and minimal heat loss downwards. Edge insulation is required to allow for expansion of the panels.

▼ Table 3.4 Key design and installation information for solid floor

Maximum heat output	Approx. 100 W/m^2
Recommended design flow temperature	50 °C
Maximum circuit length	100 m (15 mm pipe)
	120 m (18 mm pipe)
Maximum coverage per circuit	12 m^2 @ 100 mm centres
	22 m^2 @ 200 mm centres
	30 m^2 @ 300 mm centres (18 mm pipe only)
Material requirements	
Pipe	8.2 m/m^2 @ 100 mm centres
	4.5 m/m^2 @ 200 mm centres
	3.3 m/m^2 @ 300 mm centres (18 mm pipe only)
Floor plate usage	1 plate/m^2 allowing for cutting
Edging insulation strip	1.1 m/m^2
Conduit pipe	2 m/circuit

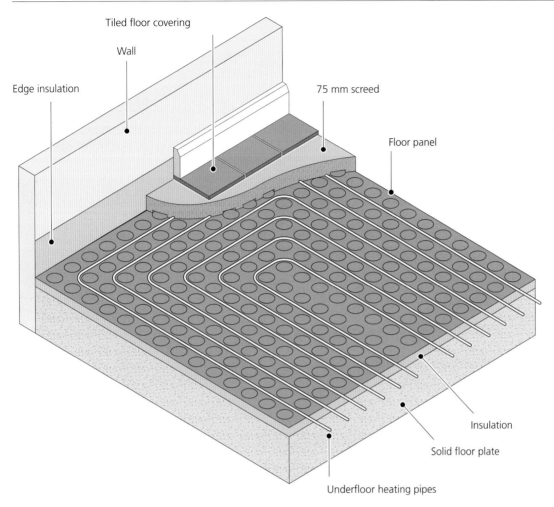

▲ Figure 3.23 Solid floor underfloor heating installation

Suspended timber floor

This system is designed for use under timber suspended floors. It uses aluminium double heat spreader plates to transmit heat evenly across the finished floor surface.

This system is suitable for any timber suspended floor with joist widths up to 450 mm. The heat plates are simply fixed to the joists using small flat headed nails or staples. A layer of insulation must be placed below the plates to prevent the heat penetrating downwards.

Where the pipework must cross the joists, the joists must be drilled in accordance with the Building Regulations.

▼ Table 3.5 Key design and installation information for suspended floor

Maximum heat output	Approx. 70 W/m^2
Recommended design flow temperature	60 °C
Maximum circuit length	80 m (15 mm pipe)
Maximum coverage per circuit using a double spreader plate	17 m^2 @ 225 mm centres
Material requirements	
Pipe	4.5 m/m^2 @ 100 mm centres
Heat spreader plates	2 plate/m^2

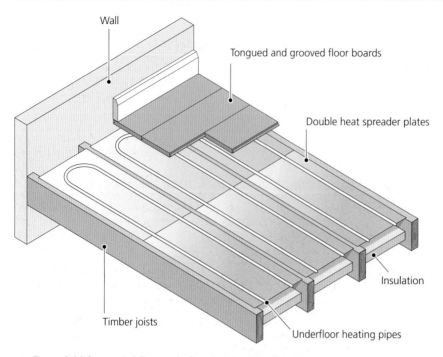

▲ Figure 3.24 Suspended floor underfloor heating installation

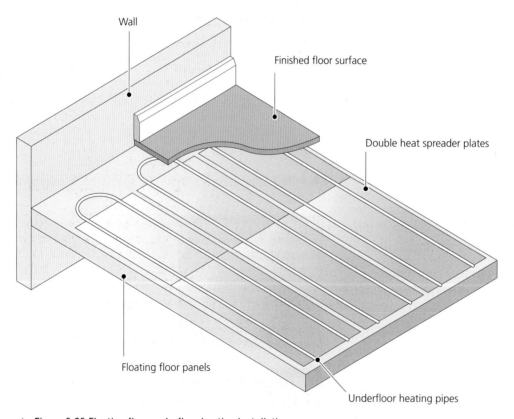

▲ Figure 3.25 Floating floor underfloor heating installation

Floating floor

This system is designed for use where a solid floor installation is not suitable due to structural limitations. It can be installed directly on to finished concrete or timber floors.

The pipework is laid on top of 50 mm thick polystyrene panels, each having a thermal transmittance of 0.036 Wm²K. The insulation has pre-formed grooves that the pipe clips into after the heat spreader plates have been fitted. The insulation is not fixed and 'floats' on the top of the sub-floor. The finished flooring can then be laid directly on to the top of the pipework, completing the 'floating' structure.

▼ Table 3.6 Key design and installation information for floating floor

Maximum heat output	Approx. 70 W/m²
Recommended design flow temperature	60 °C
Maximum circuit length	80 m (15 mm pipe)
	100 mm (18 mm pipe)
Maximum coverage per circuit using a double spreader plate	28.5 m² @ 300 mm centres (15 mm pipe)
	30 m² @ 300 mm centres (18 mm pipe)
Material requirements	
Pipe	3.1 m/m² @ 300 mm centres
Single heat spreader plates	3 plate/m²
Floating floor panel	1 panel/1.4 m²

Commissioning an underfloor heating system

Before commissioning takes place, it is recommended that the mixing valve and all other valves and pipework in the heating circuits are thoroughly flushed with water to remove flux and debris before the final filling and venting takes place.

Initial system fill

1 Close the isolating valves on the flow and return manifolds.
2 Connect a hosepipe to the flow manifold drain-off point, with the other end connected to the mains cold water supply.
3 Connect a second hose to the return manifold drain-off point. The other end must discharge over a drain gully.
4 Turn on the mains cold water and open each circuit of the underfloor system in turn until the water is running smoothly through the system and any air has been removed.
5 Turn off the water, close the drain points and remove the hoses.
6 Connect a hydraulic test pump and pressurise the system to 6 bar. Leave for a period of 1 hour.
7 Once the pressure test is complete, reduce the pressure to 3 bar.
8 Now the over screed can be laid.

Note: The screed should be allowed to dry thoroughly before the heat is turned on.

System balancing

1 Set the correct temperature at the manifold mixing/blending valve. Ensure that the boiler is operating and the correct temperature is being supplied at the mixing valve. Open the manifold isolating valves. Adjust the mixing valve as required and check the temperature of the water being supplied to the circuits using the dial thermometers fitted to the manifold.
2 Ensure that the manifold pump is set to a suitable speed.
3 Open the actuators to each circuit and adjust the flow rate through each circuit to the manufacturer's flow rate recommendations, using the flow meters on the return manifold. Repeat this process for each circuit.
4 Check the operation of each circuit actuator by operating the individual room thermostats.

2 DESIGN TECHNIQUES FOR CENTRAL HEATING SYSTEMS

The main purpose of central heating is to provide thermal comfort conditions within a building or dwelling.

The factors which affect the selection and design of central heating systems for dwellings

The factors to consider when discussing comfort are:

- **Humidity** – the amount of moisture there is within the environment. Ideal conditions for humans require 40 to 60 per cent humidity. Anything below 40 per cent can make the eyes and throat very dry. Above 60 per cent makes the atmosphere very damp and uncomfortable.
- **Air changes** – the amount of air movement (not velocity) within the building. Air movement is important because it replaces used air with fresh air, which is needed for breathing. Air changes, however, lead to heat loss. Every time an air change occurs, the fresh, cold air requires heating up. Air changes account for the biggest heat loss when calculating the fabric heat loss from a building.
- **Air temperature** – air temperatures between 16 and 22 °C, dependent upon the type of activity being carried out, age of occupants and the level and quality of clothing. Air temperature at feet level, not greater than 3 °C below that at head level. Room surface temperatures not above the air temperatures.
- **Air velocity** – this is the speed at which the air travels within the building. If it travels too fast then a draught will be felt by the occupants; if there is no movement, then the air changes will not be satisfied. Airflow past the body is horizontal and at a velocity of between 0.2 and 0.25 m per second. A variable air velocity is preferable to a constant one.
- **Activity within the building** – applies to the type of work that is being carried out within the building. The more activity that is carried out by the occupants, the less heat will be required so temperatures may have to be adjusted to suit.
- **Clothing** – relates to the type of clothing worn by the occupants of the building. Obviously, the more clothes that are worn, the warmer the person will be. Elderly people tend to feel temperature variations more.
- **Age and health** – a major factor in heating design. The age of the occupants will have a direct effect on the type of the systems we install. Older people feel the cold more than young people, which may mean that designs will have to be modified, especially when it comes to the temperatures of key rooms such as lounge and bedroom.

The criteria used when selecting heating system and component types

All too often, installers will install what they know and not what is best for the property or the customer. Heating design requires a careful consideration of specific criteria if the heating system is to fulfil its potential to the customer as an efficient and economical system.

- **Customer's needs** – the overriding concern when designing any central heating system is that the customer requirements are satisfied. However, legislation will also need to be taken into account as the system must comply with the requirements of the regulations in force.
  Consultation with the customer is most important and the design needs to be approved with them before work can begin.
- **Building layout and features** – the positions of key components and appliances are often dependent on the layout and the features of the property. Many designers will look at the plans of the property and assess the best methods of heating the space. Consultation with the building owner is important as they may have their own ideas of what they want within a particular space or room.

- **Building size and the suitability of system** – whilst small domestic properties do not present many design problems with regard to pipe sizing, pipework layout and routing, larger systems may require careful planning utilising a multi-zoning approach, complete with separate timing and temperature control for each zone and may well present a necessity for a low loss header to cope with the load of each of the separate zones. The greater the number of heat emitters, the greater the possibility that a low loss header would need to be installed. Multiple boiler installations may also feature in larger properties and these would require careful and considered design practice if the system is to be economical and energy efficient when completed.

- **Energy efficiency** – in many cases, the controls that are installed on the system will determine whether a system is energy efficient or not. New controls such as night setback, delayed start and weather compensation will assist in making the most of good system design and this will help in creating an energy efficient and economical system. Again, zoning will help in this regard by limiting the heat in areas which are not occupied whilst maintaining a comfortable environment within the dwelling.

- **Environmental impact** – new technology developed in recent years means that systems no longer have to rely on carbon-rich fuels to maintain a warm living environment. Both carbon neutral fuels, such as biomass systems, and low and zero carbon options, such as air and ground source heat pumps means that the opportunity to lower the carbon fuel usage is available. Biomass is particularly good when a communal heating system serving a number of dwellings is being considered.

- **Occupancy and purpose of the building** – heating design must always consider what the building is to be used for, the number of occupants in the building and how often the building will be occupied. You must consider:
  - how many occupants there will be and whether they will be sedentary or physically active – activity leads to heat gain
  - what heat gains can be expected from processes, machinery and electrical equipment within the building

  - if all areas of the building need the same heating requirements or if there are areas with specific requirements.

These factors may determine or even limit the heating options available. The expected occupancy patterns (such as occupied 12 hours per day and no weekend occupancy) and future building usage and adaptability may also influence heating system choice.

- **Fuel availability** – not all areas of the UK have mains gas supply. In some rural areas, LPG, oil and even solid fuel/biomass are used for heating systems. Again, these choices may limit the heating system choice, especially where solid fuel is considered. Other considerations are:
  - access for fuel delivery
  - availability of the chosen fuel
  - delivery times
  - legislation and regulation of the local area.

- **Cost** – the bigger and more complex the system, the more it will cost. To a large degree, regulations dictate the type of system that is installed with regard to zoning and control, but other factors such as heat emitter type, boiler type, size and location and system design, all play a significant role in the cost of materials and labour usage costs.

> ## KEY POINT
> Where a conflict exists between the customer requests and the legislation, it must be explained to the customer that their wishes cannot be fulfilled as the system would not meet the requirements of the regulations.

IMPROVE YOUR ENGLISH

A successful consultation process means you obtain all the necessary information in order to design a system that fulfils the individual's needs. Clear and structured communication is vital if the system is to perform to the necessary requirements of the occupants. Remember to ask questions about the structure of the building such as cavity wall insulation. Assumptions made on the designer's part could mean that the system does not live up to expectations. Also, offer choices in boiler make and model, radiator design etc. Find out where you can install pipework and where the customer does not want pipework. These are all points that make a good design into a superb installation.

VALUES AND BEHAVIOURS

Utilising 'green' technology to its full advantage reduces carbon emissions and, as a consequence, the carbon footprint of the building. Other technological advances in solar heating and micro-CHP systems have also widened the scope for significant savings, not only to the customer, but to the environment as a whole.

INDUSTRY TIP

There are two further documents in the Approved Document L range. These are:
- L2A Conservation of fuel and power in new buildings other than dwellings
- L2B Conservation of fuel and power in existing buildings other than dwellings.

You would consult these documents if you were designing heating installations in commercial or industrial properties.

Information sources required for design work on central heating systems

The installation of central heating systems is governed strictly by various regulations, British Standards and recommendations.

The regulations

- The Building Regulations:
  - Approved Document L 2010 Conservation of fuel and power
    - L1A Conservation of fuel and power in new dwellings
    - L1B Conservation of fuel and power in existing dwellings
  - Approved Document F Ventilation
  - Approved Document J Combustion appliances and fuel storage system
- The Water Supply (Water Fittings) Regulations 1999
- The Gas Safety (Installation and Use) Regulations
- **BS 7671** The IET Wiring Regulations.

INDUSTRY TIP

Visit the Government's Building Regulation pages for access to all relevant Approved Documents, at www.gov.uk/government/collections/approved-documents

The British Standards

- **BS EN 12828:2003** Heating systems in buildings. Design for water-based heating systems.
- **BS EN 12831:2003** Heating systems in buildings. Method for calculation of the design heat load.
- **BS EN 14336:2004** Heating systems in buildings. Installation and commissioning of water-based heating systems.
- **BS EN 1264.1:2021** Floor heating. Systems and components. Definitions and symbols.
- **BS EN 1264.2:2021** Water-based surface embedded heating and cooling systems. Floor heating. Prove methods for the determination of the thermal output using calculation and test methods.
- **BS EN 1264.3:2021** Water-based surface embedded heating and cooling systems. Dimensioning.
- **BS EN 1264.4:2021** Water-based surface embedded heating and cooling systems. Installation.
- **BS EN 442:2003** Specification for radiators and convectors.

The recommendations

- The Domestic Building Services Compliance Guide 2021.
- The Central Heating System Specifications (CHeSS) 2008. This publication offers advice for compliance with good practice and best practice for the installation of central heating systems.
- Chartered Institute of Building Services Engineers (CIBSE) Domestic Heating Design Guide 2021. This was produced to assist heating engineers to specify and design wet central heating systems.

Manufacturer's technical instructions

Central heating systems and components must be installed, commissioned and maintained strictly in accordance with the manufacturer's instructions.

If the manufacturer's instructions are not available or have been misplaced, most manufacturers now have the facility to download the instructions from the company website.

Manufacturer's installation and maintenance instructions were covered in Chapter 1, Cold water systems, planning and design.

Verbal and written feedback from the customer

Verbal and written customer feedback was covered in depth in Chapter 1, Cold water systems, planning and design.

Taking measurements of building features in order to carry out design calculations

Central heating system calculations require diverse types of data to allow the correct design of central heating systems. Data needs to be taken and recorded and can be sourced from:

- visiting the site and taking the information direct from the building using a tape measure or a laser rule
- taking the information from working drawings and plans using a scale rule.

The type of data needed to successfully design heating systems:

- the length, width and height of the room
- the number of external walls and their construction
- the number of internal walls and their construction
- the construction of the floor
- the construction of the ceiling or roof
- the size and type of the windows
- the type and size of the doors
- the minimum outside temperature
- the internal temperature of the room
- the exposure of the building
- any relevant manufacturer's data such as radiator sizing sheets
- CIBSE guides.

The principles of heat loss and heat gain in dwellings

In this section, we will investigate the heat loss and heat gain of a building. Both principles rely on the steady state transfer of heat through the fabric of the building itself.

Heat loss

Heat loss from a building is the reason why we need central heating systems in our homes, offices, shops and factories. The heat the building loses will have to be replaced with more heat to maintain a comfortable temperature within the building itself. The principles of heat loss are known as 'steady state thermal characteristics' of a building. Heat loss from a building is measured in watts and occurs two ways:

- through the building fabric
- due to ventilation.

Heat loss through the fabric of the building

The thermal transmittance of heat from a building to the outside is known as U-values. U-values express, for the purposes of calculation, the rate of heat transfer through the building structure (its walls, floors, ceilings, roofs and windows) and because the construction of buildings varies so much due to different materials and construction methods, the U-values vary too. This means that on occasion the U-value for a particular building will need to be calculated from scratch.

IMPROVE YOUR MATHS

The calculation of a U-value for, say, an outside wall, is based upon the heat loss of each element or material used in the wall. Each element will have its own heat loss measured in watts per metre kelvin (W/mK) known as a K-value.

The units used to express U-values are watts per m² kelvin (W/m²K). This means that if a wall, for example, had a U-value of 1.0 W/m²K, for every degree of temperature difference between the air on the surface inside the wall and the air on the surface outside, 1 watt of heat would pass through any m². So, it follows that the smaller the U-value, the better the wall is at keeping the heat in.

Most materials have published K-values for the rate of thermal conductance through them and this is measured under specific conditions. These are then sub-divided into the required thickness of the material being used to obtain its R-value. This can be very confusing so let's take a look more closely:

K-value = The thermal conductivity of a material per metre (W/mK)

$$R\text{-value} = \frac{\text{Thickness of the material (mm)}}{\text{K-value (W/mK)}}$$

$$U\text{-value} = \frac{1}{\text{Total of all R-values for the wall (W/m}^2\text{K)}}$$

So, if an outside wall was constructed from 100 mm brick with 50 mm of cavity, 100 mm of thermalite block, 50 mm of mineral wool and finished with 12.5 mm plaster board, to calculate the U-value of the wall, the procedure is as follows.

Outside surface resistance 0.040 m² K/W (Rso)

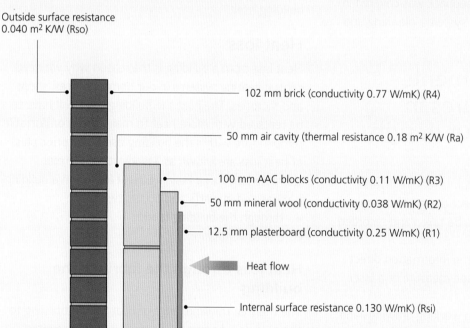

102 mm brick (conductivity 0.77 W/mK) (R4)

50 mm air cavity (thermal resistance 0.18 m² K/W (Ra)

100 mm AAC blocks (conductivity 0.11 W/mK) (R3)

50 mm mineral wool (conductivity 0.038 W/mK) (R2)

12.5 mm plasterboard (conductivity 0.25 W/mK) (R1)

Heat flow

Internal surface resistance 0.130 W/mK) (Rsi)

▲ Figure 3.26 Calculating U-values

K-values for the wall construction:

Brick = 0.77 W/mK

Cavity = 0.18 W/mK

Thermalite block = 0.11 W/mK

Mineral wool insulation = 0.38 W/mK

Plaster board = 0.25 W/mK

Because each material is of a specific thickness (such as 102 mm brick), the thickness of the material must be divided by the K-value to obtain the R-value:

Brick = 0.102 m ÷ 0.77 W/mK = 0.132 (R₄)

Cavity = 0.050 m ÷ 0.18 W/mK = 0.277 (Rₐ)

Thermalite block = 0.100 m ÷ 0.11 W/mK = 0.909 (R₃)

Mineral wool insulation = 0.050 m ÷ 0.038 W/mK = 1.315 (R₂)

Plaster board = 0.0125 m ÷ 0.25 W/mK = 0.050 (R₁)

There are two other elements that need to be added to the list. These are constants and never change.

These are air resistances to heat loss and are different for the inside and outside faces of the building.

Outside air resistance = 0.040 W/mK (R_{so})

Inside air resistance = 0.130 W/mK (R_{si})

The calculation of the U-value now looks like this:

$$\frac{1}{R_{si} + R_1 + R_2 + R_3 + R_a + R_4 + R_{so}} = W/m^2K$$

$$\frac{1}{0.130 + \frac{0.0125}{0.25} + \frac{0.050}{0.038} + \frac{0.100}{0.11} + \frac{0.050}{0.18} + \frac{0.102}{0.77} + 0.040}$$

$$= W/m^2K$$

$$\frac{1}{0.130 + 0.05 + 1.31 + 0.909 + 0.277 + 0.132 + 0.040}$$

$$= \frac{1}{2.848} = 0.35 \ W/m^2K$$

U-value = 0.35

So, the U-value for the outside wall is 0.35. This can now be used in heat loss calculations for the building.

INDUSTRY TIP

U-values express the rate of heat transfer through any element of a building – walls, roofs, floors and windows.

Fortunately, some U-values are published so this long calculation is not often needed. Approved Document L1A/B 2010 of the Building Regulations, Conservation of fuel and power, has traditionally set U-values and this is still the case where extensions to existing properties are concerned. For new buildings, however, a more holistic approach has been taken to prove energy efficiency. Whole building calculations that take a detailed look at the carbon emissions now have to be conducted. U-values are still included but they form only a small part. For new dwellings up to 450 m² standard assessment procedures (SAPs) are used to perform these calculations. For larger buildings the simplified building energy model (SBEM) is used.

Glass and glazing

Glass has a fairly high thermal conductivity of around 1.05 W/mK. However, the reason why so much heat is lost through the glass is that it is the thinnest part of a building. The heat loss can only be reduced by increasing the thickness of the glass or using it to form cavities like double and triple glazing. Table 3.7 shows the U-values for glazing.

▼ Table 3.7 Glazing U-values

Construction	U-value for stated exposure		
	Sheltered	Normal	Severe
Single window glazing	5.0	5.6	6.7
Double window glazing with air space			
25 mm or more	2.8	2.9	3.2
12 mm or more	2.8	3.0	3.3
6 mm or more	3.2	3.4	3.8
3 mm or more	3.6	4.0	4.4
Triple window glazing with air space			
25 mm or more	1.9	2.0	2.1
12 mm or more	2.0	2.1	2.2
6 mm or more	2.3	2.5	2.6
3 mm or more	2.8	3.0	3.3
Roof glazing light	5.7	6.6	7.9

Heat gain

Just as buildings lose heat during the winter months, they can also gain heat when the surrounding temperature is higher than that inside.

During the summer months, when the outside temperature is higher, a building will gain heat in two ways:
- from direct solar radiation, in other words direct sunlight
- from the surrounding warm air.

The passage of heat from outside to inside is almost the exact same principle as for heat loss. The difference is that it is not uniform. Different parts of the building will gain more heat due to the length of exposure to the Sun's solar radiation. In simple terms, the Sun rises in the east, swings out to the south during the day whilst gaining height until it reaches a peak height at noon, before slowly setting in the west. This height to angle ratio is called an 'azimuth' and is crucial when calculating solar gain for air conditioning and cooling purposes. The solar effect even exists in the winter when the angle of the Sun is much lower. On average, a south-facing wall in London will gain about 900 watts of heat per square metre (m²) every hour during the summer and around 300 watts in the winter. This reduces for east and west-facing walls as they are not exposed to the Sun for as long a period. A north-facing wall gains very little solar radiation because it never feels the heat of the Sun.

INDUSTRY TIP

The intensity of the Sun during the summer explains why solar hot water heating works so well in the summer. Solar hot water panels can give as much as 60 per cent of a household's yearly hot water needs, free of charge, just from the heat of the Sun!

Heat gain also occurs because of the warmth of the air. This is known as sol-air temperature and occurs because the Sun has warmed the air surrounding the dwelling. The same principle occurs inside the dwelling when heat is transferred from one room to another when the two rooms have different internal temperatures. If one room is at 21°C and an adjoining room is at 18°C then heat will be gained by one and lost by the other. The rate at which this heat transfer takes place and the amount of heat lost or gained will depend on the U-value of the wall.

Dwellings also gain heat from other sources, such as electrical equipment and lighting, as these give off heat when they are on. A 150 watt bulb, for instance will give off just that: 150 watts of heat. Human beings also contribute to heat gains in a building. A human will emit around 115 watts of sensible heat and around 50 watts of latent heat when at rest. This increases with physical exercise.

INDUSTRY TIP

Sensible and latent heat were discussed in Book 1, Chapter 3, Scientific principles.

IMPROVE YOUR MATHS

Heat loss due to ventilation – air change rate

Approved Document F of the Building Regulations sets the provisions for the ventilation of a building. They set out the regulations required to restrict the build-up of moisture and pollutants that would otherwise present a danger to health. The flow of air through the building results in heat loss, the lost heat having to be replaced. Ventilation rates are usually quoted as 'air changes per hour', which are defined as the volume of air moving through the room every hour divided by the volume of the room itself. The replacement air will be heated by the central heating system and is calculated by multiplying the volume of the room by the air change rate by the temperature rise and by the ventilation factor. Thus, the loss of heat due to air change can be calculated by:

Ventilation heat = room volume (m³)
loss (watts)   × air change rate (qty)
       × temperature difference (°C)
       × ventilation factor (W/m³K)

The ventilation factor is taken as the specific heat of air at 20°C, which is 0.33 W/m³K and is used to calculate the heat loss to the air changing within the rooms due to infiltration or mechanical ventilation.

The specific mass (density of air) at 20°C is 1.205 kg/m³ and the specific heat capacity of air at 20°C is 1.012 kJ/kgK. Therefore, the quantity of heat required to raise unit volume through 1 kelvin is:

$$1.012 \times 1.205 = 1.219 \text{ kJ/m}^3$$

In order to be able to apply this heat loss of air by volume, the unit of kilojoules needs to be converted to joules and from hours to seconds because 1 watt is equal to 1 joule per second. Therefore:

$$\frac{1.219 \times 1000}{3600} = 0.3386 \text{ J/s/m}^3\text{K or } 0.3386 \text{ W/m}^3\text{K}$$

Hence the air change U-value of 0.33 W/m³K.

The heating requirements of rooms in dwellings when designing a central heating system

Now that the U-values have been determined for the heat loss calculations, the information needed can be collated to allow the calculations to be performed. The information that is required is listed below:

- room size
- temperature required – indoor to outdoor
- air change rate.

The room size

The size of the room may seem like an obvious piece of information but if the room is measured inaccurately then the calculations will also be inaccurate. It follows that the bigger the room, the greater the heat loss. The heat emitter positioning also becomes troublesome as the room size increases. In most cases it is better to consider the use of two or more strategically positioned heat emitters in larger rooms rather than using just one large one. This subject will be discussed later in this chapter.

Key information is required about each room before heat loss calculations can be performed, such as:

- the length, width and height of the room
- the number of external and internal walls and their construction
- the number and type of internal walls and party walls
- the type of floor and ceiling
- the type and size of the windows and doors.

The information required may be noted during a site visit or it may be taken from working drawings and plans.

The room temperature and the outside temperature

Internal design temperatures

The design temperatures within dwellings are based upon the type and usage of the room. Internal design temperatures should be chosen to ensure satisfactory comfort conditions. In order to achieve a comfortable living condition within the room and to enable the heat loss calculations to be completed, it is necessary to select a design room temperature that needs to be achieved when the central heating system is operating. There are three aspects that must be considered to achieve a comfortable condition in a room:

- temperature
- humidity
- ventilation rate.

The temperature of the heated space during the winter at which humans feel comfortable falls within a temperature range of 19 and 23 °C when wearing normal clothing. Some people may disagree when subjected to temperatures in the higher or lower end of the acceptable range. The exact comfort temperature often results from the physical condition of the individual. Because of these varying temperature ranges, exact room temperatures need to be discussed and agreed with the customer before any design work commences.

Too much heat in a room can be controlled by thermostatic means but a shortfall in temperature is rather more difficult to assess and may result in a costly re-design of the heat losses. Design temperatures become even more critical if the customer is to be the permanent resident within the dwelling, as is the case with domestic dwellings. If the customer is an architect or a developer, temperatures may have been specified as part of the design brief or specification.

Special considerations may be needed where the customer is elderly or infirm. The table below illustrates the risk that the elderly and infirm face during the winter months.

▼ Table 3.8 The effects of low temperatures on the elderly and infirm

Temperature (°C)	Effects
24	Top end of the comfort condition
21	Recommended living condition
<20	Below 20 °C, the risk of death begins
18	Recommended bedroom temperature
16	Resistance to respiratory diseases is weakened
12	More than two hours at this temperature, the risk from raised blood pressure, heart attack and strokes
5	Significant risk of hypothermia

For normal circumstances recommended temperatures are given in Table 3.9.

▼ Table 3.9 Design room temperatures

	Lounge/sitting room	Living room	Dining room	Kitchen	Breakfast room	Kitchen/breakfast room	Hall	Cloak room	Toilet	Utility room	Study	Games room	Bedroom	Bedroom/en-suite	Bed sitting room	Bedroom/study	Landing	Bathroom	Dressing room	Storeroom
Design temperature (°C)	21	21	21	18	21	21	18	18	18	18	21	21	18	18	21	21	18	22	21	16

The temperatures listed in the table represent normal living conditions and working conditions. For properties that are designed for the elderly or the infirm, these temperatures should be increased by one or two degrees for each given application and usage. It will also be seen that the temperatures vary depending on the intended use of the room. This is simply because the body temperature of an individual and their subsequent comfort levels will vary depending on their activity (such as resting, sleeping, bathing, etc.). Hence the wide range of temperatures quoted.

Some building services engineers design heating systems for residential buildings with a constant temperature throughout, irrespective of the room type or usage. This single temperature approach reduces the amount of calculations required, as it uses the 'whole house' assessment method but this can lead to over-sizing of the heating system and an unnecessary increase in overall energy usage.

External design temperatures

The successful design of any central heating system is based on the fact that the dwelling is maintained at a certain specified temperature based upon the prevailing (at the time) external temperature. It follows that calculations must be based on a realistic external temperature that can be expected for the region during the coldest months. Research shows that the temperatures vary greatly in the UK with the south-eastern corner being the mildest throughout the year and the north of Scotland being the coldest.

INDUSTRY TIP

For further information on climatic conditions in the UK, reference should be made to CIBSE guides A and J, and local meteorological data.

The UK often sees temperatures between −2 and −6°C during the winter and, on occasions, it has been known to drop as low as −15°C or lower. However, it would be uneconomical to design a central heating system using −15°C as it occurs so infrequently. The resultant over-sizing would encourage the boiler to 'hunt' on its thermostat and increase the risk of boiler breakdown. Therefore, a more logical external design temperature should be used based upon the lowest average temperature for the location. Another possibility for determining the external design temperature is to use the lowest two-day mean temperature, which has been registered ten times over a twenty-year period.

Table 3.10 shows the recommended base temperatures for the UK but local knowledge can also be applied, as severe exposure conditions are not always obvious.

▼ Table 3.10 Recommended UK base temperatures

Type of building	Exposure	Base design temperature (°C)
House and multi-storey buildings with solid intermediate floors up to and including the 4th floor: England and Wales	Normal, sheltered in towns and cities surrounded by other buildings	−1
House and multi-storey buildings with solid intermediate floors up to and including the 4th floor: Scotland, Northern England and Northern Ireland	Normal, sheltered in towns and cities surrounded by other buildings	−3
Single-storey houses	Normal	−3
Houses in coastal areas or at high altitude including exposed rural areas	Exposed	−4
Multi-storey buildings with solid intermediate floors up to and including the 4th floor and single-storey in coastal or exposed rural areas	Exposed	−5

Air change rates

It is necessary to prevent the air in the room from becoming stale and to prevent the onset of moisture problems and mould growth. As the air change occurs, the heat in the room is lost by warm air leaving the room and cold air entering.

Air enters the building because of the poor seal in the building structure, usually through airbricks, ventilators, flues and chimneys. Flues and chimneys often cause excessive air change because warm air escapes through the flue, due to the fact it is less dense. As cold air enters, it forces out more warm air. This effect is known as the stack effect.

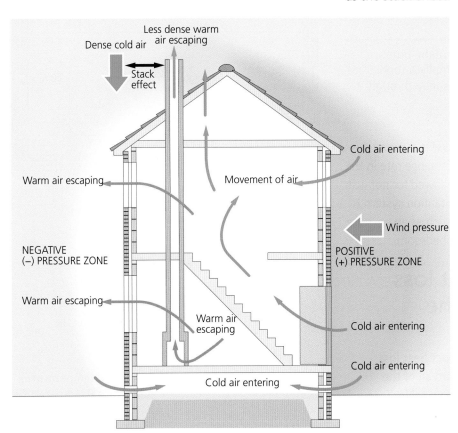

▲ Figure 3.27 The stack effect

▼ Table 3.11 Air change rates

	Lounge/sitting room	Living room	Dining room	Kitchen	Breakfast room	Kitchen/breakfast room	Hall	Cloak room	Toilet	Utility room	Study	Games room	Bedroom	Bedroom/en-suite	Bed sitting room	Bedroom/study	Landing	Bathroom	Dressing room	Storeroom
Air change rate	1.5	1.5	1.5	2.0	2.0	2.0	2.0	2.0	2.0	1.5	1.5	1.5	1.0	2.0	1.5	1.5	2.0	2.0	1.5	1.0

The **air change rates** listed below are for modern buildings. When calculating heat losses for older dwellings, there is a case for increasing the rates to allow for ill-fitting doors and windows because these will affect the heat loss from the building.

The exact amount of air infiltration is difficult to assess and because of this, exact design temperatures are often difficult to predict. The air changes listed in Table 3.11 are arrived at by methods that are verifiable or by demonstrable means and can be considered accurate for new and well-maintained buildings.

KEY TERM

Air change rate: a measure of how many times the air within a defined space (normally a room or a house) is replaced per hour, usually through natural ventilation.

Where mechanical ventilation is installed in a room, it is advisable to allow for the increased air change rate in the heat loss calculations. This should be allowed for not just in the room where the mechanical ventilation system is fitted but also any connecting rooms such as an en-suite bathroom with an extractor fan and a bedroom.

Calculating the heat loss from a dwelling – the tabulation method

The easiest and quickest way to conduct heat loss calculations is by using a table to record all of the figures. Each room would require its own table. Take a look at the example in Table 3.12 on page 228. You will see that it is divided into various columns. Not all of these columns will be used for every room.

Each room needs to be dealt with separately. We will look at a simple example which should help in understanding how the heat losses are put together.

The simple room shown in Figure 3.28 has four outside walls and heat loss through both the floor and roof. Intermediate rooms in a large building may not have these losses unless there is a temperature difference between the rooms.

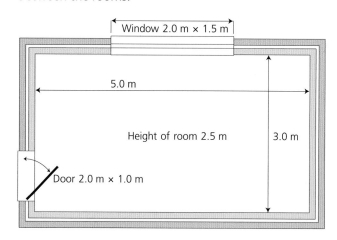

▲ Figure 3.28 Plan of a simple room

The U-values required are as follows:

External walls:	0.35
Floor:	0.25
Roof:	0.25 (flat roof)
Windows:	2.9
External door:	2.9 (the same as the windows)
Air change factor:	0.33
Temperature difference:	21 (internal temperature) −−3 (external temperature) = 24 °C

The first point is that there are no internal walls to worry about. As all walls are external the calculation is quite straightforward. So, the calculations revolve around the following elements:

1 Calculate the air change heat loss.
2 Calculate the external wall heat loss.
3 Calculate the glazing heat losses.
4 Calculate the floor and roof heat loss.
5 Calculate any adjustments due to exposure and intermittent heating loads.

IMPROVE YOUR MATHS

First, let's concentrate on Point 1.

Point 1 – air change heat loss

The room is 5 m × 3 m × 2.5 m. This gives a volume of 37.5 m^3. When this figure is multiplied by the temperature difference (24 °C), the number of air changes (2) and the air change factor (0.33), the total becomes 594 watts. This means that because there are two air changes every hour, the room will lose 594 watts of heat during these changes:

$$37.5 \times 24 \times 2 \times 0.33 = 594 \text{ watts}$$

Now look at Point 2.

Point 2 – external wall heat loss

This can be a little involved. If we laid all of the external walls flat, we would end up with the following:

$$5 + 5 + 3 + 3 = 16 \text{ m}$$

This figure must now be multiplied by the height of the wall:

$$16 \times 2.5 = 40 \text{ m}^2$$

But this figure is not much use to us as it is. There are windows and doors that have a greater heat loss than the wall so these MUST be deducted BEFORE we calculate the external wall heat loss. The window and door heat loss will be dealt with separately. So first calculate the area of the windows and doors:

$$\text{Window: } 2 \text{ m} \times 1.5 \text{ m} = 3$$
$$\text{Door: } 2 \text{ m} \times 1 \text{ m} = 2$$
$$= 5 \text{ m}^2$$

This figure can now be deducted from the external wall total:

$$40 - 5 = 35 \text{ m}^2$$

So, the heat loss from the external walls is:

$$35 \times 24 \times 0.35 = 294 \text{ watts}$$

Progress to Point 3.

Point 3 – glazing heat loss

This part of the calculation is made easy by the fact that we have already calculated the areas of the glazing, so it's a straightforward calculation. Also, the same U-value is used for both window and door because they are made from the same material. This

isn't always the case and you must check before doing the calculation:

$$\text{Window heat loss: } 3 \times 24 \times 2.9 = 208.8 \text{ watts}$$
$$\text{Door heat loss: } 2 \times 24 \times 2.9 = 139.2 \text{ watts}$$

The penultimate stage, Point 4.

Point 4 – heat loss from the floor and roof/ceiling

Because these use the same area, the calculation, again, becomes very straightforward. The area of the room is:

$$5 \text{ m} \times 3 \text{ m} = 15 \text{ m}^2$$

So, now multiply all of the figures together for both floor and ceiling:

$$\text{Floor: } 15 \times 24 \times 0.25 = 90 \text{ watts}$$
$$\text{Ceiling/roof: } 15 \times 24 \times 0.25 = 90 \text{ watts}$$

So, the total heat loss for the room is the sum of all of the calculated elements:

$$594 + 294 + 208.8 + 139.2 + 90 + 90 = 1416 \text{ watts}$$

Point 5, the last stage, is the calculation of any adjustments.

Point 5 – calculating adjustments

We are assuming that the room is in an exposed location. This would mean an increase in the total of 10 per cent in case of severe weather. Furthermore, we are also assuming that the heating will be on intermittently. In other words, it will only be on at certain times of the day or when external controls call for heat such as a frost stat. This will result in an extra 15 per cent increase in the total to cope with the intermittent heating patterns:

$$\text{Exposed location @ 10\% } = 1416 \times 0.10 = 141.6 \text{ watts}$$
$$\text{Intermittent heating load @ 15\% } = 1416 \times 0.15$$
$$= 212.4 \text{ watts}$$
$$= 1416 + 141.6 + 212.4 = 1770 \text{ watts}$$

Therefore, the heat loss from the room when all adjustments have been made is 1770 watts.

The calculation is probably more understandable when viewed in a table format. Take a look at Table 3.12. It must be realised, however, that because all of the walls were external and the structure was a single-storey, there is no data for internal wall structures or where there is another room above. This will be dealt with next.

▶ Table 3.12 Heat loss calculation table for Figure 3.28

Design temperature (°C)	21	Outside temperature (°C)	−3	Design temperature difference (°C)	24		
Room volume	**Room dimensions**			Volume of room (m³)	Air change factor (W/m³K)	Temperature difference (°C)	Air changes per hour	Total heat loss watts (W)
	Length (m)	Width (m)	Height (m)					
Total room	5	3	2.5	37.5	0.33	24	2	594

Fabric heat loss	Length (m)	Width (m)	Fabric area (m²)	U-value (W/m³K)	Temperature difference (°C)	Gain	Loss	Total heat loss watts (W)
Floor	5	3	15	0.25	24		90	90
Ceiling/roof	5	3	15	0.25	24		90	90
External glazing	2	1.5	3	2.9	24		208.8	208.8
External doors	1	2.0	2	2.9	24		139.2	139.2
External wall	16	2.5	40					
External wall (minus glazing)	40 − (3.0 + 2.0) = 35		35	0.35	24		294	294
Internal wall 1								
Internal wall 2								
Internal wall 3								
Internal wall 4								
Party wall								
Design heat loss (total watts for all elements)								1416

Additional factors	Y/N	% add	Total
Exposed location	Y	10	141.6
Intermittent heating	Y	15	212.4
Grand total heat loss for room (W)			1770

IMPROVE YOUR MATHS

Heat loss and heat gain from the internal walls and rooms above and below

The first example we looked at was a building with four external walls. Now we must advance to a room that contains both external and internal walls.

The method of calculation is exactly the same, but with addition of extra walls that may mean a loss of heat from our example rooms or, indeed, may give a heat gain! Take a look at the example rooms shown in Figure 3.29 and Table 3.13.

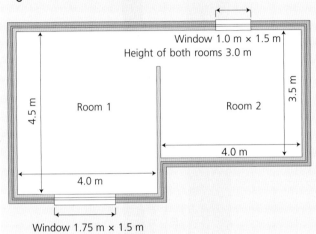

Window 1.0 m × 1.5 m
Height of both rooms 3.0 m
4.5 m
Room 1
Room 2
3.5 m
4.0 m
4.0 m
Window 1.75 m × 1.5 m

▲ Figure 3.29 Plan of Rooms 1 and 2

▼ Table 3.13 Room data for Rooms 1 and 2

Element	Room 1	Room 2
Temperature	18 °C	21 °C
Outside temperature	−3 °C	−3 °C
Temperature difference	21 °C	24 °C
Number of exterior walls	4	3
Number of interior walls	1	1
U-value of exterior walls	0.35	0.35
U-value of interior walls	1.72	1.72
U-value of the glazing	2.9	2.9
U-value of the floor	0.25	0.25
U-value of the roof/ceiling	0.16	0.16
Number of air changes	1	2

The rooms are of normal exposure.
The heating is intermittent.

We will deal with Room 1 first. The first thing to note is that the data states that there are four outside walls

and one internal wall. This due to the fact that Room 2 is slightly shorter in length than Room 1 and leaves a small amount of wall as an external structure. We cannot ignore this and so it must be treated as a small external wall.

The dimensions for the external wall of Room 1 are:

$$4.5 + 4 + 4 + 1 = 13.5 \text{ m}$$

Multiply by the height of the room:

$$13.5 \times 3 = 40.5 \text{ m}^2$$

The window size is:

$$1.75 \text{ m} \times 1.5 \text{ m} = 2.625 \text{ m}^2$$

Deduct the window from the wall:

$$40.5 - 2.625 = 37.875 \text{ m}^2$$

Now using the formula previously shown, we can enter the details on the table and calculate the external wall heat loss (see Table 3.14). We can also calculate the heat loss for the glazing.

Look at the internal wall dimensions. The wall is 3.5 m in length including the door opening and 3 m high. Some designers will work out the heat loss for the door, but for this chapter it will be treated as part of the wall for simplification. The most important factor here is that the room temperatures are different. The adjoining room requires a warmer temperature at 21 °C. This means that Room 1, at 18 °C, will GAIN heat from the Room 2 next door and so any heat gain must be DEDUCTED from our calculation and NOT added. The temperature difference between Room 1 and Room 2 is 3 °C. Take a look at Table 3.14.

When the calculation for the internal walls is conducted, it will be seen that there is a gain of 54.18 watts. This is highlighted in red to remind us to deduct it and not add it. If the temperatures were the same, then the internal wall would be ignored purely because there is neither a heat loss nor a heat gain. The rest of the table can now be completed.

Now look at Room 2. The heat gain that was seen in Room 1 now becomes a heat loss in Room 2 because Room 1 is a cooler 18 °C and so this must be allowed for in the heat loss calculations. All other factors remain the same and so the heat loss for the room can be calculated in the normal way.

→

▶ Table 3.14 Heat loss calculation table for Figure 3.29, Room 1

Design temperature (°C)	18	Outside temperature (°C)		-3	Design temperature difference (°C)		21	
Room volume	**Room dimensions**			**Volume of room (m³)**	**Air change factor (W/m³K)**	**Air changes per hour**	**Temperature difference (°C)**	**Total heat loss watts (W)**
	Length (m)	Width (m)	Height (m)					
Total room	4.5	4	3	54	0.33	1	21	374.22
Fabric heat loss	Length (m)	Width (m)	Height (m)	Fabric area (m²)	U-value (W/m³K)	Temperature difference (°C)	Gain	Loss
Floor	4.5	4		18	0.25	21		94.5
Ceiling/roof	4.5	4		18	0.16	21		60.48
External glazing	1.75		1.5	2.625	2.9	21		159.86
External doors								
External wall	13.5		3	40.5				
External wall (minus glazing)	40.5 – 2.625			37.875	0.35	21		278.38
Internal wall 1	3.5		3	10.5	1.72	3	54.18	–54.18
Internal wall 2								
Internal wall 3								
Internal wall 4								
Party wall								
Design heat loss (total watts for all elements)								913.26

Additional factors		Y/N	% add
Exposed location		N	10
Intermittent heating		Y	15
		Total	136.99
Grand total heat loss for room (W)			1050.25

▶ Table 3.15 Heat loss calculation table for Figure 3.29, Room 2

Design temperature (°C)	21	Outside temperature (°C)	−3	Design temperature difference (°C)		24	
Room volume	**Room dimensions**			Volume of room (m³)	Air change factor (W/m³K)	Temperature difference (°C)	Air changes per hour	Total heat loss watts (W)

Room volume	Length (m)	Width (m)	Height (m)	Volume of room (m³)	Air change factor (W/m³K)	Temperature difference (°C)	Air changes per hour	Total heat loss watts (W)
Total room	4.0	3.5	3	42	0.33	24	2	665.28

Fabric heat loss	Length (m)	Width (m)	Height (m)	Fabric area (m²)	U-value (W/m³K)	Temperature difference (°C)	Gain	Loss
Floor	4.0	3.5		14.0	0.25	24		84
Ceiling/roof	4.0	3.5		14.0	0.16	24		53.76
External glazing	1.0		1.5	1.5	2.9	24		104.4
External doors								
External wall	11.5		3	34.5				
External wall (minus glazing)	34.5 − 1.5 = 33.0			33.0	0.35	24		277.2
Internal wall 1	3.5		3.0	10.5	1.72	3		54.18
Internal wall 2								
Internal wall 3								
Internal wall 4								
Party wall								
Design heat loss (total watts for all elements)								1238.82

Additional factors	Y/N	% add	Total
Exposed location	N	10	
Intermittent heating	Y	15	185.82
Grand total heat loss for room (W)			1424.64

From the table, it can be seen that the heat gain from Room 1 is exactly the same as the heat loss in Room 2. The total heat loss for the room is greater simply because the air change rate was doubled in Room 2 and the temperature difference between the internal temperature and external temperature was higher.

These two examples show only heat loss. They do not tell us the size of heat emitter each room requires. The totals here require further work to convert them into heat emitter emissions and these will be discussed later in the chapter.

IMPROVE YOUR MATHS

The diagram in Figure 3.30 shows the floor plan for a simple single-storey dwelling situated in an exposed location. Using the knowledge you have gained, calculate the heat losses for each room when the external temperature is −5 °C. All other factors and U-values remain as the examples previously attempted.

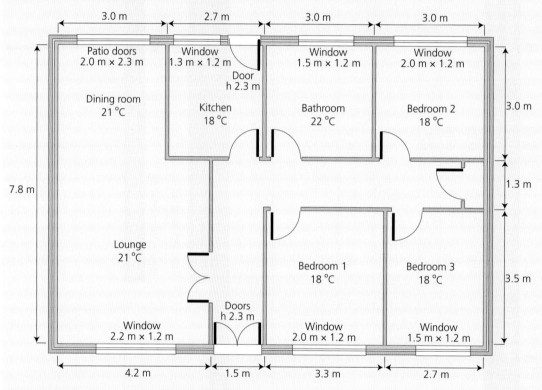

▲ Figure 3.30 Plan of a simple single-storey dwelling

A blank heat loss calculation table is provided (Table 3.16). You will need one table for every room.

▼ Table 3.16 Blank heat loss calculation table

Design temperature (°C)	Outside temperature (°C)			Design temperature difference (°C)				Total heat loss watts (W)
Room volume	**Room dimensions**			Volume of room (m³)	Air change factor (W/m³K)	Temperature difference (°C)	Air changes per hour	
	Length (m)	Width (m)	Height (m)					
Total room								
Fabric heat loss	Length (m)	Width (m)	Height (m)	Fabric area (m²)	U-value (W/m³K)	Temperature difference (°C)	Gain	Loss
Floor								
Ceiling/roof								
External glazing								
External doors								
External wall								
External wall (minus glazing)								
Internal wall 1								
Internal wall 2								
Internal wall 3								
Internal wall 4								
Party wall								
Design heat loss (total watts for all elements)								

Additional factors	Y/N	% add	Total
Exposed location	Y	10	
Intermittent heating	Y	15	
Grand total heat loss for room (W)			

The methods of sizing the central heating pipework and central heating circulating pumps

There are three very different methods for sizing the pipework and circulators for central heating systems and each one has its advantages and disadvantages:

- **Longhand calculations** – this is by far the most accurate of all heating design methods, as U-values can be calculated to suit the building structure where the installation is to take place. This ensures that an economical and environmentally friendly design of the building heating system can be guaranteed. Longhand calculations involve the use of external data such as the tables for flow rate and resistance published by the Chartered Institute of Building Services Engineers (CIBSE) in the CIBSE design guide C.

  Circulating pump size is fairly simple to calculate as most of the work is done during the pipe sizing calculation. It should be remembered that the size of the circulator will be based around the resistance in the index circuit. This will be discussed later in this chapter.

- **Computer aided central heating design programs** – computer design programs have become more widespread over the last few years as they offer not only an accurate sizing method, but also the ability to print out the finished design complete with calculations for presentation to the customer. Pipesizing computer programs are widely used by building services engineers for the design of heating, air conditioning, lighting and hot and cold water services.

- **Mears calculator** – the Mears calculator is a circular slide rule that enables the calculation of central heating systems. The key functions of the calculator are:
  - calculation of heat loss
  - calculation of heat emitter size and output
  - calculation of hot water load
  - calculation of boiler size
  - calculation of heating pipework size.

  The calculator is usually used as an on-site tool because it allows the quick calculation of heat losses and heat emitter size, but it can be a little inaccurate for modern dwellings as the U-values it uses are higher than modern U-values.

The different models available allow the calculation of single dwelling heat losses, conservatory heat losses and the heat loss from industrial heating systems.

KEY POINT

The index circuit of a central heating system is the circuit which has the biggest resistance to flow and NOT the circuit with the biggest heat load. The index circuit is always taken as the circuit with the longest run of pipework because the circulating pump will have to overcome the resistance due to length of pipe.

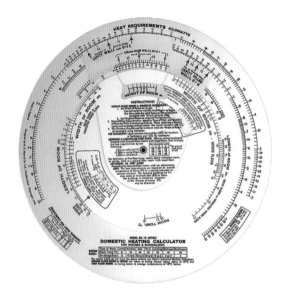

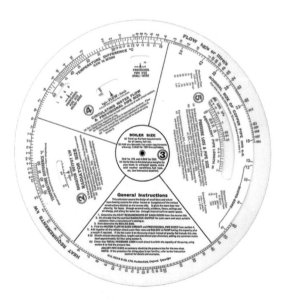

▲ Figure 3.31 The Mears calculator

The selection criteria for central heating boilers

The selection of central heating boilers is often based on experience, efficiency and cost. Many installers will install what they are familiar with simply because of the cost and the ease of installation. However, boiler selection should be based upon the suitability of the boiler in relation to the system on which it is installed. The points that should be considered are:

- **Space heating load** – will the boiler satisfy the space heating load? Or, put more simply, will it warm the dwelling to the design temperatures? It is uneconomical to oversize a boiler. It may well heat the system but the water content may be higher, which will require more heat, therefore more fuel and in consequence will cost more to run and give off more CO_2 emissions than would otherwise be necessary. Similarly, if the boiler is undersized, then the boiler will struggle to meet the demands of the system. This will mean that the boiler is working far too hard and could fail to reach the desired temperature as quickly as it should. Boiler cycling would increase and fuel usage would be high.
- **Hot water heating load** – will the boiler satisfy the hot water demand? Combination boilers are not always the best choice, especially where hot water demand is considered. They may well satisfy the heating requirements but will they deliver the amount of hot water required at the desired flow rate and temperature? The 'one boiler fits all' approach is the wrong approach to take, especially where multiple hot water outlets are to be supplied or long distances exist between the appliance and the hot water outlets.

  When storage of hot water is considered, then the kW required to heat it must firstly be calculated properly and then added to the total space heating load, otherwise the boiler will fail to meet the total heat demand, especially where a hot water priority system is installed, such as the Honeywell W-plan system.
- **Heat loss from pipework** – uninsulated heating pipes will lose heat. The amount of heat loss will depend on the material that the pipe is made from and the temperature of the water. On average, a copper 15 mm pipe will lose around 40 watts per metre (W/m) when uninsulated and 8 watts

if the pipe is insulated and the water is at 75 °C. An uninsulated pipe distributes the heat loss into its surrounding and, as such, will help to warm the dwelling. However, the heat loss contributes to the water losing heat, so by the time it reaches the heat emitter, it has already cooled from when it had left the boiler. In most cases, the heat loss from the pipes is acceptable as insulating all of the heating pipes could lead to the boiler overheating and the undesirable operation of the high limit thermostat on the boiler due to an insufficient temperature difference between the flow and return heating pipes as they leave and return from and to the boiler. As a rough guide, a 5 to 10 per cent allowance can be made on the boiler size for uninsulated pipes.
- **Factors for intermittent heating** – central heating systems for domestic dwellings are normally operated on time clocks and programmers that allow for the system to run at specific times of the day. When the heating system is off, the rooms will lose heat very quickly creating a noticeable drop in temperature. Because of this, an increase of around 15 per cent on the total heating load is advisable to counter the effects of intermittent heating. This would amount to a slight increase in heat emitter size across all rooms ensuring a quick heat up of the room to bring the rooms back to the desired temperature in cold weather.

VALUES AND BEHAVIOURS

All too often, the customer is not told of the disadvantages with combination boilers. It's good practice for you to ensure they are well informed on the positive and negatives of all systems available to them.

Sizing expansion vessels for sealed heating systems and feed and expansion cisterns for open vented systems

Water, when it is heated, expands. The amount of expansion will depend on the temperature of the water. At atmospheric pressure, water is at its greatest density at 4 °C. At this temperature, 1 m³ of water has a mass of 1000 kg. From this point forwards as the water temperature increases, 1 m³ will lose density. At 100°C, 1 m³ of water has a mass of 958 kg or an expansion rate of 4 per cent. The densities of water at various temperatures and pressures are shown in Table 3.17.

▼ Table 3.17 The densities of water at various temperatures and pressures

Temperature (°C)	Density (kg/m³)	Specific volume 10⁻³ (m³/kg)
0 (ice)	916.8	
0.01	999.8	1.00
4 (maximum density)	1000.0	
5	999.9	1.00
10	999.8	1.00
15	999.2	1.00
20	998.3	1.00
25	997.1	1.00
30	995.7	1.00
35	994.1	1.01
40	992.3	1.01
45	990.2	1.01
50	988	1.01
55	986	1.01
60	983	1.02
65	980	1.02
70	978	1.02
75	975	1.03
80	972	1.03
85	968	1.03
90	965	1.04
95	962	1.04
100	958	1.04

The expansion of the water in a central heating system, if not accommodated, will lead to an increase in system pressure and possibly component or appliance failure as a result. In a sealed heating system, the expansion of water is accommodated in an expansion vessel;

in an open vented system, the feed and expansion cistern accommodates the expansion of water. Both of these vital parts of the installation will need to sized correctly.

Sizing an expansion vessel

There are several methods for sizing expansion vessels. All methods must take into account the volume of cold water in the system and the amount by which it will expand in order to reach its design temperature. The CIBSE method is shown below.

If the system volume is known, expansion vessels can be sized with the following formula:

$$V = \frac{eC}{1 - \dfrac{P_1}{P_2}}$$

V = the total volume of the expansion vessel

C = the total volume of water in the system in litres

P₁ = the fill pressure in bars absolute (gauge pressure + 1 bar)

P₂ = the setting of the pressure relief valve + 1 bar

e = the expansion factor that relates to the maximum system requirements

Expansion factor 'e'	Temperature (°C)
0.0324	85
0.0359	90
0.0396	95
0.0434	100

'e' can be found from the formula:

$$e = \frac{d1 - d2}{d2}$$

Where:

d_1 = density of water at filling temperature (kg/m³)

d_2 = density of water at operating temperature (kg/m³)

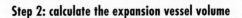

IMPROVE YOUR MATHS

Sizing an expansion vessel – CIBSE method: example 1

A sealed central heating system has a total water volume of 600 litres. The pressure of the water main is 1.5 bar and the pressure relief maximum pressure is 6 bar. The system is designed to operate at a maximum temperature of 80 °C, which means the expansion factor will have to be calculated. The fill temperature of the water is 10 °C.

Calculate the expansion factor using:

$$e = \frac{d_1 - d_2}{d_2}$$

Calculate the expansion vessel volume using:

$$V = \frac{eC}{1 - \frac{P_1}{P_2}}$$

Step 1: calculate the expansion factor 'e'

The temperature of the fill water is 10 °C with a density of 999.8 kg/m³. The maximum operating temperature is 80 °C with a density of 972 kg/m³. Therefore the 'e' factor is:

$$\frac{999.8 - 972}{972} = 0.0286$$

Step 2: calculate the expansion vessel volume

V = the total volume of the expansion vessel

C = 600 litres

$P_1 = 1.5 + 1$

$P_2 = 6 + 1$

e = 0.0286

$$V = \frac{eC}{1 - \frac{P_1}{P_2}}$$

Therefore, the expansion vessel volume is:

$$V = \frac{0.0286 \times 600}{1 - \frac{2.5}{7}}$$

$$= \frac{17.16}{1 - 0.357}$$

$$= \frac{17.16}{0.643}$$

$$= 26.68 \text{ litres or } 4.4\%$$

So, the expansion vessel volume is: 26.68 litres or 4.44% of total system volume.

IMPROVE YOUR MATHS

A sealed central heating system has a total water volume of 400 litres. The pressure of the water main is 1 bar and the pressure relief maximum pressure is 4 bar. The system is designed to operate at a maximum temperature of 85 °C, which means the expansion factor will have to be calculated. The fill temperature of the water is 4 °C.

Calculate the expansion factor using:

$$e = \frac{d_1 - d_2}{d_2}$$

Calculate the expansion vessel volume using:

$$V = \frac{eC}{1 - \frac{P_1}{P_2}}$$

Sizing a feed and expansion cistern

Sizing a feed and expansion cistern for open vented systems is quite straightforward. If the expansion of water from 4 to 100 °C is 4 per cent, then an allowance of 4 per cent of the total system volume can be used.

Therefore, if an open vented central heating system has a total system volume of 400 litres, the expansion of water is:

$$400 \times 0.04 = 16 \text{ litres}$$

This can be checked using the following formula:

$$V = C\left[(v_1 \div v_0) - 1\right]$$

Where:

V = required expansion cistern volume (litres)

C = water volume in the system (litres)

v_0 = specific volume of water at initial (4 °C) temperature (1 m³/kg)

v_1 = specific volume of water at operating (85 °C) temperature (0.968 m³/kg)

$V = 400 \times [(1.0 - 0.968) - 1] = 13.223$ litres

IMPROVE YOUR MATHS

Calculate the expansion volume of water for a central heating system containing 600 litres of water. The operating temperature is 90 °C and the initial water temperature is 4 °C.

Calculating the size of central heating components

Earlier in this chapter we looked at how to calculate the heat loss from a dwelling. This was only the first step in designing a successful heating system. Further calculations are required, and these are listed below:

- heat emitter size
- hot water heating load
- boiler size
- pipe size
- pump size.

These follow a logical pattern and will be discussed in turn.

Calculating heat emitter size

Calculating the size of the heat emitter is not just a case of performing the heat loss calculations and selecting a heat emitter size. Adjustments have to made using correction factors. Radiator catalogues list radiator outputs but these outputs were obtained in a controlled laboratory environment using tests laid down by various British Standards. Whilst these tests give accurate outputs in laboratory conditions, the requirement for the designer is to adjust these figures based upon **BS EN 442**.

In July 1997, according to **BS EN 442**, all radiators manufactured in the EU had to undergo specific tests based upon a flow temperature of 75 °C and a return of 65 °C in a test room with a temperature of 20 °C. In addition, the flow and return connections were at the same end. This is known as top, bottom, same end, or TBSE. The pipe arrangements that are common are:

- top, bottom, same end or **TBSE**
- top, bottom, opposite end or **TBOE**, which is the most efficient method of connecting a radiator
- bottom, bottom, opposite end or **BBOE** (the most common arrangement in domestic heating systems).

Most domestic systems in the UK are designed with flow and return temperatures and pipework arrangements that are different from those that were used under the test conditions of **BS EN 442**. This means that to obtain the correct output, correction factors have to be applied.

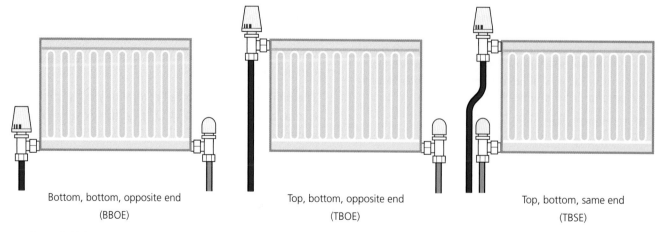

Bottom, bottom, opposite end (BBOE)

Top, bottom, opposite end (TBOE)

Top, bottom, same end (TBSE)

▲ Figure 3.32 Radiator connections

IMPROVE YOUR MATHS

The correction factor is based upon the mean water temperature, or MWT. This is half of the sum of the flow and return when added together. For example, if a condensing boiler is fitted to a central heating system with a flow temperature of 82 °C and a return of 60 °C, then the MWT is:

$$82 + 60 \div 2 = 142$$

$$142 \div 2 = 71\,°C$$

From this, we now deduct the temperature of the room, say a lounge at 21 °C:

$$71 - 21 = 50\,°C\,\Delta t$$

In Table 3.18, it can be seen that an adjustment factor of 0.798 must be applied to all radiators installed in rooms that require 21 °C. Similarly, a factor of 0.858 must be applied to all radiators to rooms of 18 °C and a factor of 0.778 to those rooms that require 22 °C.

A further adjustment factor is required because the radiators are connected BBOE. The BBOE factor is 0.98.

Therefore, the calculation of the output from the radiator looks like this.

Figure 3.33 shows a simple single-storey dwelling. The heat loss calculations were performed using a Mears calculator, model number 15: Domestic Central Heating Calculator.

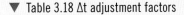

▼ Table 3.18 Δt adjustment factors

°C	Factor
45	0.700
46	0.719
47	0.739
48	0.758
49	0.778
50	0.798
51	0.818
52	0.838
53	0.858
54	0.878
55	0.898
56	0.918
57	0.938
58	0.959
59	0.979
60	1.000
61	1.021
62	1.041
63	1.062
64	1.083
65	1.104

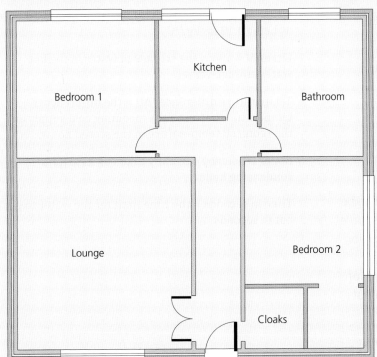

Lounge: L 6 m × W 5 m
21 °C
2 air changes
Window: H 1.5 m × W 2 m
Heat loss: 2400 watts

Bathroom: L 4 m × W 3 m
22 °C
2 air changes
Window: H 1.5 m × W 1.7 m
Heat loss: 1300 watts

Bed 1: L 4.5 m × W 4.5 m
18 °C
1 air change
Window: H 1.5 m × W 2 m
Heat loss: 1400 watts

Hall: L 6 m × W 1.5 m
18 °C
1 air change
Door: H 2.2 m × W 1.2 m
Heat loss: 960 watts

Bed 2: L 6 m × W 4 m
18 °C
1 air change
Window: H 1.5 m × W 2 m
Heat loss: 1700 watts

Cloaks: L 2 m × W 2 m
18 °C
1 air change
Heat loss: 460 watts

Kitchen: L 3 m × W 3 m
18 °C
2 air changes
Door: H 2.2 m × W 1.2 m
Heat loss: 860 watts

▲ Figure 3.33 Plan of a simple single-storey dwelling

The heat losses from the dwelling are as follows:

- Lounge: 2400 watts
- Bedroom 1: 1400 watts
- Bedroom 2: 1700 watts
- Kitchen: 860 watts
- Bathroom: 1300 watts
- Cloaks: 460 watts
- Hall: 960 watts.

The lounge at 21 °C has a heat loss of 2400 watts, so, if the factors are applied:

$$2400 \div 0.898 \div 0.98 = 2727 \text{ watts}$$

▼ Table 3.19 Heat emitter final sizes

Room	watts + adjustment factors	Final emitter size (watts)
Lounge @ 21 °C	2400 ÷ 0.898 ÷ 0.98 = 2727	2727
Bedroom 1 @ 18 °C	1400 ÷ 0.858 ÷ 0.98 = 1665	1665
Bedroom 2 @ 18 °C	1700 ÷ 0.858 ÷ 0.98 = 2021	2021
Kitchen @ 18 °C	860 ÷ 0.858 ÷ 0.98 = 1022	1022
Bathroom @ 22 °C	1300 ÷ 0.778 ÷ 0.98 = 1705	1705
Cloaks @ 18 °C	460 ÷ 0.858 ÷ 0.98 = 547	547
Hall @ 18 °C	960 ÷ 0.858 ÷ 0.98 = 1141	1141
Total for the entire system		**10828**

Therefore, a radiator with an output of 2727 watts is required and NOT 2400. If a radiator of 2400 watts was chosen, with a Δt of 50 °C, the radiator would only give out the following:

$$2400 \times 0.798 \times 0.98 = 2112 \text{ watts}$$

What this means is that even though the manufacturer's data states that the output of the radiator is 2400 watts, this is with water with a Δt of 75 °C and NOT 50 °C as is required so the radiator would be only 88 per cent of the required size.

The final heat emitter sizes can now be calculated.

IMPROVE YOUR MATHS

You should have, by now, calculated the heat loss for the single-storey building previously shown. Calculate the heat emitter/radiator sizes for the building from the heat losses previously calculated.

Calculating hot water heating load

The hot water heating load has been discussed previously, but it is of sufficient importance to be discussed again here.

The heat input into a hot water storage cylinder can be calculated based upon the efficiency of the heat source. If a boiler is the main source of heat for the generation of hot water, then the efficiency of the boiler is required. This is one of the most important factors for improving energy efficiency for the selection of a hot water storage vessel.

The formula for calculating heat input is as follows:

$$\frac{\text{SHC} \times \text{litres of water} \times \text{temperature difference } (\Delta t) \times \text{boiler efficiency}}{\text{Time in seconds} \times 100}$$

Where:

SHC = specific heat capacity of water. This is taken as being 4.19 kJ/kg/°C

Litres of water = the storage of water in the hot water storage vessel

Temperature difference (Δt) = the difference in temperature between the incoming cold supply and the required temperature of the stored hot water

Boiler efficiency = usually taken as 93% for a condensing boiler

Time in seconds = the time limit for the water to get hot in seconds

IMPROVE YOUR MATHS

Calculating hot water heating load: example

A hot water storage cylinder has a capacity of 240 litres. The cylinder is required to be at 60 °C within 2 hours. The temperature of the incoming water is 5 °C. What is the kW required?

$$\frac{\text{SHC} \times \text{litres of water} \times \text{temperature difference } (\Delta t) \times \text{boiler efficiency}}{\text{Time in seconds} \times 100}$$

SHC = 4.19

Litres of water = 240 litres

Δt = 55 °C

Time in seconds = 7200 seconds (2 hrs)

$$\frac{4.19 \times 240 \times 55 \times 93}{7200 \times 100} = 7.14 \text{ kW}$$

IMPROVE YOUR MATHS

A hot water storage cylinder has a capacity of 190 litres. The cylinder is required to be at 65 °C within 2 hours. The temperature of the incoming water is 5 °C. What is the kW required?

$$\frac{\text{SHC} \times \text{litres of water} \times \text{temperature difference } (\Delta t) \times \text{boiler efficiency}}{\text{Time in seconds} \times 100}$$

SHC = 4.19

Litres of water = 190 litres

Δt = 60 °C

Time in seconds = 7200 seconds (2 hrs)

Calculating total boiler size

In previous sections, the totals for both heat emitter output and hot water input were calculated. Now, these must be added together to find out what the boiler size will be:

- Radiator output: 10828 watts or 10.828 kW
- Hot water input: 7.14 kW

When these are added together a total of 17.968 kW is obtained. However, this does not consider the fact that there will be heat loss from the pipework. If all of the pipework is to be surface mounted in the rooms on walls and skirting boards, then no allowance need be made because the heat loss will contribute directly to the heating of the room but, typically, much of the pipework will be under floors and in roof spaces. In this instance it is usual to allow around 10 per cent to 15 per cent extra to take into account heat loss from pipework and further extensions to the system. The extra percentage of heat allowance will also ensure a quick water heat up but caution must be exercised if boiler oversizing is to be avoided.

So, taking the 17.968 kW and adding 15 per cent will give:

17.968 × 0.15 = 2.7

17.968 + 2.7 = 20.668

This can be rounded up to 21 kW. A suitable boiler can now be sourced from the various manufacturers' literature.

Calculating central heating pipe sizes

Pipe sizing central heating systems is not calculated on the amount of heat required by any one room in kW. It is calculated on the amount of water containing heat that will flow down a pipe in kilograms per second or kg/s.

As the heated water moves along the pipe, it will encounter frictional resistance. In other words, the movement of water will be slowed by the size of the pipe, the smoothness of the internal surface of the pipe, the number of changes of direction and restrictions, such as radiator valves and other fittings. To counteract the resistance to flow, a central heating circulating pump is used to force the water around the system. In any system, the resistance will always be greater at the beginning of the circuit than at the end because the resistance, measured in pascals, diminishes with length. The greatest resistance to flow will be in the longest circuit. If the pump used will overcome the resistance in the longest circuit, it will always circulate around any other circuit, simply because the resistance is less. For this reason, pump sizing is always calculated from the longest circuit. This is known as the index circuit. This will be discussed in the next section.

IMPROVE YOUR MATHS

So, let us work through the pipe sizing process. Take a look at the drawing in Figure 3.34.

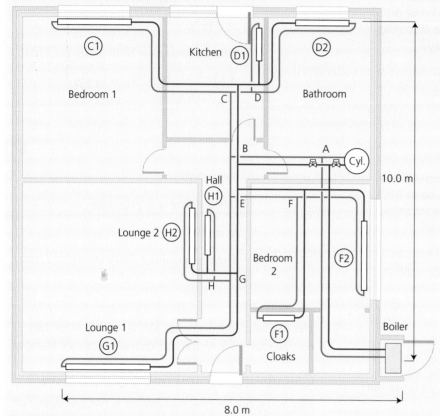

Index circuit	Length	F + R
Boiler to A	6 m	12 m
A to B	3 m	6 m
B to E	2 m	4 m
E to G	2 m	4 m
G to G1	10 m	20 m

Other circuits

A to Cyl.	1.5 m	3 m
B to C	1.5 m	3 m
C to C1	7 m	14 m
C to D	1.5 m	3 m
D to D1	5 m	10 m
D to D2	8 m	16 m
E to F	2 m	4 m
F to F1	6 m	12 m
F to F2	7 m	14 m
G to H	1.5 m	3 m
H to H1	5 m	10 m
H to H2	7 m	14 m

▲ Figure 3.34 Heating system

Figure 3.34 shows a heating system that has been designed to heat the radiators from the heat losses of the single-storey dwelling completed in Figure 3.33. As can be seen, the pipework has been split into sections with each section being given a letter or letter/number combination. This is so that pipe sizing can be calculated by section. These sections are then entered on to a table to simplify the pipe sizing process and to provide a visual reminder of progress. You will see that the table is divided into various columns.

- Column 1: this is the section of pipe that is being sized.
- Column 2: this is the total heat requirement for the section.

- Column 3: this is the percentage of the heat required being allowed for heat loss from the pipe. Typically, this is between 5 per cent and 20 per cent depending on the length of the pipe run. As a rough guide, add 10 per cent for a 20 m run and 5 per cent up to 10 m run but never less than 5 per cent for any section.
- Column 4: this is the total heat requirement with the heat allowance added.
- Column 5: this is the flow rate through the pipe in kg/s and needs to be calculated.
- Column 6: the pipe size.

▼ Table 3.20 Headings for pipe sizing table

1	2	3	4	5	6	7	8	9	10	11	12
Section	Heat required (kW)	Mains loss (%)	Total heat required (kW)	Flow rate (kg/s)	Pipe size (mm)	Length of pipe run (m)	Fittings resistance (m)	Effective length of pipework (m)	Velocity (m/s)	Pressure loss (Pa/m)	Total pascals by section

- Column 7: the actual length of pipe from the drawing.
- Column 8: this is the allowance made for fittings and changes of direction, usually around 33 per cent of the actual pipe length.
- Column 9: the presumed length of pipe once the resistance from column 8 has been added. This will be used to calculate the total resistance to flow so that the pump can be sized correctly.
- Column 10: the velocity of the water through the pipe. This should not be greater than 1.5 metres per second (m/s) or the system may be noisy. A typical velocity of between 0.5 and 1 m/s should be adequate.
- Column 11: this is the resistance per metre of the size of pipe chosen and will be used to calculate the pump size.
- Column 12: the total of the pascals per section is calculated by multiplying columns 9 and 11. This is used for pump sizing.

The pipe sizing procedure

Because the pump size is calculated from the index circuit, this is what we will be concentrating on. The index circuit is the longest circuit and follows the sections of pipework shown in Table 3.21.

▼ Table 3.21 The index circuit (see Figure 3.34)

Section	Heat required (kW)
Boiler to A	17.968
A to B	10.828
B to E	6.436
E to G	3.868
G to G1	1.364

A provisional pipe size can be estimated by looking at how much heat a pipe will carry:

28mm = 22 kW

22mm = 12 kW

15mm = 6 kW

10mm = 3 kW

Stage 1

Look at section **Boiler to A** in Table 3.21. The total boiler load is 17.968 kW (without the 15 per cent) and so the pipe size from Boiler to A will need to carry all of that heat. A 10 per cent margin for heat loss from the pipe is being added and the length of the pipe run is 12 m.

$$17.968 \times 1.10 \, (10\%) = 19.765 \text{ kW}$$

If you look at Table 3.22, you will see that these figures have been added.

As was discussed earlier, the heat required is converted from kilowatts to kg/s. The method is as follows:

$$\text{Flow rate} = \frac{kW}{SHC \times \Delta t} = kg/s$$

Where:

kW = total heat carried by the pipe

SHC = specific heat capacity of water taken as 4.19 kJ/kg °C

Δt = flow and return temperature difference

The boiler is to be a condensing boiler with a 20 °C temperature difference across the flow and return. Therefore, the calculation looks like this:

$$\text{Flow rate} = \frac{19.765}{4.19 \times 20} = 0.235 \text{ kg/s}$$

Now, we must look at the CIBSE copper pipe sizing tables for water at 75 °C.

As you can see, the CIBSE tables are divided into columns. To the far left and the far right are columns with the heading Δ**p/l**. This is the resistance and is measured in pascals, the unit of pressure. This is a vital piece of data as we cannot calculate the pump size without it and so must be entered on our Table 3.22 in column 11. Across the top of the tables are the various pipe sizes. Below each pipe size are two columns. The left column is marked q_m and is the flow rate that we have calculated. The flow rates calculated and the flow rate on the table may not be identical. In this instance, the nearest flow rate ABOVE the calculated flow rate should be used and NEVER below or the pipe will not deliver enough heat. To the right is a column marked **c**. This is the water velocity and should not exceed 1 m/s for small-bore systems and 1.5 m/s for micro-bore systems. The maximum velocity is 1.5 m/s across all systems. The velocity must be entered on to Table 3.22 in column 10. The zig-zag line also relates to velocity. To read it, follow the line upwards until you find the velocity in m/s.

Remember! The flow rate we require is 0.235 kg/s.

Now look at the chart and find the flow rate, which either matches or is slightly above. Also keep an eye on the left/right columns as the pascals should not exceed 300 pa/m for any one section or the pump, when its size is calculated, will be need a large head of pressure and this could possibly create noise in the system. Figure 3.35 is a snapshot of CIBSE table 1. The nearest flow rate is boxed for identification.

→

Pipe sizing tables 1

q_m	=	mass flow rate	kg/s
c	=	velocity	m/s
$\Delta p/l$	=	pressure loss per unit length	pa/m

Copper tubes BSEN1057 R250

Water at 75 °C

$\Delta p/l$	c	10 mm		12 mm		15 mm		22 mm		28 mm		35 mm		42 mm		c	$\Delta p/l$
		q_m	c	q_m	c	q_m	c	q_m	c	q_m	c	q_m	c	q_m	c		
35.0		0.006	0.10	0.012	0.13	0.023	0.16	0.069	0.22	0.140	0.27	0.253	0.31	0.429	0.36		35.0
37.5		0.007	0.12	0.012	0.13	0.024	0.17	0.071	0.23	0.145	0.28	0.263	0.32	0.446	0.37		37.5
40.0		0.007	0.12	0.013	0.15	0.025	0.18	0.074	0.24	0.151	0.29	0.273	0.33	0.462	0.38		40.0
42.5		0.007	0.12	0.013	0.15	0.026	0.18	0.077	0.25	0.156	0.30	0.283	0.35	0.478	0.40		42.5
45.0		0.008	0.13	0.014	0.16	0.026	0.18	0.079	0.25	0.161	0.31	0.292	0.36	0.494	0.41		45.0
47.5		0.008	0.13	0.014	0.16	0.027	0.19	0.082	0.26	0.166	0.32	0.301	0.37	0.509	0.42		47.5
50.0		0.008	0.13	0.015	0.17	0.028	0.20	0.084	0.27	0.171	0.32	0.310	0.38	0.524	0.44		50.0
52.5	0.15	0.008	0.13	0.015	0.17	0.029	0.20	0.087	0.28	0.176	0.33	0.319	0.39	0.539	0.45		52.5
55.0		0.009	0.15	0.016	0.18	0.030	0.21	0.089	0.28	0.181	0.34	0.327	0.40	0.553	0.46		55.0
57.5		0.009	0.15	0.016	0.18	0.031	0.22	0.091	0.29	0.186	0.35	0.336	0.41	0.567	0.47		57.5
60.0		0.009	0.15	0.016	0.18	0.031	0.22	0.094	0.30	0.190	0.36	0.344	0.42	0.581	0.48		60.0
62.5		0.009	0.15	0.017	0.19	0.032	0.23	0.096	0.31	0.195	0.37	0.352	0.43	0.594	0.49	0.50	62.5
65.0		0.010	0.17	0.017	0.19	0.033	0.23	0.098	0.31	0.199	0.38	0.360	0.44	0.608	0.51		65.0
67.5		0.010	0.17	0.018	0.20	0.034	0.24	0.100	0.32	0.203	0.39	0.368	0.45	0.621	0.52		67.5
70.0		0.010	0.17	0.018	0.20	0.034	0.24	0.102	0.33	0.208	0.40	0.375	0.46	0.634	0.53		70.0
72.5		0.010	0.17	0.018	0.20	0.035	0.25	0.104	0.33	0.212	0.40	0.383	0.47	0.646	0.54		72.5
75.0		0.010	0.17	0.019	0.21	0.036	0.25	0.107	0.34	0.216	0.41	0.390	0.48	0.659	0.55		75.0
77.5		0.011	0.19	0.019	0.21	0.036	0.25	0.109	0.35	0.220	0.42	0.398	0.49	0.671	0.56		77.5
80.0		0.011	0.19	0.019	0.21	0.037	0.26	0.111	0.35	0.224	0.43	0.405	0.50	0.683	0.57		80.0
82.5		0.011	0.19	0.020	0.22	0.038	0.27	0.113	0.36	0.228	0.43	0.412	0.51	0.695	0.58		82.5
85.0		0.011	0.19	0.020	0.22	0.038	0.27	0.114	0.36	0.232	0.44	0.419	0.51	0.707	0.59		85.0
87.5		0.011	0.19	0.021	0.24	0.039	0.28	0.116	0.37	0.236	0.45	0.426	0.52	0.718	0.60		87.5
90.0		0.012	0.20	0.021	0.24	0.040	0.28	0.118	0.38	0.240	0.46	0.432	0.53	0.730	0.61		90.0

▲ Figure 3.35 Snapshot of CIBSE copper pipe sizing table 1 (The Chartered Institution of Building Services Engineers (CIBSE), www.cibse.org)

Pipe sizing tables 2

q_m	=	mass flow rate	kg/s
c	=	velocity	m/s
$\Delta p/l$	=	pressure loss per unit length	pa/m

Copper tubes BSEN1057 R250

Water at 75 °C

$\Delta p/l$	c	10 mm		12 mm		15 mm		22 mm		28 mm		35 mm		42 mm		c	$\Delta p/l$
		q_m	c	q_m	c	q_m	c	q_m	c	q_m	c	q_m	c	q_m	c		
92.5		0.012	0.20	0.021	0.24	0.040	0.28	0.120	0.38	0.243	0.46	0.439	0.54	0.741	0.62		92.5
95.0		0.012	0.20	0.022	0.25	0.041	0.29	0.122	0.39	0.247	0.47	0.446	0.55	0.752	0.63		95.0
97.5		0.012	0.20	0.022	0.25	0.042	0.30	0.124	0.40	0.251	0.48	0.452	0.55	0.763	0.63		97.5
100		0.012	0.20	0.022	0.25	0.042	0.30	0.126	0.40	0.254	0.48	0.459	0.56	0.774	0.64		100
120		0.014	0.24	0.025	0.28	0.047	0.33	0.139	0.44	0.282	0.54	0.508	0.62	0.857	0.71		120
140		0.015	0.25	0.027	0.30	0.051	0.36	0.152	0.49	0.308	0.59	0.554	0.68	0.934	0.78		140
160		0.017	0.29	0.029	0.32	0.056	0.40	0.164	0.52	0.332	0.63	0.598	0.73	1.000	0.83		160
180	0.30	0.018	0.30	0.032	0.36	0.060	0.42	0.176	0.56	0.354	0.67	0.638	0.78	1.070	0.89		180
200		0.019	0.32	0.034	0.38	0.063	0.44	0.186	0.59	0.376	0.71	0.677	0.83	1.130	0.94		200
220		0.020	0.34	0.035	0.39	0.067	0.47	0.197	0.63	0.397	0.75	0.714	0.88	1.200	1.00	1.00	220
240		0.021	0.35	0.037	0.41	0.070	0.49	0.207	0.66	0.417	0.79	0.750	0.92	1.260	1.05		240
260		0.022	0.37	0.039	0.44	0.074	0.52	0.216	0.69	0.436	0.83	0.784	0.96	1.310	1.09		260
280		0.023	0.39	0.041	0.46	0.077	0.54	0.226	0.72	0.454	0.86	0.817	1.00	1.370	1.14		280
300		0.024	0.40	0.042	0.47	0.080	0.56	0.235	0.75	0.472	0.90	0.849	1.04	1.420	1.18		300
320		0.025	0.42	0.044	0.49	0.083	0.59	0.243	0.78	0.490	0.93	0.880	1.08	1.470	1.22		320

▲ Figure 3.36 Snapshot of CIBSE copper pipe sizing table 2 (The Chartered Institution of Building Services Engineers (CIBSE), www.cibse.org)

This shows that the nearest flow rate is slightly above 0.235 at 0.236 with a pipe size of 28 mm. The velocity 0.45 m/s and the pascals are 87.5 pa/m pressure drop. On the face of it this looks good BUT are there any alternative pipe sizes? Look at sheet 2, in Figure 3.36.

There is an alternative pipe size because a 22 mm pipe will also deliver the required flow rate with a velocity of 0.75 m/s (nearer to the ideal 1 m/s) but take a look at the pascals per metre. At 300 Pa/m it is at the limit of the 300 Pa/m maximum. There are several advantages to using 22 mm. It is cheaper to buy, which keeps the cost of the installation down and it is easier to work with and install. However, the pascals may present a problem later, but this will not be known until the rest of the index circuit is calculated. For the purpose of this example, 22 mm pipe will be chosen. The data can be entered into Table 3.22.

▼ Table 3.22 Pipe sizing

1	2	3	4	5	6	7	8	9	10	11	12
Section	Heat required (kW)	Mains loss (%)	Total heat required (kW)	Flow rate (kg/s)	Pipe size (mm)	Length of pipe run (m)	Fittings resistance (33%)	Effective length of pipework (m)	Velocity (m/s)	Pressure loss (Pa/m)	Total pascals by section
Boiler to A	17.968	10	19.765	0.235	22	12	1.33	15.96	0.75	300	4788
A to B	10.828	5	11.37	0.135	22	6	1.33	7.98	0.44	120	957.6
B to E	6.436	5	6.75	0.080	22	4	1.33	5.32	0.26	47.5	252.7
E to G	3.868	5	4.10	0.049	15	4	1.33	5.32	0.36	140	744.8
G to G1	1.364	10	1.50	0.017	10	20	1.33	26.6	0.29	160	4256
					Total pascals for pump sizing from the index circuit						10999.1
A to Cyl	7.140	5				3	1.33				
B to C	4.392	5				3	1.33				
C to C1	1.665	7				14	1.33				
C to D	2.727	5				3	1.33				
D to D1	1.022	5				10	1.33				
D to D2	1.705	7				16	1.33				
E to F	2.568	5				4	1.33				
F to F1	0.547	6				12	1.33				
F to F2	2.021	7				14	1.33				
G to H	2.505	5				3	1.33				
H to H1	1.141	5				10	1.33				
H to H2	1.364	7				14	1.33				

Stage 2

At the beginning of the process we allowed a percentage of the total kW for the section for heat loss from the pipe. This can now be checked to see if it is adequate.

To check to see if the 5 per cent we added to the heat required is sufficient, we perform another calculation.

If the room is to be maintained at 21 °C and the flow pipe is at 80 °C, this gives a temperature difference of 59 °C. For the purposes of the calculation, we can round this up to 60.

Therefore:

Length of run × 78 watts (see Table 3.22)

$12 \times 78 = 936$ watts

$$\text{Percentage emission} = \frac{\text{Total watts of run} \times 100}{\text{Total heat emission in watts}}$$

$$\text{Percentage emission} = \frac{93,600}{17,968} = 5.2\%$$

So, the estimation of 5 per cent heat loss from the pipework was correct therefore the pipe size is also correct.

▼ Table 3.23 Heat emission in watts per metre of pipe run

Nominal pipe size (mm)	Temperature difference of surface to surroundings (°C)				
	40	45	50	55	60
8	28	32	35	39	43
10	33	37	41	46	50
15	40	46	53	59	66
22	48	56	62	70	78
28	58	68	78	88	98
35	71	82	93	110	120
42	78	92	105	110	130
54	96	112	130	150	170
66.7	120	140	160	180	200
76.1	140	160	180	211	230

INDUSTRY TIP

There are four sheets for pipe sizing copper tubes. These documents are taken from the CIBSE concise guide. CIBSE also produce pipe sizing tables for low-carbon steel pipe and plastic pressure pipe.

Stages 1 and 2 can now be performed for the entire index circuit.

The index circuit

A to B

A to B = 10.828 + 5% = 11.37 kW

$$\text{Flow rate} = \frac{11.37}{4.19 \times 20} = 0.135 \text{ kg/s}$$

Pipe size = 22 mm

Velocity = 0.44 m/s

Pascals = 120

Percentage check

Length of run × 78

6 × 78 = 468 watts

$$\text{Percentage emission} = \frac{\text{Total watts of run} \times 100}{\text{Total heat emission in watts}}$$

$$\text{Percentage emission} = \frac{46800}{17968} = 2.6\%$$

B to E

B to E = 6.436 + 5% = 6.75 kW

$$\text{Flow rate} = \frac{6.75}{4.19 \times 20} = 0.080 \text{ kg/s}$$

Pipe size = 22 mm

Velocity = 0.44 m/s

Pascals = 120

Percentage check

Length of run × 78

4 × 78 = 312 watts

$$\text{Percentage emission} = \frac{\text{Total watts of run} \times 100}{\text{Total heat emission in watts}}$$

$$\text{Percentage emission} = \frac{31200}{17968} = 1.73\%$$

E to G

E to G = 3.868 + 6% = 4.10 kW

$$\text{Flow rate} = \frac{64.10}{4.19 \times 20} = 0.049 \text{ kg/s}$$

Pipe size = 15 mm

Velocity = 0.36 m/s

Pascals = 140

Percentage check

Length of run × 66

4 × 66 = 264 watts

$$\text{Percentage emission} = \frac{\text{Total watts of run} \times 100}{\text{Total heat emission in watts}}$$

$$\text{Percentage emission} = \frac{26400}{17968} = 1.4\%$$

Pipe sizing tables 1

q_m = mass flow rate   kg/s
c = velocity   m/s
$\Delta p/l$ = pressure loss per unit length   pa/m

| Copper tubes BSEN1057 R250 |
| Water at 75 °C |

$\Delta p/l$	c	10 mm		12 mm		15 mm		22 mm		28 mm		35 mm		42 mm		c	$\Delta p/l$
		q_m	c	q_m	c	q_m	c	q_m	c	q_m	c	q_m	c	q_m	c		
0.1								0.001	0.00	0.00	0.00	0.007	0.01	0.015	0.01		0.1
0.2								0.002	0.01	0.005	0.01	0.014	0.02	0.023	0.02		0.2
0.3								0.003	0.01	0.008	0.02	0.019	0.02	0.026	0.02		0.3
0.4								0.004	0.01	0.011	0.02	0.019	0.02	0.032	0.03		0.4
0.5						0.001	0.01	0.005	0.02	0.014	0.03	0.021	0.03	0.036	0.03		0.5
0.6						0.001	0.01	0.006	0.02	0.015	0.03	0.023	0.03	0.040	0.03		0.6
0.7						0.001	0.01	0.007	0.02	0.015	0.03	0.026	0.03	0.044	0.04		0.7
0.8						0.001	0.01	0.008	0.03	0.015	0.03	0.028	0.03	0.048	0.04		0.8
0.9						0.001	0.01	0.009	0.03	0.016	0.03	0.030	0.04	0.051	0.04		0.9
1.0						0.002	0.01	0.010	0.03	0.017	0.03	0.032	0.04	0.055	0.05	0.05	1.0
1.5				0.001	0.01	0.003	0.02	0.012	0.04	0.022	0.04	0.040	0.05	0.070	0.06		1.5
2.0				0.001	0.01	0.004	0.03	0.012	0.04	0.026	0.05	0.048	0.06	0.083	0.07		2.0
2.5				0.002	0.02	0.005	0.04	0.014	0.04	0.030	0.06	0.055	0.07	0.094	0.08		2.5
3.0		0.001	0.02	0.002	0.02	0.006	0.04	0.016	0.05	0.033	0.06	0.061	0.07	0.105	0.09		3.0
3.5		0.001	0.02	0.003	0.03	0.007	0.05	0.017	0.05	0.036	0.07	0.067	0.08	0.114	0.09		3.5
4.0		0.001	0.02	0.003	0.03	0.008	0.06	0.019	0.06	0.039	0.07	0.072	0.09	0.124	0.10		4.0
4.5		0.001	0.02	0.003	0.03	0.008	0.06	0.020	0.06	0.042	0.08	0.078	0.10	0.132	0.11		4.5
5.0		0.001	0.02	0.004	0.04	0.008	0.06	0.022	0.07	0.045	0.09	0.083	0.10	0.141	0.12		5.0
5.5		0.002	0.03	0.004	0.04	0.008	0.06	0.023	0.07	0.048	0.09	0.087	0.11	0.149	0.12		5.5
6.0		0.002	0.03	0.005	0.06	0.008	0.06	0.024	0.08	0.050	0.09	0.092	0.11	0.156	0.13		6.0
6.5		0.002	0.03	0.005	0.06	0.008	0.06	0.025	0.08	0.053	0.10	0.096	0.12	0.164	0.14		6.5
7.0		0.002	0.03	0.006	0.07	0.008	0.06	0.027	0.09	0.055	0.10	0.100	0.12	0.171	0.14		7.0
7.5		0.002	0.03	0.006	0.07	0.009	0.06	0.028	0.09	0.057	0.11	0.105	0.13	0.178	0.15	0.15	7.5
8.0		0.003	0.05	0.006	0.07	0.009	0.06	0.029	0.09	0.059	0.11	0.108	0.13	0.185	0.15		8.0
8.5		0.003	0.05	0.006	0.07	0.010	0.07	0.030	0.10	0.061	0.12	0.112	0.14	0.191	0.16		8.5
9.0		0.003	0.05	0.006	0.07	0.010	0.07	0.031	0.10	0.064	0.12	0.116	0.14	0.198	0.16		9.0
9.5		0.003	0.05	0.006	0.07	0.010	0.07	0.032	0.10	0.066	0.13	0.120	0.15	0.204	0.17		9.5
10.0	0.05	0.003	0.05	0.006	0.07	0.011	0.08	0.033	0.11	0.068	0.13	0.123	0.15	0.210	0.17		10.0
12.5		0.004	0.07	0.006	0.07	0.012	0.08	0.037	0.12	0.077	0.15	0.140	0.17	0.239	0.20		12.5
15.0		0.005	0.08	0.007	0.08	0.014	0.10	0.042	0.13	0.086	0.16	0.156	0.19	0.265	0.22		15.0
17.5		0.005	0.08	0.008	0.09	0.015	0.11	0.046	0.15	0.094	0.18	0.170	0.21	0.289	0.24		17.5
20.0		0.005	0.08	0.008	0.09	0.016	0.11	0.049	0.16	0.101	0.19	0.184	0.23	0.312	0.26		20.0
22.5		0.005	0.08	0.009	0.10	0.017	0.12	0.053	0.17	0.108	0.21	0.197	0.24	0.334	0.28		22.5
25.0		0.005	0.08	0.010	0.11	0.019	0.13	0.056	0.18	0.115	0.22	0.209	0.26	0.354	0.29	0.30	25.0
27.5		0.005	0.08	0.010	0.11	0.020	0.14	0.060	0.19	0.122	0.23	0.221	0.27	0.374	0.31		27.5
30.0		0.006	0.10	0.011	0.12	0.021	0.15	0.063	0.20	0.128	0.24	0.232	0.28	0.393	0.33		30.0
32.5		0.006	0.10	0.011	0.12	0.022	0.16	0.066	0.21	0.134	0.25	0.243	0.30	0.411	0.34		32.5
35.0		0.006	0.10	0.012	0.13	0.023	0.16	0.069	0.22	0.140	0.27	0.253	0.31	0.429	0.36		35.0
37.5		0.007	0.12	0.012	0.13	0.024	0.17	0.071	0.23	0.145	0.28	0.263	0.32	0.446	0.37		37.5
40.0		0.007	0.12	0.013	0.15	0.025	0.18	0.074	0.24	0.151	0.29	0.273	0.33	0.462	0.38		40.0
42.5		0.007	0.12	0.013	0.15	0.026	0.18	0.077	0.25	0.156	0.30	0.283	0.35	0.478	0.40		42.5
45.0		0.008	0.13	0.014	0.16	0.026	0.18	0.079	0.25	0.161	0.31	0.292	0.36	0.494	0.41		45.0
47.5		0.008	0.13	0.014	0.16	0.027	0.19	0.082	0.26	0.166	0.32	0.301	0.37	0.509	0.42		47.5
50.0		0.008	0.13	0.015	0.17	0.028	0.20	0.084	0.27	0.171	0.32	0.310	0.38	0.524	0.44		50.0
52.5	0.15	0.008	0.13	0.015	0.17	0.029	0.20	0.087	0.28	0.176	0.33	0.319	0.39	0.539	0.45		52.5
55.0		0.009	0.15	0.016	0.18	0.030	0.21	0.089	0.28	0.181	0.34	0.327	0.40	0.553	0.46		55.0
57.5		0.009	0.15	0.016	0.18	0.031	0.22	0.091	0.29	0.186	0.35	0.336	0.41	0.567	0.47		57.5
60.0		0.009	0.15	0.016	0.18	0.031	0.22	0.094	0.30	0.190	0.36	0.344	0.42	0.581	0.48		60.0
62.5		0.009	0.15	0.017	0.19	0.032	0.23	0.096	0.31	0.195	0.37	0.352	0.43	0.594	0.49	0.50	62.5
65.0		0.010	0.17	0.017	0.19	0.033	0.23	0.098	0.31	0.199	0.38	0.360	0.44	0.608	0.51		65.0
67.5		0.010	0.17	0.018	0.20	0.034	0.24	0.100	0.32	0.203	0.39	0.368	0.45	0.621	0.52		67.5
70.0		0.010	0.17	0.018	0.20	0.034	0.24	0.102	0.33	0.208	0.40	0.375	0.46	0.634	0.53		70.0
72.5		0.010	0.17	0.018	0.20	0.035	0.25	0.104	0.33	0.212	0.40	0.383	0.47	0.646	0.54		72.5
75.0		0.010	0.17	0.019	0.21	0.036	0.25	0.107	0.34	0.216	0.41	0.390	0.48	0.659	0.55		75.0
77.5		0.011	0.19	0.019	0.21	0.036	0.25	0.109	0.35	0.220	0.42	0.398	0.49	0.671	0.56		77.5
80.0		0.011	0.19	0.019	0.21	0.037	0.26	0.111	0.35	0.224	0.43	0.405	0.50	0.683	0.57		80.0
82.5		0.011	0.19	0.020	0.22	0.038	0.27	0.113	0.36	0.228	0.43	0.412	0.51	0.695	0.58		82.5
85.0		0.011	0.19	0.020	0.22	0.038	0.27	0.114	0.36	0.232	0.44	0.419	0.51	0.707	0.59		85.0
87.5		0.011	0.19	0.021	0.24	0.039	0.28	0.116	0.37	0.236	0.45	0.426	0.52	0.718	0.60		87.5
90.0		0.012	0.20	0.021	0.24	0.040	0.28	0.118	0.38	0.240	0.46	0.432	0.53	0.730	0.61		90.0

▲ Figure 3.37 CIBSE sheet 1 (The Chartered Institution of Building Services Engineers (CIBSE), www.cibse.org)

Pipe sizing tables 2

q_m = mass flow rate kg/s
c = velocity m/s
$\Delta p/l$ = pressure loss per unit length pa/m

| Copper tubes BSEN1057 R250 |
| Water at 75 °C |

$\Delta p/l$	c	10 mm q_m	10 mm c	12 mm q_m	12 mm c	15 mm q_m	15 mm c	22 mm q_m	22 mm c	28 mm q_m	28 mm c	35 mm q_m	35 mm c	42 mm q_m	42 mm c	c	$\Delta p/l$
92.5		0.012	0.20	0.021	0.24	0.040	0.28	0.120	0.38	0.243	0.46	0.439	0.54	0.741	0.62		92.5
95.0		0.012	0.20	0.022	0.25	0.041	0.29	0.122	0.39	0.247	0.47	0.446	0.55	0.752	0.63		95.0
97.5		0.012	0.20	0.022	0.25	0.042	0.30	0.124	0.40	0.251	0.48	0.452	0.55	0.763	0.63		97.5
100		0.012	0.20	0.022	0.25	0.042	0.30	0.126	0.40	0.254	0.48	0.459	0.56	0.774	0.64		100
120		0.014	0.24	0.025	0.28	0.047	0.33	0.139	0.44	0.282	0.54	0.508	0.62	0.857	0.71		120
140		0.015	0.25	0.027	0.30	0.051	0.36	0.152	0.49	0.308	0.59	0.554	0.68	0.934	0.78		140
160		0.017	0.29	0.029	0.32	0.056	0.40	0.164	0.52	0.332	0.63	0.598	0.73	1.000	0.83		160
180	0.30	0.018	0.30	0.032	0.36	0.060	0.42	0.176	0.56	0.354	0.67	0.638	0.78	1.070	0.89		180
200		0.019	0.32	0.034	0.38	0.063	0.44	0.186	0.59	0.376	0.71	0.677	0.83	1.130	0.94		200
220		0.020	0.34	0.035	0.39	0.067	0.47	0.197	0.63	0.397	0.75	0.714	0.88	1.200	1.00	1.00	220
240		0.021	0.35	0.037	0.41	0.070	0.49	0.207	0.66	0.417	0.79	0.750	0.92	1.260	1.05		240
260		0.022	0.37	0.039	0.44	0.074	0.52	0.216	0.69	0.436	0.83	0.784	0.96	1.310	1.09		260
280		0.023	0.39	0.041	0.46	0.077	0.54	0.226	0.72	0.454	0.86	0.817	1.00	1.370	1.14		280
300		0.024	0.40	0.042	0.47	0.080	0.56	0.235	0.75	0.472	0.90	0.849	1.04	1.420	1.18		300
320		0.025	0.42	0.044	0.49	0.083	0.59	0.243	0.78	0.490	0.93	0.880	1.08	1.470	1.22		320
340		0.026	0.44	0.046	0.52	0.086	0.61	0.252	0.81	0.506	0.96	0.910	1.12	1.530	1.27		340
360		0.027	0.46	0.047	0.53	0.089	0.63	0.260	0.83	0.523	0.99	0.940	1.15	1.570	1.31		360
380		0.028	0.47	0.049	0.55	0.092	0.65	0.268	0.86	0.539	1.02	0.968	1.19	1.620	1.35		380
400		0.028	0.47	0.050	0.56	0.094	0.66	0.276	0.88	0.555	1.05	0.996	1.22	1.670	1.39		400
420		0.029	0.49	0.052	0.58	0.097	0.68	0.283	0.90	0.570	1.08	1.020	1.25	1.720	1.43		420
440	0.50	0.030	0.51	0.053	0.59	0.099	0.70	0.291	0.93	0.585	1.11	1.050	1.29	1.760	1.46		440
460		0.031	0.52	0.054	0.60	0.102	0.72	0.298	0.95	0.600	1.14	1.070	1.31	1.800	1.50	1.50	460
480		0.032	0.54	0.056	0.63	0.105	0.74	0.306	0.98	0.614	1.17	1.100	1.35	1.850	1.54		480
500		0.032	0.54	0.057	0.64	0.107	0.76	0.313	1.00	0.628	1.19	1.120	1.37	1.890	1.57		500
520		0.033	0.56	0.058	0.65	0.109	0.77	0.320	1.02	0.642	1.22	1.150	1.41	1.930	1.61		520
540		0.034	0.57	0.060	0.67	0.112	0.79	0.326	1.04	0.656	1.25	1.170	1.44	1.970	1.64		540
560		0.035	0.59	0.061	0.68	0.114	0.81	0.333	1.06	0.669	1.27	1.200	1.47	2.010	1.67		560
580		0.035	0.59	0.062	0.69	0.116	0.82	0.340	1.09	0.682	1.30	1.220	1.50	2.050	1.70		580
600		0.036	0.61	0.063	0.71	0.119	0.84	0.346	1.11	0.695	1.32	1.240	1.52	2.090	1.74		600
620		0.037	0.62	0.065	0.73	0.121	0.85	0.353	1.13	0.708	1.35	1.270	1.56	2.130	1.77		620
640		0.037	0.62	0.066	0.74	0.123	0.87	0.359	1.15	0.721	1.37	1.290	1.58	2.160	1.80		640
660		0.038	0.64	0.067	0.75	0.125	0.88	0.365	1.17	0.733	1.39	1.310	1.61	2.200	1.83		660
680		0.039	0.66	0.068	0.76	0.127	0.90	0.371	1.19	0.745	1.42	1.330	1.63	2.240	1.86		680
700		0.039	0.66	0.069	0.77	0.130	0.92	0.378	1.21	0.757	1.44	1.350	1.66	2.270	1.89		700
720		0.040	0.67	0.070	0.78	0.132	0.93	0.383	1.22	0.769	1.46	1.380	1.69	2.310	1.92		720
740		0.041	0.69	0.071	0.80	0.134	0.95	0.389	1.24	0.781	1.48	1.400	1.72	2.350	1.95		740
760		0.041	0.69	0.073	0.82	0.136	0.96	0.395	1.26	0.793	1.51	1.420	1.74	2.380	1.98	2.00	760
780		0.042	0.71	0.074	0.83	0.138	0.97	0.401	1.28	0.804	1.53	1.440	1.77	2.410	2.00		780
800		0.043	0.73	0.075	0.84	0.140	0.99	0.407	1.30	0.816	1.55	1.460	1.79	2.450	2.04		800
820		0.043	0.73	0.076	0.85	0.142	1.00	0.412	1.32	0.827	1.57	1.480	1.82	2.480	2.06		820
840		0.044	0.74	0.077	0.86	0.144	1.02	0.418	1.34	0.838	1.59	1.500	1.84	2.510	2.09		840
860		0.044	0.74	0.078	0.87	0.146	1.03	0.423	1.35	0.849	1.61	1.520	1.86	2.550	2.12		860
880		0.045	0.76	0.079	0.88	0.147	1.04	0.429	1.37	0.860	1.63	1.540	1.89	2.580	2.15		880
900		0.046	0.78	0.080	0.90	0.149	1.05	0.434	1.39	0.871	1.65	1.560	1.91	2.610	2.17		900
920		0.046	0.78	0.081	0.91	0.151	1.07	0.440	1.41	0.881	1.67	1.570	1.93	2.640	2.20		920
940		0.047	0.79	0.082	0.92	0.153	1.08	0.445	1.42	0.892	1.69	1.590	1.95	2.680	2.23		940
960		0.047	0.79	0.083	0.93	0.155	1.09	0.450	1.44	0.902	1.71	1.610	1.97	2.710	2.25		960
980		0.048	0.81	0.084	0.94	0.157	1.11	0.455	1.45	0.913	1.73	1.630	2.00	2.740	2.28		980
1000		0.048	0.81	0.085	0.95	0.158	1.12	0.461	1.47	0.923	1.75	1.650	2.02	2.770	2.30		1000
1100		0.051	0.86	0.090	1.01	0.167	1.18	0.486	1.55	0.973	1.85	1.740	2.13	2.920	2.43		1100
1200		0.054	0.91	0.094	1.05	0.176	1.24	0.510	1.63	1.020	1.94	1.820	2.23	3.060	2.54		1200
1300		0.056	0.94	0.098	1.10	0.184	1.30	0.533	1.70	1.060	2.01	1.910	2.34	3.200	2.66		1300
1400	1.00	0.059	1.00	0.103	1.15	0.191	1.35	0.555	1.77	1.110	2.11	1.980	2.43	3.330	2.77		1400
1500		0.061	1.03	0.107	1.20	0.199	1.41	0.577	1.84	1.150	2.18	2.060	2.53	3.460	2.88		1500
1600		0.063	1.06	0.111	1.24	0.206	1.45	0.598	1.91	1.190	2.26	2.140	2.63	3.580	2.98	3.00	1600
1700		0.066	1.11	0.115	1.29	0.214	1.51	0.618	1.97	1.230	2.34	2.210	2.71	3.700	3.08		1700
1800		0.068	1.15	0.118	1.32	0.220	1.55	0.638	2.04	1.270	2.41	2.280	2.80	3.820	3.18		1800
1900		0.070	1.18	0.122	1.37	0.227	1.60	0.658	2.10	1.310	2.49	2.350	2.88	3.930	3.27		1900
2000		0.072	1.21	0.126	1.41	0.234	1.65	0.677	2.16	1.350	2.56	2.410	2.96	4.040	3.36		2000

▲ Figure 3.38 CIBSE sheet 2 (The Chartered Institution of Building Services Engineers (CIBSE), www.cibse.org)

G to G1

G to G1 $= 1.364 + 10\% = 1.50$ kW

Flow rate $= \dfrac{1.50}{4.19 \times 12} = 0.017$ kg/s

Pipe size $= 10$ mm

Velocity $= 0.29$ m/s

Pascals $= 160$

Percentage check

Length of run $\times 50$

$20 \times 50 = 1000$ watts

Percentage emission $= \dfrac{\text{Total watts of run} \times 100}{\text{Total heat emission in watts}}$

Percentage emission $= \dfrac{100000}{17968} = 5.5\%$

IMPROVE YOUR MATHS

Using the techniques discussed in this section, complete the pipe sizing table for the single-storey dwelling.

IMPROVE YOUR MATHS

Sizing the circulating pump

Before we can size the circulating pump, we must calculate the total pascals per section of pipe. This is completed by first finding out the effective length of pipe in column 9. It is usual to allow a 33 per cent increase in the pipework to allow for fittings resistance. This is a fairly simple task that involves multiplying the pipe length by 1.33 as shown in the index circuit table below. This will give the effective pipe length. Now, multiply the effective length by the pascals for the section. Then when all of the total pascals per section have been calculated, add the totals together as shown.

▼ Table 3.24

1	2	3	4	5	6	7	8	9	10	11	12
Section	Heat required (kW)	Mains loss (%)	Total heat required (kW)	Flow rate (kg/s)	Pipe size (mm)	Length of pipe run (m)	Fittings resistance (33%)	Effective length of pipework (m)	Velocity (m/s)	Pressure loss (Pa/m)	Total pascals by section
Boiler to A	17.968	10	19.765	0.235	22	12	1.33	15.96	0.75	300	4788
A to B	10.828	5	11.37	0.135	22	6	1.33	7.98	0.44	120	957.6
B to E	6.436	5	6.75	0.080	22	4	1.33	5.32	0.26	47.5	252.7
E to G	3.868	5	4.10	0.049	15	4	1.33	5.32	0.36	140	744.8
G to G1	1.364	10	1.50	0.017	10	20	1.33	26.6	0.29	160	4256
						Total pascals for pump sizing from the index circuit					10,999.1

The total needs to be converted to metres head. If one pascal is equal to 0.0001019977334 metres head then:

$10{,}999.1 \times 0.0001019977334 = 1.122$ metres head

You can see that the index circuit has a pressure loss of 1.122 metres head.

The next step is to find out if the boiler creates a significant pressure drop. Some low water content boilers generate a high resistance to flow through the heat exchanger and this results in a drop in pressure. This, too, is measured in metres head. Consulting the boiler manufacturer's instructions will indicate if this is the case. Any pressure drop at the boiler will also need to be added to the pressure drop across the index circuit.

Assuming there is a pressure drop at the boiler of 2 m head, then the total for the system will be:

$2 + 1.122 = 3.122$ metres head pressure drop

This must be converted to kilopascals (kPa). To convert metres head to kPa simply multiply the metres head by 9.81:

$3.122 \times 9.81 = 30.63$ kPa

You must now consult the pump manufacturer's literature to select a pump and pump speed based on your calculation. From Figure 3.39, it will be seen that a Grundfos UPS-15/50 pump will give 30.63 kPa on either speed 2 @ 0.28 l/s or speed 3 @ 0.38 l/s.

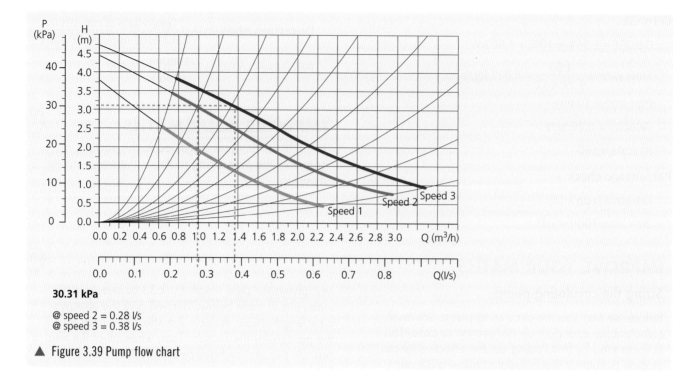

30.31 kPa

@ speed 2 = 0.28 l/s
@ speed 3 = 0.38 l/s

▲ Figure 3.39 Pump flow chart

Presenting the calculations to the customer

The methods that can be used to present the design of a central heating system and the results of calculations performed were covered previously, but it is worth revisiting the basic principle of design and calculation presentation.

Scale and not-to-scale drawings

Scale drawings and schematic drawings produced using computer aided design (CAD) packages help to show the customer what you are proposing to install to fulfil their heating requirements. This is especially important when a large installation is to be completed as it helps the customer to keep a track of what is being installed and where.

Presentation of designs and calculations

The calculation process for central heating takes time to complete and unless the calculations are set out correctly, mistakes can be made. The use of spreadsheets and tables when completing design calculations, especially for pipe sizing and fabric heat loss, is commonplace amongst most professional building services engineers. These are excellent for including with any quotation or design specification that the company wishes to present to the customer

and can be saved as either a hard copy in portable document format (PDF) or as Word or Excel documents. Many proprietary heating design packages also produce a printable report of the heat losses, heat emitters and pipework sizes. These too can be supplied to the customer at the estimation stage.

3 POSITIONING CENTRAL HEATING SYSTEM COMPONENTS

Many of the components listed below were discussed in Chapter 7 of Book 1. However, because of the importance of positioning central heating components, we will revisit them again to ensure learning and understanding.

Boilers

Boilers must always be positioned in accordance with the manufacturer instructions to maintain compliance with the Gas Safety (Installation and Use) Regulations 1998.

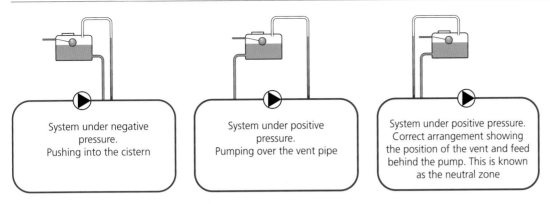

▲ Figure 3.40 The position of the circulating pump

Pumps

The circulating pump must be positioned with care to avoid design faults that could lead to problems with corrosion by aeration of the water due to water movement in the feed and expansion cistern in open vented systems (see Figure 3.40). This occurs when water is either pushed up the cold feed pipe and the open vent pipe or is circulated between the cold feed pipe and the open vent pipe.

On sealed systems, the pump should, wherever possible, be positioned on the flow pipe. In most cases, sealed system boilers and combination boilers have the pump already installed within the boiler casing.

Motorised valves

The position of motorised valves will depend upon the type of system and the type of motorised valve:

- S-plan and S-plan plus systems use 2-port zone valves, one for each circuit to be controlled. These are usually positioned in an accessible place, usually the airing cupboard.
- Y-plan systems use a 3-port mid-position valve which controls both hot water and heating circuits. Additional zoning may be achieved by installing 2-port motorised zone valves to give extra temperature control.
- W-plan systems use a 3-port motorised diverter valve. Mid-position operation is not possible with this type of valve.

Expansion vessels

Expansion vessels should be positioned on the return pipework wherever possible. If the expansion vessel is positioned on the flow pipe, it must be placed on the suction (negative) side of the circulating pump so that the pump does not force open the pressure relief valve, causing water loss.

Isolation valves

Isolation valves should be positioned at the fill point of the system for draining and maintenance. Isolation valves should also be installed either side of the pump so that the pump may be removed/replaced.

Fill points

The filling point of the system depends on the type of system installed.

- Sealed systems are fed through a temporary filling loop that connects the mains cold water supply to the return pipe on the heating system. Filling loops contain two isolation valves and must contain a double check valve type EC or type ED backflow prevention device. The filling loop **must** be removed when the system is full and commissioned.
- Open vented systems are filled via the cold feed pipe that leads from the small feed and expansion cistern in the roof space. Both the open vent and the cold feed should be positioned on the suction side of the pump and no more than 150 mm apart. Alternatively, an air interceptor can be used to connect the cold feed and open vent to the system.

Radiators

Radiators should be positioned with care to give the best heat distribution throughout the property. The customer's wishes on heat emitter position should be taken into account.

UFH manifolds

See the section on underfloor heating (page 210).

UFH pipework

See the section on underfloor heating (page 210).

Cylinders

The position of the cylinder is usually dictated by the position of the airing cupboard within the property. However, unvented hot water storage cylinders may be positioned in almost any position in the property with the only constraint being accessibility to outside for the discharge (D2) pipework.

Automatic bypass

The automatic bypass valve connects the flow pipe to the return pipe. The flow connection to the automatic bypass valve is connected immediately after the pump and connects directly to the return pipework via the automatic bypass valve. It is designed to open when all thermostatic radiator valves are closed, and the hot water circuit is satisfied. Boilers that have an electrical pump overrun circuit designed into the boiler require an automatic bypass valve to prevent boiler overheating. Most modern system and combination boilers have bypass circuits built in to the appliance.

Sealed system components

As we have already seen, sealed systems do not contain a feed and expansion cistern nor open vent pipe. Instead, these systems incorporate the following components:

- an external expansion vessel fitted to the system return
- a pressure relief valve
- the system is filled via a temporary filling loop or a CA disconnection device
- a pressure gauge.

The expansion vessel

The expansion vessel is a key component of the system. It replaces the feed and expansion cistern on the vented system and allows the expansion of water to take place safely. It comprises of a steel cylinder which is divided in two by a neoprene rubber diaphragm.

The vessel is installed on to the return because the return water is generally 20 °C cooler than the flow water and this does not place as much temperature stress on the expansion vessel's internal diaphragm as the hotter flow water. If installing the vessel on the flow is unavoidable, it should be placed on the suction side of the circulating pump in the same way as the cold feed and open vent pipe on the open vented system.

On one end of the expansion vessel is a Schrader air pressure valve where air is pumped into the vessel to 1 bar pressure and this forces the neoprene diaphragm to virtually fill the whole of the vessel.

On the other end is a ½ inch male BSP thread and this is the connection point to the system. When mains pressure cold water enters the heating system via the filling loop and the system is filled to a pressure of around 1 bar, the water forces the diaphragm backwards away from the vessel walls compressing the air slightly as the water enters the vessel. At this point, the pressure on both sides of the diaphragm is 1 bar pressure.

As the water is heated, expansion takes place. The expanded water forces the diaphragm backwards compressing the air behind it still further and, since water cannot be compressed, the system pressure increases.

On cooling, the water contracts, the air in the expansion vessel forces the water back into the system and the pressure reduces to its original pressure of 1 bar.

Periodically, the pressure in the vessel may require topping up. This can be done by removing the cap on the Schrader valve and pumping the vessel up to its original pressure with a foot pump. The operation of expansion vessels was discussed in more detail in Chapter 2.

The pressure relief valve

The pressure relief valve (also known as the expansion valve) is installed on to the system to protect against over-pressurisation of the water. Pressure relief valves are usually set to 3 bar pressure. If the water pressure rises above the maximum pressure that the valve is set to, the valve opens and discharges the excess water pressure safely to the outside of the property through the discharge pipework.

Pressure relief valves are most likely to open because of lack of room in the system for expansion due to a malfunction with the expansion vessel. This can be caused if:

● the diaphragm in the expansion vessel has ruptured allowing water both sides of the diaphragm, or
● the vessel has lost its charge of air.

▲ Figure 3.41 A pressure relief valve

The filling loop

The filling loop is an essential part of any sealed system and should contain an isolation valve at either end of the filling loop and a double check valve on the mains cold water supply side of the loop. The filling loop is the means by which sealed central heating systems are filled with water. Unlike open vented systems, sealed systems are filled directly from the mains cold water via a filling loop. The connection of a heating system to the mains cold water supply constitutes a cross connection between the cold main (fluid category 1) and the heating system (fluid category 3), which is not allowed under the Water Supply (Water Fittings) Regulations. The filling loop must protect the cold water main from backflow and this is done in two ways:

● A filling loop has a type EC verifiable double check valve included in the filling loop arrangement.
● The filling loop must be disconnected after filling creating a type AUK3 air gap for protection against backflow.

▲ Figure 3.42 The filling loop

The filling loop is generally fitted to the return pipe close to the expansion vessel and may even be supplied as part of the expansion vessel assembly.

Permanent filling connections to sealed heating systems

It is possible to permanently connect sealed heating systems to the mains cold water supply by using a type CA backflow prevention device. The type CA backflow prevention device, when used with a pressure reducing valve, can be used instead of a removable filling loop to connect a domestic heating system direct to the water undertaker's cold water supply. This is possible because the water in a domestic heating system is classified as fluid category 3 risk. A CA device can also be installed on a commercial heating system but only when the boiler is rated up to 45 kW. Over 45 kW, the water in the system is classified as fluid category 4 risk and so any permanent connection would require a type BA RPZ valve. An example of a CA backflow prevention device can be seen in Chapter 1.

The device contains an integral tundish to remove any discharge should a backflow situation occur. Under normal operation, the valve should not discharge water. However, the valve may discharge a small amount of water if the supply pressure falls below 0.5 bar or 11 per cent of the downstream pressure.

The pressure gauge

This is to allow the correct water pressure to be set within the system. It also acts as a warning of component failure or an undetected leak should the pressure begin to inexplicably rise or fall.

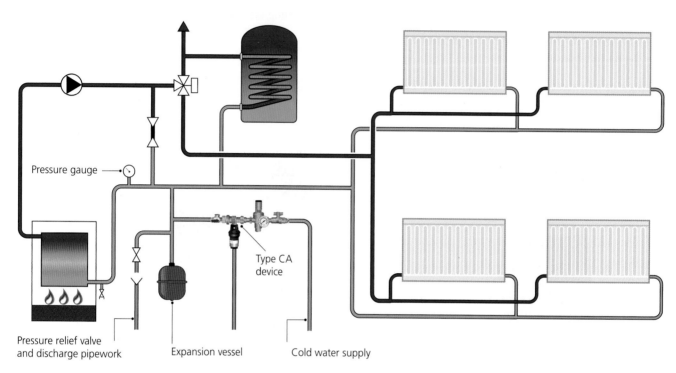

Pressure gauge

Type CA device

Pressure relief valve and discharge pipework

Expansion vessel

Cold water supply

▲ Figure 3.43 A sealed system with CA backflow prevention device

4 INSTALLATION, CONNECTION AND TESTING REQUIREMENTS OF ELECTRICALLY OPERATED CENTRAL HEATING COMPONENTS

In this part of the chapter, we will look at some of the more common wiring diagrams for domestic central heating systems:

- pumped heating systems with combination boilers
- pumped heating with gravity hot water systems
- fully pumped incorporating 3-port valves – mid-position and diverter valves
- fully pumped incorporating 2 × 2-port valves

- fully pumped incorporating hot water and multiple space heating zones
- fully pumped incorporating weather compensation, optimum start or delayed start controllers
- application of frost thermostats and boilers with pump overrun facility.

It should be remembered that electrical connections to heating systems must be through a switched fuse spur fitted with a 3 amp fuse.

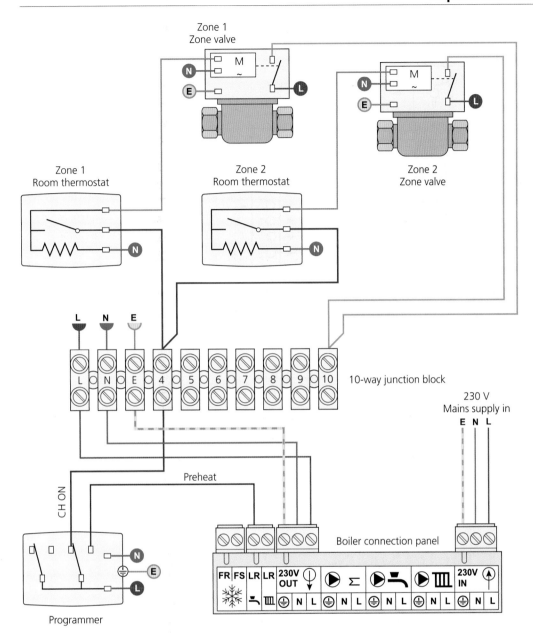

▲ Figure 3.44 Pumped heating systems with combination boilers

Pumped heating systems from a combination boiler

The wiring diagram in Figure 3.44 shows a typical combination boiler wired to separate upstairs and downstairs zone control through a wiring centre. In this particular drawing the system would have an external time clock or programmer and not a boiler integrated model.

Pumped heating with gravity hot water (C-plan)

This system is better known as the C-plan and has gravity primary circulation pipes to the cylinder from the boiler. The 2-port zone valve is operated by a cylinder thermostat to control the hot water temperature by simply preventing gravity circulation when closed. The room thermostat operates the pump. This system uses an external time clock/programmer.

C-PLAN

Gravity hot water pumped central heating

Link terminals 5-9 in the 10-way junction box

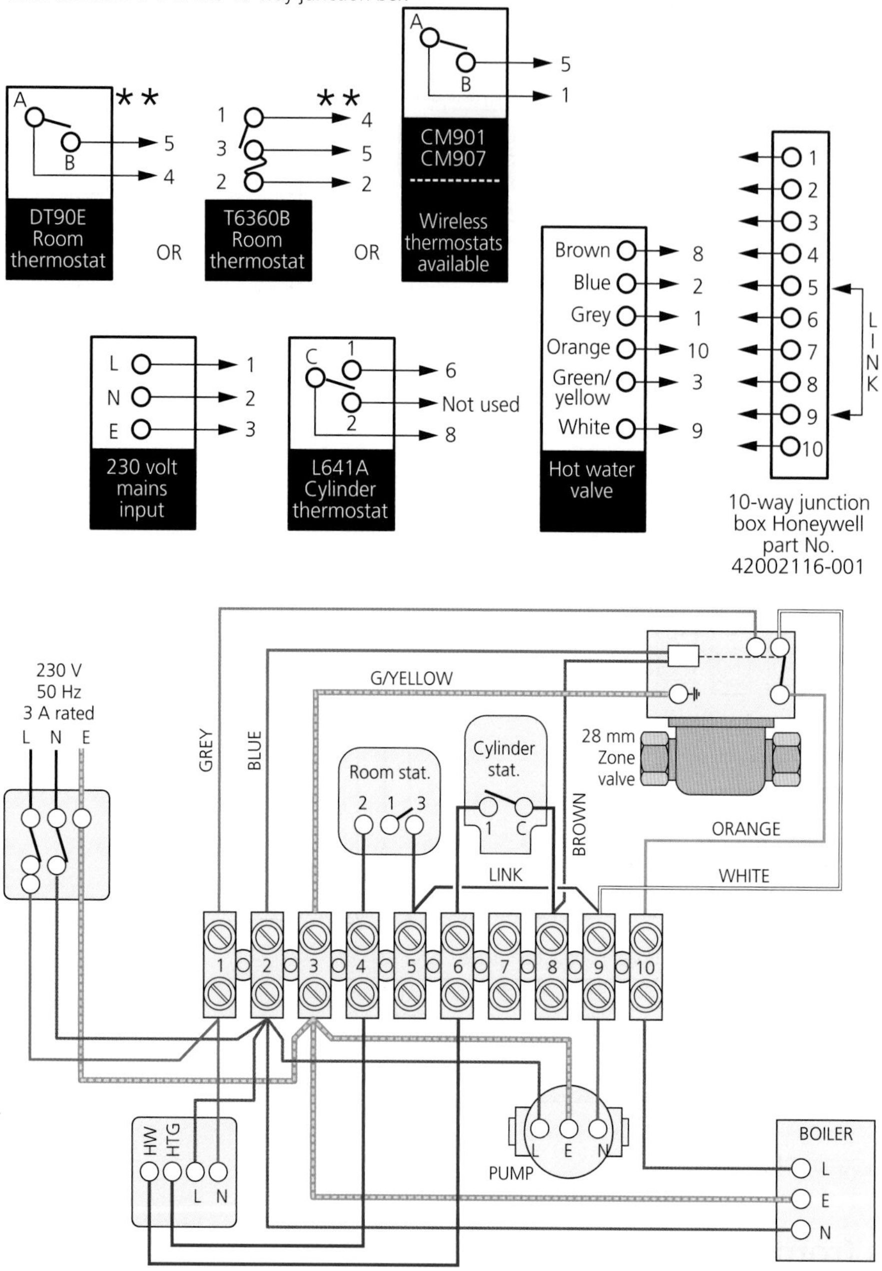

▲ Figure 3.45 Pumped heating with gravity hot water

Fully pumped incorporating 3-port valves – mid-position (Y-plan) and diverter valves (W-plan)

The two diagrams here look very similar but are two very different systems. The mid-position valve (Y-plan) allows water to be circulated to both heating and hot water circuits simultaneously because the valve will sit in the mid-position. The diverter valve (W-plan), however, will either allow flow to the hot water or central heating circuits. The W-plan system is known as a hot water priority system. In other words, the hot water storage cylinder must be satisfied before the central heating circuit will operate. The W-plan is identifiable because there are only three wires (L/N/E) from the diverter valve.

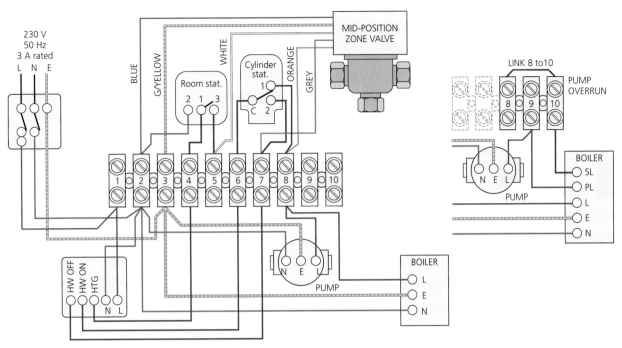

Mid-position valve (Y-plan)

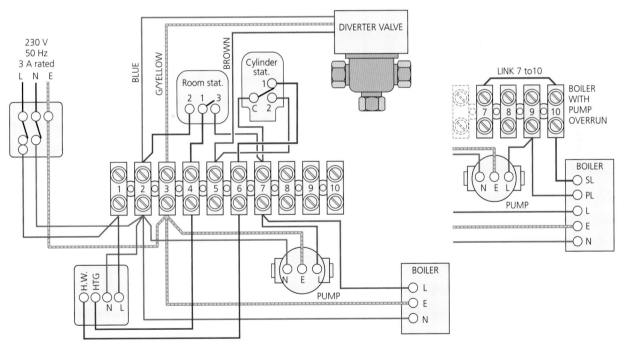

Diverter valve (W-plan)

▲ Figure 3.46 Fully pumped incorporating 3-port valves – mid-position and diverter

Fully pumped incorporating 2 × 2-port valves (S-plan)

This system is immediately identifiable because of the 2 × 2-port motorised zone valves. Some systems that include a system boiler will not require an external pump as shown in the diagram and this can be omitted.

S-PLAN

If using a 6 wire 28mm or 1" BSP V4043H on either circuit, the white wire is not needed and must be made electrically safe.

Fully pumped system only

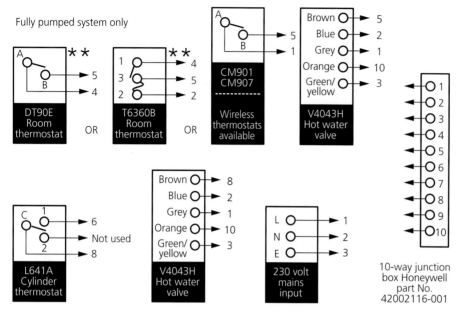

When circuit is wired as above: Complete wiring will be as line drawing below

NOTE:

It is recommended that either the 10-way junction box or sundial Wiring Centre should be used to ensure first time, fault free wiring.

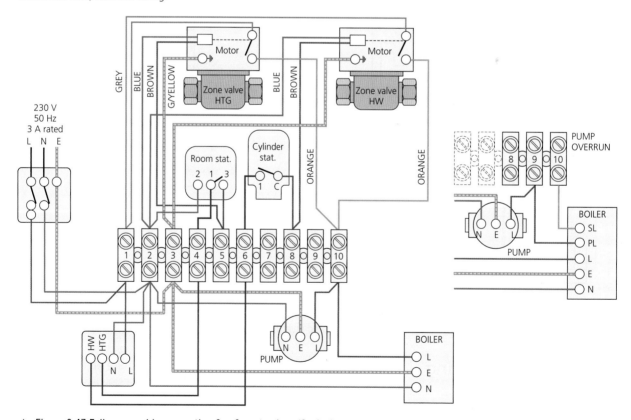

▲ Figure 3.47 Fully pumped incorporating 2 × 2-port valves (S-plan)

Fully pumped incorporating hot water and multiple space heating zones (S-plan plus)

To comply with Building Regulations Document L1, this is the recommended system for all new build domestic dwellings and refurbishments. It incorporates multiple heating zones with separate time/temperature control. Two heating zones are shown but more can be added if required.

S-PLAN PLUS

If using a 6 wire 28 mm or 1" BSP V4043H on either circuit, the white wire is not needed and must be made electrically safe.
Fully pumped system only

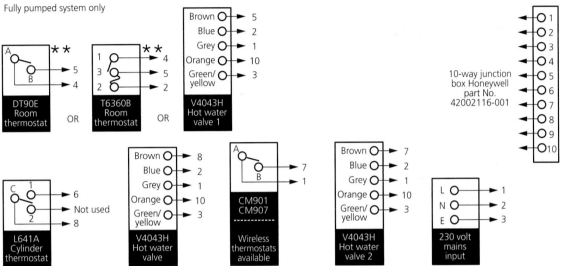

When circuit is wired as above: Complete wiring will be as line drawing below
NOTE:
It is recommended that a 10-way junction box should be used to ensure first time, fault free wiring.

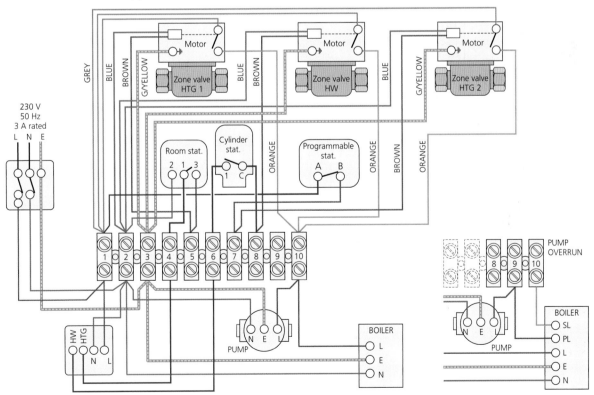

▲ Figure 3.48 Fully pumped incorporating hot water and multiple space heating zones (S-plan plus)

Fully pumped incorporating weather compensation, optimum start or delayed start controllers

The diagram in Figure 3.49 shows a fully pumped S-plan system including weather compensation. The red dashed line shows the alteration that must be made to a standard S-plan wiring arrangement to accommodate the weather compensator.

Application of frost thermostats and boilers with pump overrun

Pump overrun is a feature that some boilers require to dissipate the heat after the burner has finished firing. Pump overrun allows the hot water to be circulated away from the boiler preventing the operation of the energy cut-out. It often occurs when both the hot water and heating circuits have reached temperature and the zone valves have closed. Water is then circulated through the automatic bypass valve to keep the boiler below the high limit temperature.

The diagrams previously shown indicate the wiring arrangement of a pump overrun facility.

Frost thermostats

Frost thermostats should always be fitted in conjunction with a pipe thermostat. The frost thermostat will sense the temperature of the air whilst the pipe thermostat will sense the temperature of the water. Both thermostats have to be closed before the boiler will fire. When the water temperature reaches near freezing point, then the boiler will fire preventing the system from freezing. Frost thermostats will override all other controls to ensure the safety of the system.

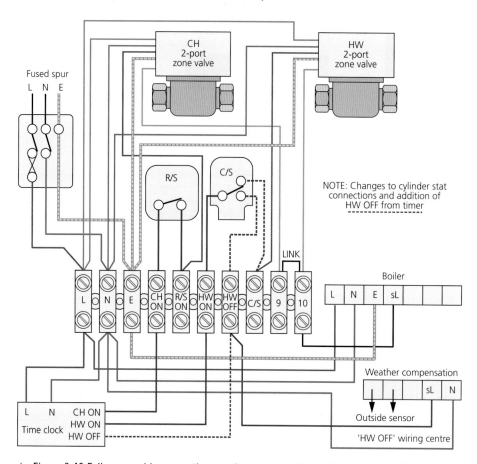

▲ Figure 3.49 Fully pumped incorporating weather compensation, optimum start or delayed start controllers

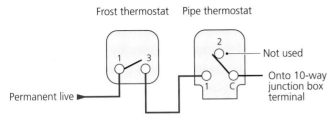

▲ Figure 3.50 Frost and pipe thermostat wiring

Electrical testing procedures

Safe isolation procedure

The Electricity at Work Regulations 1989 require that circuits engineers work on should be dead. Live work on circuits is only permitted when it is not practical for the circuit to be dead. This rule also applies to testing. However, it does not mean an engineer can work on a live circuit merely because it is inconvenient to turn the electrical supply off.

VALUES AND BEHAVIOURS

Anyone working on an electrical system should ensure that when isolating the circuit other people who could be affected are warned in advance that the supply will be turned off, and they should be allowed enough time to finish the work they are doing. For example, ensure that if working in a large building, lifts are evacuated before isolating the supply if you think that they may be affected.

HEALTH AND SAFETY

The key requirement of electrical testing procedures is to ensure the safety of others who may be affected by your actions.

Safe isolation of an electrical circuit requires that the plumber should follow the procedure stated below:
1 Firstly, identify the circuit.
2 Then if safe to do so, isolate the circuit.
3 Make sure your test voltage instrument works.
4 Ensure the test circuit is dead.
5 Test the voltage tester.
6 Lock off supply.

7 Keep the key with you.
8 Post notices in appropriate places or lock switch gear room, etc.

It is important to test your equipment before you carry out any electrical testing. This principle is the same when carrying out tests in plumbing or gas work. For example, you would test a manometer when testing sanitary pipework or a combustion analyser when commissioning the performance of a boiler. These instruments should be calibrated and tested before use. Quite often a plumber will be involved in a range of testing procedures when installing, servicing and fault-finding appliances.

Before working on any electrical appliance, it is important to identify the circuit which it is connected to and then safely isolate it. The following step-by-steps aim to help illustrate a realistic process of safe isolation.

Testing for polarity

AC electrical circuits consist of live (also known as line), neutral and earth conductors. When connections are made to AC circuits, it is vitally important the electricity flows through the live conductor to the neutral and NOT the other way around. To ensure that this is correct, a polarity test can be performed.

Polarity tests are done using a GS38 approved voltage indicator.

KEY POINT

It is important to understand that a polarity test IS A LIVE TEST so extra care is advised.

The polarity test sequence

1 Unscrew the switched fused spur from the back box to expose the electrical connections behind.
2 Using a suitable GS38 approved voltage indicator, test between live and neutral. The reading should be 230 V.
3 Test between live and earth. Again, the reading should be 230 V.
4 Test between neutral and earth. The reading should be 0 V.
5 Re-screw the switched fuse spur to the back box.

Testing for earth continuity

When we test for earth continuity, we are ensuring that the current will have a safe path to earth in the event of a fault or short circuit occurring.

Insulation and continuity testers

The use of an insulation and continuity tester covers all aspects of domestic, commercial and industrial electrical contracting, building maintenance, testing, inspection and servicing. Combining both insulation testing and continuity testing, each instrument caters for all aspects of fixed wiring installations. The instruments are useful for checking circuits that trip their protective devices.

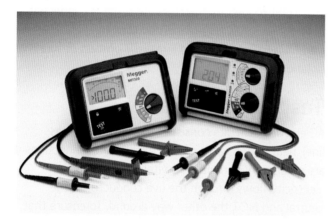

▲ Figure 3.51 Insulation and continuity testers

> **KEY POINT**
> The earth continuity test is performed WHILE THE CIRCUIT IS DEAD. Always perform the safe isolation procedure before working on electrical circuits.

The earth continuity test sequence

1 Select the circuit to be tested on the distribution board (consumer unit) and remove the live conductor from the residual current device (RCD) or circuit breaker (MCB).

2 Connect the live conductor to the earth conductor by using one of the spare earth connections on the earth bar of the consumer unit. By doing this, a circuit will be created which is half live conductor and half earth conductor.

3 Select the correct test function on the test equipment. This should usually be the low reading ohm meter function.

4 The test equipment **must** be zeroed before the test is performed. This can usually be completed by connecting the two test leads together and pressing the test button on the tester until the reading on the test equipment reads 0 ohms.

5 Unscrew the switched fused spur from the back box to expose the electrical connections behind.

6 On the switch fuse spur, touch the earth connection with the green test lead and the live connection with the brown test lead. The test reading must be less than 1 ohm.

7 At the consumer unit, place the live conductor back into the correct circuit connection.

5 COMMISSIONING CENTRAL HEATING SYSTEMS AND COMPONENTS

Testing and commissioning of central heating systems is probably the most important aspect of any heating installation as it is here that we see if the system we have installed is leak-free and performs to the requirements of the design.

Correct commissioning procedures are laid down by industry standards and manufacturer's instructions, but experience will also play a vital part in the testing and commissioning process.

In this section, we will assess the correct commissioning and testing processes that are designed to set up the system to work as the design intended.

Interpret information sources required to complete commissioning work on domestic central heating systems

Inadequate commissioning, system flushing and maintenance operations can inflict damage to even the best designed and installed central heating system. Building debris and swarf (pipe filings) can easily block pipes and these can also promote bacteriological growth. In addition, excess flux used during the installation can cause corrosion and may lead to system leakage and component failure.

It is obvious, then, that correct commissioning procedures must be adopted if the problems stated are to be avoided. There are documents that must be consulted:

- the Water Supply (Water Fittings) Regulations 1999
- British Standards **BS EN 12828, BS EN 12831, BS EN 14336** and **BS 7593**
- the Building Regulations Approved Document L1A and L1B
- the Domestic Building Services Compliance Guide
- the manufacturer's instructions of any equipment and appliances.

The Water Supply (Water Fittings) Regulations 1999

These are the national requirements for the design, installation, testing and maintenance of cold and hot water systems in England and Wales.

INDUSTRY TIP

Remember, Scotland has its own but almost identical Scottish Water Bylaws 2004.

The purpose is to prevent contamination, wastage, misuse, undue consumption and erroneous metering of the water supply used for domestic purposes. Schedule 2 of the Regulations states that:

> The whole installation should be appropriately pressure tested, details of which can be found in the Water Regulations Guide (Section 4:

Guidance clauses G12.1 to G12.3). This requires that a pressure test of 1.5 times the maximum operating pressure for the installation or any relevant part.

This document should be read in conjunction with the British Standards.

The British Standards – **BS EN 12828, BS EN 12831, BS EN 14336** and **BS 7593**

The main British Standards for the design, installation, commissioning and testing of central heating systems are:

- **BS EN 12828:2003** Heating systems in buildings. Design for water-based heating systems
- **BS EN 12831:2003** Heating systems in buildings. Method for calculation of the design heat load
- **BS EN 14336:2004** Heating systems in buildings. Installation and commissioning of water-based heating systems
- **BS 7593:2006** Code of practice for treatment of water in domestic hot water central heating systems. **BS 7593** gives guidance on the potential problems and the remedies required to maintain system efficiency in line with Approved Document L1, thereby increasing the lifespan of hot water central heating systems.

For commissioning and testing of systems, the key British Standards are **BS EN 14336** and **BS 7593**. The following sections are of relevance:

- **BS EN 14336** Section 5 Pre-commissioning checks:
  - 5.1 Objective
  - 5.2 State of the system
  - 5.3 Water tightness test
  - 5.4 Pressure test
  - 5.5 System flushing and cleaning
  - 5.6 System filling and venting
  - 5.7 Frost precautions
  - 5.8 Operational checks
  - 5.9 Static completion records
- Section 6 Setting to work
- Section 7 Balancing water flow rates
- Section 8 Adjusting of controls
- Section 9 Handover:
  - 9.1 Objective
  - 9.2 Documents for operation, maintenance and use

- 9.3 Instructions on operation and use
- 9.4 Hand over documentation
- Annex A: Guide to good practice for water tightness test
- Annex B: Guide to good practice for pressure testing
- Annex C: Guide to good practice for system flushing and cleaning
- Annex D: Guide to good practice for operational tests
- Annex E: Guide to good practice for static completion
- Annex F: Guide to good practice for setting to work
- Annex G: Guide to good practice for balancing water flow rates
- Annex H: Guide to good practice for setting of control systems.

These will be looked at a little later in the chapter.

The Building Regulations

The Building Regulations make reference to commissioning of central heating systems. These are mentioned briefly in Approved Document L1A/B. Here it states that:

(3) Where this regulation applies the person carrying out the work shall, for the purpose of ensuring compliance with paragraph F1(2) or L1(b) of Schedule 1, give to the local authority a notice confirming that the fixed building services have been commissioned in accordance with a procedure approved by the Secretary of State.

(4) The notice shall be given to the local authority–

a Not later than the date on which the notice required by regulation 16(4) is required to be given; or

b Where that regulation does not apply, not more than 30 days after completion of the work.

The Domestic Building Services Compliance Guide

The Domestic Building Services Compliance Guide 2013 (amended 2018) gives specific instructions regarding the commissioning of central heating systems. In Section 2 of the guide, its states that:

a On completion of the installation of a boiler or hot water storage system, together with associated equipment such as pipework, pumps and controls, the equipment should be commissioned in accordance with the manufacturer's instructions. These instructions will be specific to the particular boiler or hot water storage system.

b The installer should give a full explanation of the system and its operation to the user, including the manufacturer's User Manual where provided.

Manufacturer's instructions

Where appliances and equipment are installed on a system, the manufacturer's instructions are a key document when undertaking testing and commissioning procedures and it is important that these are used correctly at both installation and commissioning operations. Only the manufacturers will know the correct procedures that should be used to safely put the equipment into operation so that it performs to its maximum specification. Remember:

- Always read the instructions before operations begin.
- Always follow the procedures in the correct order.
- Always hand the instructions over to the customer upon completion.
- Failure to follow the instructions may invalidate the manufacturer warranty.

Visual inspections of central heating systems

Before soundness testing a central heating system, visual inspections of the installation should take place. British Standard **BS EN 14336** states:

- that all plant items are in accordance with the design, drawings, specifications and, where applicable, the manufacturers' instructions
- that correct installation procedures are being followed
- that the standards of installation are being met
- availability of a fuel supply and the correct installation of the flue gas removal system.

This should include the following actions:

- Walk around the installation. Check that you are happy that the installation is correct and meets installations standards.
- Check that the power supply and fuel supply i.e. gas, oil, etc. is off and cannot be inadvertently switched on. If necessary, place warning notices at key isolation points.
- Check that all radiator valves and air release valves are closed.
- Check that the drain-off valves for the system are closed.
- Check that all motorised valves have been manually opened.
- Check that the pump has been removed and a temporary section of pipe installed. This is to prevent the pump being damaged by any debris that has found its way into the system.
- Ensure that all room and cylinder thermostats are in the off position.
- Check that all capillary joints are soldered and that all compression joints are fully tightened.
- Check that enough pipe clips, supports and brackets are installed and that all pipework is secure.
- Check that the equipment (boiler, pump, motorised valves, expansion vessels, etc.) are installed correctly and that all joints and unions on and around the equipment are tight.
- Check that cisterns and tanks are supported correctly and that float operated valves are provisionally set to the correct water level.

The initial system fill

The initial system fill is always conducted at the normal operating pressure of the system. Again, the British Standard **BS EN 14336** is very clear:

> The heating system shall be water tight and tested for leakage. … This test may be an independent test or a combined test for water tightness and pressure verification.

The system must be filled with fluid category 1 water direct from the water undertaker's mains cold water supply. It is usual to conduct the fill in stages so that the filling process can be managed comfortably. There are several reasons for this:

- Filling the system in a series of stages allows the operatives time to check for leaks stage by stage.

Only when the stage being filled is leak free should the next stage be filled.

- Air locks from cistern-fed, open vented systems are less likely to occur as each stage is filled slowly and methodically. Any problems can be assessed and rectified as the filling progresses without the need to isolate the whole system and initiate a full drain down.
- Where a sealed system has been installed, the system should be filled until the pressure gauge reaches the normal fill pressure, usually 1 bar.

Once the system pressure has stabilised and the cistern (if an open vent system) is full, the furthest radiator on the index circuit can be opened and the radiator bled of air. Where sealed systems are concerned, this will cause the pressure to drop. The system should then be recharged with water to normal operating pressure.

> **KEY POINT**
> Do not be tempted to overfill the system just for the sake of filling the system with water, as this can often cause the pressure relief valve to open and discharge water.

Working back towards the boiler, open and fill the downstairs radiators first. This will ensure that airlocks do not occur on any pipework work drops to the lower radiators. Once the downstairs circuit is full, then the upstairs circuit can be filled working from the furthest radiator back towards the boiler.

If the system has been connected to the heat exchanger coil of a hot water storage vessel, open and vent the air from any automatic air valves.

When the system has been filled with water, the system should be allowed to stabilise and any float operated valves should be allowed to shut off and a check made to see where the water line is to ensure there is room for the expansion of water. Where a sealed system is installed, the water pressure should be topped up and the system allowed to stabilise. The system will then be deemed to be at normal operating pressure.

Once the filling process is complete, another thorough visual inspection should take place to check for any possible leakage. The system is then ready for pressure testing.

Pressure testing procedures

Pressure testing can commence when the initial fill to test the pipework integrity has been completed. Again, on large systems, this is best done in stages to avoid any possible problems.

The requirements of the British Standards

BS EN 14336:2004, Heating systems in buildings. Installation and commissioning of water-based heating systems, is very specific regarding the testing of central heating systems. It states:

> The heating system shall be pressure tested to a pressure at least 30 per cent greater than the working pressure for an adequate period, as a minimum of 2 hours duration. A suggested method is given in Annex B.

Annex B of BS EN 14336

B.2.2 Hydraulic pressure testing

B.2.2.1 Preparations

When preparing a hydraulic pressure testing, the following procedure should be applied:

1 Blank, plug or seal off all open ends.
2 Remove and/or blank off vulnerable items, fittings and plant pressure switches and expansion bellows.
3 Close all valves at the limits of the test section of the pipework. Plug the valves if they are not tight, or could be subjected to vibration or tampering.
4 Open all valves in the enclosed test section.
5 Check that all high points have vents, and that these vents are closed.
6 Check that the testing pressure gauge or manometer is functioning, has the correct range and has been recently calibrated.
7 Check that there are adequate drain cocks, a hose is available and that it will reach from the cocks to the drain.
8 Assess the best time to start the test in view of the duration required after completion of all the preliminaries.

B.2.2.2 During tests

For a hydraulic pressure testing, the following procedure should be applied:

1 When filling the system with water or other liquid, 'walk' the system continuously checking for leaks by the noise of escaping air or signs of liquid leakage.
2 Release air from high points systematically up through the system.
3 When the system is full of water, raise the pressure to test pressure and seal.
4 Should the pressure fall, check that stop valves are not letting and then 'walk' the system again checking for leaks.
5 When satisfied that the system is sound, have the test witnessed by, for example, the clerk of works, the client's representative, and obtain relevant signatures.

The requirements of the Water Regulations

In most domestic situations, pressure testing of central heating systems follows closely the requirements of Water Supply (Water Fittings) Regulations. The procedure for this is given in British Standard **BS EN 806**. These documents suggest that the method of testing should relate to the pipework materials installed and give specific tests for both systems with metallic pipes and those with elastomeric (plastic) pipes. This subject was covered in detail in Chapter 1.

Planning the test

Before the test is conducted, a risk assessment should be carried out. Pressure testing involves stored energy, the possibility of blast and the potential hazards of high velocity missile formation due to pipe fracture and fitting failure. A safe system of work should be adopted and a permit to work sought where necessary. Personal protective equipment should also be used.

The following factors should be carefully considered:

- Is the test being used appropriate for the service and the building environment?
- Will it be necessary to divide the vertical pipework into sections to limit the pressures in multi-storey buildings?
- Will the test leave pockets of water that might cause frost damage or corrosion later?
- Can all valves and equipment withstand the test pressure? If not, these will need to be removed and temporary pipework installed.

- Are there enough operatives available to conduct the test safely?
- Can different services be inter-connected as a temporary measure to enable simultaneous testing?
- How long will it take to fill the system using the available water supply?
- When should the test be started when the size of the system is considered?
- Preparation should also be taken into account.

Preparing for the test

- Check that the high points of the system have an air vent to help with the removal of air from the system during the test. These should be closed to prevent accidental leakage.
- Blank or plug any open ends and isolate any valves at the limit of the test where the test is being conducted in stages.
- Remove any vulnerable equipment such as the circulating pump and components and install temporary piping.
- Open the valves within the section to be tested.
- **Important!** Check that the test pump is working correctly and that the pressure gauge is calibrated and functioning correctly.
- Attach the test pump to the pipework and install extra pressure gauges if necessary.
- Check that a suitable hose is available for draining down purposes.

The hydraulic test procedure

1 Using the test pump, begin to fill the system. When the pressure shows signs of rising, stop and walk the route of the section under test. Listen for any sounds of escaping air and visually check for any signs of leakage.

2 Release air from the high points of the system or section and completely fill the system with water.

3 When the system is full and free of air, pump the system up to the required test pressure.

4 If the pressure falls, check that any isolated valves are not passing water and visually check for leaks.

5 Once the test has been proven sound, the test should be witnessed and a signature obtained on the test certificate.

6 When the pressure is released, open any air vents and taps to atmosphere before draining down the system.

7 Refit any vulnerable pieces of equipment, components and appliances.

The flushing requirements of central heating systems

Again, British Standard **BS EN 14336** is very specific regarding the flushing requirements of central heating systems. It states that:

> Systems shall, if necessary, be cleaned and/or flushed. If the system is not to be used immediately, consideration shall be given to whether the system is to be left full or empty.

In this section of the chapter, we will look at the requirements of **BS EN 14366** system flushing, paying particular attention to:

- cold and hot flushing
- system additives:
  - cleansers
  - neutralisers
  - corrosion inhibitors
- power flushing.

Cold and hot system flushing

- Cold flushing – after the system has been filled with water and checked for leakage, the system should be drained and completely emptied. This will remove much of the debris left over from the installation process, such as copper filings, that may otherwise foul the moving parts of the system or become lodged in crucial sections of the system such as the boiler heat exchanger. Once the system has drained, those components of the system removed prior to testing, such as the circulating pump, can be re-fitted and the system re-filled with fresh water following the same procedure stated on page 265.
- Hot flushing – a hot flush is conducted to remove any excess jointing compounds, flux residues and oils that may be present after the installation has been completed. The hot water separates these from the pipework and components.

With the system now full, the system should be run to its maximum operating temperature. Before this can take place, it is important to remember:

- The electricity supply to the system must be tested and switched on. Always check to ensure that the correct size fuse of 3 amps has been used in the switched fused spur.

- Ensure that the fuel supply to the boiler has been tested in line with the regulations in force and that it has been turned on.
- Ensure that all radiator valves are open to allow water circulation to take place.
- Ensure that all thermostats are calling for heat.

When the system has reached temperature, it should be switched off and completely isolated from the electricity and gas/oil supplies to prevent accidental operation of the system and drained down whilst the water is still hot.

Using central heating system cleaners, neutralisers and inhibitors

British Standard **BS EN 14336** makes recommendations regarding chemical cleaning. It states that:

> For chemical cleaning, the following procedure should be applied:
>
> a Chemical cleaning should be preceded by flushing with frequent sample testing as necessary;
>
> b The system shall be completely flushed and water filled with or without inhibitor, in accordance with the specification;
>
> c Where the whole system is not being chemically cleaned at the same time, it is recommended that the isolating valves be locked in order to avoid pollution from untreated sections.

Chemical cleanser

Central heating systems can become contaminated with mineral oils that are used to protect steel components, such as radiators, from corrosion during the manufacturing process and excess flux residues from the installation process. If the oil is not removed, then component failure becomes a greater risk because the oil attacks the rubber parts that are present in components such as motorised valves and thermostatic

radiator valves. Mineral oils can also lead to eventual pump failure. Flux residues lead to corrosion of the pipe and fittings, especially in those systems using copper tubes. The risk of failures is eliminated by the use of a chemical cleanser which is administered in accordance with the recommendations in **BS 7593:2006**. The cleanser should then be circulated around the system for a period of one hour with the boiler on, after which the system should be drained down and the system flushed to remove the cleanser until the water runs clear.

▲ Figure 3.52 System cleaner

Central heating inhibitor

Untreated central heating systems are prone to electrolytic corrosion where metal deposits form a thick, black oxide sludge which blocks boiler heat exchangers, radiators and fouls circulating pumps causing them to seize. In severe cases, sludging can result in component and even boiler failure.

In some areas of the UK, where temporary hard water exists, boilers and the associated system pipework are susceptible to scaling by the calcium carbonate deposits (limescale) present in the water. Again, this can cause major problems with loss of boiler efficiency, boiler noise and component failure.

System corrosion and scaling is prevented by adding a central heating inhibitor to the system water via the feed and expansion cistern in open vented systems or via a radiator in sealed systems. Approved Document L of the Building Regulations stipulates that where the water undertaker's cold water main exceeds 200 ppm of calcium carbonate, the feed water must be chemically treated to reduce the rate of limescale accumulation.

To be completely effective, the inhibitor must be administered at the correct dosage. If not enough inhibitor enters the system, then its protection will be diminished. A 1-litre bottle of inhibitor is enough for a ten radiator system with a water content of 100 litres.

One radiator = 10 litres. Double panel radiators count as two radiators.

▲ Figure 3.53 System inhibitor

Central heating neutralisers

Older central heating systems may require de-sludging and de-scaling using an acid-based cleanser to remove any hard, encrusted deposits that have formed on the inside of the pipework, radiators and components. Any

water containing an acid is harmful to the environment and causes major problems for water undertakers when discharged down a drain or sewer. A neutraliser, administered through either the F and E cistern, injected through a radiator or via a power flushing unit, will pacify the effects of the cleanser on the inside of the system, neutralising its effects. The system should be thoroughly flushed and the water tested to ensure that the system is free of both acid cleaner and neutraliser. An inhibitor can then be put into the system once the system has been flushed through.

▲ Figure 3.54 System neutraliser

Power flushing

When replacing boilers, or dealing with blocked pipework or radiators, a power flush may be required to remove any sludge within the system. In most cases, where a new boiler is being installed, a power flush is required as part of the warranty.

INDUSTRY TIP

Manufacturer's warranties are void if power flushing is not carried out during new boiler installations.

Power flushing involves using a high-powered pump to circulate strong, often acid based cleaning chemicals and de-sludging agents through the system. These powerful chemicals strip the old corrosion residue from the system, ensuring that the system does not contain sediment which may be harmful to new boilers, controls and valves.

The power flushing unit is connected to the heating system, often by removing the pump or a radiator.

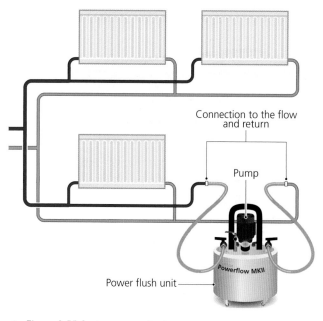

▲ Figure 3.55 System power flush

After power flushing is complete, the system may have an inhibitor added to the system water to keep the system free from corrosion.

▲ Figure 3.56 Power flush pump

Balancing a central heating system

> ### KEY POINT
> Balancing a central heating system is one of the most important parts of the commissioning process.

Balancing means adjusting all of the various thermostats, thermostatic valves and the circulating pump speed to give the desired temperatures in every room in the dwelling, whilst maximising the efficiency of the system, thus saving money and energy.

All radiators (except one) will have a thermostatic valve, usually on the flow and a lockshield balancing valve on the other. The idea is to achieve similar temperature drop of 10 °C across all radiators on the system. If the pump rate is set too high, then the temperature drop will be less than this on a correctly balanced system, so it is important that the circulating pump speed is set correctly.

VALUES AND BEHAVIOURS

Before beginning the balancing process, it is a good idea to record the temperatures and valve settings of each radiator. If anything should go wrong, then this reference point will help in starting the process again.

Where zone valves are fitted, it is especially important that these valves are open during the balancing operation.

Balancing a central heating system can be quite a lengthy process. A simple chart or table helps to record all the necessary information (see Table 3.25). Here, each radiator in turn can be adjusted and the thermostatic radiator valve and lockshield valve settings, flow and return temperatures and the vital temperature difference can all be recorded.

▼ Table 3.25 Example of the record chart

Room	Radiator	Reading	TRV setting	Lockshield valve setting	Flow temperature	Return Temperature	Temperature difference
		Original					
		1					
		2					
		3					
		4					

The table shows the record for just one room. Each room would require the same table.

The balancing process

1 Record the initial TRV and lockshield valve settings of all radiators. Also record the room thermostat temperature, the boiler temperature setting and the pump speed.
2 Open all zone valves.
3 Open all TRVs fully.
4 Open all lockshield valves fully. It is important that you record how many turns it takes to open the lockshield valves from fully closed to fully open.
5 Set the room thermostats to maximum.
6 While the system is off and cold, bleed all radiators to remove any air.
7 Check and reset, if necessary, the system pressure.
8 Turn on the boiler.
9 As the system begins to warm up, visit each radiator in turn and check which is the flow and return of each radiator. Make a note of which radiators get hot first.
10 While the system is warming, turn down the lockshield valves of the radiators that are heating up the quickest half way so that the cooler radiators catch up. This gives an approximate balance.
11 Now let the system stabilise for about an hour.
12 Using an infrared thermometer, record the first set of temperatures on the chart. Don't forget to calculate the temperature difference. This should be about 10 °C but, as this is only the first set of figures it is unlikely to be so.
13 On the radiators with the smallest temperature difference, close the lockshield valves a little, recording how many turns were made. The radiators with the highest temperature difference should be left fully open.

14 Repeat 11, 12 and 13 until all radiators are fully hot and the temperature differences are as near to 10 °C as possible.
15 Adjust the boiler temperature to give a flow temperature of 80 °C. Then, adjust the pump speed to give a temperature difference of 10 °C across the radiator flow and returns. Caution must be exercised here as altering the pump speed may mean that points 12 and 13 may need to be repeated to give the correct temperature difference.
16 Adjust all TRVs to give the desired temperatures in all rooms. This is best done over several days as rooms often take time to warm up and cool down.
17 Now, set the room thermostats to the desired temperatures.
18 Turn the boiler down to the lowest setting needed to maintain the temperatures required. It can always be increased later if needed.
19 Measure and record the flow and return temperatures at the boiler. This information should be kept with all other system records to assist in any fault-finding procedures in the future.

Dealing with defects with central heating systems that are discovered during commissioning

Commissioning is the part of the installation where the system is filled and run for the first time. It is now that we see if it works as designed. Occasionally problems will be discovered when the system is fully up and running, such as:
- Systems that do not meet correct installation requirements. This can take two forms:
  - Systems that do not meet the design specification – problems such as incorrect flow rates and

temperatures are quite difficult to deal with. If the system has been calculated correctly and the correct equipment has been specified and installed to the manufacturer's instructions, then problems of this nature should not occur. However, if the pipe sizes are too small in any part of the system, then flow rate and temperature problems will develop almost immediately downstream of where the mistake has been made. In this instance, the drawings should be checked and confirmation with the design engineer that the pipe sizes that have been used are correct before any action is taken. It may also be the case that too many fittings or incorrect valves have been used causing pipework restrictions. Another cause of flow rate and temperature deficiency is the incorrect set-up of equipment and balancing processes. In this instance, the manufacturer's data should be consulted and set-up procedures followed in the installation instructions.

- Poor installation techniques – installation is the point where the design is transferred from the drawing to the building. Poor installation techniques can cause:
  - Noise – incorrectly clipped pipework can often be a source of nuisance within systems running at high pressures because of the noise that it can generate. Incorrect clipping distances and, often, lack of clips and supports can put strain on the fittings and cause the pipework to reverberate throughout the installation, even causing fitting failure and leakage. To prevent these occurrences, the installation should be checked as it progresses and any deficiencies brought to the attention of the installing engineer. Upon completion, the system should be visually checked before flushing and commissioning begins.
  - Leakage – water causes a huge amount of damage to a building and can even compromise the building structure. Leakage from pipework if left undetected causes damp, mould growth and an unhealthy atmosphere. It is, therefore, important that leakage is detected and cured at a very early stage in the system's life. It is almost impossible to ensure that every joint on every system installed is leak

free. Manufacturing defects on fittings and equipment and damage sometimes cause leaks. Leakage due to badly jointed fittings and poor installation practice are much more common, especially on large systems where literally thousands of joints have to be made until the system is complete. These can often be avoided by taking care when jointing tubes and fittings, using recognised jointing materials and compounds and using manufacturers' recommended jointing techniques.

- Poor balancing techniques – as we have seen, the correct balancing of a system is vital if the system is to achieve the correct room temperatures in the dwelling. Incorrect balancing often means that a well designed and installed system is incapable of achieving its true efficiency potential. A poorly balanced system will often lead to:
  - insufficient room temperatures
  - increased running costs
  - increased CO_2 emissions
  - differing flow and return differentials.
- Defective components and equipment – defective components cause frustration and cost valuable installation time. If a component or piece of equipment is found to be defective, do not attempt a repair as this may invalidate any manufacturer's warranty. The manufacturer should first be contacted as they may wish to send a representative to inspect the component prior to replacement. The supplier should also be contacted to inform them of the faulty component. In some instances where it is proven that the component is defective and was not a result of poor installation, the manufacturer may reimburse the installation company for the time taken to replace the component.

Commissioning records for central heating systems

Commissioning records for large central heating systems should be kept for reference during maintenance and repair and to ensure that the system meets the design specification. Typical information that should be included on the record is as follows:

- the date, time and the name(s) of the commissioning engineer(s)
- the location of the installation

- the types and manufacturer of equipment and components installed
- the type of pressure test carried out and its duration
- the type and quantities of any inhibitors and de-scalers administered
- the temperatures of the flow and return from the boiler
- the balancing procedure and resulting temperature differences of the heat emitters/radiators
- the expansion vessel pressure.

IMPROVE YOUR ENGLISH

Clear and thorough completion of written records during initial installation will ensure the best course of action can be followed later in the event of any repair work. Ask yourself – will this information be clear and **legible** for future engineers?

KEY TERM

Legible: readable.

Examples of commissioning reports can be found in Annexes A to H of British Standard **BS EN 14336**. The records should be kept in a file in a secure location.

Notification of work carried out

At all stages of the installation from design to commissioning, notification of the installation will need to be given so that the relevant authorities can check that the installation complies with the regulations and to ensure that the installation does not constitute a danger to health.

Notification must be given to the local building control office – under Building Regulations Approved Document L, central heating systems are notifiable to the local authority building control office. Building Regulations approval can be sought from the local authority by submitting a building notice. Plans are not required with this process so it's quicker and less detailed than the full plans application. It is designed to enable small building works to get under way quickly. Once a 'building notice' has been submitted and the local authority has been informed that work is about to start, the work will be inspected as it progresses. The

authority will notify if the work does not comply with the Building Regulations.

Building Regulations Compliance certificates

From 1 April 2005, the Building Regulations demanded that all installations must be issued with a Building Regulations Compliance certificate, issued by the local authority. This is to ensure that all Building Regulations relevant to the installation have been followed and complied with.

Hand over to the customer or end user

When the system has been tested and commissioned, it can then be handed over to the customer. The customer will require all documentation regarding the installation and this should be presented to the customer in a file, which should contain:

- all manufacturers' installation, operation and servicing manuals for the boilers, heat emitters and any other external controls, such as motorised zone valves, pumps and temperature/timing controls fitted to the installation
- the commissioning records and certificates
- the Building Regulations Compliance certificate
- an 'as fitted' drawing showing the position of all isolation valves, drain-off valves and strainers, etc. and all electrical controls.

The customer must be shown around the system and shown the operating principles of any controls, time clocks and thermostats. Emergency isolation points on the system should be pointed out and a demonstration of the correct isolation procedure in the event of an emergency. Explain to the customer how the systems work. Finally, point out the need for regular servicing of the appliances and leave emergency contact numbers.

The Energy Related Products Directive and central heating boilers

The Energy Related Products Directive (ErP) sets new efficiency levels for many energy related products. Its aim is to reduce carbon emissions from products that consume energy. It came into effect on 26 September 2015. The customer has no responsibility under ErP.

ErP has two parts:

- Ecodesign – this implemented new regulations for manufacturers. Appliances such as boilers and heat pumps have to meet directed energy efficiency targets, and this resulted in manufacturers designing more efficient boiler models to meet the new criteria.
- Energy labelling – all energy related products must display a label showing the appliance energy rating. This is banded from A to G. High efficiency boilers are banded as 'A'.

  Although the responsibility for the appliance banding lies with the manufacturer, the installer may be required to provide more information by assessing the overall efficiency of the entire system. The ratings for the installed heating system and the heating controls have to be combined with the appliance efficiency to give an overall system efficiency.

INDUSTRY TIP

For more information on the ErP Directive, visit: www.tuv-sud.co.uk/uk-en/activity/product-certification/european-approvals/ce-marking-gain-access-to-the-european-market/erp-directive

Components of a gas central heating boiler

Here we will look at some of the components of central heating boilers. It is impossible to look at all the components in this book, so some of the more common ones are covered:

- heat exchanger
- water-to-water heat exchanger
- diverter valve
- gas valve (multi-functional control)
- condense trap
- air pressure switch.

The position of these components is shown on the cutaway diagram of a combination boiler (Figure 3.57).

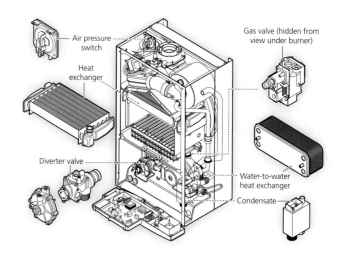

▲ Figure 3.57 Combination boiler cutaway

Heat exchanger

Heat exchangers transfer the heat from the burner into the heat transfer medium. With boilers, the heat transfer medium is water. Common materials include cast iron, stainless steel, aluminium and copper.

The heat exchanger is often located above the burner, and the products of combustion pass through it to heat the water, although with modern condensing boilers the heat exchanger is circular, with the burner passing through the middle. Here, a gas/air mixture is blown by a fan through the burner, where it is ignited, heating the heat exchanger. Irrespective of the method, the products of combustion are then directed into the flue system.

The cooler return water typically enters at the bottom of the heat exchanger and the heated flow water exits from the top.

Water-to-water heat exchanger

Often called plate heat exchangers, these are constructed from a number of corrugated plates – the more plates that are sandwiched together, the better the heat transfer. The plates provide two separated circuits for water to flow through. One circuit is connected to the primary system and is heated by the water in the primary heat exchanger. The other circuit is connected to the mains cold water supply.

As the cold water flows through the heat exchanger, it is warmed instantly by the heated plates from the primary circuit, thereby providing instant hot water to the taps.

Diverter valve

In a combination boiler, the diverter valve is an electrically operated valve that provides a means of changing the direction of the flow of heated water produced by the primary heat exchanger.

When the boiler is in heating mode, the diverter valve allows water to flow to the heating circuit. When a hot tap is opened, a flow switch activates the diverter valve to alter the flow of water through the plate heat exchanger to warm the flowing water to the hot tap.

Gas valve (multi-functional control)

The gas valve, or to give its correct name, the multi-functional control, is a valve that incorporates many components in one unit:
- a filter
- a thermoelectric safety shut off valve – this shuts down the boiler in the event of pilot failure (if applicable)

- a gas regulator – this regulates the gas to the correct pressure for the burner
- a solenoid valve – this opens when the appliance calls for heat
- a gas inlet pressure test point
- a gas burner pressure test point
- a pilot flame adjustment screw (if applicable).

There are many designs of multi-functional control, and most are based on two specific methods of operation:
- controls fitted to boilers that have a permanent pilot flame and incorporate a thermoelectric flame failure device (thermocouple)
- controls that utilise some form of electronic ignition and flame-ionisation flame failure system.

Condense trap

A condense trap, or condensate trap, is fitted to condensing boilers to collect the condensate that forms as a part of the boiler's operation, and to allow a path for the condensate to evacuate from the boiler to the outside of the building via a drain or soakaway. Condense traps are usually made from moulded plastic, and often incorporate a siphon to maintain some water in the trap at all times. The body of the trap has an inlet from the boiler heat exchanger/combustion chamber and an outlet to the drain.

Air pressure switch

The air pressure switch uses pressure differential caused by the running of the flue fan to activate a small micro-switch. Inside the air pressure switch is a small diaphragm that moves due to the movement of air caused when the fan starts. When the diaphragm moves, it operates the small micro-switch that activates the ignition sequence on the boiler. The air pressure switch is usually connected to the fan and the flue by rubber tubing.

⑥ DIAGNOSING AND RECTIFYING FAULTS IN CENTRAL HEATING SYSTEMS AND COMPONENTS

Central heating systems, if installed and commissioned in accordance with the manufacturer's data, should last at least 30 years without any major problems developing. Faults occur because of wear and tear on the system, but occasionally, faults develop many years later because of poor installation, commissioning or poor servicing techniques. A simple omission, such as not administering inhibitor to the system at the commissioning stage, will, over time develop into a major problem with black oxide sludge that could, in severe cases, mean the system has to be de-commissioned, removed and re-installed.

In this part of the chapter, we will look at fault-finding procedures for central heating systems, the causes of faults and, of course, their remedies.

The periodic servicing requirements of central heating systems

Central heating systems, like other plumbing systems in the home, require a certain amount of periodic maintenance to ensure a continued and efficient operation. Most servicing techniques centre around the appliance and its correct operation. In reality, though, this is only part of the story. For a central heating system to be properly serviced, there are many points and NOT just the boiler that need to be looked at:

1 Talk to the customer and ask if there have been any problems with the system.
2 The boiler is the centre of the system. Without it, none of the heat emitters will work. The boiler must be thoroughly serviced in accordance with the manufacturer's instructions by a gas safe registered engineer. The flue gases must be analysed for the correct CO_2 content and the appliance case checked to ensure a correct seal.
3 The system must be run and the temperatures of the flow and return at the boiler checked to ensure that the temperatures and subsequent temperature difference across the flow and return are within manufacturer's data. This is crucial for condensing appliances as the boiler will not go into condensing mode without the correct temperature difference.
4 Linked with point 2, check the correct operation of the boiler thermostat to ensure it operates with acceptable limits. It should be checked at minimum and maximum settings to ensure that the boiler shuts down satisfactorily.
5 Every radiator must be checked to ensure that it achieves the correct temperature and that the temperature difference across the flow and return is within $10\,°C \pm 1\,°C$. Over time, as deposits within the system form, openings of valves become narrower and this affects the balancing of the system. The system may need a complete re-balancing procedure depending on its age and condition.
6 Linked to point 4, if the system balancing is out, check the condition of the water. Initiate a test to ensure that the system has inhibitor in it by using a test kit like the example below. If too much sediment is detected, the system may need a complete power flush.
7 Check the operation of the room thermostat and cylinder thermostat. Take a water temperature reading and, if possible an air temperature reading.
8 Check the system for signs of leakage. Repair any leaks that may be present on the system.
9 The customer should receive a boiler servicing report on the checks carried out and their findings. The benchmark service log book should also be completed by a suitably qualified gas safe registered engineer.

IMPROVE YOUR ENGLISH

Communicate with your customer in the first instance – they will be your first source of information on any developing or persistent problems.

Obtaining details of system faults from end users

When identifying faults that have occurred on central heating systems, the customer can prove an invaluable source of information as they can often describe when and how the fault first manifested itself and any characteristics that the fault has shown.

IMPROVE YOUR ENGLISH

Verbal discussion with the customer often results in a successful repair without the need for extensive diagnostic tests. The customer should be asked:

- The immediate history of the fault.
  - When did it first occur?
  - How did they notice it?
  - What characteristics did it show?
- Did they notice any unusual discharge of water around the storage vessel or an increase or decrease in flow rate or pressure? This may well indicate the type of component failure that has taken place.
- Did they attempt any repairs themselves? If so, what did they do? This is important because if repairs have been attempted, they may well have to be undone to successfully diagnose the problem.
- What was the result of the fault? Again, an important aspect because it can often indicate where the fault lies. For instance, if the customer has noticed a drop in flow rate or pressure, this might indicate a blockage, a blocked strainer or scale growth.

How to use manufacturer instructions and industry standards to establish the diagnostic requirements of central heating system components

When attempting to identify faults with central heating systems, the most important document that can be consulted is the manufacturer's instructions. In most cases, these will contain a section on fault finding that will prove an invaluable source of information. Fault-finding using manufacturer's instructions usually takes three forms:

- known problems that can occur and the symptoms associated with them
- methods by which to identify the problem in the form of a flow chart. These usually follow a logical, step-by-step approach, especially if the equipment has many parts that could malfunction, such as a pressure reducing valve or an expansion vessel
- the techniques required for replacement of the malfunctioning component.

A replacement parts list will also be present for those components that can be replaced. When ordering parts,

it is advisable to use the model number of the equipment and the parts number from the replacement parts list. This will ensure that the correct part is purchased.

The associated British Standards may also be consulted as it contains important information regarding minimum flow rates required by certain appliances. Again, this should be used in conjunction with the manufacturer's instructions.

KEY POINT

Remember! Manufacturer's instructions always take precedence over the British Standards and Regulations.

The routine checks and diagnostics performed on central heating system components as part of a fault-finding process

Routine checks on components and systems can help to identify any potential problems that may be developing within the system as well as keeping the system operating to its maximum performance and within the system design specification. Checks performed can include:

- **Checking components for correct temperatures** – these are important checks, simply because they can indicate whether a component has started to fail and will require replacement or whether the system will require rebalancing. Those components that are pressure and temperature related such as expansion vessels and thermostats are particularly vulnerable and susceptible to failure.
- **Rust spots on radiators** – this is usually a sign that the radiator has rusted through due to excessive system corrosion. Do not touch the rust spot as it could start to leak. It should be pointed out to the customer.
- **Excessive system air in the radiators** – in actual fact, it may not be air at all. It is probably hydrogen that has collected in the top of the radiator due to electrolytic corrosion. It usually occurs regularly in one or two radiators.
- **Cleaning system components** (including dismantling and reassembly).

- **Checking for correct operation of system components:**
  - **Thermostats** – can be checked using a thermometer in the hot water flow once the thermostat has shut off and against the surface of the radiator. This will indicate whether the thermostat is operating at the correct temperature.
  - **Thermostatic radiator valves** – should be checked to ensure that they have not stuck in the closed position. TRVs that stick could indicate either potential TRV failure or system debris holding the valve closed.
  - **Motorised valves** – should be activated to ensure that they are operating in line with the system thermostats.
  - **Pumps** – these should be checked using the manufacturer's commissioning procedures to ascertain whether the pump is performing as the data dictates. A slight fall in performance is to be expected with age. Check to ensure:
    - there no signs of damage or wear and tear on the pump
    - there are no signs of leakage from the pump
    - that there are no unusual noises or vibrations when the pump is operating.
  - **Timing devices** – time clocks can be checked to see if they activate at the correct time and that any advance timings such as 1-hour boost buttons, work correctly. The time display should be checked against the correct time of the check and any alterations to the time made.
  - **Expansion vessels** – should be checked for the correct pressure using a portable Bourdon pressure gauge. The type used to check tyre pressures is ideal for this. Any signs of water leakage should be investigated. Always refer to the manufacturer's instructions for the correct charge and pre-charge pressures.
  - **Feed and expansion cisterns** – check that the float operated valve has not stuck in the off position and that the vent pipe terminates above the water line. Check that the feed pipe is clear of any debris.
  - **Gauges and controls** – gauges are notorious for requiring replacement or re-calibration as they often display an incorrect pressure. They should be replaced as necessary.
- **Checking for correct operation of system safety valves:**
  - **Pressure relief valves** – can be checked by twisting the top and holding the valve open for 30 seconds. Always ensure that the valve closes completely and that the water stops without any drips.
- **Checking for blockages in heat emitters** – if a radiator only gets hot at the top and down each side, it is probably blocked with black oxide sludge. Black oxide sludge collects in low spots on the system, especially radiators and circulating pumps. Being metallic in nature, the sludge is also attracted to any electrical component that emits a magnetic field. Radiators can be removed and flushed out but this will only be a temporary solution. The problem is system deep and requires further investigation. It could even be the system design that is causing the fault. One solution is to initiate a full system power flush using a sludge remover and then treat the system with inhibitor but the root cause of the problem must be investigated if a total cure to the problem is to be found.

> **HEALTH AND SAFETY**
>
> Caution should be exercised when checking for system air, especially when bleeding the radiator as hydrogen is highly flammable!

Should any components require replacing, they should be replaced with like-for-like components or, if this is not possible, check with the manufacturers that the part is approved for use with the system.

Methods of repairing faults in central heating system components

Repairing system components should be undertaken using the manufacturer's servicing and maintenance instructions, as these will contain the order in which the component should be dismantled and re-assembled. As with all components, there will be occasions when it cannot be repaired and replacement is the only option. Some of the components that may be repaired and/or replaced are listed in Table 3.26.

▼ Table 3.26 Central heating components

Component	Known faults	Symptom	Repair
Filling and venting – open vent systems	Blocked cold feed pipe to the system or blocked air separator	System not filling after drain down due to a blockage of sediment	The affected section of pipe or the air separator must be removed and replaced.
	System discharging water from the feed and expansion cistern overflow when the system is heating up	Water level in the F and E cistern too high and will not accommodate the expansion of water	Lower the water level and reset the float operated valve to shut off at a lower level.
	Air locks in the system	Usually occur at high spots on the installation pipework	Fit air release valves at all high spots in the pipework.
		Often occurs with older systems where the feed and vent pipes are combined	The feed and vent pipes must be separated or the problem will re-occur every time the system is drained down.
	Water pumping over the feed and expansion cistern	Often occurs when the pump, vent and feed are arranged incorrectly	Check that the pump, vent and cold feed pipes are arranged – vent/cold feed/pump (VCP) working away from the boiler.
Filling and venting – sealed systems	Service valve to the filling loop passing water	Service valve worn	Replace service valve.
	Rising pressure causing the pressure relief valve to discharge water	Service valve open slightly or worn (see fault above)	Check the service valve, replace as necessary and remove the filling loop.
		Pin hole in the hot water plate heat exchanger causing the cold water to pass through to the heating system (combination boilers only)	Replace the plate heat exchanger.
	Radiators will not vent	No pressure	Top up the pressure at the filling loop and retry venting procedure.
Pumps	Worn/broken impeller	Motor working but water not being pumped	No repair possible. Replace the pump.
	Burnt out motor	Voltage detected at the pump terminals but pump not working	No repair possible. Replace the pump.
	Cracked casing	Water leaking from the pump body	No repair possible. Replace the pump.
	Faulty capacitor	Slow starting pump	Replace the capacitor if possible. Check manufacturer's instructions.
Expansion vessels	Pressure loss due to faulty Schrader valve	No pressure in the expansion vessel. Water discharging from the pressure relief valve during water heat up	Pump air into the expansion vessel using a foot pump and check the Schrader valve with leak detector fluid. Check for bubbles. Replace Schrader valve as necessary.
	Ruptured bladder/diaphragm	Water discharging from the Schrader valve. Water discharging from the pressure relief valve on water heat-up	It is possible to replace the bladder/diaphragm of some accumulators. Check the manufacturer's instructions.
Expansion (pressure) relief valve	Water dripping intermittently when the water is being heated	Usually an indication that the expansion vessel has lost its air charge or internal expansion bubble has disappeared	Check and recharge the expansion vessel or internal air bubble.
Motorised valves	Motorised valve not activating when thermostats calling for heat	Faulty actuator or faulty motor	1 Replace valve actuator head. 2 Replace valve motor.

→

▼ Table 3.26 Central heating components (continued)

Component	Known faults	Symptom	Repair
	Valve not shutting off	As most motorised valves close when de-energised, the problem is most likely to be a seized valve or broken valve spring	Check the operation of the valve with the manual lever: 1 If the valve moves freely, then the valve spring is broken, so replace the valve head. 2 If the valve will not move, the valve is seized and must be replaced.
	Valve leaking from below the actuator head	Valve spindle seal has worn	Replace the valve.
Cylinder thermostats	Hot water too hot	System thermostat is not operating at the correct temperature	Check the temperature of the hot water with a thermometer against the setting on the thermostat. Replace the thermostat as necessary.
	No hot water	System thermostat not operating	Check the thermostat with a GS38 electrical voltage indicator for correct on/off functions. Replace as necessary.
Room thermostats – hard wired	Thermostat not activating the heating – radiators cold	Dirt/dust on the sensors	Follow safe isolation procedure and clean the dust from the thermostat.
		Room thermostat not working. Faulty thermostat	Check the thermostat with a GS38 electrical voltage indicator for correct on/off functions. Replace as necessary.
Room thermostats – wireless	Thermostat not activating the heating – radiators cold	Loss of radio connection between the thermostat and the boiler	1 Follow manufacturer's instruction to re-establish the signal. 2 Check and replace thermostat batteries.
		Room thermostat not operating correctly	Replace thermostat.
Room thermostat – Wi-Fi	Thermostat not activating the heating – radiators cold	Loss of internet connection	Check internet connection – contact internet provider.
High limit thermostat	No hot water	Usually an indication that the system thermostat has malfunctioned and the high limit thermostat has activated to isolate the heat source	Check the main system thermostat and reset the high limit thermostat.
Pressure (Bourdon) gauges	Sticking pressure indicator needle	Gauge not reading the correct pressure and does not move when the pressure is raised or lowered	No repair possible. Replace the gauge.
Weather compensation, delayed and optimum start controls	As these controls often contain multiple sensors, it is recommended that the manufacturer's installation instructions are referred to whenever possible. Alternatively, contact the manufacturer's technical help support line.		
Radiator/heat emitter	Radiator/emitter is blocked with black oxide sludge	Emitter heating at the top and sides only. Middle of the emitter is cold	1 Remove radiator and flush through with clean water. 2 Undertake a full, chemical system power flush to remove the system sludge build-up.
Thermostatic radiator valve (TRV)	Stuck valve	Radiator not getting hot	● May be possible to release the valve by gently working the pin at the top of the valve underneath the temperature sensor head up and down. ● Replace the TRV.

Safe isolation of central heating systems or components during maintenance and repair

Repair and maintenance tasks on central heating systems, appliances and valves are essential to ensure the continuing correct operation of the system. The term used when isolating a water supply during maintenance operations is 'temporary decommissioning'. There are basically two types:

- planned preventative maintenance
- unplanned/emergency maintenance.

When a maintenance task involves isolating the central heating system, a notice should be placed at the point of isolation stating 'system off – do not turn on' to prevent accidental turn on of the system. Where key components such as the expansion vessel, pressure relief, motorised valves and thermostats are found to be faulty, then the system should be isolated and temporarily decommissioned until replacement parts are obtained and fitted. If a component requires removal and replacing, it is always a good idea to cap off any open ends until the new component is installed. When removing radiators fitted with thermostatic radiator valves, the TRV should be shut off using the proprietary cap that came with the valve.

Where the equipment also uses an electrical supply, safe isolation of the electricity supply is vital and the safe isolation procedure should be followed and the fuse/supply locked off for safety (see Chapter 5, Electrical principles).

A record of all repairs and maintenance tasks completed will need to be recorded on the maintenance schedule at the time of completion, including their location, the date when they were carried out and the type of tests performed. This will ensure that a record of past problems is kept for future reference.

Where appliance servicing is carried out, the manufacturer's installation and servicing instructions should be consulted. Any replacement parts may be obtained from the manufacturers.

Procedures for carrying out diagnostic tests to locate faults in central heating system components

With central heating components, there are simple diagnostic tests that can be performed to identify exactly what the fault is. Here, we will look at the basic tests that can be performed on some of the common components and controls:

- replacement of circulating pumps
- sealed heating system components
- control components.

Replacement of circulating pumps

The first time that the customer notices that the pump isn't working is when the radiators fail to get hot or the boiler makes unusual noises. In most cases, the boiler will shut down on the high-limit energy cut out because the hot water is not being circulated away.

If pump failure is suspected:

1 Check that the rest of the system is operating (such as the room and cylinder thermostats operate the motorised valves, etc.).
2 Turn off the switched fuse spur.
3 Remove the centre pump bleed screw. This will expose the pump impeller shaft. Check that it rotates freely with a small screw driver.
   a If it does, the pump is not seized up. Now go to step 4.
   b If it does not, then free the shaft by rotating it several times and try the system again. In most cases the pump will now operate satisfactorily.
4 With the electricity on, check that there is electricity at the pump live and neutral terminals with a suitable test lamp or GS38 proving unit.
5 If 230 V is detected, it can be assumed that the pump is faulty and will need to be replaced.

KEY POINT

DO NOT use a multimeter to check for voltage. EXERCISE CAUTION AS THE SYSTEM WILL BE LIVE!

Replacing the circulating pump

1 Ensure that the new pump will fit. Some older pumps were larger and a pump extension might need to be used.
2 Isolate the heating system from the electricity supply using the safe isolating procedure as detailed in Chapter 5.
3 Remove the electrical cover and disconnect the live, neutral and earth wires.
4 Isolate the pump using the gate valves either side.
5 Placing a bowl under the pump to catch water, break open the pump unions. If the pump has been in for some years, this may prove difficult. If the unions will not move, try tapping them all the way around with a small hammer to loosen them.
6 Carefully remove the pump, taking note of the direction of flow.
7 Clean the pump valves where the washer sits to ensure a good water tight joint when the new pump is installed.

8 Install the new pump, ensuring that the direction of flow is correct. If possible, install the pump with a slight upwards angle towards the bleed screw as this helps with venting the air and prevents excessive wear on the pump bearings. Do not forget the rubber pump washers between the pump and the valves. Do not use any jointing compound unless this is necessary.
9 Reconnect the live, neutral and earth wires to the new pump. Ensure that this is done correctly. Replace the terminal box cover on the pump.
10 Turn on the pump valves and bleed the pump of air. Check for leaks.
11 Re-instate the electricity supply and test.

Sealed heating system components

Sealed heating system components occasionally malfunction. Fortunately, these are very easy to diagnose. Sealed heating system components are:

- the expansion vessel
- the filling loop
- the expansion relief valve
- the pressure gauge.

The expansion vessel

▼ Table 3.27 Expansion vessel fault finding

Fault	Probable cause	Recommended solution
No air charge in the expansion vessel	Faulty Schrader valve	Recharge the vessel with air and check the Schrader valve with leak detection fluid. If the valve is leaking, then replace the expansion vessel.
Water detected at the Schrader valve	Expansion vessel full of water due to a ruptured membrane/diaphragm in the expansion vessel	Replace the membrane if possible (check the manufacturer's instructions). If not replace the expansion vessel with one of similar capacity.

If a faulty expansion vessel is diagnosed, the system should be isolated and temporarily decommissioned until a replacement vessel is obtained and fitted.

The filling loop

Filling loops must be disconnected from the system in line with the Water Supply (Water Fittings) Regulations. However, some installers choose to ignore this fact and the filling loop is left connected to the system. This constitutes a cross connection between a fluid category 1 fluid and a fluid category 3 and, as such, is a breach of the Water Regulations. After any fault-finding process, the filling loop MUST be disconnected.

▼ Table 3.28 Filling loop fault finding (this assumes that the filling loop has been left connected)

Fault	Probable cause	Recommended solution
System pressure rising constantly	Faulty service valve	Replace the service valve and disconnect the filling loop.
Water detected when the filling loop is removed	See above. Check the double check valve as this may be passing water under back pressure	See above. Replace double check valve and leave the filling loop disconnected after refilling.

Expansion (pressure) relief valve

▼ Table 3.29 Expansion relief valve fault finding

Fault	Probable cause	Recommended solution
Water discharges from the pressure relief valve	**Intermittently** Expansion vessel has lost its air charge	• Recharge the air bubble by draining down, or; • check and recharge expansion vessel as necessary. See expansion vessel fault finding chart.
	Continually 1 Pressure relief valve has dirt under the valve seat 2 Pressure relief valve seating damaged	Replace pressure relief valve as required.

Pressure gauge

If a fault with the pressure gauge is suspected, simply tap the front of the gauge to see if the needle moves. If the gauge continues to give a false reading, replace the pressure gauge.

Control components

Control components include motorised valves, thermostats and time clocks. These are best diagnosed through the system type. In other words, the type of system may dictate the diagnostics that are performed.

> **KEY POINT**
>
> Remember the fault-finding rule with controls – first, check that you have wired the system correctly! Only start suspecting component faults when you are sure that the system is correctly wired.

Cylinder thermostat

1 Make sure that the terminals have been wired correctly:
  a Terminal C (common) is the left-hand terminal.
  b Terminal 1 is the middle terminal.
  c Terminal 2 is the right-hand terminal.
2 Disconnect terminals 1 and 2 while the checks are taking place; this will prevent false readings due to backfeed. The cylinder thermostat is faulty if:

  a terminal 1 does not become live when calling for hot water, or
  b terminal 2 does not become live when satisfied (terminal C must be live in both cases).

Room thermostat

1 Make sure the terminals have been wired correctly:
  a Disconnect terminal 3 while the checks are taking place. This will prevent false readings due to backfeed.
  b Remove wire from terminal 3.
  c Make sure terminal 1 is live.
  d Turn the room thermostat to call for heat, if live is not detected on terminal 3 then the thermostat is faulty.

Mid-position valve – Y-plan system

The valve is faulty if the valve does not operate as described in the following checks. The checks should be made in the correct order, from 1 to 6.

Valve is open for heating only:

1 a Switch off the electricity supply.
  b Disconnect grey and white wires from appropriate junction box terminals.
  c Reconnect both grey and white wires to permanent live terminal in junction box.
2 a Switch on the electricity supply. The valve should move to the fully open heating position at port A.

b   The motor should stop automatically when port A is open. The valve should remain in this position as long as electricity is applied to white and grey wires.

c   With port A fully open, the orange wire becomes live to start pump and boiler. This can be checked by feeling port A heating outlet is getting progressively hotter.

Valve is open for domestic hot water only:

3   a   Switch off the electricity supply.

b   The valve should now automatically return to open the domestic hot water port B.

c   Heating port A should close.

4   a   Isolate grey and white wires and tape over them to make them safe.

b   Remove cylinder stat wire from terminal 6 in the wiring centre box and connect to permanent live terminal.

c   Switch on the fused spur.

d   Cylinder thermostat must be set to call for heat, pump and boiler should start.

Valve is open for both domestic hot water and heating:

5   a   Switch off the electricity supply.

b   Replace the cylinder thermostat wire to terminal 6.

c   Isolate and make safe the grey wire by taping it over and connect the white wire to the permanent live.

d   Switch on the electricity supply. The motor should now move to the mid-position and stop automatically.

e   Cylinder thermostat must be set to call for heat.

f   Both ports A and B are should now be open for hot water and heating and the boiler and the pump should start. This can be checked by feeling port A heating outlet and B hot water outlet to see if they are getting progressively hotter.

6   a   Switch off the electricity supply.

b   Reconnect white and grey wires to their junction box terminals.

c   If check 5 completes satisfactorily then the problem is not the mid-position valve. The fault is elsewhere in the circuit.

2-port motorised zone valve – S-plan and S-plan plus

Zone valves on S-plan systems are faulty:

- if the motor fails to operate when live is applied to the brown wire and neutral to the blue wire (motor can be viewed with valve cover removed) – the motor should stop automatically when the valve is fully open and will stay in this condition as long as live is applied to the brown wire; the valve automatically closes under the spring return when live is removed from the brown wire
- the orange wire only becomes live after the valve has fully opened (make sure the grey wire is live)
- if the boiler continues to run when the cylinder thermostat and/or room thermostat is satisfied and/or the clock is in **off** position.

2-port motorised zone valve – C-plan

Zone valves on C-plan systems are faulty:

- if the motor fails to operate when live is applied to the brown wire and neutral to the blue wire (motor can be viewed with valve cover removed) – the motor should stop automatically when the valve is fully open and stay in this condition as long as live is applied to the brown wire; the valve automatically closes under the spring return when live is removed from the brown wire
- the orange wire only becomes live after the valve has fully opened (make sure the grey wire is live)
- if the boiler and pump continue to run when the cylinder and room thermostats are satisfied and the clock is in **off** position.

Check 1:

- Isolate the electricity supply.
- Disconnect the brown wire to the valve and terminate safely by taping it over.
- Disconnect the white wire and reconnect it to the permanent live terminal at the wiring centre.
- Disconnect pump live connection at the wiring centre and re-connect it to the permanent live terminal.
- Switch on the electricity supply. Valve should remain in the closed position. The orange wire should become live and the boiler should fire.

Check 2:

- Isolate the electricity supply.
- Restore the white wire and pump live connections to their original positions at the wiring centre.
- Connect the brown wire to the permanent live terminal at the wiring centre.
- Ensure that the grey wire is connected to the permanent live.
- Switch on the electricity supply. The valve should now open. When fully open, the orange wire should become live to fire boiler.
- Isolate the electricity supply.
- Restore the brown wire to its original terminal on the wiring centre.

If these checks complete satisfactorily, the problem is not on valve but elsewhere in circuit. It should be noted that a 28 mm or 1-inch BSP motorised valve is required for the C-plan system.

The programmer

The programmer should only be suspected as faulty:

- after a check that any links required are in place
- after a check that the programmer has power to the correct terminals
- after a check that the programmer timing is set up correctly
- if a 230 V reading does not appear at heating ON terminal when heating only is selected either on continuous or timed
- if 230 V reading does not appear at hot water ON terminal when hot water only is selected either on continuous or timed
- if 230 V reading does not appear on hot water OFF terminal with hot water off on programmer.

Alternative methods of wiring arrangements

The easiest method of wiring a domestic central heating system is using a plug-and-play-type system. Controls are simply connected to the system using a special low-voltage plug and socket system. Controls such as the room thermostat and motorised valve are plugged into a base unit which is mounted in a convenient location close to them.

Or alternative wireless components are becoming more popular and have advantages over wired controls when it comes to positioning.

SUMMARY

We have seen that central heating systems are often very complex in nature and require much planning both in their design, calculation and installation. We have investigated the new technology that is emerging and the problems that arise when things go wrong either through a fault developing or through poor installation techniques.

The way we live our lives means that a modern, efficient and economical central heating system is now considered a 'must have' for any modern home. We, as installers, need to keep our skills up to date so that we are best able to cope with the changes that are happening within the industry. We need to be able to respond to the new methods of installation, to innovative ideas and, more importantly, to the wishes of the customer.

Test your knowledge

1 A house which has a floor area of 162 m^2 should comply with which of the following control principles:

   a Controlled under one zone with one timeclock or programmer

   b Split into two zones and controlled with one timeclock or programmer

   c Controlled by one zone valve providing all radiators are fitted with thermostatic control

   d Split into two zones, each zone having independent time control

2 Identify the component in the image below:

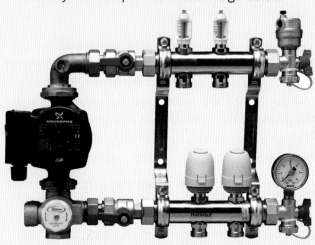

   a Underfloor heating manifold

   b Low loss header

   c 'H' Frame

   d Magnetic filter system

3 What is the most suitable backflow prevention device to be installed to the filling loop between the heating circuit and the mains cold water in a domestic situation?

   a Single check valve

   b AUK air gap

   c Double check valve

   d RPZ valve

4 When designing a central heating system for a new dwelling, which of the following British standards would be most relevant for reference?

   a BS 1710

   b BS EN 806

   c BS EN 14336

   d BS 7670

5 The term 'U-value' has a unit of measure taken as:

   a W/m^2 K

   b KW/m^3

   c J/sec

   d J/m^3 sec

6 The image below shows a wiring diagram for a heating system. What type of heating system layout does it represent?

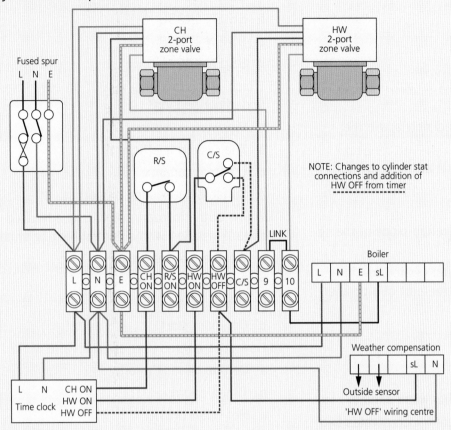

a C-plan

b S-plan

c W-plan

d Y-plan

7 Select the correct formula to determine the required flow rate (Q) in kg/s through heating pipes from the options below:

a $kW/\Delta p \times 9.81$

b $kW \times SHC/\Delta p$

c $SHC/3600 \times \Delta t$

d $kW/SHC \times \Delta t$

8 Identify the pattern of the underfloor heating pipework in the image below.

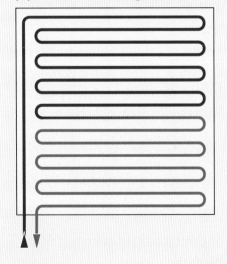

a Snail

b Bifilar

c Linear

d Series

9 What is the most efficient method of connecting radiators to pipework?

   a BOE

   b TBOE

   c BBOE

   d TBSE

10 Where is the best place for an automatic bypass valve to be installed?

   a On the flow pipe immediately above the boiler

   b On the primary return after the circulator

   c Between the flow and return pipes near to the hot water cylinder

   d Between the flow and return pipes of the index radiator

11 Where does the water system water flow to on the 'B' connection of a three-port mid-position valve?

   a The boiler

   b The hot water cylinder

   c The radiators

   d Both the heating and hot water circuits

12 Which central heating system incorporates two two-port valves to control the system water?

   a S-plan system

   b Y-plan system

   c C-plan system

   d W-plan system

13 On a central heating system with a conventional boiler, what is the recommended temperature difference between the flow and return pipework?

   a 25 °C

   b 20 °C

   c 15 °C

   d 10 °C

14 What type of central heating system is filled using a filling loop?

   a Y-plan plus system

   b W-plan system

   c Sealed system

   d Open vented system

15 What standard outlines the commissioning procedure for a central heating system?

   a BS EN 806

   b BS EN 14336

   c BS EN 12828

   d BS EN 442

16 Which of the following is NOT an advantage of underfloor heating?

   a Pipework is hidden

   b The system is efficient

   c Low maintenance

   d Suitable for existing properties

17 Which of the following is NOT an advanced control installed with a central heating system?

   a Weather compensation

   b Automatic by-pass

   c Delayed start

   d Optimum start

18 How does a three-port divertor valve work?

   a Central heating priority

   b Timed priority

   c Hot water priority

   d Equal system pressure

19 What is the principal way in which a radiator heats up a home?

   a Convection

   b Radiation

   c Infrared

   d Conduction

20 You are called to a customer's property, where a radiator fitted with a TRV is not getting hot. What could be the problem?

   a The circulator has failed

   b The boiler has not fired

   c The regulating pin in the head is stuck

   d There is blockage in the return pipe

21 What would be the primary symptom if a circulator were positioned immediately prior to the open vent pipe?

a Pumping over

b Over heating

c Restricted heating

d Positive pressure

22 What is required to be installed either side of a circulator?

a Air release valve

b Isolators

c Zone valves

d Lockshield valves

23 When replacing a synchron motor, what should be your first course of action?

a Release the motor housing and pull the tab connectors off

b Isolate the electrics and check the supply is dead

c Disconnect the wires, noting the connections

d Isolate the water for the boiler

24 To install an oil fired boiler, the competent person is required to be registered with which organisation?

a OFTEC

b NICEIC

c HETAS

d Gas Safe

25 Which statement best describes a programmer?

a A switched timer that controls the boiler

b A controller that is designed to directly activate the pump and boiler

c A switched timer that independently controls the hot water and heating systems

d A timer that activates the zone valves

26 Using the formula below, calculate the size of an expansion vessel required for a large domestic central heating system containing 200 litres of water.

$$V = \frac{eC}{1 - \frac{P1}{P2}}$$

Where:

V = the total volume of the expansion vessel

C = 200 l

P1 = 1.5 + 1

P2 = 6 + 1

e = 0.0324

27 Calculate the air change heat loss of a dining room measuring 4 m × 4 m × 3 m, when the outside temperature is −5 °C and the internal temperature is 20 °C.

28 What function does a weather compensator provide?

29 Where would a low loss header typically be used?

30 What is the basic difference between an S-plan system and an S-plan plus system?

31 Describe the function of a low loss header in a central heating system.

32 List the advantages of an underfloor heating system compared to a conventional central heating system.

33 Describe why air changes are a factor when selecting and designing a heating system.

34 Complete the following table for the faults and repairs that may occur on a central heating circulating pump.

Fault	Symptom	Repair
Broken impellor		
Burnt out motor		
Cracked case		
Faulty capacitor		

Answers can be found online at www.hoddereducation.co.uk/construction.

CHAPTER 4

ADVANCED SANITATION AND RAINWATER SYSTEMS

Plumbers all over the country will be working on sanitation and rainwater systems which need to be installed correctly, work efficiently and remove waste hygienically. This means a good understanding is required about the fundamental design and workings of these systems so that an informed discussion can take place with the customer over the choice of materials, components and layouts. It will also mean that the customer's and installer's health and safety will be paramount whilst working on the system at height or underground; whilst using some heavy materials and due to the nature of the foul water.

A well-designed, installed and maintained system will result in a satisfied customer who will be much more likely to offer repeat trade.

By the end of this chapter, you will have knowledge and understanding of the following areas of advanced sanitation and rainwater systems:
- the legislation relating to the installation and maintenance of sanitation and rainwater systems
- system layouts of sanitation systems
- design techniques for sanitation and rainwater systems
- commissioning sanitation systems and components
- servicing and maintenance of sanitation and rainwater systems
- diagnosing and rectifying faults in sanitary systems and components.

You can revisit Book 1 to remind yourself of the basics of sanitation and rainwater systems. In Book 1, Chapter 8, we covered:
- understanding layouts of gravity rainwater systems
- installing gravity rainwater systems
- understanding maintenance and service requirements of gravity rainwater systems
- decommissioning rainwater and gutter systems and components
- performing a soundness test and commission rainwater, gutter systems and components.

In Book 1, Chapter 9, we covered:
- sanitary pipework and appliances used in dwellings
- installing sanitary appliances and connecting pipework systems
- service and maintenance requirements for sanitary appliances and connecting pipework systems
- the principles of grey water recycling.

1 THE LEGISLATION RELATING TO THE INSTALLATION AND MAINTENANCE OF SANITATION AND RAINWATER SYSTEMS

Knowing and understanding the correct sources of information is paramount to any work you undertake. Following the relevant legislation will not only lead to good design and proper installation of a system, but it makes sure the installation is within the law. Building Regulations are governed by law which means they are mandatory and must be followed. British Standards and Codes of Practice outline Building Regulations in slightly more practical terms, and if you follow British Standards fully, then you will have covered everything required in Building Regulations.

Drainage design to the European code **BS EN 12056**

The correct sizing of discharge pipework is essential to maintain the **equilibrium** of the traps within a system, which will prevent the ingress of foul smells through the loss of trap seals.

KEY TERM

Equilibrium: when referring to drainage systems this relates to keeping the air pressure even within the system, so that any negative pressure or pressure fluctuation does not cause any trap seal loss and therefore the ingress of foul air in to the property.

When designing any drainage system, it is important to discuss the customer's needs trying to incorporate them in to the design. The building layout may also pose problems with any proposal, but most of these can be overcome with careful planning and design which would incorporate all current standards and regulations. Regular and good communication with the customer is important outlining any options and variations to the original design at the outset of any work, as changes later in the installation could be costly and could lead to disputes.

IMPROVE YOUR ENGLISH

Regular and good communication with the customer is important, outlining any options and variations to the original design at the outset of any work, as changes later in the installation could be costly and could lead to disputes.

A full survey of the proposed job is essential to understand what type of system would be most suitable. A visual inspection of an existing installation may show up concerns or areas where the system does not comply with current standards. There could also be problems with previous workmanship, such as incorrect gradients causing backflow or trap seal loss; excessive pipe runs; insufficient clipping also causing problems for the installation.

▲ Figure 4.1 Plumber talking with customer

▲ Figure 4.2 An example of poor practice of pipework installation

Presenting design calculations and quotations

IMPROVE YOUR MATHS

When a Level 3 plumber is qualified, they may take more of an active role in discussing the pricing for jobs in more detail.

Along with the written quotation, there may be a request for some working drawings which would take the form of well-drawn, but not to scale drawings. If scaled drawings were required, these may have to be drawn professionally.

INDUSTRY TIP

The name, address and contact details of your company should be clearly stated on any quotation, so often company headed paper is used which gives a professional appearance. If the company belongs to any professional bodies like the Chartered Institute of Plumbing and Heating Engineering, then the logo of the body would also be included.

CIPHE
ciphe.org.uk

▲ Figure 4.3 Chartered Institute of Plumbing and Heating Engineering logo

The customer's name and address should also be clearly shown on the quotation as well. The make and model of appliances and fittings should be clearly identified on the quotation. The start and end dates for a quoted job could be outlined along with any possible disruption to the customer like existing appliances or rooms that may be out of service for a period of time.

It is advisable for a plumber to be part of a certification scheme when installing a new system or when extending an existing system. If they are not a member then the local authority will need to be informed prior to the commencement of work.

> **KEY POINT**
> You should have read Book 1, Chapter 9, which covers all the basic information for drainage.

Regulations and Standards

Before we go too deeply into the sanitation and rainwater systems it would be useful to outline the regulations associated with these areas as a reminder. More detail can be found in Book 1, Chapter 9.

The Building Regulations Approved Document H

Each Building Regulation starts with a different letter and sometimes with an additional number, such as Building Regulations Approved Document H, which deals with the following:

- H1 Foul water drainage
- H2 Waste water treatment systems
- H3 Rainwater drainage
- H4 Building over sewers
- H5 Separate systems of drainage
- H6 Solid waste storage.

The Building Regulations Approved Document F

This gives information on ventilation requirements for dwellings and states that a bathroom should have an intermittent rate of extraction of 15 l/s. It also gives advice on the ventilation requirements for other rooms.

The Building Regulations Approved Document G

This covers the hygiene for sanitary conveniences and washing facilities. It also includes bathrooms and section 3 (G3) deals with hot water.

BS EN 12056

This offers guidance and advice when installing any sanitary system and is broken into five parts:

- Part 1 General information and performance requirements
- Part 2 Sanitary pipework including layout calculations
- Part 3 Roof drainage and layout calculations
- Part 4 Waste water lifting plant
- Part 5 Installation and testing, instructions for operation maintenance and use.

▲ Figure 4.4 Bathroom suite

② SYSTEM LAYOUTS OF SANITATION SYSTEMS

> **KEY POINT**
> Return to Book 1, Chapter 8 to remind yourself of the layout of rainwater systems.

British Standards are one of the most useful sources of information when designing and installing a sanitation system. The standard allows you to look at various designs used in different situations outlining their individual design criteria.

The different types of systems (BS EN 12056 Part 2)

Book 1, Chapter 9 has already gone into some detail about the four main types of soil stack systems used in the UK.

System I – single stack system with partly filled branch discharge pipes

This is the primary ventilated stack (Figure 4.5). This means the sanitary appliances are connected to partly filled branch discharge pipes which are designed with a filling degree of 0.5 (50 per cent) and are connected to a single discharge stack.

System II – single stack system with small bore branch discharge pipes

This is the secondary ventilated stack (Figure 4.6). When sanitary appliances are connected to small discharge pipes, the small bore discharge pipes are designed with a filling degree of 0.7 (70 per cent) and are then connected to a single discharge stack.

System III – single stack system with full bore branch discharge pipes

When sanitary appliances are connected to full bore discharge pipes, the full bore branch discharge pipes are designed with a filling degree of 1.0 (100 per cent) and each branch pipe is separately connected to a single discharge stack. (This is based on UK practice.)

System IV – separate discharge stack system

This is the ventilated discharge branch stack (Figure 4.7). Drainage systems type I, II, III may also be divided into black water stack serving WCs and urinals and then a grey water stack serving other appliances.

> **HEALTH AND SAFETY**
> Always wear the correct PPE when working on foul water systems to protect yourself from infections and diseases, such as Weil's disease.

Ventilation of sanitary systems

As outlined in Book 1 and earlier in this chapter, for any sanitation system to work efficiently the equilibrium of the trap seals MUST be maintained to prevent trap seal loss due to negative pressure or pressure fluctuations. Equilibrium of pressure is created by good design, correct commissioning and regular maintenance.

When selecting the above ground drainage system, you must consider:
- size of the property
- number of outlets
- lengths of waste run
- occupancy.

This information should be used in line with **BS EN 12056**.

There are four types of sanitation systems used in the UK:
- primary ventilated stack system
- secondary ventilated stack system
- ventilated branch discharge system
- stub stack.

All these systems have been outlined in detail in Book 1.

> **KEY POINT**
> On a primary ventilated stack system, the soil pipe acts as the ventilation pipe, hence the name soil and vent pipe – SVP.

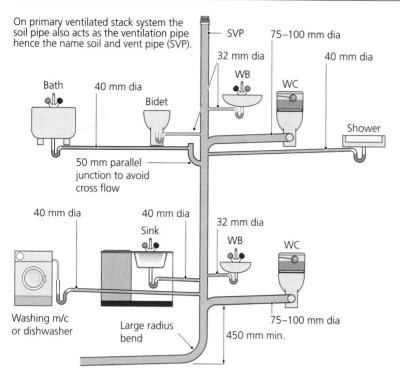

On primary ventilated stack system the soil pipe also acts as the ventilation pipe hence the name soil and vent pipe (SVP).

SVP

75–100 mm dia

32 mm dia

40 mm dia

WB

WC

Bath

40 mm dia

Bidet

Shower

50 mm parallel junction to avoid cross flow

40 mm dia

40 mm dia

32 mm dia

Sink

WB

WC

Washing m/c or dishwasher

Large radius bend

75–100 mm dia

450 mm min.

▲ Figure 4.5 Primary ventilated stack system

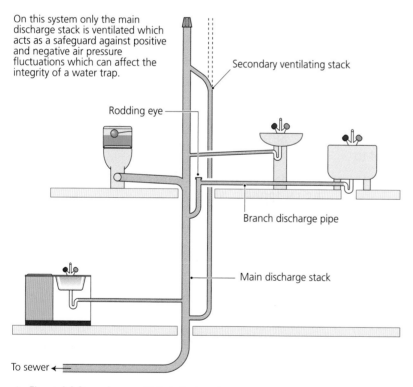

On this system only the main discharge stack is ventilated which acts as a safeguard against positive and negative air pressure fluctuations which can affect the integrity of a water trap.

Secondary ventilating stack

Rodding eye

Branch discharge pipe

Main discharge stack

To sewer ◄

▲ Figure 4.6 Secondary ventilated stack system

Water trap seals are protected from induced or self siphonage because a branch ventilating pipe is located not more than **750 mm** from any appliance.

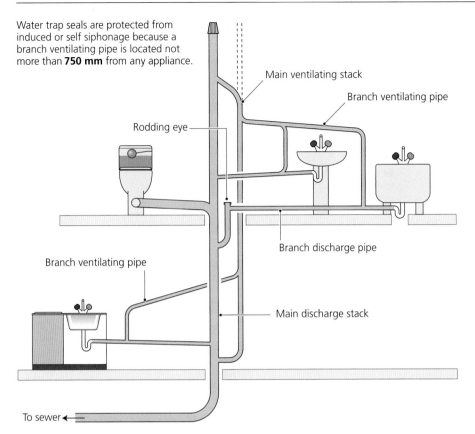

▲ Figure 4.7 Ventilated branch discharge system

Ventilation is required if the highest connection of an appliance from the invert from the drain exceeds 2 metres or if the distance between the crown of the WC connection and the drain inverts exceeds 1.3 metres.

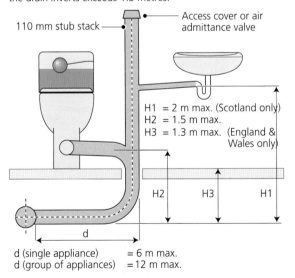

H1 = 2 m max. (Scotland only)
H2 = 1.5 m max.
H3 = 1.3 m max. (England & Wales only)

d (single appliance) = 6 m max.
d (group of appliances) = 12 m max.

▲ Figure 4.8 Stub stack system

Primary ventilated stack system

The design of this system relies on the soil pipe acting as the ventilation pipe and so there are no separate ventilation pipes. This can only be achieved by following the guidelines laid down in **BS EN 12056**:

- All sanitary appliances must be closely grouped to the discharge stack.
- All appliances, as far as possible, should be fitted with a 'P' trap or waste valve, with a discharge pipe diameter equal to that of the trap.
- Bends and branches should be avoided and the gradient must be kept to a minimum, normally 2½–5°.
- The vertical discharge stack must be installed as straight as possible and incorporate a long radius bend or 2 × 45° bends at the base (the radius should be twice the diameter of the pipe). This will avoid compression.

Close grouping of appliances is achieved by adhering to the pipework **dimensions and distances** from appliance to soil stack.

Dimensions and distances of appliances and waste pipework from the soil stack.

32 mm waste	1.7 m
40 mm waste	3.0 m
50 mm waste	4.0 m
100 mm waste	6.0 m

Waste pipes

All of the measurements above will be drastically reduced if bends are introduced to change direction or the gradient on the branch discharge pipework is too steep.

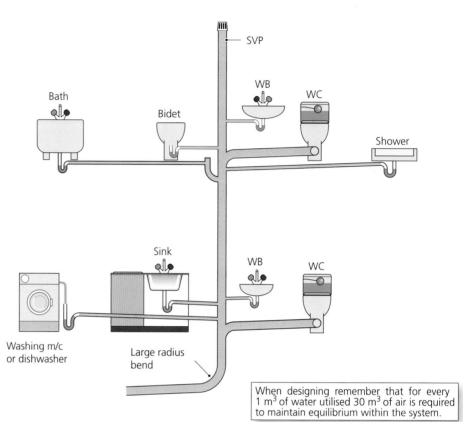

When designing remember that for every 1 m³ of water utilised 30 m³ of air is required to maintain equilibrium within the system.

▲ Figure 4.9 Primary ventilated stack system

Close grouping of appliances is achieved by adhering to the pipework **dimensions and distances** from appliance to soil stack.

Dimensions and distances of appliances and waste pipework from the soil stack.

32 mm waste	1.7 m
40 mm waste	3.0 m
50 mm waste	4.0 m
100 mm waste	6.0 m

Waste pipes

All of the measurements above will be drastically reduced if bends are introduced to change direction or the gradient on the branch discharge pipework is too steep.

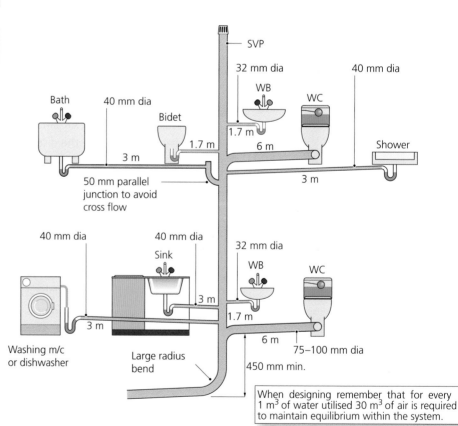

When designing remember that for every 1 m³ of water utilised 30 m³ of air is required to maintain equilibrium within the system.

▲ Figure 4.10 The maximum branch pipework diameters, lengths, gradients and trap seals on a ventilated stack system

KEY POINT

Normally the distance from a soil stack for a 32 mm appliance (such as a basin) branch pipe is 1.7 m, but the length of permissible pipe run will decrease if the gradient is increased.

▼ Table 4.1 The effects of gradient on sanitary pipework

Branch connections		
Pipe size (mm)	Maximum length (m)	Approximate gradient (mm/m)
32	1.7	22
32	1.1	44
32	0.7	87
40	3.0	Between 18 and 80
50	4.0	Between 18 and 80
100	6.0	Minimum 18

Whenever the pipe lengths shown in Table 4.1 are exceeded, the waste pipe diameter should be increased to the next size up.

By using the graph in Figure 4.11, showing 32 mm waste pipe, it can be seen that if the gradient changes

from 22 mm/m to 44 mm/m, the permissible length of discharge pipework reduces.

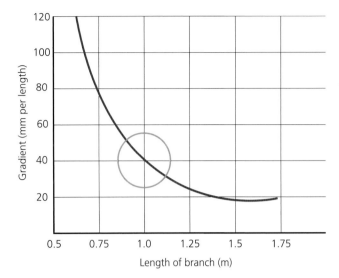

▲ Figure 4.11 Gradient graph for a 32 mm wash basin waste pipe

KEY POINT

The graph clearly shows that if the gradient of fall is increased, then the length of waste pipe is reduced accordingly.

▼ Table 4.2 **BS EN 12056** – maximum permissible pipe lengths

Appliance	Diameter (mm)	Maximum pipe length (m)	Pipe gradient (mm per m run)	Maximum number of bends	Maximum drop (m) vertical pipe
Wash hand basin or bidet	32	1.7	18 to 22	None[1]	None
	32	1.1	18 to 44	None[1]	None
	32	0.7	18 to 87	None[1]	None
	40	3.0	18 to 44	2[1]	None
Bath or shower	40	3.0[2]	18 to 90	No limit	1.5
Kitchen sink	40	3.0[2]	18 to 90	No limit	1.5
Domestic washing machine or dishwasher	40	3.0	18 to 44	No limit	1.5
WC with outlet up to 80 mm diameter	75	No limit	18 min.	No limit[4]	1.5
WC with outlet over 80 mm diameter	100	No limit	18 min.	No limit[4]	1.5
Bowl urinal[4]	40	3.0[3]	18 to 90	No limit[4]	1.5
Trough urinal	50	3.0[3]	18 to 90	No limit[4]	1.5
Slab urinal[5]	65	3.0[3]	18 to 90	No limit[4]	1.5
Food waste disposal unit[6]	40 min.	3.0[3]	135 min.	No limit[4]	1.5
Sanitary towel disposal unit	40 min.	3.0[3]	54 min.	No limit[4]	1.5
Floor drain	50 to 100	3.0[3]	18 min.	No limit	1.5
Branch serving 2 to 4 wash basins	50	4.0	18 to 44	None	None

→

▼ Table 4.2 **BS EN 12056** – maximum permissible pipe lengths (continued)

Appliance	Diameter (mm)	Maximum pipe length (m)	Pipe gradient (mm per m run)	Maximum number of bends	Maximum drop (m) vertical pipe
Branch serving several bowl urinals[4]	50	3.0[3]	18 to 90	No limit[4]	1.5
Branch serving 2 to 8 WCs	100	15.0	9 to 90 (21.5)	2	1.5
Up to 5 wash basins with spray taps[7]	32	4.5[3]	18 to 44	No limit[4]	None

1 Excluding the 'connection bend' fitted directly or close to the trap outlet.

2 If no longer than 3 m, this may result in noisy discharge, and there will be an increased risk of blockage.

3 Should ideally be as short as possible to limit deposition problems.

4 Sweep bends should be used: not 'knuckle' bends.

5 For up to 7 people: longer slabs should have more than one outlet.

6 Includes small potato-peeling machines.

7 Wash basins must not be fitted with outlet plugs.

ACTIVITY

When you are next on site, identify the design of soil stack used for various styles of buildings including larger and non-residential properties; or as a learner group walk past nearby buildings and carry out a similar survey. Photograph them and discuss their installation on return.

These drainage systems will connect in to differing underground systems according to the location and age of the property. The design of the underground system can be identified by a site survey or by contacting the local authority. The three main types of underground drainage systems are fully outlined in Book 1, Chapter 9:

- separate system
- partially separate system
- combined system.

Air admittance valves (AAV) and their importance

There are alternative ways which allow air into a system to keep the equilibrium. Air admittance valves can be installed in certain situations to allow this. The British Standard for air admittance valves is **BS EN 12380**.

An air admittance valve or AAV, is a device fitted to the top of sanitary pipework which allows air to enter the system, but does not allow foul smells to escape. Its purpose is to maintain the equilibrium in the system and balance out any suction effects which could cause a trap seal to be lost. The mechanism of the AAV returns to the closed position after it has been opened by a negative pressure created within the stack.

Operating principle of an AAV

The valve incorporates a sealing diaphragm which lifts and allows air to be drawn into the system when subject to negative pressure. Once the negative pressure has ceased – equilibrium – the pressure on the diaphragm returns to the closed position thereby preventing any escape of foul air into the building. Some AAVs have a spring to return the diaphragm to the closed position. The AAV is designed to open and close spontaneously when required allowing a supply of air to adequately ventilate the system to ensure a smooth discharge.

An AAV enables ventilating pipes to be terminated inside a building, which eliminates the need for a vent pipe to penetrate a roof and so offers flexibility in the design of new systems. Bear in mind that an AAV is NOT a substitute for a ventilation stack; wherever an AAV is installed the system at some point must have an open vent to outside air – a ventilating pipe should be provided at or near the head of each main drain.

- An AAV should not be used if the soil stack provides the only ventilation to a septic tank or cesspool.
- An AAV should not be used on a system with no open ventilation on an existing system.
- An AAV should not normally be used outside because of the risk of freezing due to moisture causing the AAV to stick in either the open or closed position. However, there are some AAVs available which are suitable for external use – these external AAVs can be removed to provide a secondary **rodding point** – reading the manufacturer's instructions is key.

KEY TERM

Rodding point: a place where the drain or section of drain can be accessed to clear any blockages.

Figure 4.12 shows how an AAV functions with the arrows showing the air flow.

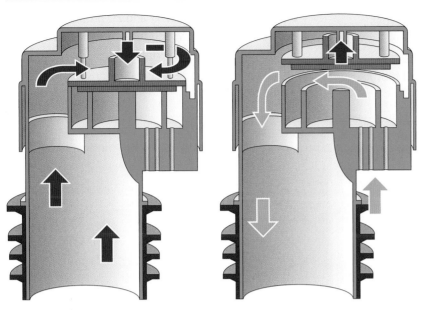

▲ Figure 4.12 AAV in action

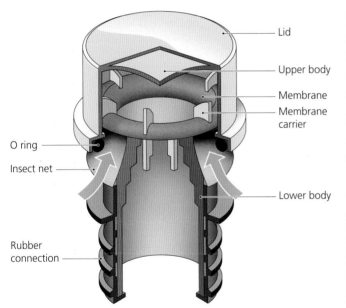

Lid

Upper body

Membrane

Membrane carrier

O ring

Insect net

Lower body

Rubber connection

▲ Figure 4.13 Cut away AAV

AAVs are generally used on discharge stacks up to 45 m or ten storeys, with an airflow capacity that equals 43 l/sec. They should always be installed vertically with a minimum of 200 mm above the highest branch connector. Preferably they should be located in a non-habitable space, such as a duct, boxing or roof space which is required to have adequate ventilation and be accessible for maintenance. If the AAV is boxed in, vents should be located at high and low positions to allow full movement of air and avoid any build-up of moisture.

The valves can also help indicate the possibility of blockages within the system by showing evidence of high water levels in a WC bowl following flushing and the consequent slow drainage of appliances located upstream from the blockage. An AAV should therefore always be installed above the flood level of the highest appliance.

It is important to remember at the design and installation stage that size, limitations and the use of AAVs are subject to national and local regulations and practice. It would be good practice to consult with the building control officer (BCO) at the local authority office.

> **KEY POINT**
>
> An air admittance valve provides a means of ventilation to a drainage system under conditions of reduced pressure when ventilating pipes are terminated inside a building in accordance with **BS EN 12056.2:2000.**

AAV installation in multi-dwelling properties

If AAVs are fitted, then the following rules apply:
- One stack in five must be ventilated to atmosphere and this should be located at the head or start of the drain run.
- Up to four properties of up to three storeys high can use AAVs.
- All multi-storey buildings will require additional ventilation if more than one property is connected to a drain which is not ventilated by a conventional stack.

- Where a drain serves more than four properties with AAVs, like a housing estate, then:
  - if there are between 5 and 10 properties ventilation is required at the head and foot of the drain run

- if there are between 11 and 20 properties then ventilation is required at the head, foot and mid-point of the drain run.

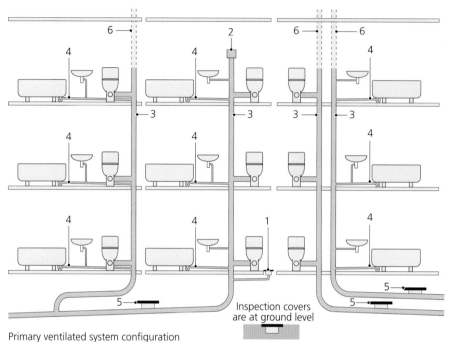

Primary ventilated system configuration
1. Floor gully   2. Air admittance   3. Stack   4. Branch discharge   5. Drain lid   6. Stack vent
                     valve   pipe   (mid and ground level)

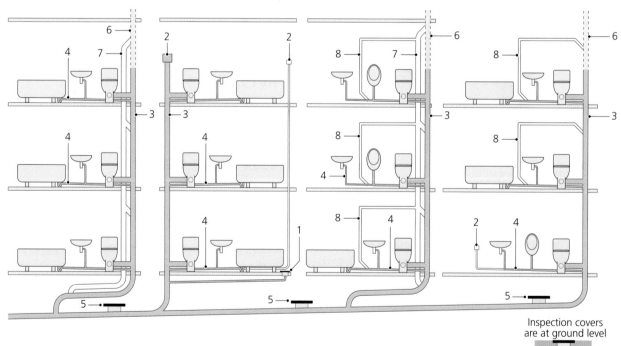

Inspection covers are at ground level

Secondary ventilated system configuration
1. Floor gully   2. Air admittance   3. Stack   4. Branch discharge   5. Drain lid   6. Stack vent   7. Ventilating stack   8. Branch
                     valve   pipe   (mid and   Ventilating pipe
                                                                        ground level)

▲ Figure 4.14 Illustrations from **BS EN 12056** to demonstrate the installation variations for AAVs

ACTIVITY

When next on site, identify the location of the vent pipe for various styles of buildings, including multi-occupancy; or as a learner group walk past nearby buildings and carry out a similar survey. Have a look to see if you can locate any AAVs, photograph them, taking note of the manufacturer. Go online and look up the installation instructions and confirm the correct installation.

Different types of AAVs

There are many different applications and pipe sizes for AAVs which give the plumber a variety of options when specifying and installing.

The working principle and layout features for foul tanks in sanitary systems

Building Regulations H2 should be sourced and referenced for the design and installation requirements.

Cesspools

In rural areas many homes and buildings are self-contained; the combined waste ends up in a local cesspool, septic tank or treatment plant. These have NO connection to the public sewer system and are known as off-main. A cesspool is also known as a cesspit.

A cesspool is an underground tank that stores the sewage until the time of disposal. The design of a cesspool will incorporate an inlet pipe but no outlet pipe, as the cesspool only collects the waste until it is disposed. Older cesspools were made from brickwork, but modern ones are made from glass-reinforced plastic (GRP) – fibre glass.

Cesspools must be constructed so they are watertight to prevent any leakage of foul water or the ingress of surrounding groundwater.

KEY POINT

Ingress in this context describes the action of, or fact that, foul water from the cesspool has entered the groundwater and contaminated it.

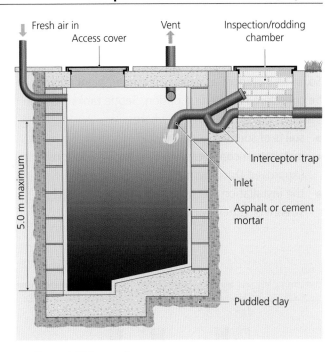

▲ Figure 4.15 Cesspool

It is important that customers are informed that cesspools require emptying on a regular basis to prevent the unit from overflowing. This process must be carried out by a drainage contractor who will use the principle of mechanical suction to draw the waste into the vehicle. Older cesspools were designed with an overflow, but these now no longer conform to current requirements.

Leakage is another problem which is more common with brick-built units, due to the fabric and structure breaking down, which would lead to the ingress of groundwater and the leakage of foul water. This will result in foul smells and pollution of the surrounding area. The use of cesspools is no longer an option in most instances.

Septic tanks

A septic tank is a multi-chamber storage tank allowing liquid and solid waste to separate. The liquid is allowed to flow out of the tank and be disposed of separately. The sewage enters the settlement chamber, allowing the solid waste (sludge) to sink to the bottom and the liquid to rise to the surface. The surface liquid is in contact with oxygen and the organic material starts to break down biologically. This liquid still contains some sewage but only in small enough particles to be carried away through the discharge outlet and into the ground (soakaway).

Basic septic tanks only partly treat the sewage, and so many parts of the UK prohibit the installation of septic tanks as they discharge effluent of low quality. So, in all instances a sewage treatment plant should be considered as the first option. Septic tanks may be installed subject to consent where:

- soil is of suitable porosity
- installation complies with Building Regulations (Approved Document H)
- the installation will not contaminate any ditch, stream or other watercourse.

▲ Figure 4.17 A section showing the inside of a basic septic tank

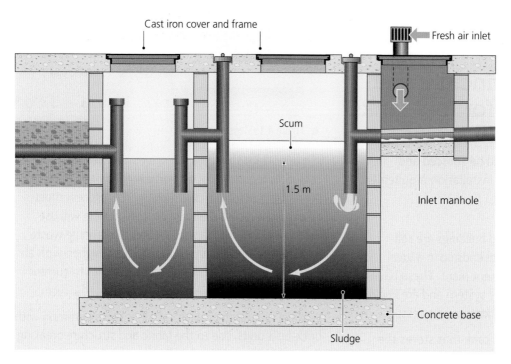

▲ Figure 4.16 Septic tank

It is very important to consult with the building control officer from the local authority prior to any installation. A proper survey of the area and ground type will also need to take place; this will include a soil porosity test. The following procedure will need to be adopted:

1 Excavate a hole 300 mm × 300 mm × 300 mm below the proposed invert level of the land drain.
2 Fill the hole with water and allow it to drain away overnight.

3 Refill with water and observe the time taken to drain from 75 per cent full to 25 per cent full (which is 150 mm of water).
4 Divide the time by 150 (150 mm).
5 The answer will give the average time in seconds (Vp) required for the water to drop 1 mm.
6 Repeat the exercise two more times in at least two trial holes.

▼ Table 4.3 A chart used to help installers work out the suitability of land drainage

Time taken to fall 150 mm (mins)	Equivalent value of Vp (secs/mins)	Alpha* Overall length of drain run required (m)			Large capacity Overall length of drain run required (m)		
		2,800 litres *4 people*	2,800 litres *10 people*	2,800 litres *14 people*	2,800 litres *22 people*	2,800 litres *30 people*	2,800 litres *38 people*
20	8	18	40	56	88	120	152
30	12	24	60	84	182	180	228
40	16	32	80	112	176	240	304
60	24	48	120	168	264	360	466
120	48	96	240	338	528	720	912
180	72	144	360	504	782	1,080	1,368

The above is based upon 500 mm trench width, 2 m apart.

For lengths above 100 m it is acceptable to use 1 m wide trench and half the length.

*Alpha is a type of Klargester septic tank, available in three sizes.

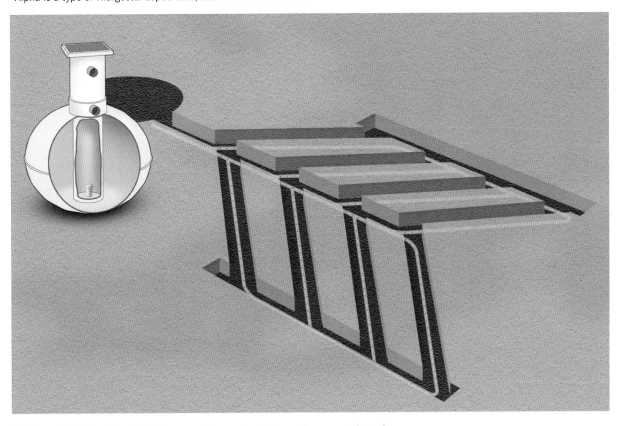

▲ Figure 4.18 A section of land prepared to receive discharge from a septic tank

The average time for drain away can be calculated by referencing the table to work out the length of drain run required for various capacity tanks. Drainage field disposal should only be used on test values between 15 and 100. Where the Vp value is outside these figures, technical advice should be sort from the supplier.

Reed bed

A reed bed is only required when local water authorities request a better quality of effluent than that being discharged from a standard unit.

> **KEY POINT**
>
> A reed bed is a natural filtration process which is more aesthetically pleasing. Oxygen from effluent is transferred from leaves down through the reed stem and out via its root system in the gravel bed.

The advantages of reed beds are:
- they satisfy the new Building Regulations
- they improve effluent quality for existing works
- they are very low maintenance
- they are easy to install
- they are aesthetically pleasing and environmentally friendly
- they significantly improve the discharge from a treatment plant.

A sewage treatment plant is ideal for a location where there will be a discharge or sub-surface irrigation or to a suitable watercourse (which has been approved by the regulator) and where a septic tank will not meet the required standard.

▲ Figure 4.19 Reed beds

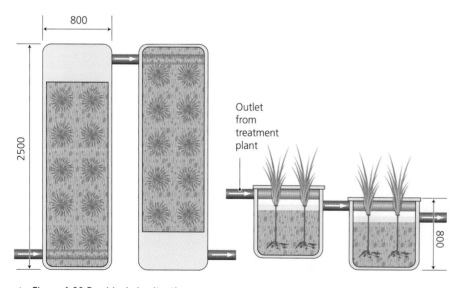

▲ Figure 4.20 Reed beds in situation

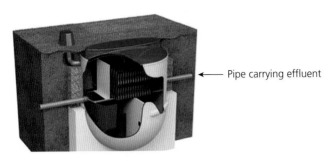

▲ Figure 4.21 Sewage treatment plant below ground

← Pipe carrying effluent

INDUSTRY TIP

Employers are now requiring plumbers to have a hepatitis A jab when they are working on foul water systems.

WC macerators

Macerators can offer the plumber many options when installing sanitary appliances in remote locations. They offer a solution where access to the main soil stack is not practical from a conventional gravity appliance.

▲ Figure 4.22 WC macerator

KEY POINT

If a macerator is installed, Building Regulation Part G requires that a gravity WC MUST be located in the same building.

Gravity fall from appliances

A pump/macerator unit does not suck in the incoming waste, which means the waste water has to enter the unit by gravity. This is especially important when installing a low trap such as in a shower. If the fall from the outlet of the appliance to the inlet of the macerator is not sufficient (less than 1:40) then problems such as backing up of soil waste could occur in the shower tray. It is therefore important to read through and follow the installation instructions carefully to measure the minimum height of the shower trap above floor level.

Vertical rise pipework

It is always important to run any vertical outlet pipework as near to the macerator unit connection as possible. This is to ensure the maximum discharge capability through the outlet pipework. This is important as long horizontal pipe runs add unwanted frictional loss and resistance to the mechanical pumping operation.

The outlet pipework must rise at least 300 mm before any horizontal pipework is fitted.

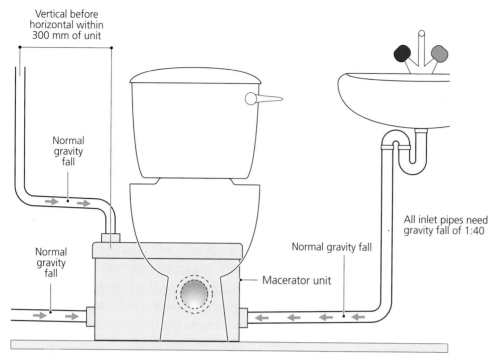

Vertical before
horizontal within
300 mm of unit

Normal
gravity
fall

Normal
gravity
fall

Normal gravity fall

All inlet pipes need
gravity fall of 1:40

Macerator unit

▲ Figure 4.23 Vertical rise of outlet pipework

Discharge pipework

Copper or solvent weld plastic pipes are both suitable for discharge pipework from a macerator unit.

If plastic pipe is used it MUST be solvent weld and NOT push fit. A continuous run of Hep$_2$O pipe can be used, but it is important to ensure that it is regularly supported to avoid any sagging or dipping in the pipe run, which would lead to poor performance.

INDUSTRY TIP

With any macerator installation, always read the installation instructions carefully.

For most macerator units, 22 mm discharge pipe is suitable. However, if a horizontal run exceeds 10–12 metres, most manufacturers advise that the discharge pipe should be increased to 32 mm, which would allow for easier drainage.

Installation factors for discharge pipework

- Always use smooth bends and not knuckle type 90° elbows.
- If using copper pipe, pull machine bends rather than using fitting where possible.
- If using plastic solvent weld pipe, use two 45° bends in series to achieve a 90°.
- If using plastic solvent weld pipe, ensure you wipe off all excess solvent cement.
- Remember to deburr cut edges to prevent the build-up of effluent at that point.

Failure to follow these points could result in a narrowing of the pipe bore which will inevitably lead to blockages.

INDUSTRY TIP

Installation factors or criteria are important and key points to consider at the design stage.

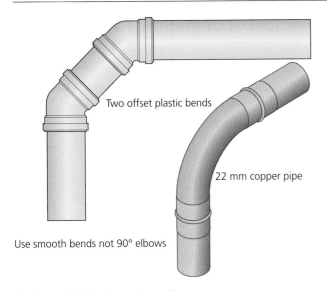

Two offset plastic bends

22 mm copper pipe

Use smooth bends not 90° elbows

▲ Figure 4.24 Discharge pipework

Discharge pipe underground to a manhole

Copper and plastic pipes are suitable for above and below ground installations provided that the following criteria are adhered to:

- The 1:200 minimum fall is laid correctly.
- The ground above is not subject to heavy wheel loads.
- The pipe is insulated to protect against frost.
- If the pipe is laid within 450 mm of the ground level, either concrete casing or paving slabs should protect the pipe.

Installation options

It is normal practice to position the macerator unit directly behind and connected to the WC. However, it is possible to position the unit a short distance away from the WC (for example, the other side of a wall). The extension should be no longer than 150–200 mm long – extensions longer than this tend to cause blockages. Always check for leaks at every joint and never install the unit underneath the floor.

Macerators are often installed behind a panel for aesthetic reasons. If this is the case, make sure the unit

is accessible for future maintenance. Full access to all jubilee clips securing the pipework is important.

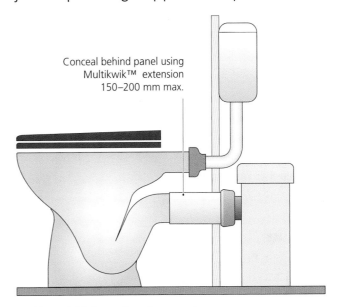

Conceal behind panel using Multikwik™ extension 150–200 mm max.

▲ Figure 4.25 Connections to macerator

INDUSTRY TIP

Installation options need to be considered at the design stage and are often outlined in the manufacturer's instructions. The instructions may offer two or three installation options that could be considered.

Electrical connections

Full details of how to connect the macerator to the electrical supply will be outlined in the manufacturer's instructions. The macerators require an unswitched fused spur connected with a 5 amp fuse. The connection can be to a spur which is connected to the ring main circuit.

HEALTH AND SAFETY

Safe isolation is essential if you are working on any electrical appliance. The full safe isolation procedure is outlined in Book 1, Chapter 1.

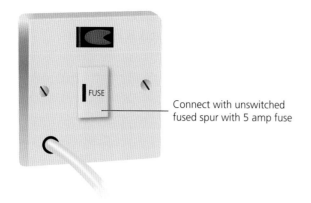

Connect with unswitched
fused spur with 5 amp fuse

▲ Figure 4.26 Fused spur

Principle of operation

The WC discharges into the macerator drum. A float
or diaphragm switch rises and starts the motor. The
motor has stainless steel blades attached to the
rotating spindle which breaks up any solids entering the
macerator drum area and suspends the solid in liquid.
This transformed mass is then positively pumped to the
outlet at a higher part of the unit and into a vertical
pipe through a non-return valve into the discharge
pipe. The discharge pipe then runs horizontally with a
minimum of 1:200 fall to the soil stack. Figures 4.27
to 4.29 show the mechanical sequence of operation.

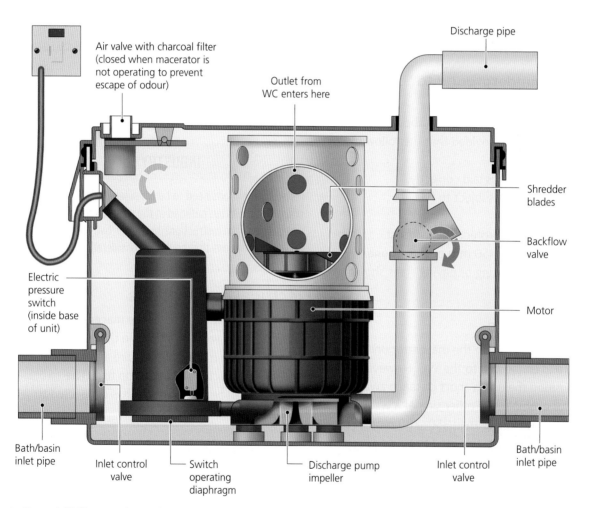

▲ Figure 4.27 Macerator in starting position

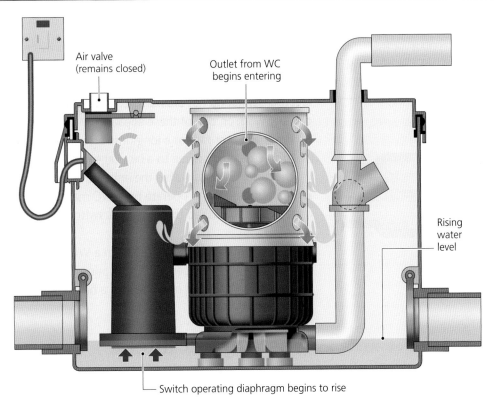

Air valve
(remains closed)

Outlet from WC
begins entering

Rising
water
level

Switch operating diaphragm begins to rise

▲ Figure 4.28 Macerator process beginning as WC discharges

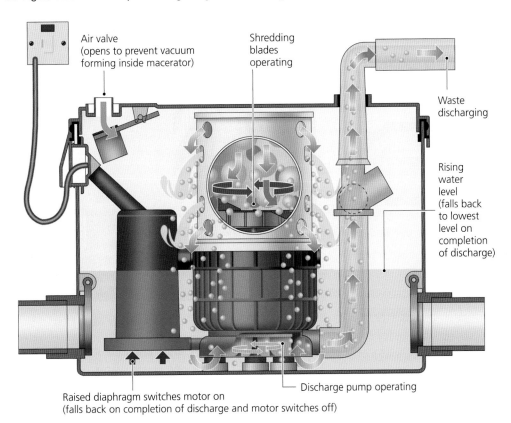

Air valve
(opens to prevent vacuum
forming inside macerator)

Shredding
blades
operating

Waste
discharging

Rising
water
level
(falls back
to lowest
level on
completion
of discharge)

Raised diaphragm switches motor on
(falls back on completion of discharge and motor switches off)

Discharge pump operating

▲ Figure 4.29 Macerator shredding in process

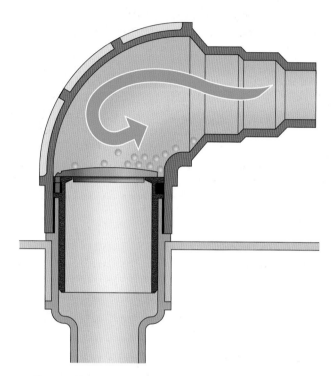

▲ Figure 4.30 An external non-return valve – no discharge

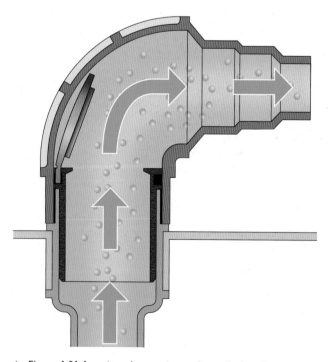

▲ Figure 4.31 An external non-return valve – discharging

Macerator faults

The manufacturer's instructions will usually provide a flow chart that will help with fault finding. Before any work starts on a macerator, it is imperative that safe electrical isolation has been achieved before you carry out any work on one of these appliances.

- If a macerator ran for a long time after flushing, the problem could be due to a blockage in the area of the pump located within the unit.
- If a pump failed to operate on a WC macerator after it had been flushed then the float switch could have a fault.
- If there was a slow discharge from a WC pan connected to a macerator with a pump working correctly, this could be caused by a blocked carbon air filter.

Waste water lifters

Pumping stations

Compact pump systems for small domestic waste water applications are suitable in situations where foul drainage by gravity is not an option. Larger domestic pumping stations are recommended for 8–13 people for the removal of sewage effluent. These units are fitted with an alarm in the event of high fluid levels.

▲ Figure 4.32 Installation option for a waste water lifter

The waste water lifting units are quite innovative in their methods of removing waste water, which negate the use of huge tanks. These devices can take waste water from basement height and lift the waste into the gravity pipe under the basement ceiling. This then fills the septic tank which has the same effective capacity but much smaller volume. This option offers versatility when specifying and installing therefore impacting on the overall cost of a project. These units can reliably discharge domestic waste water with a pH4–pH10 value containing fibres, textiles and faeces.

BS EN 12056 Part 4 gives guidance on the types of waste water lifting plant and discharge pipework.

▼ Table 4.4 Minimum size of discharge pipe in accordance with BS EN 12056.4

Type of wastewater lifting plant	Minimum size of discharge pipework
Non-macerating faecal lifting plant to prEN 12050-1	DN 80
Macerating faecal lifting plant to prEN 12050-1	DN32
Faecal-free lifting plant to prEN 12050-2	DN32
Non-macerating faecal lifting plant for limited applications to prEN 12050-3	DN25
Macerating faecal lifting plant for limited applications to prEN 12050-3	DN20

If the building is located far away from the main sewer, for example in a rear housing situation or when the use of gravity is impossible, a pumping station could provide an ideal solution against having to excavate the land, which would be costly.

▲ Figure 4.33 Pipework outlet to a remote drain where no gravity drain is available

The installation of a waste water lifter can be below or on the same finished floor level of a property. The discharge pipe enters the soil stack and forms a **backflow loop**, as shown in Figure 4.34. The vent pipe must discharge in accordance with **BS EN 12056.1** for faecal lifting plants to above roof level to avoid the ingress of foul smells in to the property.

KEY TERM

Backflow loop: prevents backflow. See Figure 4.34, where the backflow loop rises above the mains sewer and drops into the top of the pipe.

Lifting units

AAVs cannot be installed on the discharge pipework. These lifting units are designed to lift waste water from below the sewer level in a property and then discharge into the sewer system. They combine pumps, tanks and controls in one self-contained unit to ease installation and maintain reliable operation. They include cabling, sensors and hoses. These units have a sloping tank bottom design to ensure that dirt and solids are guided towards the pump at all times. The aim is to minimise sedimentation and reduce the need for tank cleaning.

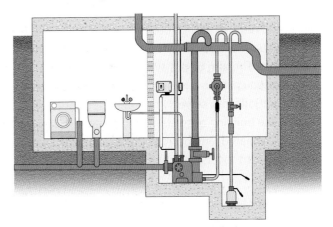

▲ Figure 4.34 A waste water lifter

▲ Figure 4.35 A waste water lifter in a commercial installation

An inspection cover enables a plumber to check the valves and remove foreign bodies; the design ensures that adequate access to the built-in non-return valve is available. The aim of this style of unit is to keep construction costs to a minimum by making the need for excavations into sewers and drain-pipe connections outside the building obsolete. The digging of pump pits can be eliminated, and basement areas can then be converted into toilets and utility rooms with minimal inconvenience.

Waste water lifters (Figure 4.35) incorporate a permanent level control which triggers an alarm in the event of high water levels and incorporates a control unit (Figure 4.36). As water is discharged into the sump, the water level rises and triggers the internal pump to start. As the water level decreases, the pump stops.

KEY POINT

The permanent level control is a device which triggers an alarm in the event of high water levels.

The unit also incorporates a piezo-pressure sensor that detects waste water levels inside the tank without the need for any moving parts or submersion. The tanks are generally made from non-adhesive polyethylene (PE) and designed to withstand a column of water 5.0 m and water temperatures of up to 50 °C and up to 90 °C for short periods of time.

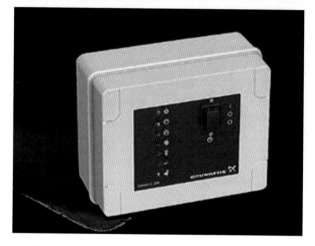

▲ Figure 4.36 Control unit

Sink waste disposal unit

These units are installed under kitchen sinks and need a ready-made hole 89–90 mm in diameter, made in the sink to fit the unit, this is usually positioned in the cutlery bowl. A standard 40 mm trap will fit on to the outlet of the waste disposal unit.

These units dispose of waste food and cooking products discharging them into the drainage system. The cutting or grinding blades can deal with a large range of food matter including bones.

HEALTH AND SAFETY

Always outline the safe use of these appliances to the customer, which are explained in the manufacturer's instructions.

▲ Figure 4.37 A waste disposal unit fitted to the base of a sink unit

The process turns anything in the unit into a paste solution and water flushes it into a drain via a 40 mm waste outlet. The electric motor that turns the rotor where the blades are attached is located at the base of the unit. The motor on the unit should be connected to an electrical supply via the correctly sized fuse spur, typically 10 amp, but will be specified in the manufacturer's instructions. Manufacturers often require the circuit to be protected by a 30 mA RCD.

▲ Figure 4.38 Fused spur

HEALTH AND SAFETY

A residual current device or RCD protects the customer and engineer from a live–earth fault. The standard that relates to this appliance is found in **BS EN 60335.2,** Specification for safety of household and similar electrical appliances. An RCD is a life-saving device which is designed to prevent you from getting a fatal electric shock if you touch something live, such as a bare wire. It can also provide some protection against electrical fires. RCDs offer a level of personal protection that ordinary fuses and circuit-breakers cannot provide. An RCD is a sensitive safety device that switches off electricity automatically if there is a fault. The RCD is designed to protect against the risks of electrocution and fire caused by earth faults. For example, if you cut through the cable when mowing the lawn and accidentally touched the exposed live wires or a faulty appliance overheats causing electric current to flow to earth. It constantly monitors the electric current flowing through one or more circuits it is used to protect. If it detects electricity flowing down an unintended path, such as through a person who has touched a live part, the RCD will switch the circuit off very quickly, significantly reducing the risk of death or serious injury.

ACTIVITY

Use manufacturer's instructions and the internet to compare the advantages and disadvantages between waste water lifters and macerators in a domestic and then a commercial installation.

HEALTH AND SAFETY

Care should be taken when servicing these appliances; it is important to ensure you are fully trained and aware of the manufacturer's servicing procedures. The units will be supplied with a special release tool for the blades, should they become trapped. The device must have some form of accessible thermal trip cut-off device that will turn off the unit in the event of blades jamming or no water flow.

KEY POINT

The device MUST have some form of accessible thermal trip cut-off device that will turn off the unit in the event of blades jamming or no water flow.

Wet rooms

A wet room is a waterproofed (tanked) area which is usually equipped with a walk-in shower, and therefore the room itself becomes the enclosure. A wet room can be fitted out with standard or specialist equipment to suit the customer's requirements, so it is very important to discuss with the customer prior to designing a wet room to get a complete understanding of what is required. The drain point is inset into the floor area which has a gently sloping floor draining the water towards the drain point (this is instead of the shower tray).

▲ Figure 4.39 Open plan wet room

VALUES AND BEHAVIOUR

A wet room is quite often favoured by people with physical impairments as it offers easy access. As this can be such a personal and individual consideration, good and sensitive communication skills are required. Go online and search for specialist wet room installations and use these websites to create a list of important questions to ask a prospective customer.

Its spacious layout can facilitate more contemporary designs and because of the free floor area it is also suitable for underfloor heating.

Any house or flat can be made suitable for a walk-in wet room, as their installation requires them to be **tanked**, which is equivalent of completely sealing the floor and shower walls to ensure there is no water leakage from the wet room area.

KEY TERM

Tanking: a process used to ensure that a wet room area installation is completely leak free.

The waterproof membrane is a unique feature in a wet room and installing it is usually carried out by a specialist contractor and can take several days to complete.

INDUSTRY TIP

There are now manufacturers who offer wet rooms in a 'kit' form for ease of installation. The customer's budget and requirements will indicate which installation is chosen.

INDUSTRY TIP

The waterproof membrane, if used, would be installed by a specialist contractor to ensure watertight wet rooms. This comes in the form of sheet material and or liquid compounds which are sealed to the floor and walls.

Low-level grates allow the collection and removal of water from the showering area with the aid of a built-in fall (slope) to the drain at floor level. There is quite a process involved in installing the drainage system for a wet room. The following illustrations give some indication, but in all cases the manufacturer's instructions MUST be read and followed at all times.

The built-in fall is designed to drain the water towards the outlet. There are two main options for the location of

the grate which will accommodate the flow of waste water. The trap is connected under the grate and then flows to the soil stack. Figures 4.41 and 4.42 show one style of installation method and design for wooden floors which is adapted to suit solid floors as well.

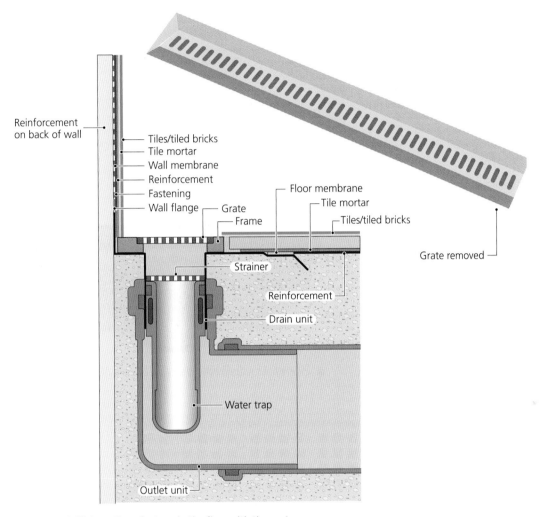

▲ Figure 4.40 A section of a trap in the floor with the grate

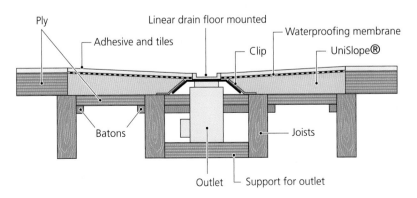

▲ Figure 4.41 Cross section showing waterproof membrane centre outlet

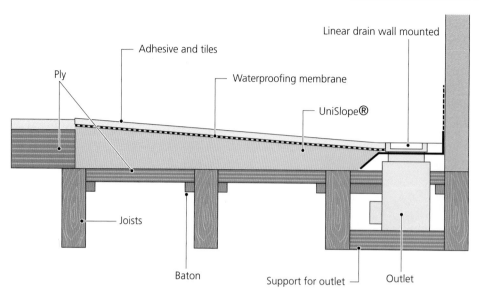

▲ Figure 4.42 Cross section showing waterproof membrane end outlet

Grates are designed to assist even discharge and full removal of waste water and often use a collection gulley. The device shown in Figure 4.43 gathers waste water and houses the trap which is located under the grate system.

Special traps are included in the design and provide easy access. The waste outlet is often discreetly located to enhance the aesthetic appearance of the installation.

▲ Figure 4.43 A drain collection gulley with outlet

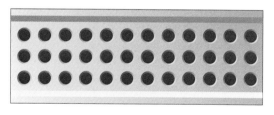

Classic

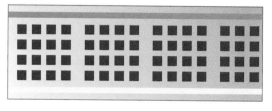

Square

Column

Stripe

▲ Figure 4.44 Four grate styles for wet rooms

③ DESIGN TECHNIQUES FOR SANITATION AND RAINWATER SYSTEMS

First, we will address design techniques of sanitation systems, followed by rainwater systems later in this section.

Selection and design of sanitary systems

A full survey of a proposed installation is essential to ascertain what type of system and appliances would be suitable. A visual inspection of an existing system would outline the condition and design of the current system. It may even show up areas that do not comply with current standards. We have outlined that all systems and installations should comply with:

- **BS EN 12056 Parts 1–5**
- Building Regulations Approved Document Part H.

They should also comply with:

- **BS 8000.0:2014** Workmanship on building sites. Code of practice for above ground drainage and sanitary appliances
- The Water Supply (Water Fittings) Regulations 1999, which covers the supply of hot and cold water to sanitary appliances which must be protected against backflow or back siphonage.

There are a lot of other individual standards for pipework, fittings and appliances that have been outlined in Book 1.

On new-build installations, such as a housing development, the position and design of the soil and vent pipe along with the positioning and style of appliances has already been determined by the architect. Any preparation work or disruption should be agreed with the relevant trades or customer before work starts.

Design considerations about materials need to be made. Plastic is commonly used on modern builds, but cast iron could be chosen for an older property. A full survey to confirm what design the underground system is would need to be carried out – this would highlight any changes that may need to be made and also avoid any incorrect connections to the underground system.

IMPROVE YOUR ENGLISH

In order to ensure you fully understand the customer's objectives and expectations for the installation, discussions should be had at an early stage. It would be useful to ask the following questions:

- What is the size of property?
- What is the style of property?
- What is the age of property?
- What style of appliances?
- How many appliances?
- What is the position of appliances?
- How many people are living in the property?
- What are the potential uses of appliances?
- What existing installations are present?

All of these are important factors to be considered.

Spacing requirements

Building Regulations Approved Document Part M aims to ensure that buildings are accessible and useable to people regardless of disability, age or gender. They should be able to gain access within a building and use its facilities as a visitor, living in the property or if the building is a place of work. This mainly relates to non-domestic buildings and new domestic buildings, but when an existing property undergoes a significant alteration or has an extension then consideration is required.

Plumbers may be asked to install a 'Doc M' pack into a building. This means that the sanitary equipment installed will aid people with physical impairments by the addition of bars and levers so that they can use the facilities.

When installing bathroom appliances, consideration MUST be made to allow for the minimum space requirements for each appliance for personal use and to supervise the bathing of children. **BS EN 6465.2:2017** gives advice on these space requirements.

> ### KEY POINT
> **BS EN 6465.2:2017** gives advice on appliance space requirements. There must also be a minimum amount of appliances within a property based on the number of people occupying the property, and this information is detailed in **BS 6465.1:2006+A1:2009**.

▼ Table 4.5 Appliances required per dwelling

Sanitary appliances	Number per dwelling	Additional notes
WC	One for up to 4 people, two for 5 people or more	There should be a wash basin adjacent to every WC in every property
Wash basin	One	
Bath or shower	One for every 4 people	
Kitchen sink	One	

It is not always possible to achieve the recommended dimensions, especially when dealing with small bathrooms. The British Standard therefore makes allowances for overlaps of the appliance space. These overlaps also apply to a cloakroom and downstairs WC.

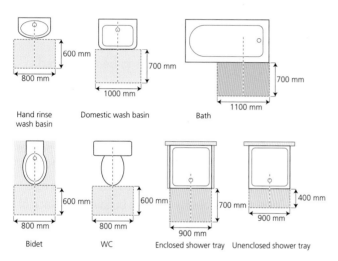

▲ Figure 4.45 Provision of space for sanitary installations

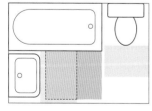

In this layout, the activity space of both the bath and the wash basin overlap. The space for the WC usage is not affected.

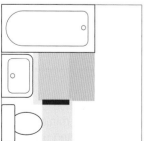

In this layout, the activity space of the bath, wash hand basin and WC all overlap. The overlap is shown by the red rectangle on the drawing.

This one is the most common of all bathroom layouts.

▲ Figure 4.46 Overlap – provision in bathrooms

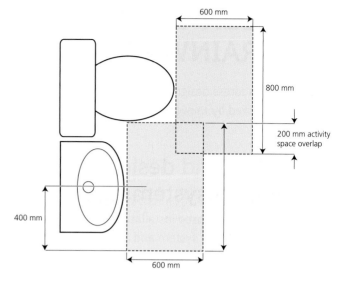

▲ Figure 4.47 Overlap – provision in a downstairs WC

Fire stop arrangements

Whenever a soil or vent pipe penetrates a wall or a ceiling an intumescent collar is fitted around the soil pipe to prevent the risk of smoke and fire spreading between areas.

KEY POINT

An intumescent collar is a device used to fit around a plastic soil or waste pipe penetrating a wall or ceiling. The collar contains material that will expand and fill the void when the pipe melts because of the presence of extreme heat from a fire. This therefore acts as a smoke and fire stop to prevent them spreading for a set period of time.

▲ Figure 4.48 An intumescent collar

BS EN 12056.1 Section 5.4.1 states that, where pipes pass through walls, floors and ceilings, subject to specific fire resistance requirements, special precautions should be taken in accordance with national and local regulations and practice.

A **PVCu** pipe will melt if exposed to excessive heat and the space the pipe once occupied could create a venturi effect for fire, smoke and fumes. When exposed to such conditions the intumescent material in the collar will become a carbonaceous char and will expand then compress the hot plastic pipe and close the void to prevent the spread of fire, smoke and fumes.

KEY TERM

PVCu (unplasticised poly vinyl chloride): a common material used in rainwater guttering and pipework systems.

Whenever an intumescent collar is installed, it must be checked to confirm that it is fit for use and in date. It is very important to refer to the manufacturer's instructions to make sure it is installed correctly.

HEALTH AND SAFETY

The effect of fire spreading through a void associated with internal pipework which penetrates a fire-resistant wall or floor can be identified in the following descriptions:
- products of toxic gases and smoke
- the addition of fuel to total load of the fire
- the risk of fire spreading along the pipework
- the reduction of the fire resistance of the building materials that have been penetrated with the pipe.

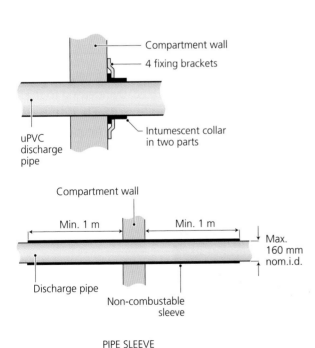

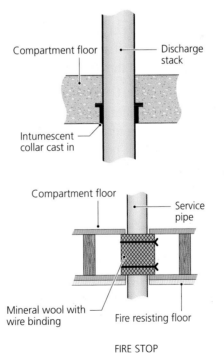

▲ Figure 4.49 Building Regulations Approved Document B3 – internal fire spread

The Building Regulations Approved Document B3 gives an example, showing a range of different applications of fire stopping.

If a material, such as a timber suspended floor, is penetrated by an incorrectly installed heating appliance flue pipework which did not have the required air gap, then the adjacent material could become **pryolised** and its ignition temperature could be significantly reduced, and as a result create a weak point where fire could spread.

KEY TERM

Pryolised: when a material begins to decompose due to elevated temperatures.

Calculating the size of sanitary pipework in accordance with BS EN 12056

BS EN 12056.2 explains that water flowing into discharge stacks will cause air pressure fluctuations and suction can occur below discharging branch connections and offsets. This can cause water seal loss by induced siphonage from the appliances connected to the stack. Back pressure or positive pressure can occur about bends and offsets in the stack which can cause foul air to be blown through the trap water seal and potential loss of seal.

The following identifies some causes for loss of seal:
- the flow load which relates to the total number of appliances connected to the stack, how they are distributed on each floor and their frequency of use
- the height and diameter of the stack size must be calculated to accommodate the height of the building and the amount of appliances connected to it
- the design of pipe fittings, the shape and size of branch inlets and radius of the bend and the base of the stack connecting the system to the drain
- changes in direction in the wet portion of the stack

- the provision, or lack of a ventilating pipe
- surcharging of the drain
- provision, or lack of an interceptor trap.

Therefore, when sizing drainage systems we must refer to **BS EN 12056** and the following pipework dimensions (Table 4.6) in relation to appliances in the UK.

▼ Table 4.6 Pipework dimensions showing waste pipe diameter and appliances

Waste pipe size (mm)	Appliance
32	Wash basin, bidet, drinking fountain
40	Sink, bath, shower, urinal, sanitary towel macerator
50	Food waste disposal unit, multiple appliances
100	Toilet

Houses of single occupancy rarely need sizing, as a 100 mm stack is required for one WC. Even houses with one WC can be connected to a 100 mm vertical stack. Sizing is however required for flats, halls of residence, commercial buildings and public entertainment venues where discharge units are used to help calculate a stack size for a specific installation.

KEY POINT
Discharge units are used to calculate the size of a soil stack.

The following step-by-step example will help to show how stack sizes can be calculated. By referring to the tables and formulas extracted from **BS EN 12056**, discharge units (DU), flow rates (Q_{ww}), frequency factors (K) and hydraulic capacities (Q_{max}) can be calculated and result in the calculation of the correct size of discharge stack needed.

▼ Table 4.7 Abbreviations for the formula used for sizing

Discharge units	DU
Flow rates	Q_{ww}
Frequency factor	K
Hydraulic capacity	Q_{max}

Calculating stack sizes step-by-step
Step 1

▼ Table 4.8 Discharge units (DU)

Appliance	System I (DU, l/s)	System II (DU, l/s)	System III (DU, l/s)	System IV (DU, l/s)
Wash basin, bidet	0.5	0.3	0.3	0.3
Shower without plug	0.6	0.4	0.4	0.4
Shower with plug	0.8	0.5	1.3	0.5
Single urinal with cistern	0.8	0.8	0.4	0.5
Urinal with flushing valve	0.5	0.3	–	0.3
Slab urinal	0.2*	0.2*	0.2*	0.2*
Bath	0.8	0.6	1.3	0.5
Kitchen sink	0.8	0.6	1.3	0.5
Dishwasher (household)	0.8	0.6	0.2	0.5
Washing machine up to 6 kg	0.8	0.6	0.6	0.5
Washing machine up to 12 kg	1.5	1.2	1.2	1.0
WC with 4.0 l cistern	**	1.8	**	**
WC with 6.0 l cistern	2.0	1.8	1.2–1.7***	2.0
WC with 7.5 l cistern	2.0	1.8	1.4–1.8***	2.0
WC with 9.0 l cistern	2.5	2.0	1.6–2.0***	2.5
Floor gully DN50	0.8	0.9	–	0.6
Floor gully DN70	1.5	0.9	–	1.0
Floor gully DN100	2.0	1.2	–	1.3

* Per person
** Not permitted
*** Depending on type (valid for WCs with siphon flush cisterns only)
*** Not used or no data

Use Table 4.8 (discharge units) to select the total number of appliances running in to the stack. System III is based on the British above ground sanitation system as stated earlier. The next stage is to add up the discharge units (DU) for these appliances.

Step 2

▼ Table 4.9 Typical frequency factors (K)

Usage of appliances	K
Intermittent use, e.g. in dwelling, guesthouse, office	0.5
Frequent use, e.g. in hospital, school, restaurant, hotel	0.7
Congested use, e.g. in toilets and/or showers open to the public	1.0
Special use, e.g. laboratory	1.2

Use Table 4.9 to select the frequency factor for the use of the appliances.

Step 3

Q_{ww} is the expected flow rate of waste water in a part or in the whole drainage system where only domestic sanitary appliances are connected to the system.

Where:

$$Q_{ww} = K\sqrt{\Sigma DU}$$

Q_{ww} = waste water flow rate (l/s)

K = frequency factor

ΣDU = sum of discharge units

Use the formula to work out the waste water flow rate.

Step 4

▼ Table 4.10 Hydraulic capacity (Q_max) and nominal diameter (DN)

Stack and stack vent	System I, II, III, IV Q_max (l/s)	
DN	Square entries	Swept entries
60	0.5	0.7
70	1.5	2.0
80*	2.0	2.6
90	2.7	3.5
100**	4.0	5.2
125	5.8	7.6
150	9.5	12.4
200	16.0	21.0

* Minimum size where WCs are connected in system II.

** Minimum size where WCs are connected in system I, III, IV.

Use the Q_{ww} (litres per second) to size the pipe from Table 4.10. We tend to use swept entry fittings in the UK, but as you can see there is also a column for sizing square entry.

▼ Table 4.11 Hydraulic capacity (Q_max) and nominal diameter (DN) for secondary vent systems

Stack and stack vent	Secondary vent	System I, II, III, IV Q_max (l/s)	
DN	DN	Square entries	Swept entries
60	50	0.7	0.9
70	50	2.0	2.6
80*	50	2.6	3.4
90	50	3.5	4.6
100**	50	5.6	7.3
125	70	7.6	10.0
150	80	12.4	18.3
200	100	21.0	27.3

* Minimum size where WCs are connected in system II.

** Minimum size where WCs are connected in system I, III, IV.

If the soil and waste system includes secondary ventilation, then Table 4.11 should be used.

IMPROVE YOUR MATHS

How to calculate the size of a discharge stack

If a congested football stadium is used as an example, we can end up with a calculation as shown. Once the total discharge units are added up, then the square root of that number is applied and multiplied by the frequency factor.

Use Table 4.8:

50 × 6 l WCs: 50 × 1.7 = 85

100 × wash basins: 100 × 0.3 = 30

Total discharge units = 115

Use formula from Step 3:

Square root of 115 = 10.7

Use Table 4.9:

Frequency factor = 1.0

1.0 × 10.7 = 10.7 litres per second flow rate

Use your answer and refer to Table 4.10.

10.7 l/s is the calculated flow rate and we will be using swept entries. Refer to the swept entry column, going down the column you choose the figure that is equal to or above the calculated figure. So, going down the column 7.6 l/s is too small, then 12.4 l/s is slightly above our calculated flow rate, so this is the figure we would use.

Reading across the table, 12.4 l/s will mean we use 150 mm soil and vent pipe.

KEY POINT

The frequency factor is a variable that should be used when determining the pipework system flow rate based on the frequency of use of the sanitary appliances for different building functions.

IMPROVE YOUR MATHS

How to calculate the size of a discharge stack for a single occupancy

The previous example related to a commercial installation, but the same process can be applied to a domestic property. As a Level 3 qualified plumber, you will need to apply this method to achieve a well-designed sanitary installation.

For example, we have a property that has two (6 l) WCs, two baths, two wash basins, one shower with plug, one bidet, one kitchen sink, one (6 kg) washing machine, one domestic dishwasher.

Use Table 4.12 to work out the required soil stack size.

▼ Table 4.12

Appliances	Discharge units
2 × 6 l WCs	1.7 × 2 = 3.4
2 × baths	1.3 × 2 = 2.6
2 × wash basins	0.6 × 2 = 1.2
1 × shower with plug	1.3
1 × bidet	0.6
1 × kitchen sink	1.3
1 × 6 kg washing machine	0.6
1 × domestic dishwasher	0.2
TOTAL	11.2

The square root of 11.2 is 3.35, and the frequency factor for this type of property is 0.5.

$$3.35 \times 0.5 = 1.75 \text{ litres per second}$$

If swept entries are used again, you can see that 100 mm diameter soil stack would be more than adequate.

IMPROVE YOUR MATHS

Using the tables provided, calculate the size of a discharge stack serving two flats, each containing one (6 l) WC, one bath, one shower without plug, one (6 kg) washing machine and a dishwasher.

Design and installation of advanced rainwater systems

It is so important to design and install an effective rainwater system on properties. There are many environmental issues that need to be considered and if the system is not suitable the integrity of the property could be affected. For larger buildings and constructions there are companies that specialise in rainwater systems that allow large roof areas to remove large volumes of water effectively and efficiently.

> **KEY POINT**
>
> It is important to revisit the information about basic rainwater systems in Book 1, Chapter 8 to remind yourself of the fundamental criteria concerning the choice and installation of guttering systems.

Designing rainwater systems

When designing a rainwater system, a survey of the property must be carried out followed by discussion with the customer concerning their requirements and choice. If a system is to be replaced, it is important to ask the customer how the existing installation has performed and as to whether the customer has any concerns or has had any problems with it.

IMPROVE YOUR ENGLISH

Remember, customers will usually be able to provide you with invaluable information about any work you undertake. By asking the right questions, you can ascertain how the current installation has performed and what issues have been experienced by the customer to help you make improvements to the new system design.

When working on a new build or alteration, it is good practice to select some manufacturers' brochures and show the customer so they are able to select the design that is aesthetically pleasing to them. This will also give an opportunity for the plumber to outline suitable designs and considerations to the customer.

VALUES AND BEHAVIOURS

The customer will appreciate being consulted at the design stage – you can use visual aids such as manufacturers' brochures to enable them to make selections and preferences that can then be integrated into your design.

When carrying out the survey of the property, it would be wise to establish the type of drainage for the premises for the property and plan your system around whether it is a combined, separate or partially separate system and whether the local authority has any preference for connections. If a partially separate system is installed, check for any soakaways on existing remote downpipes.

Sometimes a new build extension will require a rainwater system. This could be connected to an existing system already installed on the premises. In this situation a recalculation for the gutter size and design may be required, to accommodate gutter size and number of outlets for the additional flow of rainwater.

Legislation relating to design and installation of advanced rainwater systems

There are many documents which need to be referred to when designing and installing a rainwater system. This is because there are a range of restrictions in legislation to ensure that the water is efficiently collected and safely discharged from the property.
- Building Regulations Approved Document H3 Rainwater drainage
- **BS EN 12056.3:2000** Rainwater system design, outlet position and rainfall intensity calculations
- the manufacturer's instructions installation guide
- **BS EN 12200.1:2000** Rainwater pipe systems

Calculating the size of a gutter

To assess the suitability of a gutter system to drain the roof on a building efficiently, the following factors need to be taken in to account:
- the effective roof area to be drained
- the local rainfall intensity
- the flow characteristics of the gutter system
- the number and position of downpipes.

BS EN 12056.3 measures rainfall intensity in litres per second per metre squared ($l/s/m^2$) for a two-minute storm event. The standard also has maps that give details of the rainfall intensity in areas around the country.

The angle of the roof is a vital consideration when designing a gutter system. The steeper the roof angle the faster the rain will flow off the roof, the shallower the roof incline the slower the rain will flow off the roof.

BS EN 12056.3 gives the following formula to work out the **effective roof area**.

KEY TERM

Effective roof area: different to the actual size of the roof area. In effect it is the plan view area of the roof.

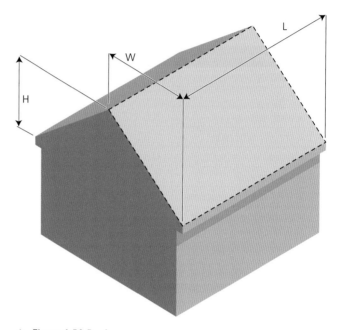

▲ Figure 4.50 Roof area

To calculate effective maximum roof area (allowance for wind), use the following formula:

$$\left(W + \frac{H}{2}\right) \times L$$

Where:

W = horizontal span of the roof slope

H = height of the roof pitch

L = length of the roof

IMPROVE YOUR MATHS

Calculating roof area

Example 1

Calculate the area if a roof is 12 m long, 7 m wide and 3 m high.

Solution:

$$\left(7+\frac{3}{2}\right) \times 12 = (7 + 1.5) \times 12 = 8.5 \times 12$$
$$= 102 \text{ m}^2 \text{ effective roof area}$$

IMPROVE YOUR MATHS

Calculating roof area

Example 2

Calculate the area if a roof is 11 m long, 6.5 m wide and 4 m high.

Solution:

$$\left(6.5+\frac{4}{2}\right) \times 11 = (6.5 + 2) \times 11 = 8.5 \times 11$$
$$= 93.5 \text{ m}^2 \text{ effective roof area}$$

The size of the roof directly relates to the size of the guttering required, so if the size of the roof increases so does the guttering and discharge pipework. This principle also applies to the angle of the roof as the angle affects the flow of rainwater and therefore the speed at which the rainwater enters the gutter.

The rainwater system must therefore be accurately designed to adequately manage the predicted rainfall. Table 4.13 shows the factor which should be added to the roof area once the pitch of the roof is known.

▼ Table 4.13 Multiplication factor for all roof pitches

Type of surface	Design in (m²)
Flat roof	Plan area of relevant portion
Pitched roof at 30°	Plan area of portion × 1.29
Pitched roof at 45°	Plan area of portion × 1.50
Pitched roof at 60°	Plan area of portion × 1.87
Pitched roof over 70° or any wall	Elevation area × 0.5

The next stage is to include the factor for the correct roof pitch which can be selected from Table 4.14.

▼ Table 4.14 Roof pitch and factors

Roof pitch	Factor
10°	1.088
12.5°	1.111
15°	1.134
17.5°	1.158
20°	1.182
22.5°	1.207
25°	1.233
27°	1.260
30°	1.288
32.5°	1.319
35°	1.350
37.5°	1.384
40°	1.419
42.5°	1.459
45°	1.500
47.5°	1.547

For roofs of 50° and above, a factor of 1.600 may be used.

IMPROVE YOUR MATHS

Calculating using roof pitch factor

By using the figures from calculating roof area example 1, the multiplication factor for the pitch of the roof (as shown in Table 4.14) can be included in the new calculation.

Calculate the area if a roof is 12 m long, 7 m wide and 3 m high.

Solution:

$$\left(7+\frac{3}{2}\right) \times 12 = (7 + 1.5) \times 12 = 8.5 \times 12$$
$$= 102 \text{ m}^2 \text{ effective roof area}$$

If the roof pitch is 45° then the roof area should be multiplied by a factor of 1.5.

$$102 \text{ m}^2 \times 1.5 = 153 \text{ m}^2$$

The manufacturer's chart can now be used to select an appropriate gutter size and style for the installation.

The area of a flat roof should be considered as the total plan area. When a roof is more complex with a range of spans and pitches, then each individual area should be calculated and added together.

In an alternative calculation offered by Building Regulations, the length and span are multiplied then multiplied again by a factor for a given roof pitch:

Width × length × (design factor for a given roof angle)

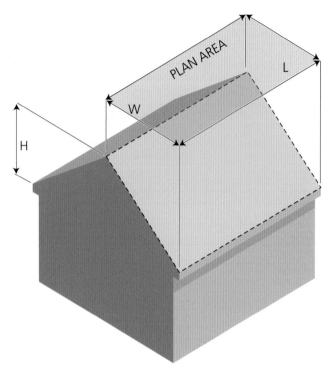

▲ Figure 4.51 Roof plan

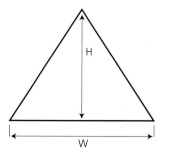

▲ Figure 4.52 Elevation of a roof with a 75° pitch

If the width of the roof is 9 m and the height is 3.5 m then: 9 × 3.5 = 31.5 m

Multiply with the factor given for elevations from Table 4.13 (this is 0.5). Therefore, 31.5 × 0.5 = 15.75 m² = roof area.

Working out the gutter size

The size of the gutter is dependent upon the size of the roof area to be drained. The effective roof area for flat roofs is different from that for pitched roofs and needs to be calculated.

Table 4.15 is from Building Regulations Approved Document Part H and shows that for pitched roofs, a greater 'run-off' or flow will be experienced and must be allowed for.

If the angle of the roof is known then we can calculate using the factor for the pitch as shown in Table 4.14.

IMPROVE YOUR MATHS

Calculating roof area using alternative formula

What is the effective area if a roof is 6 m long, 4 m wide and with a pitch of 45°?
Solution:
The design factor for a 45° pitch is 1.5 taken from the table for roof pitch and factors. If the plan area of a pitched roof measures 6 m × 4 m, then the effective area will be:

6 × 4 × 1.5 = 36 m²

Elevation of a roof

When calculating the effective area from an elevation, different criteria are used. The elevation area of the roof is the width × height.

IMPROVE YOUR MATHS

Calculating gutter size

In the calculating roof area using alternative formula example, 36 m² was calculated as the effective area of the roof which had an angle of 45°, so by using Table 4.15, a gutter size of 100 mm diameter. A half round gutter can be chosen along with a 63 mm diameter downpipe.

▼ Table 4.15 Gutter sizes and outlet sizes

Maximum effective roof area (m²)	Gutter size (mm diameter)	Outlet size (mm diameter)	Flow capacity (litres/sec)
6.0	–	–	–
18.0	75	50	0.38
37.0	100	63	0.78
53.0	115	63	1.11
65.0	125	75	1.37
103.0	150	89	2.16

Refers to nominal half round eaves gutters laid level with outlets at one end sharp-edged. Round-ended outlets allow smaller downpipe sizes.

ACTIVITY

Locate manufacturers online and use the calculating roof area examples 1 and 2 to choose a gutter size for half round, ogee and square line.

Gutter flow capacity

BS EN 12056.3 requires that only 90 per cent of the gutter flow should be relied upon. It is also recommended by many manufacturers that gutters be fixed at this level as this enables the gutter to be fitted as high as possible while still ensuring that the correct position to the roof's edge is maintained.

Careful consideration needs to be given to the position of the rainwater outlets. Centre outlets are more efficient than end outlets as the area that can be drained is almost double, with the result of reducing the number of downpipes required on the system, saving money and time.

The fall of the gutter

Although some manufacturers say that gutters should be installed level, generally good practice will follow a fall of 1–3 mm/m. This is interpreted as a fall of 1:600 (25 mm for every 15 m). This fall greatly increases the flow capacity.

Another reason for the fall of 1:600 is so that the gutter will not fall too low at the end of the run. Brackets should be installed at a maximum of 1.0 m intervals so that it does not sway when filled with water, but holds the fall.

Selecting a gutter type

There are a range of different gutter profiles and gutter materials produced today to suit a variety of installations. Plastic guttering is by far the most popular material used in domestic installations. Cast iron, cast aluminium and extruded aluminium are also available. The expansion and contraction of plastic guttering is a major installation factor.

Outlet positions

This part of the guttering system acts as the connection between the gutter run and the downpipe

and is commonly positioned above the location of the ground level gulley leading to the sewer. The gulley positions can be located either by a visual survey or by referring to the property layout drawings.

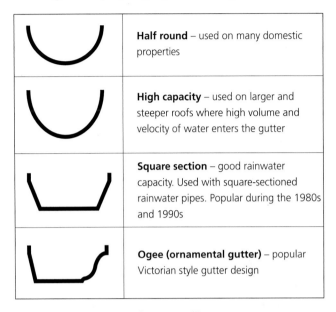

	Half round – used on many domestic properties
	High capacity – used on larger and steeper roofs where high volume and velocity of water enters the gutter
	Square section – good rainwater capacity. Used with square-sectioned rainwater pipes. Popular during the 1980s and 1990s
	Ogee (ornamental gutter) – popular Victorian style gutter design

▲ Figure 4.53 Drawings of gutter profiles

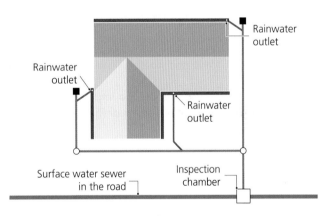

▲ Figure 4.54 Building layout drawing

The outlets, either a running outlet or a stop end outlet, will always denote the lowest point of the gutter run. The system will be more effective with more outlets as this will allow for a better and balanced flow of water from the roof. Placing an outlet centrally rather than at an end of a run will mean that the outlet will accommodate a greater area of roof water run-off. Figure 4.55 shows this positioning.

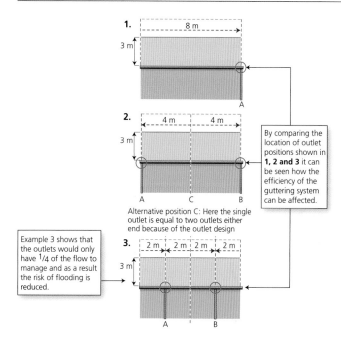

▲ Figure 4.55 Outlet positions on a roof

By dividing the expected flow rate from the roof by the flow rate of the outlet, you can work out how many outlets should be installed. When doing this, always refer to the manufacturer's literature for their particular flow rate.

ACTIVITY

Looking at Figure 4.54, discuss the important design criteria and possible repositioning of the outlets.

Changes in direction

Changes in direction are unavoidable as guttering follows the roof profile, but they will affect the flow capacity of the gutter if the change in direction is greater than 10°. A single 90° angle will affect the flow rate by around 15 per cent, so the more there are in a single run, the less efficient the flow rate will be.

When designing a gutter system, ensure you:
- try to install straight gutter runs which offer the maximum flow rate
- do not locate outlets near to a change in direction
- apply the maximum fall **ratio** (1:600) where there are lots of bends
- install larger gutters where there are lots of changes in direction.

KEY TERM

Ratio: 1:600 means 1 mm fall for every 600 mm length of gutter.

IMPROVE YOUR MATHS

Work out the maximum fall for a gutter length of 3.2 m.

4 COMMISSIONING SANITATION SYSTEMS AND COMPONENTS

KEY POINT

Commissioning of rainwater systems was covered in Book 1, Chapter 8, so refer back to remind yourself of information such as rainwater testing and maintenance.

Like all newly installed systems sanitation systems will need to be commissioned. In other words, make sure it works and performs as expected. Some manufacturers will outline a recommended commissioning procedure, but all will follow this basic outline.

The correct procedures must be followed and there are several documents that must be consulted:
- **BS EN 12056.2:2000**
- Building Regulations Approved Document H
- Building Regulations Approved Document F
- Building Regulations Approved Document G
- the manufacturer's instructions.

Visual inspection

It is important to check that all the connections are properly fitted, such as push fit fittings are completely engaged and that solvent welded components are secure. It is essential that none of these components leak. The discharge pipework fall needs to be inspected and tested to confirm correct discharge.

Clips are often overlooked, but must be inspected to make sure the pipework is supported sufficiently, and the clips are anchored securely.

Water levels in WC cisterns should be inspected and adjusted. Appliances need to be inspected to make sure they have been secured correctly, that there is a seal, if required, around the appliance to prevent water leaking out. The waste connections need to be inspected to make sure they are sealed and secure, that they in turn are connected to the correct trap and the trap is filled with water. When walking the system, make sure the swept tee connections are facing the correct direction to aid the flow of water.

Once the visual inspection is complete, the system can then be tested for leaks.

Soundness test

In the case of larger buildings or multi-storey buildings, this test may be required to be carried out in stages or floor by floor. The soundness test is carried out in accordance with **BS EN 12056.2:2000** to ensure there are no leaks, as this will result in the ingress of foul water and odours.

Air test preparation

1 Inspect all the traps to ensure they are filled with water.
2 Seal the soil stack with drain plugs or bags. The top seal will have a test nipple.
3 A manometer or U gauge is filled with water and connected to a test hose.
4 The test hose is connected to both a hand pump and the test nipple via a Y piece connector.
5 Inspect water level in manometer. It should read zero.
6 Gently squeeze the hand pump until the water level in the manometer reaches 38 mm.
7 Close air inlet valve on pump.
8 Wait for 3 minutes.
9 After 3 minutes if there is no pressure drop, the system is sound.
10 If there is a drop in pressure, leak detection fluid should be used to trace the leak. The leak should be rectified and the system retested.

> **INDUSTRY TIP**
>
> Always test your equipment before you use it. You do not want to find that the system is sound, but the leak is on the test equipment!

Performance test

Once a successful soundness test has been completed a **performance test** can be carried out.

> **KEY TERM**
>
> **Performance test**: carried out on a sanitary system to ensure that after simultaneous operation of appliances connected to the same soil stack, the trap depths remaining should be at least 25 mm deep.

The test requires appliances to be filled with water and simultaneously discharged.

The procedure is:
1 Fill all traps with water.
2 Check the depth of each trap using a matt black dip stick. Record each reading.
3 Fill each appliance up to the overflow level.
4 Discharge simultaneously and flush the WCs.
5 After the test, check the depth of each trap using a matt black dip stick. Record each reading.
6 Ensure no trap seal is less than 25 mm deep.

The performance test is required to be carried out three times – one immediately after the other.

The performance test is designed to test the system under the worst conditions and therefore see if the trap seal is sucked out, leaving the trap vulnerable, allowing foul air into the property.

It is important to check if the pipework conveying the water is either too long or too short. A plug of water accumulates while it is discharging and this in turn creates a partial vacuum, behind the plug, which sucks the water from the trap. This happens frequently where a wash basin has been installed due to the speed of discharge and the small diameter of discharge pipe. This is known as self or induced siphonage.

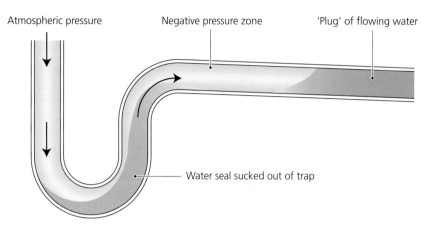

Atmospheric pressure   Negative pressure zone   'Plug' of flowing water

Water seal sucked out of trap

▲ Figure 4.56 Self siphonage in action

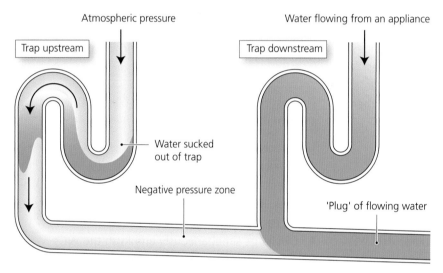

Atmospheric pressure   Water flowing from an appliance

Trap upstream   Trap downstream

Water sucked out of trap

Negative pressure zone

'Plug' of flowing water

▲ Figure 4.57 Induced siphonage in action

The second type of siphonage the performance test could highlight is induced siphonage, which occurs to a passive trap (one not in use) when water is discharged from other appliances connected to the same discharge branch. Again, a plug of water accumulates while it is discharging and this in turn creates a partial vacuum behind the plug, which sucks the water out from the passive trap. This can happen when a series of wash basins are connected to the same discharge branch, hence good practice suggests that if multiple basins are connected to the same discharge branch the pipe diameter should be increased from 32 mm to 50 mm in diameter.

Record keeping

Commissioning records, especially for larger systems, should be kept for reference during maintenance and repair to ensure future efficiency of the system.

Typical information that should be included is:
- date, time and names of commissioning engineer
- location
- manufacturers' names and components
- soundness test results
- performance test results.

These tests may need to be overseen by building control and therefore be signed off by them.

IMPROVE YOUR ENGLISH

Well written, detailed commissioning records will help improve the approach and process of any future installations. They will also be used in the future as reference material if any problems or faults occur in the system.

Handover

When a system has been tested and commissioned, it can then be handed over to the customer. The customer will require all the documentation for the installation, which should include:
- all manufacturers' instructions
- commissioning records and certificates
- Building Regulations compliance certificates
- an as-fitted drawing showing the position of access points.

The customer must be then shown around the system and shown the operating principle of the system and any appliances such as macerators, waste water lifters, sink waste disposal unit and showers. It is good practice to advise the customer about any future maintenance or self-maintenance considerations. The customer will also need to know where to isolate any electrical appliances should an emergency arise.

VALUES AND BEHAVIOURS

Upon completion of any new installation, you should take time to adequately advise the customer about how to maintain their new system or appliances. Take time to demonstrate correct operation and answer any questions they have.

⑤ SERVICING AND MAINTENANCE OF SANITATION AND RAINWATER SYSTEMS

Routine maintenance

The manufacturer's instructions, along with the handover to the customer, should outline any maintenance considerations. Maintenance procedures for different appliances are as follows.

Macerator maintenance

After installation, a demonstration on how the unit operates and is fully isolated needs to be carried out. The manufacturer's instructions will often provide a flow chart which will help with fault finding.
1 Remove fuse from the spur.
2 Lock and label the spur.
3 Check the size of fuse is correct.
4 Visually inspect that pipework has been installed correctly.
5 Inspect for sufficient clipping.
6 Inspect gradient of at least 1:40 fall from appliances to the unit.
7 Any filters, non-return valves and seals need inspecting.
8 If a macerator is faulty follow the manufacturer's flow chart.

INDUSTRY TIP

If the macerator runs for a long time after flushing, the system or unit could be blocked.

If the macerator fails to run after a discharge, check the fuse or float operated switch.

If the macerator cycles, it could be due to a faulty non-return valve.

Waste water lifter maintenance

BS EN 12056.4 covers the maintenance on these units. Always communicate with the customer to see if there have been any problems. The manufacturer's instructions are important to follow for both servicing and fault finding as designs vary. When working on these units it is important to wear the correct PPE. To carry out a service or any maintenance on these units, the engineer will require additional training.
1 Check safe isolation.
2 Ensure that the motor rotates in the correct direction.
3 Operate valves to makes sure they open and close fully.

4 Inspect the switching and setting controls in the collection tank.

5 Inspect for water tightness.

6 Carry out a functional test on the non-return valve.

7 Inspect the pipework support and clipping.

8 Test the motor protection switch.

9 Check oil levels if required.

10 Ensure correct warning light function.

11 Check correct voltage.

12 Inspect hand pump if fitted.

Sink waste disposal unit maintenance

These units are installed on specific sinks with 89–90 mm waste holes. Care should be taken when servicing and maintaining these units. It is important to be trained and fully aware of the manufacturer's procedures. The units are supplied with a specific tool to free the grinding blades should they become blocked. The device will have a thermal cut-off device that will trip in the event of the blades jamming. It is important to inspect the earth bonding, RCD and fuse sizes of these units.

⑥ DIAGNOSING AND RECTIFYING FAULTS IN SANITARY SYSTEMS AND COMPONENTS

Fault diagnosis and repairs to sanitation systems tend to be quite simple and can normally be rectified quite easily. Any fault on a system will generally be described by a customer outlining the symptoms. This can be followed up by a visual inspection which may need to backed up with manufacturer's instructions. Listen to the customer describe the symptom which will help in the diagnosis. The customer will need to be informed of your findings, costs, time scale and any disruption that may occur. With all work like this a risk assessment will need to be made and the correct PPE will need to be worn.

Addressing faults in sanitation systems

- **Leaks** – these will become obvious as the system is used and appliances discharged. Over time, moss and mould will start to show and the possibility of a foul smell may occur. The system will need to be inspected and a soundness test may need to take place to find the leak. Replacement of the seal or part will need to take place.

- **Blockages** – this will be evident by a slow discharge from an appliance or a back-up of fluid in the appliance. Care will need to be taken with the rectification of blockages. Look for any local access caps that could be removed to allow drain rods access. The use of correct chemicals is an option, but this will need to be carried out with care, after fully reading the manufacturer's instructions. It could be as simple as just removing a trap and clearing it, or as complicated as replacing a section of the system.

- **Inadequate support** – this will be evident with the sagging of discharge pipes or easy movement of pipework. Broken supports or additional supports can be installed.

- **Broken pipework** – this will be very evident with major leaks and smells. A section of the system will need to be replaced and tested.

- **Incomplete systems** – this will only be found on new builds where contractors have been changed. A visual inspection of the property will need to take place and discussion with your supervisor and the site manager will need to take place.

- **Incorrect fall** – the required fall of 2½–5° will give the correct self-cleaning discharge. If the fall is too shallow, debris can remain in the discharge pipe, creating build-ups and then blockages. If the fall is too steep, this can create siphonage and pull the trap seal out. A visual inspection of the system will need to be made and an assessment made. This could mean changing the design and fall of the system or installing an anti-vac trap.

Addressing faults in rainwater systems

- **Leaks** – these may not be obvious to start with until it rains, but over time there may be black/green moss growing on the side of the building. Leaks are caused either by a seal being damaged or a crack in the system. This can be rectified by replacing the seal or a section of the system.

- **Blockages** – these are identified by the gutter overflowing when it is raining and causes a nuisance to the customer. Blockages can be caused by something as simple as a tennis ball, leaves, etc. blocking the outlet. The blockage will need to be removed and the customer advised as to what the cause was. If there are over-hanging trees, regular leaf clearance or the use of gutter guards should be suggested.
- **Inadequate support** – a visual inspection of the system will show the sagging of the gutter or easy movement of the downpipe. This can cause pooling of rainwater and movement of components on windy days, which causes additional stress being put on the system. Broken supports will need to be replaced and additional supports could be fitted if required.
- **Broken gutter or downpipes** – this will be evident by a visual inspection, or severe leaks when it rains. Damage like this will normally have been caused by a physical knock. This could be from an overhanging branch or ladder, a wheelbarrow or car, etc. A section will need to be replaced and the customer advised about care in the future.
- **Incomplete systems** – this will generally only happen on a new build if contractors have been changed. A visual inspection of the property will need to take place, as well as a discussion with your supervisor and the site manager.
- **Incorrect fall** – the symptom will not necessarily be obvious to the customer, but should be evident at the commissioning stage of the installation. The rainwater will flow either too fast or too slow. In the worst scenario, the fall could be the wrong way. A recalculation of the 1:600 fall will need to be made and any necessary adjustments made to the installation.
- **Lack of provision for expansion** – in the warmer weather this could cause buckling of the system, and in cooler weather this could cause a section to come loose. A visual inspection of each connection will need to be made to ensure the gutter has been inserted up to the expansion point and not the stop. Necessary adjustments will need to be made.

SUMMARY

Having looked at the broad subject of sanitation and rainwater systems, it shows the importance of design, installation and maintenance. There is a high responsibility on disposing of waste products safely and hygienically, which can be affected if not enough planning, thought and care have taken place at the early stages prior to and during installation.

With growing pressure on world resources, especially water, the saving of wholesome water through the re-use of household water by installing grey water recycling systems or through the installation of rainwater harvesting systems is becoming a prominent issue. As plumbers we need to play a responsible part in this.

Test your knowledge

1 Which system does the UK primary ventilated stack come under being outlined by **BS EN 12056**?

   a System I – 50% capacity

   b System II – 70% capacity

   c System III – 100% capacity

   d System IV – black water and grey water systems

2 Where can an air admittance valve be used?

   a Below the spillover level of a basin

   b In a loft area if insulated

   c At the base of a soil stack

   d Where a discharge branch enters a soil stack

3 Why must cesspools and septic tanks be inspected for soundness?

   a To stop foul air entering the building

   b To allow controlled access of air into the system

   c To prevent the system from overflowing

   d To prevent foul water entering the surrounding ground

4 Who should be consulted with at the design stage if there are any doubts over installing an air admittance valve?

   a Building control officer

   b The customer

   c The site manager

   d A colleague

5 Which Building Regulation states that if a WC macerator is installed in a customer's property, a gravity WC must also be installed?

   a G

   b H

   c A

   d P

6 Which type of fitting should **not** be used on the discharge pipe from a macerator?

   a Solder ring

   b Push fit

   c Compression

   d Solvent weld

7 Just prior to the connection of the discharge pipe from a pumping station to the sewer what **must** be installed?

   a Backflow stop

   b Backflow enhancer

   c Backflow loop

   d Backflow restrictor

8 What does an RCD protect you from?

   a Electric shock

   b Exposure to foul water

   c Toxic gases

   d Build-up of pressure

9 Which one of the following is important at the design stage prior to installing a wet room?

   a Gaining planning permission

   b Talking to the building control officer

   c Soundness testing

   d Talking to the customer

10 What item must be installed correctly to collect and remove the flow of water in a wet room?

   a Vent grill

   b AAV

   c Low level grill

   d Back up loop

11 Which part of **BS EN 12056** outlines the different types of soil stack systems?

   a 1

   b 2

   c 3

   d 4

12 What is the maximum height above the invert level that a WC can be connected to a stub stack?

   a 2.5 m

   b 2.0 m

   c 1.5 m

   d 1.0 m

13 According to **BS EN 12056**, what gradient should the discharge branches of a solid stack be normally kept between?

a 0–2.5°

b 2.5–5°

c 3–6°

d 5–10°

14 What is the recommended discharge branch size for a kitchen sink?

a 20 mm

b 32 mm

c 40 mm

d 50 mm

15 Which appliance does not have an overflow?

a Shower

b Bath

c Sink

d Basin

16 What is the maximum number of WCs that can be attached to a 100 mm discharge branch?

a 2

b 5

c 8

d 12

17 Which phrase best describes a septic tank?

a An underground tank that stores the sewage until the time of disposal

b An underground tank that allows the sewage to settle

c A multi chamber tank allowing the liquid and solid waste to separate

d A multi chamber tank storing the sewage until the time of disposal

18 Which one of the installation factors is NOT associated with a WC macerator?

a Always use smooth bends

b Remember to re-burr any cut edges

c Gravity fall fills the macerator unit

d Only copper pipe is used for the discharge

19 Which item is the correct electrical connection for a macerator unit?

a Switch fused spur with 10 amp fuse

b Switch and luminated fused spur with 10 amp fuse

c Unswitched fused spur with 13 amp fuse

d Unswitched fused spur with 5 amp fuse

20 When would a waste water lifter be used?

a When draining by gravity is not an option

b When the discharge into the system is high

c When the maximum flow capacity of the system reaches 75%

d When a system includes a surface water discharge

21 What safety cut-out device is incorporated into a sink waste disposal unit?

a Over load cut-out

b Flow switch

c Thermal cut-out

d Non-return valve

22 Which standard outlines the minimum space requirements for each sanitary appliance?

a Building Regulations part M

b BS EN 6465

c BS EN 806

d BS 8000

23 According to **BS EN 12056**, what should be fitted when a pipe goes through a ceiling or floor to another property?

a Non-return valve

b Air admittance valve

c Water trap

d Intumescent collar

24 What size of waste pipe is required as a discharge if multiple hand wash basins are connected together?

a 32 mm

b 40 mm

c 50 mm

d 100 mm

25 What is the formula for working out the effective roof area when designing a rainwater system?

a $\left(W + \dfrac{H}{2}\right) \times L$

b $\left(L + \dfrac{W}{2}\right) \times H$

c $\left(H + \dfrac{L}{2}\right) \times W$

d $(H + L) \times + \dfrac{W}{2}$

26 Which Building Regulation Approved Document outlines foul water drainage?

27 Which pipe on a primary ventilated stack system acts as the vent?

28 Describe the function of a permanent level control in a lifting unit.

29 When designing a primary ventilated stack for a domestic property, the installation must always be designed to operate at what pressure?

30 List and describe the four types of sanitation systems outlined in **BS EN 12056**.

31 List the regulative and guidance documentation that would need to be referenced when designing and installing a soil stack at a domestic property.

32 State the advantages of a reed bed if connected to a septic tank.

33 Describe why a macerator might be installed behind a WC in a bathroom.

34 Describe the use and function of an intumescent collar.

Answers can be found online at www.hoddereducation.co.uk/construction.

CHAPTER 5

ELECTRICAL PRINCIPLES

Plumbing is a very diverse craft which often encompasses electrical applications. For example, a plumber's work could include the installation of an immersion heater for a hot water cylinder, a central heating control system, fitting electric showers or checking bonding to gas or water systems. Many components or equipment used for plumbing installations require an electrical supply or control circuit.

In addition, working in the building services industry requires the use of electrical plant and machinery, which can also involve risk. Whether you are installing electrical equipment or not, you will be working on or around electricity and the purpose of this chapter is to give you an understanding of the risks involved as well as an understanding and awareness of some of the electrical equipment and circuits you will be working on. It will show you how to connect electrical equipment to circuits and check the connections are safe.

By the end of this chapter, you will have a knowledge and understanding of:
- the principles of electricity
- the legislation relating to electrical work and the control of plumbing and domestic heating systems
- the types of electrical system and layouts
- performing pre-installation activity prior to undertaking electrical work on plumbing and domestic heating
- safe isolation procedures
- safe installation and testing of electrical equipment
- diagnosing faults and safe repair of electrical work.

Remember, some of the basic electrical safety and principles were covered in Book 1, Chapter 11, Electrical principles and processes for building services engineering. This covered:
- electrical supplies used in domestic plumbing systems
- components used in electrical installations and basic electrical tasks
- electrical tests and procedures for safely isolating supplies
- identifying critical safety faults on electrical components.

KEY POINT

It is important to note that the knowledge gained from this chapter does not make you a qualified electrician, but opportunities do exist for you to take further training in your career to enable you to install and commission a defined scope of electrical systems and circuits.

1 THE PRINCIPLES OF ELECTRICITY

Electricity is a flow of electrons that cannot be seen or smelt. It flows around a circuit where its energy makes things work. When electricity is simply quantified, we use a number of terms, represented by certain symbols (see Table 5.1).

▼ Table 5.1 Common electrical terms

Term	Unit of measurement	Symbol used	What it is
Current	amperes or amps for short (A)	I	In simple terms, it is the quantity of electricity. To use a simple analogy, if two litres of water was to flow through a pipe, that is the quantity of water. If two amps was to flow in a circuit, this is the quantity of electricity. The quantity of electricity has an impact on the size of cables needed, as cables that are too small for the intended current may get too hot, which in turn will result in damage and potential fire risks.
Resistance	ohms (Ω)	R	It is an opposition to current flow which can impede the current. If a pipe was kinked, water would be restricted and less water flows. If a circuit has too much resistance, this too can resist current flow which causes heat. Resistance may also be present in appliances that use or consume the electrical energy, such as immersed water heater elements that have a set resistance and when the current passes through the element, creates heat. Anything carrying electricity has resistance such as cables, etc. but the idea is to keep the resistance low in cables so current can flow without creating too much heat.
Voltage	volts (V)	V	Voltage is, in simple terms, the pressure that pushes current around a circuit. If a circuit has too much resistance, perhaps due to length or cable size, voltage will be lost in the circuit which could result in equipment not working correctly. In the UK, the standard supply voltage for most dwellings is 230 V. In reality, if the voltage was measured in many dwellings, it could be anywhere between 217 V and 253 V. Another term used to describe voltage is potential.
Potential difference	volts (V)	V	This is the difference in voltages between two conductive points. As an example, parts connected to the electrical earth in a building, such as metal gas pipes, would in normal circumstances be at 0 volts. A live connection on a 230 V circuit would have a voltage of 230 V. If the voltage was measured between the two parts using a volt meter, the meter would read the potential (voltage) difference which is 230 V. If, however, the pipe became live due to a fault and had a voltage present of 220 V, the potential difference the meter would actually read is 10 V.
Power	watts (W)	P	Most electrical equipment will have a power rating based on the voltage it is intended to be connected to. This would in turn dictate the amount of current needed from the supply.
Energy	kilowatt hours	kWh	This is the unit used to measure the amount of electricity consumed by an electrical installation and used to bill the consumer for what they have used. Although energy is generally measured in joules, one joule is one watt of power in one second. If joules were used to bill the quantity of electricity used, most electricity bills would have usage as millions of joules. To make this much easier to understand, energy usage is measured in kilowatts used per hour.

IMPROVE YOUR MATHS

How many watts of power are consumed by an electric water heater using 90,000 joules in 30 seconds?

KEY POINT

Both line and neutral conductors are **live** as they both carry current.

Basic circuit principles

In simple terms, current flows from a source of energy, such as a battery or transformer, around a circuit, through a load, such as a heater element and back to its source. The current passes through a **line** conductor up to the load and through the **neutral** as a return path back to the source.

KEY TERMS

Line: the **conductor**, having brown coloured **insulation**, which is normally connected to terminals marked L.

Neutral: the conductor, having blue coloured insulation, which is normally connected to terminals marked N.

Conductor: this is the part of a cable which current passes through. In most cables, this is made of copper and should have a low resistance.

Insulation: the material that covers the conductor, and should have a high resistance stopping current flow. Insulation is intended to stop current from leaking from one conductor into another conductor or person which could in turn cause an electric shock.

The load will normally cause the voltage to be consumed so the line conductor has a voltage of 230 V and the neutral 0 V, giving a measured potential difference across the load of 230 V when measured on the load L and N terminals.

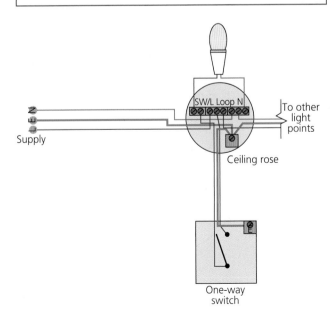

▲ Figure 5.1 A basic electrical circuit

Ohm's law and power

Having an understanding of Ohm's law and how power is calculated will help when considering cable size, checking circuits or undertaking fault diagnosis. Ohm's law and power were introduced in Book 1, Chapter 3, Scientific principles.

Ohm's law

The relationship between voltage, current and resistance can be demonstrated using Ohm's law where:

$$\text{Voltage} = \text{current} \times \text{resistance}$$

or

$$V = I \times R \text{ or } I = \frac{V}{R} \text{ or } R = \frac{V}{I}$$

So, if a circuit had a voltage of 230 V and a connected load having a resistance of 120 Ω, the current drawn by the load would be:

$$\text{as } I = \frac{V}{R}$$

then:

$$\frac{230}{120} = 1.92 \text{ A}$$

Alternatively, Ohm's law could be used to work out voltage loss due to cable conductor resistance.

IMPROVE YOUR MATHS

Worked example: calculating Ohm's law

Let's assume a line conductor was connected to a 230 V supply and the conductor had a resistance of 3 Ω. If the load draws a current of 2 A, how much voltage would be lost by the conductor resistance?

As:

$$V = I \times R$$

then:

$$V = 2 \times 3 = 6\ V$$

So, if the supply was 230 V and 6 V was lost due to the conductor resistance, only 224 V would be present on the load line terminal.

Power

The power of a load is measured in watts. This value dictates the amount of current drawn by the load depending on the circuit or supply voltage as:

$$\text{Power (P)} = \text{voltage (V)} \times \text{current (I)}$$

or:

$$I = \frac{P}{V}$$

IMPROVE YOUR MATHS

Worked example: calculating power

So, if an immersion heater element has a power rating of 3 kW and is connected to a 230 V supply, how much current is drawn by the element?

As a kilowatt is 1000 watts then 3 kW is 3000 W so:

$$I = \frac{P}{V} \text{ so } \frac{3000}{230} = 13.04\ A$$

So, the immersion heater circuit conductors would need to carry 13.04 A.

KEY POINT

It is always a good exercise to know how much current certain loads draw as this has an impact on the switches or devices selected to control them. For example, if the immersion heater above needed to be controlled by a time clock switch, the switch must be capable of switching at least 13.04 A. If the time clock switch selected was only capable of switching 10 A, the switch would burn out very quickly.

ACTIVITY

Using a catalogue, find the electrical power rating for the items listed below then calculate the amount of current they would draw from a 230 V supply:
- central heating circulating pump
- electric shower
- electric instantaneous water boiler
- motorised valve.

Electric shock

Very small current values can cause **electric shock**, and values around 0.05 A (50 mA) can cause death. The human body responds to different current values as the body uses electrical signals to make muscles move. Coming into contact with external currents can override the body signals causing involuntary operation of muscles. If the current is an alternating current (AC), this could cause parts of the body, such as the heart muscle, to operate at speeds equal to the electrical supply frequency.

KEY TERM

Electric shock: where a current flows through the human body and causes an accident or injury as a result.

The electrical supply frequency for AC in the UK is 50 hertz, meaning the current alternates direction 50 times in one second. An alternating current is effectively switching on and off twice in one alternation or cycle so this has the effect of turning the current on and off twice in one cycle or 100 times a second.

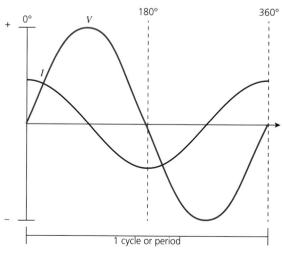

▲ Figure 5.2 Alternating current sine wave

If an AC current exceeding 0.05 A was able to flow through the heart, this could cause the heart to pump 100 times a second and it will not take long for the heart to stop functioning in a short time period causing arrest.

If a person was to come into contact with direct current (DC), which flows in one direction only, this would cause muscles to contract and stop. If someone came into contact with a direct current of approximately 0.02 A, this could cause arm muscles to lock making it impossible to let go of the live part. There is a saying that 'you stick to DC.'

The body responds in various ways to an electrical current flowing through it as shown in Table 5.2.

▼ Table 5.2 How current affects the body

Current in milliamperes (mA)	Effect on the body
10–12	A slight tingling sensation is experienced
15	Pain increases rapidly above this current
20–25	Hard to let go (DC) or the hand can be knocked away from the point of contact (AC)
50	Death has been recorded at this current
Above 120	Rapid and irregular beats of heart muscles. Breathing may be paralysed. Fatal effects are likely

In order for current to enter the body, a voltage is required to break down skin resistance. For an average person in average conditions the voltage is typically 50 V AC or 120 V DC.

Changing conditions can reduce the body resistance, meaning lower voltages can cause current to enter the body causing electric shock. These changes in conditions could be:

- humid locations such as bathrooms causing the body to sweat lowering skin resistance
- being outdoors standing on the earth, meaning more points of contact with earth – more contact points lower body resistance as there are more paths for current to take
- confined conducting locations, such as laying on a metal duct, as the multiple points of contact with earthed parts completely reduces body resistance.

There are two ways a person can receive an electric shock:

- **directly**, by touching a part which is meant to be live such as a live conductor
- **indirectly**, by touching a metallic part which isn't normally live but has become live due to a fault in the system.

Protection against direct contact

In order to protect persons against electric shock by direct contact with live parts, two methods are used in all common electrical installations:

- **Insulation of live parts**, such as the insulations around a copper conductor – as insulation must have a high resistance to stop current from leaking, the insulation must be suitable for the intended voltage. If the insulation isn't suitable for the intended voltage, the voltage will break down the resistance allowing current to flow. If insulation is damaged or deteriorating, this too could cause electric shock as live parts may be exposed to touch.
- **Barriers and enclosures**, such as junction boxes, socket-outlet fronts and boxes or switch fronts and boxes – the rule is that all barriers and enclosures must only be accessible by the use of a tool or key, such as a screwdriver. One exception to this rule is a ceiling rose type light point and a lampholder. This is because the lampholder would normally have a lamp inserted giving protection and a ceiling rose is installed on a ceiling which is a generally inaccessible position.
  In addition, all barriers and enclosures housing live electrical parts must have suitable IP protection such as **IP2X** for all surfaces and **IP4X** for all accessible horizontal top surfaces.

KEY TERMS

IP2X: meaning that there is no hole in the barrier or enclosure greater than 12.5 mm in diameter, which provides 'finger protection', meaning no person can insert their finger and touch live parts.

IP4X: meaning that there is no hole in the barrier or enclosure greater than 1 mm in diameter stopping parts from falling into the enclosure.

② THE LEGISLATION RELATING TO ELECTRICAL WORK AND THE CONTROL OF PLUMBING AND DOMESTIC HEATING SYSTEMS

Although working on electrical systems is not restricted by a licence to practise, there are many statutory and non-statutory regulations which address the dangers of working on, or installing electrical components and systems. As soon as somebody decides to work on or install electrical equipment or components, they are duty-bound to work in accordance with all electrical related regulations regardless of the nature of the work. Failure to comply with any of the regulations could lead to, as a minimum, prosecution. Failure to comply with regulatory procedures could result in death.

Legislation and information

Under the Health and Safety at Work etc. Act, many statutory regulations have been created by HSE to promote electrical safety. These are described below. Remember, when working on or around electricity, ignorance to the law is not a defence.

Electricity at Work Regulations 1989

The Electricity at Work Regulations 1989 (EAWR) are made under the HSW Act. This means that breaches in the EAWR will result in action being taken under the HSW Act.

INDUSTRY TIP

You can access the Electricity at Work Regulations 1989 at: www.legislation.gov.uk/uksi/1989/635/contents/made

In addition to the duties imposed by the HSW Act, the EAWR impose duties on **duty holders** in respect of systems, electrical equipment and conductors, and work activities on or near electrical equipment under their control. Managers of mines and quarries are also included, despite mines and quarries having other special regulations.

The EAWR cover the principles of electrical safety that apply to work activities and systems, including anything that influences them, equipment, isolation and safety systems, and the **competency** of those working with electricity.

KEY TERMS

Duty holder: any person or organisation holding a legal duty under the Health and Safety at Work etc. Act 1974.

Competency: the degree to which a person has the ability to complete something.

The EAWR apply beyond those situations that we traditionally associate with the dangers of electricity, including voltages outside the scope of **BS 7671** The IET Wiring Regulations. The EAWR cover battery operated systems to extra high voltage transmission supplies.

The HSE produces a memorandum of guidance (HSR25), which is free and provides guidance on how each regulation should be interpreted. Each regulation covers a specific topic, as follows:

- Regulation 4 – systems, work activities and protective equipment
- Regulation 5 – strength and capability of electrical equipment
- Regulation 6 – adverse or hazardous environments
- Regulation 7 – insulation, protection and placing of conductors
- Regulation 8 – earthing or other suitable precautions
- Regulation 9 – integrity of referenced conductors
- Regulation 10 – connections
- Regulation 11 – means for protecting from overcurrent
- Regulation 12 – means for cutting off the supply and for isolation
- Regulation 13 – precautions for work on equipment made dead
- Regulation 14 – work on or near live conductors
- Regulation 15 – working space, access and lighting
- Regulation 16 – persons to be competent to prevent danger and injury.

Electricity Safety, Quality and Continuity Regulations 2002

The Electricity Safety, Quality and Continuity Regulations 2002 (ESQCR) impose requirements regarding the installation and use of electric lines and the apparatus of electricity suppliers, including provisions for connection with earth. The safety aspects of these regulations are administered by the HSE. All other aspects are administered by government and the industry.

The ESQCR may impose requirements, usually on supply companies or those associated with the supply of electricity, in addition to those of the EAWR. Designers of installations have a responsibility to ensure they meet the ESQCR.

Provision and Use of Work Equipment Regulations 1998

The Provision and Use of Work Equipment Regulations 1998 (PUWER) ensure that work equipment, such as electric drills and cutters, as constructed or adapted, is suitable for the task and safe to use, taking into account all the risks of using it in a specific work environment. This applies to any equipment, machinery, appliance, apparatus or tool for use at work.

For example, in order to be fit for purpose, construction site equipment needs to cope with a wide range of weather conditions including wet weather. Equipment also needs to be robust enough and sufficiently protected to withstand mechanical impact or abrasion.

It should help to reduce to as low a level as possible, the risk of electric shock through external influences, including from reduced low voltage systems such as the 110 V system used on UK construction sites. This may mean that additional protection must be provided.

Non-statutory regulations, published guidance and codes of practice

Guidance on the legal requirements of certain statutory regulations and how to interpret them, and approved codes of practice (ACoPs), are published or approved by the HSE and the relevant secretary of state. These documents often contain a copy of the legislation they are giving guidance on. They have a special legal status in law. Where a duty holder does not follow the guidance or code of practice, they must show that their course of action or omission is no less safe.

Not all documents using the term 'regulations' are statutory instruments. For example, **BS 7671** The IET Wiring Regulations are non-statutory. They form the **national standard** in the UK for low voltage electrical installations. They deal with the design, selection, erection, inspection and testing of electrical installations operating at a voltage up to 1000 V AC. Work undertaken in accordance with **BS 7671** is almost certain to meet the requirements of the Electricity at Work Regulations 1989. **BS 7671** is supported by a number of guidance notes published by the Institute of Engineering and Technology (IET).

Other non-statutory rules include the rules and requirements of regulating bodies such as the NICEIC, the National Association of Professional Inspectors and Testers (NAPIT), the Electrical Contractors' Association (ECA) and trade unions such as Unite. Individuals and enterprises that want to belong to these organisations must comply with the relevant organisation's rules. In certain circumstances, compliance with such rules, such as those for competent person schemes (CPSs), assists in compliance with specific statutory regulations or requirements.

BS 7671 The IET Wiring Regulations

BS 7671 is the basis for all other guidance documentation, such as the IET On-Site Guide or the eight IET guidance notes (GNs). **BS 7671** is broken down into seven parts with further appendices providing information relating to the seven parts. The parts are in a logical order relating to the design and installation of an electrical installation. They are:

- Part 1 – Scope, object and fundamental principles, which sets out what the standard covers (scope), why it is setting out (object) and what risks need addressing (fundamental principles).
- Part 2 – Definitions, which provide detailed descriptions of technical words or phrases, including abbreviations and symbols, used in the document.
- Part 3 – Assessment of general characteristics, which details what a designer must assess prior to designing an electrical installation and before any equipment is installed.

- Part 4 – Protection for safety, which provides detail on methods which may or must be used for protecting against particular risks electricity may pose such as electric shock, fire or overcurrent.
- Part 5 – Selection and erection gives detail on what particular equipment can be used and how it should be installed.
- Part 6 – Inspection and testing, which gives criteria for what should be inspected and tested for every new or existing electrical installation.
- Part 7 – Special installations or locations sets out requirements where a particular location provides greater risk. In these locations, all relevant previous regulations still apply but, because of the risk, some regulations need changing to overcome the risk.

IET On-Site Guide

The purpose of the IET On-Site Guide is to provide guidance on the design, installation and inspection and testing of standard installations such as those in dwellings, shops and small offices. By following the guidance given, the need for complex design calculations is reduced.

IET Student's Guide to the Wiring Regulations

Whilst this guide is intended for students learning about electrical installations, the title should not put off electricians from using this guide as it simplifies a lot of the technical aspects of **BS 7671** and provides good guidance on the electrical industry, basic design procedures, good working practices and common calculations.

IET guidance notes

The IET guidance notes (GNs) are intended to provide more detailed guidance on aspects of **BS 7671** but the numbering of the eight GNs does not necessarily reflect the parts of **BS 7671**. The eight guidance notes are:

- GN1 – Selection and erection
- GN2 – Isolation and switching
- GN3 – Inspection and testing
- GN4 – Protection against fire
- GN5 – Protection against electric shock
- GN6 – Protection against overcurrent
- GN7 – Special locations
- GN8 – Earthing and bonding.

The IET also provide specific guidance documents on aspects of electrical installation work, such as:

- Guide to consumer units
- Considerations for DC installations
- Installation of electric vehicle charging installations
- In-service inspection and testing of electrical equipment
- Guide to the Building Regulations
- Guide to emergency lighting installations.

INDUSTRY TIP

The IET also publish a comprehensive suite of Codes of Practice known as IET Standards. A full list of these can be found on the IET website: search for 'IET electrical'.

Diagrams and manufacturers' documents providing electrical information

Numerous sets of drawings are produced to communicate information about individual systems or collections of systems within an electrical project. They include:

- plans/layout drawings
- schematic (block) diagrams
- wiring diagrams
- circuit diagrams.

Drawings usually use British Standard symbols. However, in order to represent the wide variety of materials and equipment available, non-standard symbols may also be used. A legend or key explains them.

Plans or layout drawings

Plans or layout drawings are used to locate individual systems within the overall project and give an indication of the scale of the project. In addition, there may be drawings to show specific fixing, assembly and/or completion details. This is often the case where complex construction, lifting or use of a crane is required. These details may be provided in elevation, plan or both. They will then become part of the contractor's method statement when the project goes into the construction phase.

Schematic diagrams

Schematic diagrams, sometimes referred to as block diagrams, can serve many purposes, but are primarily provided to show the overall functionality of a system, including interfaces and operational requirements. This means they are intended to show a sequence of control and connection but not necessarily scale or actual positions in the wiring. For example, if a three-plate lighting system was drawn as a wiring diagram, the light would be wired before the switch even though the switch controls the light. A schematic would show the switch before the light as a sequence of control. Schematic or block diagrams are typically used to show distribution systems and circuits in large electrical installations having many remote distribution boards.

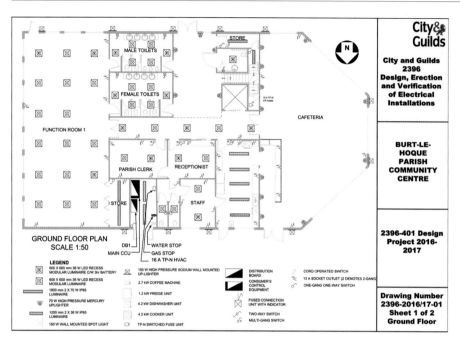

▲ Figure 5.3 Plan drawing of a building

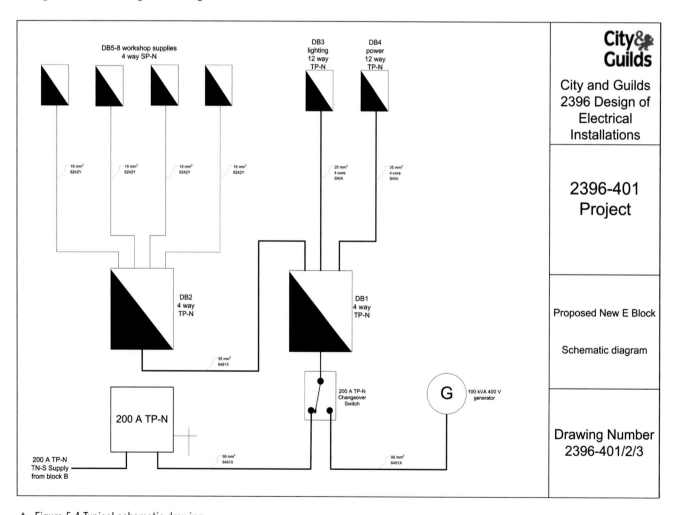

▲ Figure 5.4 Typical schematic drawing

Wiring diagrams

Wiring diagrams are generally provided to show in detail how a system or collection of systems is put together. This type of drawing shows locations, routing, the length of run and types of systems cabling. These types of diagram are sometimes mistakenly referred to when it is actually a circuit diagram that is required.

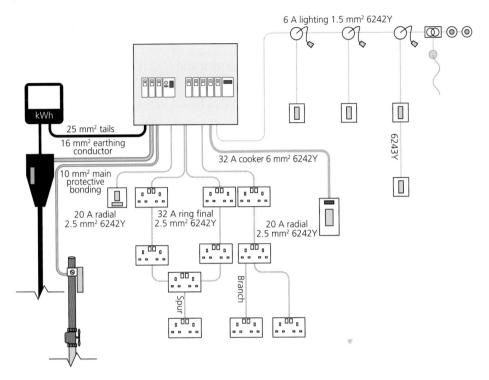

▲ Figure 5.5 Typical wiring drawing

Circuit diagrams

Circuit diagrams contain information on how circuits and systems operate. They can be provided as detailed layouts, although some information such as length of run may be omitted for clarity. In most instances, these diagrams are used for **diagnostic** purposes so that designers, installers and maintainers understand how their actions may influence a particular component or arrangement.

KEY TERM

Diagnostic: concerned with identifying problems.

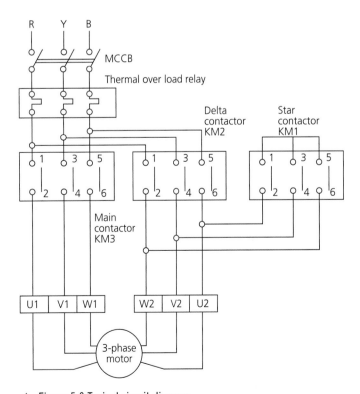

▲ Figure 5.6 Typical circuit diagram

Manufacturer's information

When installing any electrical equipment or appliances, the manufacturer's instructions supersede all other requirements.

Manufacturer's information normally comes in two forms:

- installation and maintenance information
- user information.

Installation and maintenance information contains instructions on how to install and commission the equipment. It will normally state what size fuse is required to protect the equipment, as well as recommended minimum cable size and type.

It is essential for an installer to follow the manufacturer's instructions when fitting electrical appliances. By reading them in advance it will help greatly with the selection of the product to ascertain whether or not it is suitable for a specific application.

Manufacturer's literature will also give fault-finding flow charts and a spare parts list for anyone trying to diagnose a fault with an existing appliance.

While products must comply with the relevant standards of safety and performance, sometimes there are extraordinary applications or special functions which are covered solely by a manufacturer, provided the product is installed in accordance with their directions, hence the importance of a compliance certificate and benchmark log book.

User information is limited to the daily operation of an appliance and the installer during the handover should give demonstrations and explanations to the customer when commissioning is complete.

③ THE TYPES OF ELECTRICAL SYSTEM AND LAYOUTS

It is important to understand how electrical systems are arranged in order to safely work on parts of them.

Typical electrical installation systems for domestic premises

An electrical installation is made up of many parts and in certain ways to make them safe and functional. These include:

- consumers' control unit (CCU) or distribution board (DB)
- protective devices found in the CCU
- circuits
- earthing and bonding to achieve automatic disconnection of the supply (ADS) in the event of a fault to earth
- current using equipment and specialist controls (many of these components will also be covered in other chapters).

Consumers' control unit (CCU)

Generally, the first unit in an installation, where the supply is split into circuits, is called the consumers' control unit (CCU). This may sometimes be referred to as a distribution board, or DB, and sometimes simply as a consumer unit (CU). The CCU will have protective devices inside. They control and protect the circuits and could include:

- circuit breakers (CB)
- residual current breakers with overload (RCBO)
- fuses.

The CCU will also have a main switch which can be used to isolate the entire installation. In a domestic dwelling, the main switch must be double-pole meaning it isolates both line and neutral.

INDUSTRY TIP

You may hear a CCU called many different things such as fuse boards, DBs or distribution boards. Distribution boards (DB) are often the name given to further units on the end of a distribution circuit. A distribution circuit is a circuit that supplies a distribution board whereas a final circuit is one that supplies current using equipment such as lights and heaters or socket outlets.

Arc fault detection devices (AFDDs)

BS 7671 now requires arc fault detection devices (AFDDs) to be installed to protect circuits supplying socket outlets that have a rating up to 32 A in specified buildings. These building are those considered higher risk installations such as:

- purpose-built student accommodation
- houses of multiple occupation (HMO)
- care homes
- buildings considered higher risk such as high-rise residential buildings.

AFDDs are recommended for other types of premises.

AFDDs monitor the supply to the circuit being protected and can detect both series and parallel arcing. Series arcing is generally caused by a loose or failing connection creating an arc where the electricity jumps the gap between the connections. Parallel arcing is where insulation breaks down or has been damaged, meaning arcing takes place between conductors in a circuit.

Electrical arcs or sparks can lead to fires if allowed to continue so AFDDs are designed to disconnect the circuit on detection, reducing the risk of fire.

AFDDs can be combined as part of a circuit breaker or RCBO, or installed as a standalone device.

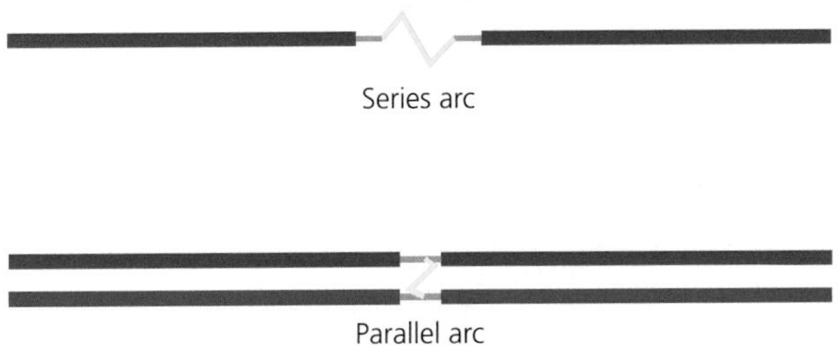

Series arc

Parallel arc

▲ Figure 5.7 Types of arc

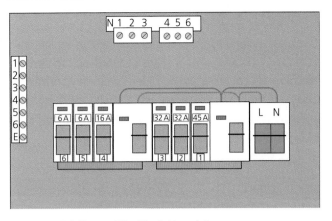

▲ Figure 5.8 Six-way DB with eight modules

Some CCUs/DBs are known as split-way boards. This means that some of the modules can be protected by one **residual current device (RCD)** or switch and the others by another RCD or switch. Split-way boards should always have one main switch that can isolate all circuits. The advantage of split-way boards is that any fault on one circuit will only trip part of the installation protected by one of the RCDs.

KEY TERMS

Residual current device (RCD): a sensitive device which trips, cutting current from a circuit, should a very small fault occur between any live conductor and earth. They are intended to give maximum protection against electric shock and can be identified by the fact they look like circuit breakers but have a test button located on them.

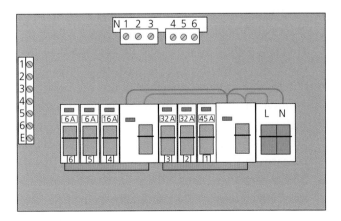

▲ Figure 5.9 Split-way board showing three ways protected by one RCD and three ways protected by another. In total, the DB is 12-module

Protective devices

There are several types of protective devices which are fitted in CCUs and what is there depends on the age of the electrical installation. Older CCUs may contain fuses but more modern ones will probably have circuit breakers (CB) or circuit breakers having a combined RCD (RCBOs).

▼ Table 5.3 Standard ratings for protective devices

Current rating (A)	Device type
5	Rewireable fuse, older cartridge fuses
6	Circuit breakers, RCBOs, cartridge fuses
15	Rewireable fuse, older cartridge fuses
20	Circuit breakers, RCBOs, cartridge fuses, rewireable fuses
30	Rewireable fuse, older cartridge fuses
32	Circuit breakers, RCBOs, cartridge fuses
40	Circuit breakers, RCBOs, cartridge fuses
45	Rewireable fuse, older cartridge fuses

Fuses

Fuses have been a tried and tested method of circuit protection for many years. A fuse is a very basic protection device where a high enough fault current causes the fuse element to heat up to the point where the element is destroyed and cuts the current from the circuit. Once the fuse has 'blown' (when the element of the fuse has melted or ruptured), the fuse needs to be replaced.

- **BS 3036 semi-enclosed rewireable fuses** – in older equipment, the fuse may be just a length of appropriate fuse wire fixed between two terminals. These devices are now becoming uncommon as electrical installations are rewired or updated. One of the main problems associated with rewireable fuses is the overall lack of protection and, in some cases, the presence of asbestos in the fuse carrier.
- **BS 88 cartridge fuses** – these are barrel shaped fuses which have metallic ends, like a plug fuse, and normally clip into a carrier. Like rewireable fuses, once these blow, they need to be replaced, which is done in a very user-friendly way. As a result, and

as electrical installations are updated, the CCU is normally replaced with one capable of having circuit breakers or RCBOs.

▲ Figure 5.10 Rewireable fuses

▲ Figure 5.11 Cartridge fuse

Circuit breakers and RCBOs

These devices are much more user friendly as they have the ability to be reset if tripped. These devices work on a magnetic effect where a high enough current causes an electromagnetism to automatically switch off the device in event of a predetermined fault current. Overload currents, which are smaller than fault currents, are detected by a thermal trip, such as a bi-metallic strip, which causes the device to trip.

The difference between a circuit breaker and RCBO is that an RCBO also incorporates a residual current device (RCD) which will cause disconnection if a fault to earth is detected as small as 30 mA, in the case of an RCBO having a residual trip setting of 30 mA, which is the most common setting for domestic type installations.

The physical difference between a CB and RCBO is an RCBO has a test button incorporated on it. It is very important that this test button is pushed at least once every six months as this keeps the tripping mechanism free from sticking and that means the device will trip quickly should very small fault currents occur to earth. With only one or two rare exceptions, all circuits in domestic type electrical installations must have RCD protection and this is either provided with individual RSBOs protecting each circuit or the use of split-way CCUs where RCDs protect several circuits.

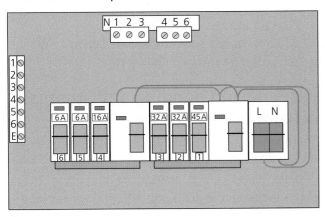

▲ Figure 5.12 Split-way CCU with RCDs protecting a group of circuits

Circuits

Circuits can be categorised in the following ways:
- **Lighting circuits**, which are relatively low power circuits with an almost constant load when lights are operating.

- **Power circuits**, which are provided for socket-outlet circuits or other appliances such as immersion water heaters or showers. They require larger protective devices due to the demand for current and, in most instances, require RCD protection due to the additional risk presented by the portable equipment connected to the socket outlets.

Lighting circuits

Lighting circuits are generally rated at 6 A but can, in some installations, be rated at 10 A or 16 A. As the name suggests, lighting circuits are intended to supply lighting points, but in some cases they may also supply some very small power equipment such as bathroom fans or shaver supply units.

Power circuits

Power circuits generally supply socket outlets, but may also supply individual appliances. Power circuits may be wired in two ways: ring-final and radial.

Ring-final circuits

Ring-final circuits are the traditional means of wiring socket-outlet circuits within the UK and are rated at 32 A. The reason for their use was to provide a high number of conveniently placed outlets adjacent to the loads. The circuit load is shared by two sets of cables run in a ring formation. This also assists in improving voltage drop as the conductors being in parallel reduce the overall resistance.

Ring-final circuits are able to supply an unlimited number of outlets serving a maximum floor area of 100 m². Permanent loads, such as immersion heaters, should NOT be connected to a ring-final circuit. Ring-final circuits supplying kitchens should be arranged to have the loads equally distributed around the circuit. The plugs and fused connection units, commonly called fused spur units, are normally fitted with 3 A or 13 A fuses.

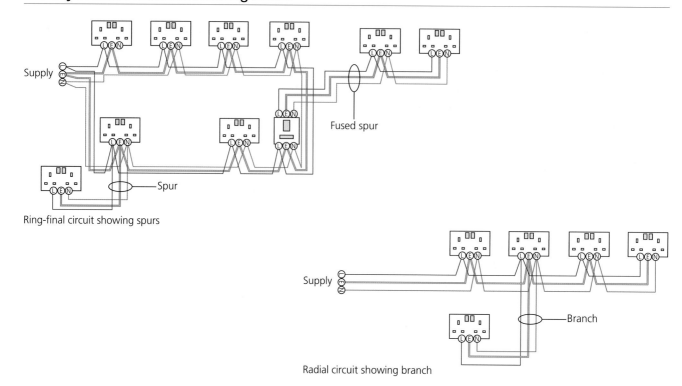

Ring-final circuit showing spurs

Radial circuit showing branch

▲ Figure 5.13 Ring-final circuit showing spurs and radial circuit showing branch

With technology reducing the consumption of appliances and equipment, the need for a ring-final circuit is being questioned as they do have several disadvantages. If a ring-final circuit becomes open circuit, this may not necessarily be detected by the user as power will still be distributed to all socket outlets. Circuit conductors may then become overloaded leading to the possibility of fire. As a result, more and more electricians are wiring socket-outlet circuits as radial type circuits to avoid the risk of unidentified open ring circuits.

Radial circuits

Radial circuits may be selected to supply multiple socket outlets and they are also chosen as a means of supplying individual appliances or dedicated fixed appliance circuits such as circuits to supply showers or immersion water heaters. Radial circuits can be any rating depending on the cable size. Typical ratings include:

- 16 A for immersion heaters, dedicated boiler circuits or single appliances such as dishwashers

- 20 A for socket-outlet circuits
- 32 A for cookers and showers rated up to 7 kW
- 40 A for showers rated up to 9 kW.

Earthing and bonding and automatic disconnection of supply

There are many green and yellow protective conductors within an electrical system, but they have different purposes such as earthing or bonding. They play a major part in achieving ADS in the event of a fault to earth. If a fault exists between line and earth, there is a major risk of electric shock as someone may be in contact with metallic parts. In order to reduce the risk of electric shock, circuits are protected using a system called automatic disconnection of supply (ADS).

Automatic disconnection of supply

In order for ADS to protect persons against the risk of electric shock circuits must have protective devices, such as a fuse, circuit breakers or RCBO, and protective conductors such as earthing and, where necessary, bonding. These conductors must provide a low resistance path to earth so should any metal part become live due to a fault, Ohm's law dictates that the low resistance path provided by the protective conductors will ensure there is a high fault current. The high fault current will disconnect the device protecting the circuit very quickly; hopefully before anyone touches the part and gets an electric shock.

Regulations for electrical circuits, known as **BS 7671** (which will be covered later in this chapter), state that most low power circuits, such as those in domestic installations, must disconnect within 0.4 seconds.

As currents as low as 0.05 A can cause electric shock (as discussed earlier in this chapter) fault currents may as well be hundreds of amperes as they are likely to kill anyway. What saves a person's life is the disconnection of the circuit's supply in a very quick time and this will only happen with high fault currents.

To give an illustration of this, Table 5.4 shows the amount of current needed under fault conditions to disconnect the various ratings of **type B circuit breakers** in the specified disconnection time of 0.4 seconds.

▼ Table 5.4 Current causing effective disconnection of type B circuit breakers

Device rating (A)	Current needed to cause automatic disconnection to occur in 0.4 seconds (A)
6	30
10	50
16	80
20	100
32	160

KEY TERM

Type B circuit breakers: the most sensitive type of CB and should be the types used to protect circuits in domestic type installations. Other types are type C intended for motors and transformers and type D for very specialised machines such as welding equipment or medical equipment.

As the current needed to cause disconnection of a protective device is high, the resistance of the whole fault circuit needs to be low. This resistance is known as the total earth fault loop impedance (Z_s). The diagram in Figure 5.14 shows the total earth fault current path

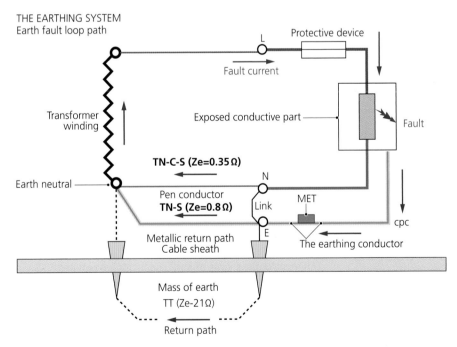

THE EARTHING SYSTEM
Earth fault loop path

The values of ohms shown are maximum values the electrical distribution network operator tries to achieve for the supply impedance.

▲ Figure 5.14 The total earth fault current path

for the three common earthing arrangements, which starts at the **sub-station transformer**, through the supply network line conductor, into the electrical installation and through the circuit line conductor. The current then flows through the fault onto earthed parts and back through the cpc (circuit protective conductor) to the main earthing terminal (MET). From this point, it flows through the installation earthing conductor which then connects to the supply network earth back to the transformer.

KEY TERM

Sub-station transformer: a piece of equipment which is owned by the electricity distribution network operator (DNO) and is used to step down large distribution voltages of 11000 V to 230 V for supplies into houses. Sub-stations are sometimes located behind panel fences and can serve up to 100 houses or more depending on its size. Sometimes they are located on poles where they serve one or two houses in more rural locations. In cities and large towns, they are normally located in brick or concrete structures.

IMPROVE YOUR MATHS

Use Ohm's law and values from Table 5.4 to work out what the earth fault impedance needs to be to cause the following circuit breakers to disconnect when the voltage is 230 V:

- 10 A
- 20 A
- 32 A.

Earthing

The green and yellow earthing conductors in each circuit are called circuit protective conductors (cpc) and must be present in all parts of a circuit. Every exposed metallic part in an electrical circuit must be connected to the cpc to ensure the metal part is connected to the earth path should a fault occur. This will ensure the low resistance earth path creates a high fault current causing quick disconnection.

As a general rule, the cpc should ideally be the same cross-sectional area (csa) as the circuit line conductor. This however isn't always the case, and cables such as the flat PVC twin and cpc cables commonly used in domestic electrical circuits in the UK have a cpc

reduced in csa. In these situations the length of the circuit should be restricted as the smaller csa cpc may end up having a resistance too high. The Institute of Engineering Technology (IET) produce a publication called the IET On-Site Guide, and Table 7.1 in the guide gives maximum circuit lengths based on the type and rating of protective device, how the cable is installed and csa of the circuit line and cpc.

ACTIVITY

Have a look at Table 7.1 in the IET On-Site Guide and see the vast range of cable sizes used depending on the circuit rating and protective device.

Using the table, determine the smallest cable combination (line/cpc) permitted for circuits protected by:

a 20 A circuit breakers

b 32 A RCBO

c 6 A **BS 3036** fuses.

Bonding

Main protective bonding conductors (MPB), sized in accordance with Section 4 of the IET On-Site Guide must be installed connecting the MET and:

- metallic installation service pipes such as gas, water, oil
- metallic exposed structural steelwork of a building rising from the ground
- other **extraneous conductive parts**.

KEY TERM

Extraneous conductive parts: metallic parts of a building structure or services that have a low resistance path to the general mass of earth but do not form part of the electrical system.

INDUSTRY TIP

Although various sizes of MBP conductor are required depending on the type of supply and installation, most domestic dwellings will be protected using 10 mm² conductors.

Where the MPB conductor is connected to the incoming service pipework, the connection must be made within 600 mm of any stop valve or meter or point of entry to the building but before any branch pipe is installed.

Connection is made to the pipework using a specific bonding clamp which complies with **BS 951**, Electrical earthing. Clamps for earthing and bonding. Different clamps have colour coding on them for installing in different conditions. They may be:
- red for dry conditions
- blue for damp locations.

INDUSTRY TIP

BS 7671 does state that services entering a building having an insulated section **do not require** the connection of a main protective bonding conductor, so always seek the advice of an electrician as connecting a bond to a part that is not extraneous may create a danger that wasn't there.

In some situations, **supplementary bonding** may need to be installed, but this is not normally needed where circuits have RCD or RCBO protection.

The purpose of MPB is to ensure **equal potential**. Should a fault occur in an electrical installation, the cpc for the circuit will become live to 230 V, as will the MET and any services or parts connected by MPB. If equal potential is present between the electrical installation and extraneous parts, current cannot flow between, especially through a person, meaning the risk of electrocution is reduced.

KEY TERMS

Supplementary bonding: where a bonding conductor is installed either between pipes or from a socket outlet or other accessory to a pipe. Unlike MPB, supplementary bonding doesn't come from the MET. The minimum csa of cable permitted for supplementary bonding is 4 mm² where the cable is in free air. Supplementary bonding is sometimes also called cross-bonding.

Equal potential: where the voltage between any two parts is within safe touch voltage levels, usually 50 V AC but dependent on the location.

Electrical accessories

There are several types of electrical accessories or junctions that are commonly used to connect electrical equipment used in the plumbing industry. These can be:
- socket outlets
- fused connection units
- double pole switches
- junction boxes
- wiring centres.

Socket outlets

Although socket outlets are common place in all buildings, they are normally provided for portable appliances. It is not good practice to use a socket outlet to connect permanently fixed equipment such as boiler systems, immersion heaters or other equipment but that doesn't mean you will not see it sometimes.

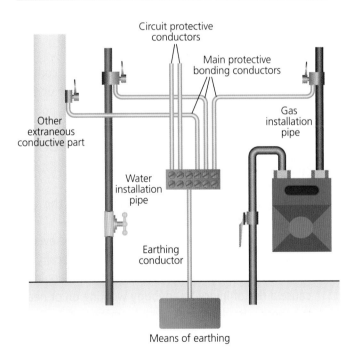

▲ Figure 5.15 Protective conductor arrangement for earthing and bonding

Fused connection units

It is far better practice to use fused connection units (FCU) to supply permanently fitted appliances and systems providing they have a total current demand below 13 A as this is the maximum fuse size that can be fitted to a fused connection unit.

There are several common types, those that have a flexible cable outlet on the front plate or those that do not. FCUs can be switched or unswitched and some have indicators on them to show they are switched on.

FCUs are the most common method of providing an electrical supply to a boiler system from a main electrical circuit.

▲ Figure 5.16 Typical FCU showing a flex outlet

Double pole switches

Double pole switches are not fused so they should only be fitted on specifically rated and designated circuits such as those supplying immersion heaters. Some can have flexible cable outlets as well as indicator lights to show when they are on.

Junction boxes

Junction boxes are not used to supply equipment directly from an electrical main circuit but could be used to connect equipment to the control wiring of a boiler control circuit providing the circuit already has adequate fused protection such as an FCU.

▲ Figure 5.17 A joint box

Wiring centre

Wiring centres are normally manufacturers specific boxes having terminals inside that are specifically labelled for certain equipment such as boiler circulating pumps, boilers or programmers.

Motor control equipment

It would be rare for a plumber to work on motor control equipment such as a direct-on-line (DOL) starter unit. You may come across one controlling swimming pool pumps as these larger pumps may require specific overload protection which is incorporated in the DOL unit. The purpose of the starter is to provide:

- a means of starting the motor
- a means of stopping the motor, especially in the case of an emergency
- undervoltage protection, which means, in the event of a power cut, the motor does not automatically restart when the power comes back on; this could cause a risk of injury to anybody working on or near the motor should it unexpectedly restart
- overload protection should the motor jam – if this happens, the overload will trip the power to the motor protecting the motor from overheating.

▲ Figure 5.18 DOL starter

Current using equipment and specialist controls

When working on plumbing systems, there are a number of electrically powered items of equipment or control devices that you may need to install and connect to an electrical supply. These include:

- shower pumps
- instantaneous showers
- jacuzzi bath or hot tubs
- macerators
- boiler control devices
- optimisers
- home automation (Wi-Fi enabled equipment).

Shower pumps

These are normally installed under a bath or in a void. They usually come fitted with a flexible cable which requires connection to a fused spur connection unit giving the pump a permanent electrical supply. The pump is controlled with an internal flow switch which switches on the pump when somebody turns on the shower tap, creating water flow. When the tap is turned off, this stops the water flow and the pump switches off.

Instantaneous showers

As these appliances heat up the cold water to produce hot water, they usually require a large electrical supply direct from the consumer unit.

Most electric showers are rated between 7 and 10 kW. This means that a 10.5 kW shower requires a 45 A electrical circuit to supply it which in turn will likely need a large 10 mm² cable to supply it, depending on the cable installation conditions. These circuits should be installed by an electrician. To allow mechanical works to be carried out on the shower, the circuit should contain a local switch such as a pull cord switch located in the bathroom.

Jacuzzi baths or hot tubs

Jacuzzi type whirlpool baths, like shower pumps, normally come with a flexible cable for termination into a local switch fused spur connection unit as these normally only contain a small powered pump. Hot tubs however heat the water as well as pump it around the tub and as a result, may require a larger electrical supply. Always consult the manufacturer's information for further detail and where a supply larger than 13 A is required, a special dedicated circuit would need installing by an electrician.

Macerators

A macerator is a device that compacts and pumps waste water or black water (toilet waste) where no natural waste fall exists such as in basement toilets where the building's waste pipes are higher than the toilet.

These devices are usually quite low in power ratings and are typically around 5 A. They normally come fitted with a flexible cable requiring a permanent connection to a local fused spur connection unit.

Boiler/heating controls

There are a large number of devices used to control heating and hot water systems which require an electrical connection, these include:

- **Programmers** – these are dual zone time clocks that turn on the central heating or hot water at varying times of the day. These are normally wired to a wiring centre to control the room thermostat or cylinder thermostat.
- **Room thermostats** – these are used to monitor the room temperature and control the zone valve for the heating system should the room temperature fall below that set by the thermostat. Older thermostats used bimetal strips, but modern ones use electronic sensors to monitor the temperature.
- **Cylinder thermostats** – on a cylinder thermostat there is a calibrated bimetallic strip positioned at the back of the thermostat, which is firmly positioned against the bare surface of a cylinder with the aid of a tensioned wire. Once the thermostat senses the pre-set temperature it turns off the electrical supply to a zone valve. A cylinder thermostat should be installed approximately one third the distance from the base of the cylinder and be set to operate between 60 and 65 °C.
- **Zone valves** – these are electrically controlled valves which are supplied using a 5-core flexible cable which connects to a wiring centre. The valve has a small motor in it which opens and closes the valve as well as a small micro-switch which is used to control the boiler and pump when the valve opens and closes.

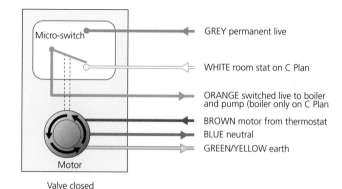

GREY permanent live

WHITE room stat on C Plan

ORANGE switched live to boiler and pump (boiler only on C Plan

BROWN motor from thermostat
BLUE neutral
GREEN/YELLOW earth

Micro-switch

Motor

Valve closed

▲ Figure 5.19 A 2-port zone valve showing the cable connection colours

Optimisers

These are more sophisticated forms of time-clock switches or programmers. They are used to control heating or hot water systems but may incorporate thermostats within them. Other forms use radio connections for remote thermostats or temperature monitors which can monitor internal and external temperatures ensuring the heating system maintains a desired temperature efficiently.

Home automation and Wi-Fi devices

With huge advances in the internet of things (IoT), home automation devices advance just as quickly. It is now very common to control all central heating and hot water systems using a smartphone connection anywhere in the world. As long as the system is able to connect to the internet using a Wi-Fi or Bluetooth connection, it can 'talk' to other devices removing the need for any control wiring. In some situations, where the devices are battery powered, the need for any wiring is completely removed.

4 PERFORMING PRE-INSTALLATION ACTIVITY PRIOR TO UNDERTAKING ELECTRICAL WORK ON PLUMBING AND DOMESTIC HEATING

Modern living is shaped by electricity. It is a safe, clean and immensely powerful source of energy and is in use in every factory, office, workshop and home in the country. However, this energy source also has the potential to be very hazardous, with a possibility of death, if it is not handled with care.

Electrical safety on site

The Electricity at Work Regulations 1989 were made under the Health and Safety at Work, etc. Act 1974 (HSW Act) and came into force on 1 April 1990.

INDUSTRY TIP

Access Electricity at Work Regulations 1989 at: www.legislation.gov.uk/uksi/1989/635/contents/made

Common electrical dangers on site

Electrical injuries can be caused by a wide range of voltages, and are dependent upon individual circumstances, but the risk of injury is generally greater with higher voltages. Alternating current (AC) and direct current (DC) electrical supplies can cause a range of injuries including:

- electric shock
- electrical burns
- loss of muscle control
- fires arising from electrical causes
- arcing and explosion.

Electric shock

It is worth remembering that electric shocks may arise either by direct contact with a live part or indirectly by contact with an exposed conductive part (for example, a metal equipment case) that has become live as a result of a fault condition.

Faults can arise from a variety of sources:

- broken equipment case exposing internal bare live connections
- cracked equipment case causing 'tracking' from internal live parts to the external surface
- damaged supply cord insulation, exposing bare live conductors
- broken plug, exposing bare live connections.

The magnitude (size) and duration of the shock current are the two most significant factors determining the severity of an electric shock. The magnitude of the shock current will depend on the contact voltage and impedance (electrical resistance) of the shock path. A possible shock path always exists through ground contact (for example, hand-to-feet). In this case:

Shock path impedance = body impedance + any external impedance

A more dangerous situation is a hand-to-hand shock path, where the current flow is through the heart. This is when one hand is in contact with an exposed conductive part (for example, an earthed metal equipment case), while the other simultaneously touches a live part. In this case the current will be limited only by the body impedance and shock currents will flow directly through the heart.

As the voltage increases, so the body impedance decreases, which increases the shock current. When the voltage decreases, the body impedance increases, which reduces the shock current. This has important implications concerning the voltage levels that are used in work situations and highlights the advantage of working with reduced low voltage (110 V) systems or battery-operated hand tools.

At 230 V, the average person has a body impedance of approximately 1300 Ω. At mains voltage and frequency (230 V–50 Hz), currents as low as 50 milliamps (0.05 A) can prove fatal, particularly if flowing through the body for a few seconds.

Electrical burns

Burns may arise due to:

- the passage of shock current through the body, particularly if at high voltage
- exposure to high-frequency radiation (e.g. from radio transmission antennas).

HEALTH AND SAFETY

Accidents do happen even when you put measures in place to remove all risks. Electricity cannot be seen, and working on a busy site could lead to exposure to electricity if others do not work safely such as an electrician not placing covers over potentially live parts. Always be familiar with the emergency procedures and accident reporting covered in Book 1, Chapter 1.

Loss of muscle control

People who experience an electric shock often get painful muscle spasms that can be strong enough to break bones or dislocate joints. This loss of muscle control often means the person cannot 'let go' or escape the electric shock. The person may fall if they are working at height or be thrown into nearby machinery and structures.

Fire

Electricity is believed to be a factor in the cause of many fires in domestic and commercial premises in Britain each year. One of the principal causes of such fires is wiring with defects such as insulation failure, the overloading of conductors, lack of electrical protection, poor connections.

Arcing

Arcing frequently occurs due to short circuit (conductive bridge between live parts) accidentally caused while working on live equipment (either intentionally or unintentionally). Arcing generates UV radiation, causing severe sunburn. Molten metal particles are also likely to be deposited on exposed skin surfaces.

Explosion

There are two main electrical causes of explosion:
- short circuit due to an equipment fault
- ignition of flammable vapours or liquids caused by sparks or high surface temperatures.

Controlling current flow

It is necessary to include devices in circuits to control current flow, that is, to switch the current on or off by making or breaking the circuit. This may be required:
- for functional purposes (to switch equipment on or off)
- for use in an emergency (switching in the event of an accident)
- so that equipment can be switched off to prevent its use and allow maintenance work to be done safely on the mechanical parts
- to isolate a circuit, installation or piece of equipment to prevent the risk of shock where exposure to electrical parts and connections is likely for maintenance purposes.

The preparation of electrical equipment for maintenance purposes requires effective disconnection from all live supplies and the means for securing that disconnection (by locking off).

Electrical supply for tools and equipment

Portable electric tools can provide valuable assistance with much of the physical effort required in electrotechnical activities. These tools can use different sources of electrical supply (mains or battery) and different means of maintaining safety in relation to that electrical supply.

Basic equipment constructions are aimed at preventing the risk of electric shock and are summarised below.

Class I

The basic insulation may be an air gap and/or some form of insulating material. External conductive parts (for example, the metal case) must be earthed by providing the supply through a three-core supply lead incorporating a protective conductor. The most important aspect of any periodic inspection/testing of Class I equipment is to check the integrity of this protective conductor. There is no recognised symbol for Class I equipment, though some appliances may show the symbol in Figure 5.20.

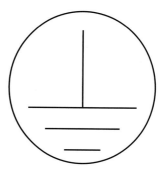

▲ Figure 5.20 This symbol may appear on Class I items

Class II

Class II equipment has either no external conductive parts apart from fixing screws (insulation-encased equipment) or there is adequate insulation between any external conductive parts and the internal live parts to prevent the former becoming live as a result of an internal fault (metal-encased equipment). Periodic inspection or testing needs to focus on the integrity of the insulation. Class II equipment is identified by the symbol shown in Figure 5.21.

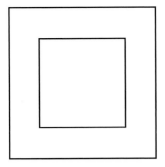

▲ Figure 5.21 This symbol may appear on Class II items

Class III

This method of protection is not designed to prevent shock but to reduce its severity and therefore make the shock more survivable. The supply (no greater than 50 V AC) must be provided from a separated extra-low voltage (SELV) source such as a safety isolating transformer or a battery. Class III equipment is identified by the diamond symbol and the safety isolating symbol by two interlinked circles, as shown in Figure 5.22.

Another way of reducing the risk of electric shock is by using a reduced low voltage system. This is *not* a Class III system but is a safer arrangement than using mains-operated (230 V) equipment because of the lower potential shock voltage. Supply is provided via a mains-powered (230 V) step-down transformer with the centre point of the secondary winding connected to earth.

 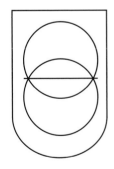

▲ Figure 5.22 This symbol may appear on Class III items

Low voltage electrical supplies for tools and equipment

The human body's impedance increases with lower touch voltages, particularly below 50 V. Although 50 V can be dangerous in certain circumstances, if the system voltage can be reduced to around this level, then the magnitude (size) of any current flow through the body will be significantly reduced.

The most common low voltage system of this type in use in the UK uses a 230/110 V double-wound step-down transformer with the secondary winding centre-tapped and connected to earth (CTE). While the supply voltage for equipment supplied from such a transformer is 110 V, the maximum voltage to earth is 55 V (63.5 V for a 110 V three-phase system). This is covered in **BS 7671** (Regulation 411.8 Reduced low voltage systems). Overcurrent protection for the 110 V supply may be provided by fuses at the transformer's 110 V output terminals or by a thermal trip to detect excessive temperature rise in the secondary winding. The latter method is generally employed for portable units. Such reduced voltage supplies may be provided:

- through fixed installations (workshops, plant rooms, lift rooms or other areas where portable electrical equipment is in frequent use)
- through small portable transformers designed to supply individual portable tools known as 110 V Centre Tapped Earth (CTE) transformers.

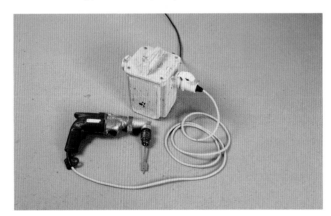

▲ Figure 5.23 Portable CTE transformer and drill

One additional advantage of using a reduced low voltage system is that this safeguard applies to all parts of the system on the load side of the transformer, including the flexible leads, as well as the tools, hand lamps, etc.

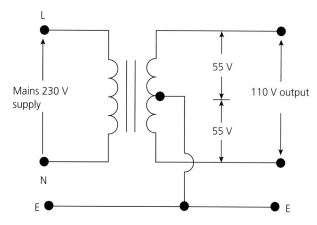

▲ Figure 5.24 Circuit diagram for a 110/55 V CTE transformer

Alternatively, cordless or battery-powered tools offer a convenient way of providing a powered hand tool without the inconvenience of using a mains supply and without the hazard of trailing power leads.

> **KEY POINT**
> PAT testing was covered in detail in Book 1, Chapter 1.

⑤ SAFE ISOLATION PROCEDURES

Isolation can be very complex due to the differing industrial, commercial and domestic working environments, some of which require experience and knowledge of the system processes. This section deals with a basic practical procedure for isolation and for the securing of isolation. It also looks at the reasons for safe isolation and the potential risks involved during the isolation process.

Safe operating procedures for the isolation of plant and machinery during both electrical and mechanical maintenance must be prepared and followed. Wherever possible, electrical isolators should be fitted with a means by which the isolating mechanism can be locked in the open/off position. If this is not possible, an agreed procedure must be followed for the removal and storage of fuse links. Regulation 12 of the Electricity at Work Regulations 1989 requires that, where necessary to prevent danger, suitable means (including, where appropriate, methods of identifying circuits) must be available for:

- **cutting off the supply** of electrical energy to any electrical equipment
- the isolation of any electrical equipment.

> **HEALTH AND SAFETY**
> NEVER rely on a standard local switch for isolation where you will be working on electrical terminals unless it can be secured in the open/off position. If the device cannot be secured as open/off, isolate the entire circuit.
>
> You should also isolate the whole circuit when you are making a connection into a fused connection unit (or similar) and the incoming supply terminals are live.

> **INDUSTRY TIP**
> Access Regulation 12 and 13 of the Electricity at Work Regulations 1989 at: www.legislation.gov.uk/uksi/1989/635/made

The aim of Regulation 12 is to ensure that work can be undertaken on an electrical system without danger. Terms used in Regulation 12 are defined in the following key terms box.

> **KEY TERMS**
> **Isolation:** this means the disconnection and separation of the electrical equipment **from every source of electrical energy** in such a way that this disconnection and separation is **secure**.
>
> **Cutting off the supply:** depending on the equipment and the circumstances, this may be no more than normal functional switching (on/off) or emergency switching by means of a stop button or a trip switch.
>
> **From every source of electrical energy:** many accidents occur due to a failure to isolate all sources of supply to or within equipment (for example, control and auxiliary supplies, uninterruptable power supply (UPS) systems or parallel circuit arrangements giving rise to back feeds).
>
> **Secure:** security can best be achieved by locking off with a safety lock (such as a lock with a unique key). The posting of a warning notice also serves to alert others to the isolation.

How to undertake a basic practical procedure for isolation

Being able to perform safe isolation is a key skill every person working with electricity needs. It is always a good idea to practise this procedure, so be sure to do this under supervision by a skilled person such as your tutor.

Gather together all of the equipment required for this task

You will need the following equipment:
- a voltage indicator which has been manufactured and maintained in accordance with Health and Safety Executive (HSE) **Guidance Note GS38**
- a proving unit compatible with the voltage indicator

- a lock and/or multi-lock system (there are many types of lock available)
- warning notices which identify the work being carried out
- relevant personal protective equipment (PPE) that adheres to all site PPE rules – this may include gloves to assist grip, hard hats due to the location and eye protection due to the risk of arcing.

INDUSTRY TIP

Return to Book 1, Chapter 1, to remind yourself of the relevant PPE.

The equipment shown in Figure 5.25 can be used to isolate various main switches and isolators. To isolate individual circuit breakers with suitable locks and locking aids, you should consult the manufacturer's guidance.

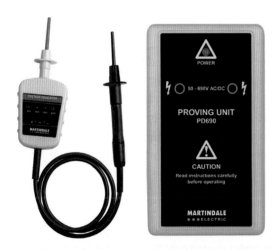

▲ Figure 5.25 Equipment used for secure isolation

KEY TERM

Guidance Note GS38: electrical test equipment for use by electricians (published by the Health and Safety Executive, HSE) was written as a guideline to good practice when using test equipment on circuits operating at voltages >50 V AC or >120 V DC or where tests use these voltages. It is intended to be followed, in order to reduce the risk of danger and injury when performing electrical tests.

The GS38 information listed below relates to the design and maintenance of approved electrical test equipment for use on electrical systems.

Probes should have:

- finger barriers
- an exposed metal tip not exceeding 4 mm; however, it is strongly recommended that this is reduced to 2 mm or less
- fuse, or fuses, with a low current rating (usually not exceeding 500 mA), or a current-limiting resistor and a fuse.

Leads should be:

- adequately insulated
- colour coded
- flexible and of sufficient capacity
- protected against mechanical damage
- long enough
- sealed into the body of the voltage detector and should not have accessible exposed conductors, other than the probe tips.

When working on or near electrical equipment and circuits, it is important to ensure that:

- the correct point of isolation is identified
- an appropriate means of isolation is used
- the supply cannot inadvertently be reinstated while the work is in progress
- caution notices are applied at the point(s) of isolation
- conductors are proved to be dead at the point of work before they are touched
- safety barriers are erected as appropriate when working in an area that is open to other people.

IMPROVE YOUR ENGLISH

Communication is key when isolating a circuit or system. You need to be clear and precise in explaining to a client what exactly is being switched off and what implications the power loss will create. As well as verbal communication, written advice before work starts is a good way to communicate to a client that disruption is likely.

ACTIVITY

Check your test instruments for any damage and for compliance with GS38.

Carry out the practical isolation

The method of isolation is outlined below.

1 **Identify** – identify equipment or circuit to be worked on and point(s) of isolation.
2 **Isolate** – switch off, isolate and lock off (secure) equipment or circuit in an appropriate manner. Retain the key and post caution signs with details of work being carried out.
3 **Check** – check the condition of the voltage indicator leads and probes. Confirm that the voltage indicator is functioning correctly by using a proving unit.
4 **Test** – using a voltage indicator, test the outgoing terminals of the isolation switch. Take precautions against adjacent live parts where necessary.
5 **Prove** – using a voltage indicator and proving unit, prove that the voltage indicator is still functioning correctly.
6 **Confirm** – confirm that the isolation is secure and the correct equipment has been isolated. This can be achieved by operating functional switching for the isolated circuit(s). The relevant inspection and testing can now be carried out.

INDUSTRY TIP

During single-phase isolation, there are three tests to be carried out:

- L – N
- L – E
- N – E

(L = line, N = neutral, E = earth)

Reinstate the supply

When the 'dead' electrical work is completed, you must ensure that all electrical barriers and enclosures are in place and that it is safe to switch on the isolated circuit.

1 Remove the locking device and danger/warning signs.
2 Reinstate the supply.
3 Carry out system checks to ensure that the equipment is working correctly.

Implications of safe isolation

When you isolate an electricity supply, there will be disruption. So, careful planning should precede isolation of circuits. Consider isolating a section of a nursing home where elderly residents live. You will need to consult the nursing home staff, to consider all the possible consequences of isolation and to prepare a procedure.

VALUES AND BEHAVIOURS

Conducting an isolation procedure can have an effect on those in the vicinity. It is good practice to consider any potential impact and how you might minimise or mitigate it. The following questions are useful:

- How will the isolation affect the staff and other personnel? For example, think about loss of power to lifts, heating and other essential systems.
- How could the isolation affect the residents and clients? For example, some residents may rely on oxygen, medical drips and ripple beds to aid circulation. These critical systems usually have battery back-up facilities for short durations.
- How could the isolation affect the members of the public? For example, fire alarms, nurse call systems, emergency lighting and other systems may stop working.

- How can an isolation affect systems? For example, IT programs and data systems could be affected; timing devices could be disrupted. In this scenario, you must make the employers, employees, clients, residents and members of the public aware of the planned isolation.

Alternative electrical back-up supplies may be required in the form of generators or uninterruptable power supply systems.

Always discuss the implications of isolation and subsequent loss of supply with the client. Good communication is key. It is best practice to talk first rather than having to argue afterwards when someone lost all their data or became stranded in a lift.

Before any isolation is carried out, you must assess the risks involved. This section deals with the practical implications and the risks involved during the isolation procedure, if risk assessment and method statements are not followed.

KEY POINT
Refer back to Book 1, Chapter 4 Planning and supervision for more on risk assessments.

Risk assessments: who is at risk and why?

If isolation is not carried out safely, what are the possible risks when performing inspection and testing tasks?

Risks to you

Risks to you might include:

- shock – touching a line conductor, e.g. if isolation is not secure
- burns – resulting from touching a line conductor and earth, or arcing
- arcing – due to a short circuit between live conductors, or an earth fault between a line conductor and earth
- explosion – arcing in certain environmental conditions, for example, in the presence of airborne dust particles or gases, may cause an explosion.

Things that can cause risk to you include:

- inadequate information to enable safe or effective installation, such as no diagrams, legends or charts
- poor knowledge of the system you are working on (and so not meeting the competence requirements of Regulation 16 of EAWR)
- insufficient risk assessment
- inadequate test instruments (not manufactured or maintained to the standards of GS38).

Even tasks that seem low risk, such as removing a bonding clamp from a water pipe, can be high risk if isolation has not been undertaken. If dangerous potentials exist between the electrical system and the water pipe, you will become the bridge between these potentials as soon as you remove the clamp. NEVER REMOVE BONDING UNLESS THE WHOLE SYSTEM IS ISOLATED.

Risks to other tradespersons, customers and clients

Risks to other tradespersons, customers and clients might include:

- switching off electrical circuits – for example, switching off a heating system might cause hypothermia (resulting from being too cold); if lifts stop, people may be trapped
- applying potentially dangerous test voltages and currents
- access to open distribution boards and consumer units
- loss of service or equipment, for example:
  - loss of essential supplies
  - loss of lights for access
  - loss of production.

Risks to members of the public

Risks to members of the public might include prolonged loss of essential power supply, causing problems, for example, with safety and evacuation systems, such as:

- fire alarms
- emergency exit and corridor lighting.

HEALTH AND SAFETY

Although safety services usually have back-up supplies such as batteries, these may only last for a few hours. Other safety or standby systems may have generator back-up, but this will also require isolation, leaving the building without any safety systems.

Risks to buildings and systems within buildings

Risks to buildings and systems within buildings might involve applying excessive voltages to sensitive electronic equipment, for example:

- computers and associated IT equipment
- residual current devices (RCDs) and residual current operated circuit breakers with integral overcurrent protection (RCBOs)
- heating controls
- surge protection devices.

There might also be risk of loss of data and communications systems.

ACTIVITY

Write down a list of the risks associated with isolation and the effects isolation can have on people, livestock, systems and buildings. You could think about:

1 Who is at risk if isolation is not carried out correctly?

2 What might happen if you need to switch off a socket-outlet circuit? For example, in a care home?

3 What must you do if you encounter a computer server that requires a permanent supply and you need to switch off the main supply to enable safe working?

⑥ SAFE INSTALLATION AND TESTING OF ELECTRICAL EQUIPMENT

When installing electrical systems, sound knowledge of the correct practices and procedures is required; these include an understanding of:

- specific electrical tools
- the risk assessment
- materials
- installing wiring systems
- terminating conductors.

Electrical tools

There are a wide range of tools you will use when working on electromechanical systems. This section will look at the ones specifically used when working on electrical systems.

Wire strippers

Wire strippers provide a safe and reliable method of removing the insulation from a wire or cable without damaging the conductor. Wire strippers come in various designs, but the common principle is that the cutting jaws only cut into the insulating material and not into the conductor.

Wire strippers can be either manually set (by a screw or dial) or automatically set. Side cutters can also be used but these often damage the conductor.

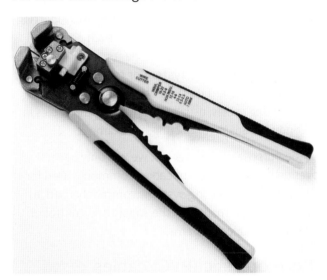

▲ Figure 5.26 Wire strippers

Cable cutters

There are a variety of different tools that can be used to cut cables, the most common being side cutters. Side cutters are probably one of the most important tools that electricians have in their tool kit. These are used for cutting cables to length, cutting sleeving and cutting nylon tie-wraps, for example. They work on a compression-force basis and are shaped so that the cutting point is along one side.

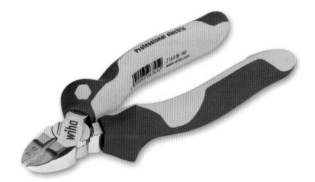

▲ Figure 5.27 Side cutters

Depending on the size of the cutters, cable sizes up to 16 mm² can be cut easily. However, larger cables require larger types of cutters. These range from cable loppers, up to hydraulic, manual pump cutters for cables up to 300 mm².

Screwdrivers

An electrician will use a selection of different sized screwdrivers. They will all have one thing in common in that they will all be of approved standards. The most common screwdrivers used include:

- terminal (3–3.5 mm)
- large flat (4–5 mm)
- pozidriv (PZ2)
- and, more recently, a consumer unit screwdriver.

▲ Figure 5.28 Electrician's screwdrivers

Electrician's knife

The electrician's knife is available in various arrangements but usually has a folding blade. The most common use of the electrician's knife is for stripping the outer sheaths of *some* cables; the blade generally has a shaped section to aid the removal of certain cable sheaths, such as on armoured or some types of mineral-insulated (MI) cables.

▲ Figure 5.29 Electrician's knives

An electrician's knife should never be used for stripping the insulation from cables or for cutting cables to length. It should always be used pointing away from the body.

> **HEALTH AND SAFETY**
>
> It is dangerous to use a disposable knife blade when stripping cables as the blade can break, creating sharp shards.

> **INDUSTRY TIP**
>
> Remember that all work, including electrical work, requires a risk assessment to be carried out before the work begins. Whether installing a new electric circulation pump or the wiring for a thermostat, the work involves risks. Refer back to Book 1, Chapter 4 for more detail on risk assessments.

Materials

When starting a task, it is important to be able to identify the materials that will be required to complete the work. As well as using your previous experience, you should be able to identify the components that are required by looking at the drawings and specifications. The project specification contains information about the type of finish that is required for the installation work and may include details of preferred manufacturers.

The drawings often include plans showing the positioning of components and accessories, as well as the routes to be followed when the wiring is being installed. This is important as none of the services being run through the installation must cause problems for other systems and services.

Installing wiring systems

There are many types of wiring systems used. A wiring system could be a cable on its own or a means of containing or managing the cable. Some of the terms you may hear include the following:

- **Wiring system** – this is the term used to describe the type of cable used and the method of supporting it. For example, single-core cable in conduit is a wiring system but, equally, so is a twin and cpc cable clipped to a wall.
- **Support system** – this is the method of supporting a cable and could include systems such as cable tray, basket or simply clips or cleats.
- **Cable management system** or **cable containment system** – this is a method of supporting and protecting cables, usually by enclosing the cables in conduit, trunking or ducting, for example.

For the purpose of this chapter, we will only be covering the installation of cables commonly used for the installation of wiring systems commonly used in domestic environments for the supply and control of electro-mechanical systems.

Thermoplastic (PVC) cables

Thermoplastic cables are commonly referred to as PVC (polyvinyl chloride) cables and they come in various shapes, sizes and forms, including:

- single-core cable
- twin and cpc flat-profile cable
- three-core and cpc flat-profile cable
- multi-core flexible cable.

In domestic installations, the most common cables are the twin and cpc, and the three-core and cpc flat-profile cables.

Cables have three main parts:

- **Sheath** – this is on the outside and holds the conductors in one cable, as well as providing minor mechanical protection to the inner conductors.

- **Insulation** – this is on the live conductors only and is used to provide basic protection against electric shock, as well as being a means to identify the conductors.
- **Conductor** – this is what carries the current and is commonly made from copper but may be made out of other materials such as aluminium.

▲ Figure 5.30 Flat-profile cables

Conductors can be formed in several ways. Conductors with a cross-sectional area (csa) up to and including 2.5 mm² can be formed out of a single piece of copper. Above 2.5 mm² csa, conductors tend to be made out of multiple strands of copper. It is important to be able to bend the cable and a solid piece of copper would be hard to bend. However, for conductors of 300 mm² csa and over, the conductors tend to be solid again; bending is not normally required. If a cable is to be bent a lot in use, for example in the flex connecting a vacuum cleaner to a plug, the conductor is normally made of many fine strands of copper, each no thicker than a hair. The reason for this is that as the conductors keep getting bent, they will eventually break. In a solid conductor, the circuit would be broken. In a conductor consisting of many strands, if one breaks, it will not significantly change the amount of current the cable can carry.

Stripping single-core thermoplastic cables

Removal of the thermoplastic sheath, or insulation, is relatively easy with thermoplastic cables.

Use of wire strippers is recommended for single-core conductors. Automatic wire strippers tend to rip the insulation from the conductor and can damage the insulation. Manual wire strippers are preferred and these come in various forms, but all work on a similar principle.

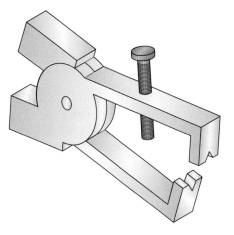

▲ Figure 5.31 Cable stripper jaws

Manual wire strippers all have two blades that cross one another like scissors.

The blades each have a notch so that they cut *around* the conductor in the middle. They also all have a means of adjustment so that different sizes of conductor can be stripped (Figure 5.31).

Before using wire strippers, they must be set to the correct size. This can be done using a scrap or off-cut piece of wire of the correct size. With the jaws of the wire strippers together, turn the adjustment screw until the hole in the jaws is just bigger than the size of the conductor to be stripped. Test the setting on the off-cut piece of wire by placing the wire strippers over the wire and squeezing the handles to close the jaws. Then slightly release the jaws and try to slide the insulation off using the wire strippers to assist.

If the wire strippers are correctly set, the insulation will come off easily and there will be no damage to the copper conductor. If the aperture is set too small, the insulation will slide off easily but the conductor may be damaged. If set too large, the insulation will not come off easily. A simple adjustment of the adjusting screw will correct these problems. Now, with the correct setting, the wires can be stripped safely.

Stripping flat-profile thermoplastic cable

Flat-profile cable is stripped in the same way as single-core cables, once the sheath has been removed. There are several ways to do this, but some methods can damage the cables.

First, identify how much cable is required within the accessory. A conductor should terminate with ease at any termination point within the accessory. As a rule of thumb, measure the diagonal size of the accessory and then add 10 per cent to allow for termination. However, if the accessory is a distribution panel, wires must be able to reach not just the required position, but also any point on the panel if the sequence of the board were to be changed.

Next, identify how much of the sheath should be removed from the cable. The purpose of the sheath is to provide some mechanical protection for the insulation on the conductors. However, too much sheath within the accessory takes up space and will put excess strain on the conductors. The sheath should be stripped back almost to where the cable enters the accessory, leaving only 10–15 mm, a thumb's width, of sheath within the accessory.

Having decided on the length of sheath required and with the cable in place, score round the sheath so that it can be removed. This can be done with a scriber or, preferably, with an electrician's knife. Care should be taken not to cut into the wires. Use a pair of side cutters to snip down the centre of the cable, splitting the line to one side and the neutral to the other.

Pull the live conductor and sheath apart, tearing the sheath as far as the score mark. Once at the score mark, bend the sheath back and snap it off, leaving a clean finish. Now the inner wires can be stripped in the same manner as single-core wires, making sure that green and yellow sleeving is applied to the bare copper cpc.

An alternative – but less safe – method of removing the sheath is to score round the sheath with an electrician's knife, then to split the sheath by running a knife along the route of the protective conductor. The risk with this method is of damaging the cpc or even the insulation of the conductors.

Stripping thermoplastic flexible cables

Stripping the actual wires of flexible cables is done in the same way as stripping single-core wires. However, the sheath is stripped in a different way from the other cables already mentioned here.

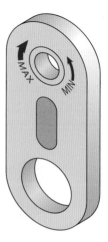

▲ Figure 5.32 Flexible cable sheath-stripping tool

The outer sheath can be removed with the use of a ringing tool, which comes in various shapes and forms, the most basic of which is shown in Figure 5.32. This tool slides over the end of the cable, to the required position, and is then rotated around the cable, cutting it slightly. The ringing tool is removed, the cable is bent to finish the cut and the sheath is then pulled off.

There may be times when a ringing tool is not available, so it is important to know how to do this without a special tool. Thermoplastic has a tendency to split when it is damaged and stretched, and this feature can be used to help strip the sheath.

Bend the cable tightly at the point where the sheath is to be removed. Using a sharp knife gently score the top of the bend, noticing the thermoplastic split open like a little mouth. Gently work the split until the inner cables are visible and then unbend, rotate and re-bend the cable at about 90° from the first point. Repeat the score until, again, the wires become visible. Taking care not to cut into the wires, repeat the bend and cut until the sheath can be removed. Once the sheath has been removed, the wires can be stripped as mentioned previously.

Terminating cables and BS 7671

Section 526 of **BS 7671** The IET Wiring Regulations is the key section relating to the terminating and connection of conductors. Below is a regulation-by-regulation discussion of the requirements.

526.1

This requires that all connections shall provide 'durable electrical continuity and adequate mechanical strength and protection'. This is a general requirement and is in keeping with requirements of the EAWR. The subsequent regulations give further information on how this can be achieved.

526.2

This requires that:
The means of connection shall take account of, as appropriate:

i   the material of the conductor and its insulation

ii the number and shape of the wires forming the conductor

iii the cross-sectional area of the conductor

iv the number of conductors to be connected together

v   the temperature attained at the terminals in normal service, such that the effectiveness of the insulation of the conductors connected to them is not impaired

vi the provision of adequate locking arrangements in situations subject to vibration or thermal cycling.

INDUSTRY TIP

The requirement for connections is that they are electrically continuous and mechanically strong. Many purists do not like the use of strip connectors (chocolate blocks) but, like anything else, used correctly they are fine. Given the very large numbers in use, they must have something going for them.

Where a soldered connection is used, the design shall take account of **creep**, mechanical stress and temperature rise under fault conditions.

There are a large number of considerations covered by this regulation. Take each of these points in turn.

KEY TERM

Solder creep: sometimes called 'cold flow', 'solder creep' can occur when the termination is under constant mechanical stress and the solder can literally move or 'creep'. Incidence of creep increases with temperature.

The material of the conductor and its insulation

Different metals or, more correctly, dissimilar metals may react with each other, resulting in corrosion of the termination. It is therefore important to make sure that compatible materials are used when terminating cables and conductors. A further discussion of corrosion takes place later in the book.

The number and shape of the wires forming the conductor

Conductors come in many formats, round or triangular, solid or stranded. It is important to select terminations that are compatible with the cable and/or conductor. Failure to use compatible parts may result in the conductor becoming loose within the termination.

When crimping lugs onto cables, it is important that the correct size lugs and crimp dies are used, to ensure that a sound mechanical and electrical connection is made. Shaped lugs are available to use with triangular and half-round conductors used in some two-, three- and four-core cables. Where flex is being terminated, it is important that the ends are treated, for example, fitting ferrules to ensure that the individual strands are not spread and are all contained within the termination.

▲ Figure 5.33 Tube lug and bell mouth lug designed to capture all strands of conductor

The cross-sectional area of the conductor

The cross-sectional area (csa) of the cable was carefully selected at the design stage, to ensure that the current-carrying capacity of the conductor was adequate for the circuit load. When terminating the cable, it is important to ensure that all strands of the conductor are contained within the terminal, to ensure that the current-carrying capacity is maintained. Terminals need to be large enough to house the conductor and to be suitably rated to carry the circuit load. Failure to meet either of these requirements could result in overheating at the termination that, in turn, may cause damage to the equipment and/or the insulation of the cable. This in turn may pose a fire risk.

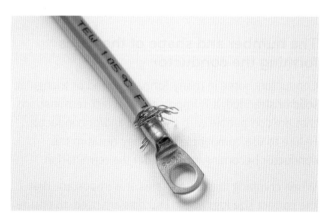

▲ Figure 5.34 Terminals that are not large enough to accommodate all the strands are problematic

The number of conductors to be connected together

The termination must be suitable for the number of conductors to be connected together. Attempting to fit more conductors into a terminal than it is designed to hold will invariably result in one or more of those conductors not being properly connected and becoming loose over time. In the case of accessories, where more than one conductor is intended to be connected, manufacturers will provide a number of linked terminals to house all of the conductors so that a sound mechanical and electrical connection is made.

The temperature attained at the terminals in normal service, such that the effectiveness of the insulation of the conductors connected to them is not impaired

The maximum operating temperature of conductors with thermoplastic insulation (usually PVC) is 70 °C, whilst the maximum operating temperature of thermosetting cables (XLPE) is 90 °C. When using thermosetting cables at 90 °C, it is important to make sure that the terminals are capable of withstanding this temperature, as the majority of electrical accessories are designed to operate at 70 °C. Other specialist cables may operate at higher temperatures, so it is important to check the maximum operating temperature of the terminals.

HEALTH AND SAFETY

The correctly sized terminal is critical as ill-fitted examples can cause a fire risk.

The provision of adequate locking arrangements in situations subject to vibration or thermal cycling

Where conductors are terminated into machinery, this can cause vibration and may adversely affect the terminations. In such cases, it is common practice to use, for the final connection, one of these options:

- a flexible cable – where added mechanical protection is required, braided flex such as SY flex can replace standard flex
- a flexible conduit – where unsheathed cables make the final connection or where additional mechanical protection is required, the cables can be housed within flexible conduit
- an anti-vibration loop in the cable – with cables such as MICC, or other cable types which are not flexible, a loop is included in the cable to allow the cable to absorb any vibration.

INDUSTRY TIP

The use of a flexible connection will also allow the motor to be moved when adjustment or alignment is required.

Thermal cycling can also harm the terminations. It is greatest when cables are run at or near their maximum operating temperature. The effects can be reduced by ensuring that terminations and connections are kept tight.

KEY TERM

Thermal cycling: heating and cooling of metal (in this case) which causes expansion and contraction and, eventually, the loosening of terminals.

526.3

Poor and loose terminations cause many fires of electrical origin. Regulation 526.3 requires every connection or joint to be accessible for inspection, testing and maintenance, with the exception of:

- joints designed to buried in the ground
- a compound-filled joint
- an encapsulated joint
- a cold tail of a floor or ceiling heating system
- a joint made by welding or soldering
- a joint made with an appropriate compression tool
- spring-loaded terminals complying with **BS 5733** and marked with the symbol (MF).

INDUSTRY TIP

From this regulation it is clear that the practice of using screw terminal junction boxes and placing these in inaccessible places, such as ceiling or floor voids, is not in compliance with **BS 7671**.

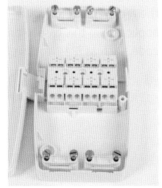

▲ Figure 5.35 Types of connection not required to be accessible

526.4

This requires that the insulation of the cable must not be adversely affected by the temperature attained at a connection. An example of this may be where a cable with an insulation temperature rating of 70 °C is connected to a bus-bar which has been designed to run at a higher temperature. In this case the insulation on the cable would be removed from the cable to a suitable distance and replaced with insulation capable of withstanding the higher temperature.

526.5

All terminations and joints in **live conductors** must be enclosed within a suitable enclosure or accessory. There are no exceptions. This requirement applies to both low-voltage and extra-low voltage connections, but sadly it is not uncommon to see poor examples of connections, especially where down-lighters are fitted.

KEY TERM

Live conductor: a conductor intended to be energised in normal service, and therefore includes a neutral conductor.

▲ Figure 5.36 Example of an acceptable connection

In harmony with Regulation 526.5 is Regulation 421.1.6, which requires that all enclosures have suitable mechanical and fire-resistant properties.

526.6

This requires that there is 'no appreciable mechanical strain in the connections of conductors.' Mechanical strain may come about due to:

- the conductor bending too tightly before entering the terminal, causing the termination to be under constant stress. Cables must be installed in accordance with the minimum bend radii, as given in the IET On-Site Guide or IET Guidance Note 1 (GN1).

- cables having no form of strain relief fitted. This can exert mechanical forces on the termination, due to the weight of the cable pulling on the termination. Cables should be fixed at the maximum distances given in the IET On-Site Guide or IET GN1. The fitting of suitable cable glands, where cables enter enclosures, provides strain relief on the terminations.
- cables terminated without any slack. Under faulty conditions large electromagnetic forces, due to high fault currents, are exerted on the cables, with the highest forces being exerted at the 'crutch point', where cables come together at the outer sheath. If there is little slack in the cables, the forces will be transferred to the terminal.

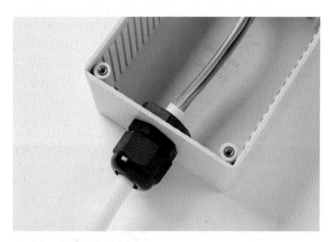

▲ Figure 5.37 Using a gland to provide strain relief

526.7

This requires that, where a joint in a conductor is made within an enclosure, the enclosure must provide adequate mechanical protection as well as protection against relevant external influences. The minimum requirements for an enclosure to meet the requirements for basic protection are that:

- the bottom sides and face meet at least IP2X or IPXXB
- for the top surface, the enclosure must meet IP4X or IPXXD.

However it may also be necessary to take into account external influences such as water ingress or dust ingress, depending on the location.

The IP Code

The IP Code is an international code specifically aimed at manufacturers of enclosures and equipment. It applies to degrees of protection provided by electrical equipment enclosures with rated voltages not exceeding 72.5 kV. The abbreviation 'IP' stands for International Protection, so the full title is International Protection Code (IP). The code is defined in IEC 60529 (**BS EN 60529**).

The three general categories of protection given in the standard are:

1 the ingress of solid foreign objects (first digit)
2 the ingress of water (second digit)
3 the access of persons to harmful electrical or mechanical parts.

When referring to the IP code in wiring regulations, 'X' is used in place of the first or second numeral, to indicate that:

1 the test is not applicable to that enclosure or
2 in the case of standards, the classification of protection is not applicable to this standard.

For example, IP2X means that protection against the ingress of solid objects must meet at least IP2 but the requirement for water ingress protection is not applicable in this case. Manufacturers will provide a full code, such as IP44, for the enclosure.

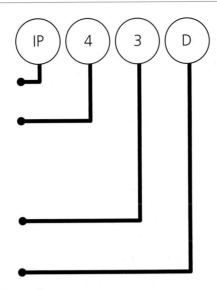

Code letters

International protection

First numeral 0–6 or letter X

Protection of persons by prevention or limiting ingress of parts of the body

Limitation of the ingress of solid objects

Second numeral 0–8 or letter X

Resistance to the ingress of water

▲ Figure 5.38 How the IP Coding system works

▼ First numeral 0–6: Ingress of solid objects

IP	Requirement	Example
0	No protection	
1	Full penetration of **50.0 mm** diameter sphere not allowed and shall have adequate clearance from hazardous parts. Contact with hazardous parts not permitted.	Back of the hand
2	Full penetration of **12.5 mm** diameter sphere not allowed. The jointed test finger shall have adequate clearance from hazardous parts.	A finger
3	The access probe of **2.5 mm** diameter shall not penetrate.	A tool such as a screwdriver
4	The access probe of **1.0 mm** diameter shall not penetrate.	A wire
5	Limited ingress of dust permitted. No harmful deposit.	Close fitting
6	Totally protected against ingress of dust.	Dust-tight

▲ Figure 5.39 The first digit of the IP code shown relates to the ingress of solid objects and the second digit relates to the ingress of liquids

INDUSTRY TIP

Do not use your own finger to test equipment. Use a proper test finger.

▼ Second numeral 0–8: Ingress of water

IP	Requirement	Example
0	No protection	
1	Protected against vertically falling drops of water	Drips from condensation or similar
2	Protected against vertically falling drops of water with enclosure tilted 15° from the vertical	

→

▼ Second numeral 0–8: Ingress of water (continued)

3	Protected against sprays to 60° from the vertical	Rainproof
4	Protected against water splashed from all directions	Outdoor electrical equipment
5	Protected against low-pressure jets of water from all directions	Where hoses are used for cleaning purposes
6	Protected against strong jets of water	Where waves are likely to be present
7	Protected against the effects of immersion between 15.0 cm and 1.0 m	Inside a bath tub
8	Protected against submersion or longer periods of immersion under pressure	Inside a swimming pool

ACTIVITY

What is meant by the code IP44?

▼ Additional letter A–D: Enhanced protection of persons

IP	Requirement	Example
A	Penetration of 50.0 mm diameter sphere up to guard face must not contact hazardous parts	The back of the hand
B	Test finger penetration to a maximum of 80.0 mm must not contact hazardous parts	A finger
C	Wire of 2.5 mm diameter × 100.0 mm long must not contact hazardous parts	A screwdriver
D	Wire of 1.0 mm diameter × 100.0 mm long must not contact hazardous parts may be protected by an internal barrier	A wire

The IP codes that you are most likely to come across are:

1 For protection against the ingress of solid objects and protection to persons. The codes are IP2X and IP4X used in relation to barriers and enclosures.

2 For protection against the ingress of water the codes are IPX4, IPX5, IPX6, IPX7 and IPX8.

3 For enhanced personal protection IPXXB and IPXXD again used in relation to barriers and enclosures.

▲ Figure 5.40 Cable entry at electrical accessory not meeting the IP code

▲ Figure 5.41 Non-sheathed cables outside of an enclosure

INDUSTRY TIP

You will find that over the years, many creatures find their way into enclosures. This might seem impossible, but insects are common and mice not unknown.

ACTIVITY

When removing knockouts from a plastic box the slot should be as tight to the cable as possible. What tool should be used to cut the slot?

INDUSTRY TIP

When cables enter the top of an enclosure, IP4X must be maintained, meaning there should be no gap larger than 1 mm.

526.8

Where the sheath of a cable has been removed the cores of the cable must be enclosed within an enclosure as detailed in 526.5. This also applies to non-sheathed cables, contained within trunking or conduit.

526.9

The group of regulations designated 526.9 relates to the connection of multi-wire, fine-wire and very-fine wire conductors.

526.9.1

To stop the ends of multi-wire, fine-wire and very-fine wire conductors from spreading or separating, this regulation requires that suitable terminals, such as plate terminals, or suitable treating of the ends be undertaken. One suitable method is to fit ferrules on the ends of the conductor. Manufacturers will almost always fit some form of ferrule to the ends of flexes so that the conductor can be terminated in a screw terminal.

▲ Figure 5.42 Ferrule fitted to pendant flex

INDUSTRY TIP

Soldering flex forms a hard mass that, when subjected to vibration, may work loose. Most appliances come fitted with a 13A plug but occasionally, a flex with soldered ends is supplied. In this situation, advice from the manufacturer should be followed.

526.9.2

Soldering or tinning of the ends of multi-wire, fine-wire and very-fine wire conductors is not permitted if screw terminals are used.

526.9.3

The connection of soldered and non-soldered ends on multi-wire, fine-wire and very-fine wire conductors is not permitted where there is relative movement between the two conductors.

Connection methods

Allowable connection methods include:

- screw
- crimped
- soldered
- non-screw compression.

Each method has its advantages and disadvantages.

Screw terminals

When cables are terminated into electrical equipment, the type of terminal used must be taken into account. Most accessories use a grub screw, which is screwed

down onto the conductor to ensure it is retained in place. Problems arise, however, when the terminal is designed to take more than one conductor, or a conductor of a larger csa than the one being installed.

The common types of screw terminal used in the accessories within electrical installations are:

- square
- circular
- moving-plate
- insulation-displacement
- pillar.

Square base terminal

The use of square terminals can be seen in accessories such as socket-outlet face-plates, where two or even three cables can be terminated in the same terminal. These terminals are designed to accept up to three cables and, where a single conductor is used, the screw can miss or damage the conductor. To minimise the potential for problems, the end of the conductor is bent over to increase the contact area available for the screw of the terminal.

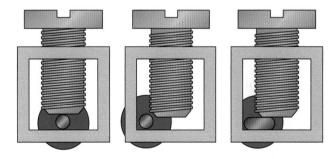

▲ Figure 5.43 Square terminal showing possible problems if conductor is not doubled over

To bend over the end of the conductor, it is necessary to remove twice the normal amount of insulation from the cable. Then a pair of long-nosed pliers is used to fold the exposed conductor in half. It is important to make sure that both sides of the bend are the same length, as shown in the centre diagram of Figure 5.44. If the return edge is too long, as shown in the left diagram of Figure 5.44 the conductor may protrude from the terminal, causing a shock hazard. If it is too short, as shown on the right diagram of Figure 5.44 the bend will be redundant.

▲ Figure 5.44 Bending over the end of a conductor: bend is too big, correct and too small

Circular terminal

Circular terminals can be seen in accessories such as light switches and ceiling roses, where single cables or small cables are terminated. They are also commonly found in consumer units and distribution boards on both the neutral and earth bars.

Circular bottom terminals are designed to ensure that the conductor is positioned directly beneath the terminal screw, so there is no need to bend the end of the conductor over.

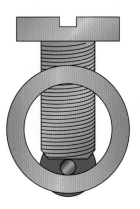

▲ Figure 5.45 Circular base terminal

Moving-plate terminal

Moving plate terminals are often used on protective devices, such as circuit breakers and fuse holders that are mounted in consumer units and distribution boards.

The option of bending over the end of the conductor depends not only on the size of the conductor, but also on the size and type of the terminal. If the terminal is the type where the bottom moves up towards the top when the screw is tightened, it is not necessary to bend over the conductor, as the terminal tightens evenly. If, however, the terminal has a plate that moves towards

the bottom of the terminal as the screw is tightened, small conductors should be doubled over. If there is any doubt, refer to the manufacturer's recommendations.

No matter what type of terminal is being used, different conductor types should never be mixed within the same terminal. If flexible cable is terminated within the same terminal as solid or stranded cable, the screw may fail to clamp on the flexible cable and only a few of the fine strands may be secured. This could result in a poor electrical connection and the wire might come loose.

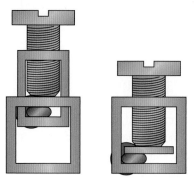

▲ Figure 5.46 Two types of moving-plate terminal

If there is no option other than to mix flexible cable and solid or stranded conductors in the same terminal, the flexible conductor must be fitted with a ferrule. This is a small, metal tube that is crimped onto the end of a flexible conductor to hold the strands together.

Terminating copper and aluminium conductors within the same terminal should also be avoided due to the electrolytic reaction between the two different metals.

Whatever types of terminal and conductor are being used, always make sure that the screw tightens on the conductor and not the insulation. To ensure this, the insulation should stop at the opening of the terminal. Take care not to stop the insulation too early, leaving the conductor exposed, with the possibility of faults occurring.

INDUSTRY TIP

Common faults with terminations are exposed conductors or screwing down onto the insulation.

Advantages and disadvantages of screw terminals

The advantages of screw terminals are:
- they are cheap to produce
- they are reliable
- they are easily terminated, with basic tools
- the terminals are reusable.

The disadvantage of screw terminals are that:
- over-tightening could result in damage to the terminal or the conductor
- under-tightening of the terminals can result in overheating and arcing
- terminals can become loose, due to movement of the conductor in use or due to mechanical vibration
- terminations need to be accessible for inspection.

Crimps

Crimps and crimp lugs come in two basic forms, insulated and uninsulated.

Insulated crimps

When using a cable crimp lug, the wire's insulation must be stripped back about 5 mm. This enables the crimp to be installed with the correct amount of conductor within the crimp and with the insulating section being sealed down onto the insulation of the wire.

▲ Figure 5.47 Insulated crimps

Crimp lugs come in three colours for the different sizes of wires:
- red – 1 to 1.5 mm² wires
- blue – 1.5 to 2.5 mm² wires
- yellow – 4 to 6 mm² wires.

The jaws of the crimping tool are shaped to apply a different crimp style and pressure to the conductor and insulation sides of the connection. The crimping tool applies the correct amount of pressure through a ratchet that cannot be defeated unless the correct amount of pressure is used, or the release button is pressed.

Once a crimp has been installed, it must be checked to ensure that the conductor of the wire protrudes from the crimped part of the lug and that the insulation has been trapped on the other end, so that no exposed conductor is showing. Bearing in mind that crimps are used in applications where vibration occurs, they should never be used on solid conductors.

ACTIVITY

It is not unknown for pliers to be used for fitting crimps, instead of the correct crimping tool. List two possible faults that could occur.

Uninsulated crimps

Uninsulated crimps are used on conductors with cross-sectional areas from 6 mm² upwards. It is important that the crimp is sized in accordance with the cross-sectional area of the conductors and is compatible with the conductor material being terminated.

Whilst a hand crimper may be suitable for smaller-sized conductors, on larger conductors a hydraulic crimper will be required. Battery-operated crimp tools are available that take all of the hard work out of crimping a lug onto a conductor.

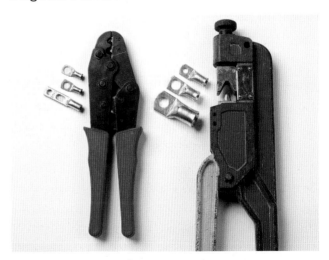

▲ Figure 5.48 Uninsulated crimps

379

Crimped connection method, step-by-step

STEP 4 The cable lug is then crimped to the cable, using a proprietary cable crimper that suits the size of the lug.

STEP 1 Select the correct size cable lug for the conductor. Cable lugs with different-sized holes are available, so be sure to always choose one that has the correct-sized hole for the connection bolt or screw.

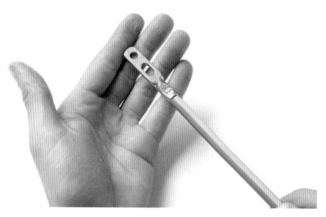

STEP 5 The conductor is now ready to be connected.

Advantages and disadvantages of crimped connections

The advantages of crimped connections are that they:
● are quick and convenient to install
● provide a secure termination
● do not need to be accessible for inspection.

STEP 2 Strip enough insulation from the cable so that the copper conductor meets the end of the cable lug, while the sheath of the cable fits tight to the base of the lug.

The disadvantages of crimped connections are that:
● special tools are required
● tools for larger sizes are expensive to purchase
● crimps cannot be reused.

INDUSTRY TIP

Crimped connections are far quicker and cleaner to make than soldered joints. They also avoid the risk of the heat source damaging the insulation or setting fire to the building.

STEP 3 Ensure that the cable reaches right to the end of the lug tube.

Soldered terminations

In the past, lugs were soldered to cables, but this has mainly been replaced by the use of crimp lugs.

Soldering is mainly used in the assembly of electronic components and equipment.

▲ Figure 5.49 Soldering is mainly used with electronic equipment

Advantages and disadvantages of soldered terminations

The advantages of soldered terminations are that:
- they provide a good electrical connection
- they offer good mechanical strength
- large numbers of connections can be made within a small area as the joint area is very small.

The disadvantages of soldered terminations are that:
- a heat source is required
- there are hazards associated with molten metals
- there may be damage to conductor insulation
- there may be damage to components.

> **ACTIVITY**
> What action should be taken if insulation is damaged during the soldering process?

Non-screw compression

Non-screw compression connectors, including push-fit connectors, have been used in many associated industries such as lighting manufacturing for a number of years and have proved to be robust and reliable, both electrically and mechanically. In recent years these have been used more and more in electrical installations and, depending on choice of connector, can be used for joining:

- solid conductors to solid conductors
- flexible conductors to solid conductors
- flexible conductors to flexible conductors.

▲ Figure 5.50 A wide range of push fit connectors is available, to cope with various cable types

Manufacturers make a range of accessories to go alongside these connectors that ensure the termination of cables can be both speedy and reliable.

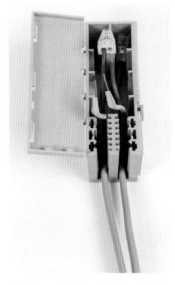

▲ Figure 5.51 Termination of PVC cables into purpose-designed box

Advantages and disadvantages of non-screw compression terminations

The advantages of non-screw compression terminations are that:
- they are quick and convenient to install
- they provide a secure termination
- they are not affected by vibration
- no special tools are required
- they are reusable
- they are maintenance free.

The disadvantage of non-screw terminations is that:

- they are generally not available for cable sizes exceeding 6 mm^2.

Proving that terminations and connections are electrically and mechanically sound

It is important that when terminations are complete, they are verified to be both electrically and mechanically sound. The procedures to follow will include both inspection and testing.

Inspection will include checks such as:

- making sure that all terminations are tight. This includes both live and protective conductors and can be accomplished by both careful scrutiny and by 'tugging' the conductor to ensure it is securely fastened in the termination
- checking visually that the electrical connection is made to the conductor rather than the insulation
- checking visually that conductive parts are not accessible to touch
- checking that the correct termination methods have been used and that they are suitable for:
  - the type of conductors being terminated
  - the environment in which the termination is to be used.

INDUSTRY TIP

Careful and thorough inspection will identify the majority of faults with terminations.

Testing will need to be carried out to ensure that:

- conductors are continuous
- there are no shorts in the conductors
- conductors are connected to the correct points.

The appropriate tests would be:

- continuity, including that of the cpc and ring final circuit conductors
- insulation resistance
- polarity and phase rotation.

It is important that correct inspection and testing procedures are followed.

The consequences of terminations not being electrically and mechanically sound

If the termination of cables and conductors is not electrically or mechanically sound, the consequences can be disastrous. The cause of terminations not being electrically and mechanically sound is usually high-resistance joints or corrosion.

The effects of high-resistance joints

The most common cause of high-resistance joints is a loose connection. When current is passed across such a joint, it heats up. This is likely to cause damage to the cable insulation and/or the connected electrical equipment. In the worst case, this may result in the overheating of adjacent material, resulting in a possible fire. It is important to make sure that cables are seated properly in the terminals and that the terminals are correctly tightened. Manufacturer's instructions need to be consulted to check whether a torque setting is given for connections, which must then be complied with.

It should be remembered that, even with a sound connection, when current is flowing, the conductors and terminations will heat up, resulting in expansion of the metal, which can lead to loosening of the terminal. This is why terminals, apart from those exempted by Regulation 526.3, must be accessible for maintenance and inspection.

Another cause of loose terminations is vibration from such things as machinery. It is important that initial terminations are correctly made and tightened and that regular maintenance is carried out to ensure that loose connections cannot occur.

The **corrosion** of terminals will also result in high-resistance joints.

KEY TERM

Corrosion: the breaking down or destruction of a material, especially a metal, through chemical reactions.

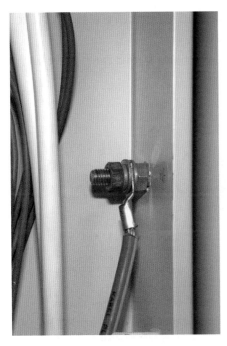

▲ Figure 5.52 Lug bolted to casing of equipment

Techniques and methods for the safe and effective termination and connection of cables

The following section will describe and illustrate the preparation of cables for connection. Various methods will be illustrated. The health and safety requirements for each method and the tools required to terminate will be discussed for each type of cable. The cables covered in this section are:

- thermosetting insulated cables, including flexes
- single and multicore thermoplastic (PVC) and thermosetting insulated cables
- PVC/PVC flat profile cable.

Terminating some of these cables requires the use of glands and shrouds as described below.

Cable glands

Cable glands are available in a range of sizes and formats and with a bewildering array of designatory letters and numbers: BW, CW, CX, CXT to name but a few. So what do all these letters mean? These tables provide the answers.

▼ First letter

Code	Definition
A1	For unarmoured cable with an elastomeric or plastic outer sheath, with sealing function between the cable sheath and the sealing ring of the cable gland.
A2	As type A1, but with seal protection degree IP66 – means 30 bar pressure
B	No seal
C	Single outer seal
E	Double (inner & outer) seal

▼ Second letter

Code	Designation of cable armouring
W	Single wire armour
Y	Strip armour used
X	Braid
T	Pliable wire armour

From the tables, these meanings can be gathered.

BW   A gland without seals suitable for single wire armour cable. SWA for indoor use.

CW   A gland with a single outer seal for single wire armour cable. SWA for outdoor use.

CX   A gland with a single outer seal for braided cables such as SY flex for outdoor use.

What is the purpose of a cable gland?

A gland can be used to:
- maintain the IP rating of an enclosure
- provide continuity of earth
- provide strain relief to terminations.

A gland is an integral part of the termination of a cable and so must be fitted correctly. Incorrect fitting could result in:
- water being allowed to enter an enclosure
- the connection to earth not being adequate and posing a shock risk in the event of an earth fault
- strain being placed on cables and the cables pulling out of terminals, creating either a short-circuit fault or an earth fault.

Shrouds

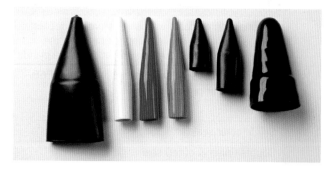

▲ Figure 5.53 A range of shrouds

What does a cable shroud do? A cable shroud can aid the process of keeping the surface of the gland clean and free from the build-up of dirt. It does not, however, necessarily improve the ingress protection (IP) rating of the cable gland. In fact, the gland will invariably have been tested and rated without the installation of a cable shroud. The shroud may provide corrosion protection to cable armour or the sheath, but it must be installed in such a way as not to trap moisture under itself and thus increase the corrosion potential. If the shroud is too loose on the cable sheath, moisture may enter the assembly and, as the fit with the gland is going to be tight, the moisture will be trapped.

Shrouds generally come in PVC or LSF (low smoke and fume) varieties, in a range of sizes to match the gland, and in a range of colours to match the sheath colour of the cable, with black being the most common.

▲ Figure 5.54 A badly fitted shroud could trap moisture

How to fit a shroud to ensure a tight fit to the cable sheath

The same method of fitting a shroud can be used with all cables.

STEP 1 Push the shroud lightly on to the cable so that a small bulge appears where the cable end is. Do not push too hard, as this will stretch the shroud and you will end up cutting in the wrong place.

STEP 2 Cut the shroud at the bulge, with a pair of side cutters or, better still, a pair of cable croppers.

STEP 3 Push the shroud onto the cable. The top of the shroud should now be a snug fit to the outer sheath of the cable. Remember, when assembling the gland and shroud combination, the shroud goes on before the gland.

The next section describes the common methods of terminating cables.

Cable entry to an enclosure

When a cable enters an enclosure, the integrity of the enclosure should not be compromised. The entry may have to meet one or more of these criteria.

- The point of entry must not cause damage to the cable. Rough edges should be removed and, as a minimum, rubber grommets should be used on all cable entries. Cable glands are a better alternative.

▲ Figure 5.55 Rubber grommet to enclosure. Grommets protect cables from rough edges

- The entry of the cable should not compromise the IP rating of the enclosure. For basic protection this is:
  - top surface IP4X – a 1 mm diameter wire will not enter
  - front, sides and bottom, a 12.5 mm diameter object will not enter.

▲ Figure 5.56 Non-compliance on cable entry

- There may be IP ratings that are applicable for the ingress of water:
  - for an enclosure outside a building, it is likely to be IPX4 splash proof
  - for an enclosure where water jets are used, the rating should be IPX5.

There may be requirements for fire protection. Where there is a fire risk due to powders or dust being present in locations such as a carpenter's workshop, the minimum IP rating is IP5X.

▲ Figure 5.57 Splash proof socket

Terminating flexes

Tools required

- Ringing tool (Method 1)
- Stripping knife (Method 2)

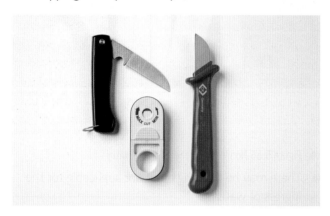

▲ Figure 5.58 Ringing tool (Method 1); stripping knife (Method 2)

Safety considerations

There is a risk of cutting your hands from use of knife. The use of gloves and eye protection is recommended.

Method

Before the flex can be terminated, the outer sheath must be removed.

Two methods of removing the outer sheath are outlined here.

Method 1

The outer sheath can be removed with the use of a ringing tool. These come in various shapes and forms, the most basic of which is shown in the diagram. This tool slides over the end of the cable to the required stripping position, and is then rotated around the cable, cutting it slightly.

▲ Figure 5.59 Flexible cable sheath stripping tool for use on flex

The ringing tool is removed, the cable is bent to finish the cut and the sheath is then pulled off.

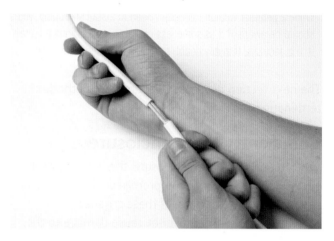

▲ Figure 5.60 Outer sheath being removed

Method 2

There may be times when a ringing tool is not available, so it is important to know how to remove the outer sheath without this tool. Thermoplastic has a tendency to split when it is damaged and pressure is applied, and this can be used to help strip the sheath.

Bend the cable into a tight bend at the point where the sheath is to be removed. Using a sharp knife, gently score the top of the bend. The thermoplastic will split open like a little mouth (Figure 5.61).

▲ Figure 5.61 Bending cable to cause split in sheath

Terminating single-core cables
Tools required

● Cable strippers.

▲ Figure 5.62 Different types of cable strippers are available

Safety considerations

There is a risk of injury from slipping with cutters or pliers. The use of gloves and eye protection is recommended.

▲ Figure 5.63 Cable stripper jaws

Method

As single-core cables do not have outer sheaths to remove because they are intended to be installed within trunking or conduit, the termination method is straightforward and requires the minimum of tools. This termination method will therefore also apply to the final connection of other cable types.

Use of wire strippers is recommended for single-core conductors. Automatic wire strippers tend to rip the insulation from the conductor and can damage it. Manual wire strippers are preferred and these come in various forms, but all work on a similar principle.

Manual wire strippers have two blades that cross one another, like scissors. Each blade has a notch so that together they cut around the conductor in the middle. In addition, wire strippers also have some means of adjustment so that different sizes of conductor can be stripped.

Before wire strippers can be used, they must be set to the correct size. This can be done using a scrap or off-cut of wire of the correct size. With the jaws of the wire strippers together, turn the adjustment screw until the hole in the jaws is just bigger than the conductor to be stripped. Test the setting on the off-cut of wire, by placing the wire strippers over the wire and squeezing the handles to close the jaws. Then slightly release the jaws and try to slide the insulation off, using the wire strippers.

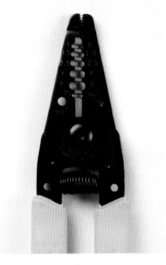

▲ Figure 5.64 Cable strippers with pre-set holes

If the wire strippers are correctly set, the insulation will come off easily and there will be no damage to the copper conductor. If the aperture is set too small, the insulation will slide off easily but the conductor may be damaged. If it is set too large, the insulation will not come off easily. A simple adjustment of the adjusting screw will correct these problems. Now, with the correct setting, the wires can be stripped safely.

Other types of wire stripper will have pre-set stripping holes so that selecting the correct hole will ensure that the depth of cut is correct every time.

Once the conductors are stripped they are ready to be connected to the electrical equipment.

387

Terminating PVC/PVC flat profile cable

Tools required

Dependent on method used:

- electrician's knife
- side cutters
- pliers.

▲ Figure 5.65 Pliers, side cutters, electrician's knife (from left to right)

Safety considerations

- Cuts to hands from use of knife
- Injury from slipping with cutters or pliers

The use of gloves and eye protection is recommended.

Method

Before the conductors can be connected, the outer sheath of the cable must be removed.

First, identify how much of the sheath should be removed from the cable. The purpose of the sheath is to provide some mechanical protection for the insulation on the conductors. Too much sheath, however, takes up space within the accessory and will put excess strain on the conductors. The sheath should, therefore, be stripped back almost to where the cable enters the accessory, leaving only 10–15 mm, a thumb's width, of sheath within the accessory.

There is more than one way to remove the outer sheath. Each has its advantages and disadvantages but, whichever method is used, care must be taken to avoid damage occurring to either the conductors or the insulation around the conductors.

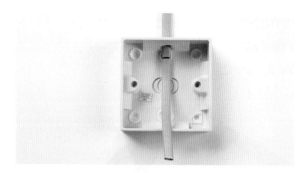

▲ Figure 5.66 Determining the point to which to strip the sheath

Method 1

Having decided on the length of sheath required, and with the cable in place, score the sheath at the point to which it is to be removed. This can be performed with an electrician's knife. Care should be taken not to cut into the cable.

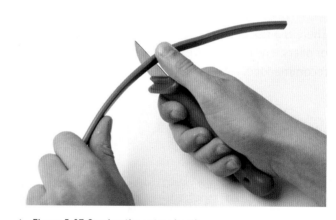

▲ Figure 5.67 Scoring the outer sheath

From the end of the cable, snip down the centre of the cable, using a pair of side cutters. Split the line to one side and the neutral to the other.

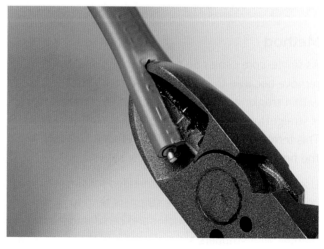

▲ Figure 5.68 Snipping the end of the cable

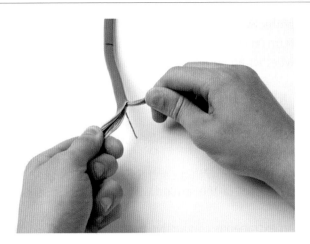 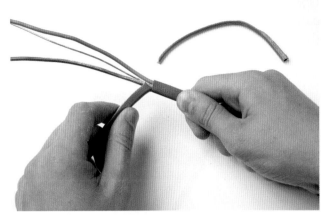

▲ Figure 5.69 Tearing down to the score mark (left) and removing the outer sheath (right)

Grip each piece of split cable, conductor and sheath and pull them apart, tearing any uncut sheath as you go. Continue up to the score mark.

Once the score mark is reached, the sheath can be removed from the wires. Then, holding the cable in one hand and with the thumb of the other hand placed on the sheath behind the score, jerk the torn sheath to make it break along the score mark, leaving a clean finish.

Method 2

Snip down the end of the cable with the side cutters, then use a sharp electrician's knife to slice down the cable. To do this, run the blade along the protective conductor. Note that this can damage the protective conductor or, even worse, the insulation of one of the live conductors if it is not done carefully.

Method 3

Snip down the end of the cable and gently pull the cpc with pliers, almost like a cheese wire, to tear the outer sheath as far as the point to which it is to be stripped. Take care not to apply too much force or to damage the cpc, nor to strip too far.

▲ Figure 5.70 Running an electrician's knife along the cpc

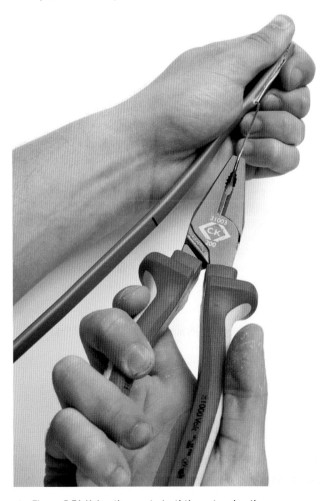

▲ Figure 5.71 Using the cpc to 'cut' the outer sheath

389

Once you have reached the desired strip position in the outer sheath, use side cutters to cut away the outer sheath.

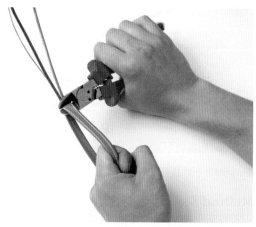

▲ Figure 5.72 Using side cutters to cut away the outer sheath

Termination

Now the inner wires can be stripped in the same manner as for single-core wires, making sure that green and yellow sleeving is applied to the bare copper cpc. The sleeving must cover all of the bare cpc, apart from a small amount at the end, enough to go into the terminal.

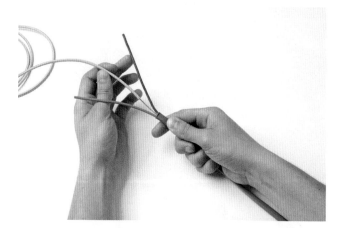

▲ Figure 5.73 Applying sleeving to bare cpc ready for final termination

The conductors can now be terminated into the electrical equipment, as described previously.

What needs bonding?

BS 7671 The IET Wiring Regulations requires that an extraneous conductive part be connected to the main earthing terminal (MET) by means of a suitably sized protective **bonding** conductor. An understanding of the term 'extraneous conductive part' helps when deciding whether an item requires bonding or not.

KEY TERMS

Bonding: a term used to describe the connection of extraneous conductive parts to the earthing system.

Earth: earth with a capital E represents the potential of the ground we stand on.

An extraneous conductive part is defined in **BS 7671** as:

- a conductive part
- liable to introduce a potential, generally Earth potential, and
- not forming part of the electrical installation.

By breaking down the definition in this way, it helps us come to a conclusion as to whether or not something requires bonding.

- A conductive part is generally metallic.
- The potential is usually taken as the mass of **Earth**, hence the capital E in 'Earth potential'.

Extraneous conductive parts would include such items as:

- metallic water installation pipes
- metallic gas installation pipes
- other metallic installation pipework and ducting
- exposed structural metalwork
- central heating and air-conditioning systems.

When considering items on the list above, always use the criteria that define an extraneous conductive part to decide whether or not the item is an extraneous conductive part in a particular situation. For example, a structural steel beam supported on brick piers would not be regarded as an extraneous conductive part and would not require bonding, but a structural steel beam in contact with the ground is an extraneous conductive part that would require bonding. Where plastic pipes enter a building, or metal pipes have insulated sections, the metallic installation pipes will not require bonding as these are not extraneous.

Metallic parts of the electrical installation are *exposed* conductive parts rather than *extraneous* conductive parts and these require earthing.

HEALTH AND SAFETY

Metal parts that are not in contact with earth cannot become live or create a circuit so they are safe. If they are bonded, this can introduce a risk that did not exist before.

▲ Figure 5.74 A clamp used for bonding extraneous conductive parts such as pipework

Installing protective bonding

When installing protective bonding to mains services, the following steps should be taken.

1 Select the correct size of protective bonding conductor.
2 Install the protective bonding conductor.
3 Make the correct choice of bonding clamp.
4 Install the bonding clamp.
5 Terminate the cable.

Cable sizes

The size of bonding conductor depends on the type of earthing system used and the electrical system to be worked on, including the type of electrical supply earthing arrangement and length of the cable run. In the majority of domestic type installations, a 10 mm^2 conductor would normally be suitable.

Installation of main protective bonding conductors

A main protective bonding conductor of the correct size should be run from the MET to the point of connection to the service to be bonded. The conductor used must be of the correct colour: green and yellow.

Support of main protective bonding conductors

Section 522 of **BS 7671** requires that all cables are correctly supported throughout their length to avoid mechanical stresses on both the cable and the terminations. Gas, water and other service pipework should not be used as a method of support. Consideration needs to be given to this aspect, especially where the bonding cable is terminated to the service pipework to ensure that the protective bonding conductor does not become disconnected from the pipe clamp.

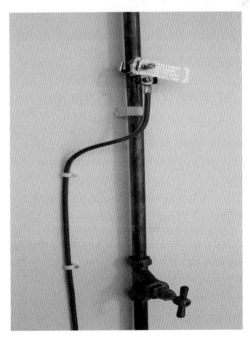

▲ Figure 5.75 Clipped bonding conductor

Point of connection

The connection of the bonding conductor to the service pipes should be:

- as near as practicable to the point of entry of the service into the building
- where practicable, within 600 mm of the service meter or at the point of entry if the service meter is external

- to the consumer's pipework – in the case of a water service, after the stopcock
- to hard metal pipework, not soft or flexible pipe
- before any branch pipework
- after any insulating sections.

The connection point is required to be accessible for future inspection and testing.

▲ Figure 5.76 Bonding to gas supply

Choice of bonding clamps

Bonding clamps must:
- meet the requirements of **BS 951**
- be suitable for the environment in which they are to be installed
- be labelled in accordance with Regulation 514.13.1 of **BS 7671** The IET Wiring Regulations.

BS 951 clamps are designed only to fit circular pipes or rods. They are not designed to fit objects of irregular shape, to be attached to steel wire armoured cable or lead-sheathed cable. They are available in different lengths and different materials to suit different pipe sizes and different installation environments.

> **HEALTH AND SAFETY**
> Always make sure all connections are electrically and mechanically sound. Bad joints are dangerous.

BS 951 clamps are designed for connection of bonding conductors of between 2.5 and 70 mm². They also come in three standard band lengths, to suit pipes with diameters of:

- 12–32 mm
- 32–50 mm
- 50–75 mm.

BS 951 clamps are available to suit different environments and manufacturers have adopted a three-colour coding system to make selection easier:
- red for dry, non-corrosive atmospheres
- blue for corrosive or humid conditions
- green for corrosive or humid conditions, and for larger sizes of conductor.

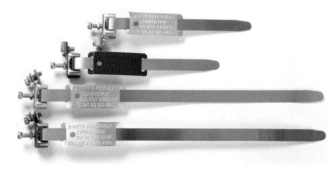

▲ Figure 5.77 Colour-coded **BS 951** bonding clamps

Some extraneous parts, such as steelwork, are of an irregular shape but still require bonding. Various clamps are available from suppliers and these must be:
- electrically durable
- of adequate mechanical strength
- suitable for the environment where they are to be installed
- labelled in accordance with Regulation 514.13.1 of **BS 7671**.

Labels for bonding clamps

BS 7671 requires that the point of connection of every bonding conductor to an extraneous conductive part has a permanent and durable label fixed in a visible position stating 'safety electrical connection – do not remove' (Figure 5.78).

▲ Figure 5.78 **BS 7671** permanent label

Installation of bonding clamps

The photograph in Figure 5.79 shows a bonding clamp as it appears when it first arrives. Note that the label is fitted this way for packaging; the slots in the label are only there for this purpose.

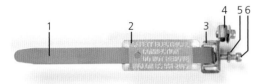

▲ Figure 5.79 A bonding clamp as supplied

The six parts of the bonding clamp are:
1. strap
2. label
3. bonding clamp body
4. cable termination point and screw
5. locking nut
6. tightening screw.

Step-by-step instructions for fitting a BS 951 clamp

Step 1 – preparing the bonding clamp for installation

- Remove the label from the strap.
- Remove the tightening screw and locking nut from the body of the clamp.
- Pass the tightening screw through the hole in the label and fit the locking nut.
- Refit the assembly to the clamp body, but leave the screw so that it is not protruding into the slot of the body.
- The clamp is now ready for installation.

▲ Figure 5.80 Preparing the bonding clamp for installation (Step 1)

Step 2 – preparing the pipe for installation of the bonding clamp

It is vitally important that the connection between the bonding clamp and the pipework is electrically sound. Tarnished pipework can be cleaned using wire wool; paint must be removed from painted pipework and the metal should be cleaned with wire wool.

INDUSTRY TIP

Even if the pipe looks clean there may be an oxide on the surface. Make sure you clean the pipe before fixing the clamp to it.

▲ Figure 5.81 Preparing the pipe for installation of the bonding clamp (Step 2)

Step 3 – installation of bonding clamp to pipe

▲ Figure 5.82 Place the body of the clamp against the cleaned section of pipework

▲ Figure 5.83 Pass the strap around the pipe and through the slot in the body of the bonding clamp

Pull the strap tight with one hand and, with the other hand, tighten the screw with a screwdriver.

Make sure that the locking nut is far enough up the tightening screw to prevent it locking against the body of the clamp.

▲ Figure 5.84 The bonding clamp should now be tight against the pipe

▲ Figure 5.85 The locking nut is now tightened against the body of the clamp

▲ Figure 5.86 The bonding clamp is now installed ready for connection of the protective bonding

Termination of bonding conductors

Regulation 526.1 of **BS 7671** The IET Wiring Regulations requires that all electrical connections are:

- electrically durable
- of adequate mechanical strength.

The ideal method of terminating the bonding conductor is by means of a cable lug of the correct size for the cable.

▲ Figure 5.87 Cable to be terminated and correct size lug

Strip enough insulation from the cable so that the copper conductor meets the end of the cable lug, while the sheath of the cable fits tight to the base of the lug.

▲ Figure 5.88 Fitting the lug to the cable

The cable lug is then crimped to the cable using a proprietary cable crimper that suits the size of the lug.

▲ Figure 5.89 Crimped lug

Connection of bonding conductor to bonding clamp

Remove the cable termination screw.

Align the cable lug hole and reinsert the cable termination screw. Tighten with a screwdriver.

▲ Figure 5.90 Connecting the bonding conductor to the bonding clamp

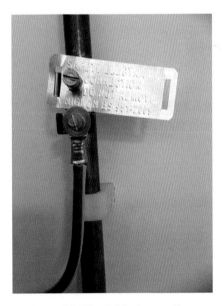

▲ Figure 5.91 The finished connection

Some of the larger bonding clamps are designed so that the cable can be terminated without the use of a cable lug. In this case, the cable is stripped back far enough to allow good metal-to-metal contact; the stripped end is inserted between the two connection screws and the screws are tightened equally to give a sound electrical connection.

▲ Figure 5.92 Bonding clamp for use without cable lug

Termination of more than one cable to a bonding clamp

Where two services are in close proximity to one another, it is acceptable to run one bonding cable to serve both services. If the bonding cable loops from one bonding clamp to another, the bonding conductor should be unbroken.

INDUSTRY TIP

There is a risk that if another trade removes a bonding conductor they may not bother to replace it. By using an unjointed bonding conductor for two services at least the end service is still bonded.

As shown in Figure 5.93, strip off enough insulation from the bonding cable to enable the cable to make metal-to-metal contact with the bonding clamp. A stripping knife would be the ideal tool to accomplish this task.

▲ Figure 5.93 Preparing cable to loop from one bonding clamp to another

▲ Figure 5.94 Parting the strands of the cable so that the connection screw of the bonding clamp can be fitted

▲ Figure 5.95 Passing the screw through the hole created

7 DIAGNOSING FAULTS AND SAFE REPAIR OF ELECTRICAL WORK

Communication and regulations

There are many faults that can occur in electrical circuits that include component failure as well as errors in design and poor workmanship in the installation of electrical systems.

Before any work begins on investigating an electrical problem it is essential that a conversation with the client is carried out as this will help in identifying the problem and enable effective faulting finding to be applied in conjunction with the manufacturer's instructions. Reference to the IET On-Site Guide may often be required and the systems and processes of testing electrical circuits can be found in Section 9. In Section 10 of the IET On-Site Guide, it states that:

> It is the test operative's duty to ensure their own safety and the safety of others whilst working through the test procedures.

The guide goes on to emphasise the importance of having a good working knowledge and experience of the correct electrical application and use of the test instruments that includes the leads, probes and accessories. It goes on to say check to ensure any test equipment is safe and suitable. It is important to check the accuracy and calibration of such devices and to make sure they are in good working order.

Testing equipment

Before any testing begins the plumber must ensure that all leads, probes and instruments are in good condition, undamaged and that they function properly. HSE Guidance Note GS38 for electrical test equipment gives guidance on safe and suitable use of electrical test equipment. The document goes on to advise that the test probe of any instrument should not in itself be a hazard to the user. The leads should be robust and have a barrier to prevent access to terminals. In addition the connectors to the instrument should be shrouded.

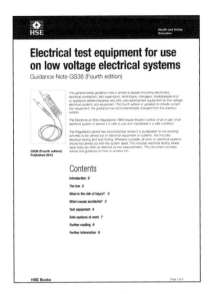

▲ Figure 5.96 HSE GS38 Guidance Note

It is important that test equipment is calibrated in accordance with the manufacturer's instructions.

▲ Figure 5.97 Metallic tips on test instrument probes (min. 2 mm; max. 4 mm; exposed conductors)

HSE Document GS38 also advises that the leads on test equipment should be suitably fused.

▲ Figure 5.98 HTC fuse 500 mA max.

It is recommended that the types of testing equipment you will use, like test lamps and voltage indicators, are marked clearly with the maximum voltage that may be tested by the device and any short time rating for the device where applicable. The rating is the maximum recommended current which should pass through the device for a few seconds. Generally these devices are not designed to be connected for more than a few seconds.

> **KEY POINT**
>
> HSE Document GS38 advises that the test probe of any instrument should not exceed 4 mm, and where practicable be reduced to 2 mm. The leads should also be insulated and each be a different colour.

Test lamp

Test lamp detectors rely on an illuminated bulb to indicate the voltage. The robust leads are well insulated and the probe has a HBC fuse not exceeding 500 mA. They are fitted with glass bulbs which should not cause any danger if the bulb is broken and the bulb can be protected by a guard.

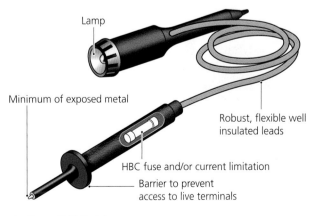

▲ Figure 5.99 Test lamp

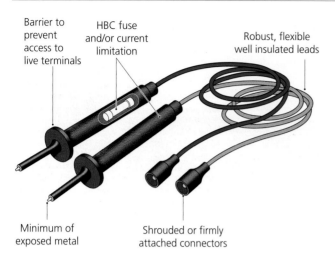

Barrier to prevent access to live terminals

HBC fuse and/or current limitation

Robust, flexible well insulated leads

Minimum of exposed metal

Shrouded or firmly attached connectors

▲ Figure 5.100 Approved test leads

Voltage tester

A voltage tester enables **AC** and **DC** voltage indication and measurement on the model shown in Figure 5.101 is between 6–1,000 volts. They often have other features incorporated into their design, such as an acoustic sounder and a visual continuity function. They can also test the function of 30 mA RCD, RCBO and MCBs.

KEY TERMS

Alternating current: flows in *both* directions
Direct current: an electrical current in which electrons flow in a *single* direction.

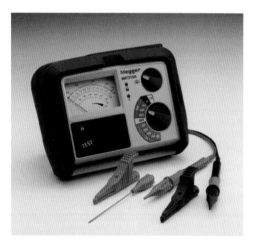

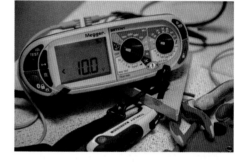

▲ Figure 5.101 Voltage testers

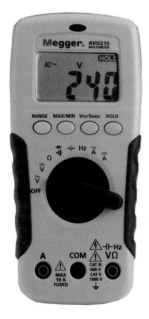

▲ Figure 5.102 Digital multimeter

Digital clampmeter

A digital clampmeter will give readings not only of current and voltage but of power quantities too.

They are ideal for use in the installation, maintenance, monitoring or for the checking of AC or DC electrical systems and equipment. It is also useful for testing applications, as they are typically on electrical systems and equipment where there is a need to measure current, volts, resistance and frequency.

The digital clampmeter is therefore intended for use while installing, maintaining, fault finding or monitoring those systems. One of their features includes being able to detect the load currents from equipment and identify a range of load currents and start-up currents to motors on pumps for heating systems. Information from tests can be transferred to a PC which can be convenient when storing and producing maintenance reports.

To carry out the test, first isolate the supply by following the safe isolation procedure and then position the clamp on the live wire only. Then, if safe to do so, turn on the supply and make a note of the reading. The meter reads and interprets the magnetic field that is being produced by the flow of current in the live conductor that is supplying the resistive component, i.e. the pump motor.

> ### KEY POINT
>
> If a customer reported a problem with nuisance tripping of a CB when the heating was on, a clampmeter could confirm a measurement of the current drawn from the pump motor to help the engineer ascertain if that component was the cause of the problem. The advantage of this method means that testing can occur without disconnection of the supply.

▲ Figure 5.103 Digital clampmeter

Using a clampmeter is a safer way to test current flow as there is no connection in series to the load and as a result the risk of electric shock is reduced.

▲ Figure 5.104 Clampmeter with no reading as power is off

▲ Figure 5.105 Reading with *Power on* shows 0.24 A being drawn, which equates to circa 60 W which is normal for a pump of this size

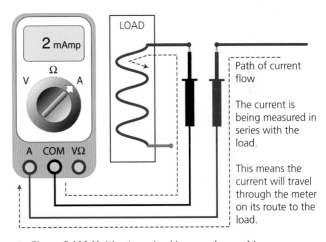

▲ Figure 5.106 Multimeters should **never** be used to measure current in electrical systems as this involves placing the meter into the circuit meaning exposure to live parts

> ### KEY POINT
>
> Remember from earlier in this chapter, if the earth fault loop impedance is low, the fault current will be high enough to disconnect the circuit quickly in the event of an earth fault.

Socket testers

Socket testers are very useful additions to a working plumber's tool kit as they can be used for initial identification of faults following socket replacement or re-wire work of any kind. Simple miss-wires or missing connections can be quickly identified prior to full **BS 7671** electrical test and inspection, which saves time during the testing and certification process. They may also be used during installation or commissioning of electrical equipment connected to 13 mA sockets. By simply plugging the device into a socket outlet, the polarity of the connections can be identified as any faults are signalled by three bright LEDs on the front of the unit and a simple diagnosis chart for typical errors is included on the label next to the LED display, as shown on the *Megger*. A typical use of a socket tester would be when a boiler or other electrical appliance requires servicing or replacing as the plumber will be aware of the existing wiring at the point where the tester is fitted.

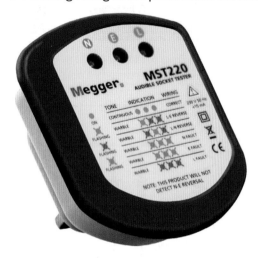

▲ Figure 5.107 Socket tester

▲ Figure 5.108 A common socket tester used in situation

Proving unit

As with any gas or water testing apparatus, it is very important that the accuracy of any such testing equipment is verified before and during use. Working on the same principle, a proving unit provides a convenient way of verifying or proving the safe operation of any two pole electrical voltage tester which is to be used for safe

isolation. When the probe tips are applied to the proving unit, it automatically begins in auto-test mode. The output voltage from the device will increase from 50 V up to around 690 V and then gradually return back to 50 V and then OFF. The voltage output is clearly indicated on the five LED indicators on the front of the unit.

▲ Figure 5.109 Proving unit   ▲ Figure 5.110 Voltage indicating device in test position on a proving unit

Always ensure a good connection is made when inserting probes into a proving device.

Types of testing – dead and live

There are two types of testing, dead and live. Although most of the electrical tests that are carried out are **dead tests** because the electrical power is off during testing. **BS 7671** specifies some examples of when **live tests** must be carried out on electrical installations, but these should only be carried out by competent persons who are electrically skilled.

KEY TERMS

Dead testing: testing is carried out on electrical components or parts of an installation when the electrical supply is disconnected.

Live testing: a test is carried out when components are live.

A visual inspection must be done by the plumber before testing can begin and is done at any point of an installation when connecting to an electrical supply. A visual inspection should ensure that the system is installed in accordance with the appropriate standards and that there are no visible signs of damage. In addition, a check should be made to ensure that conductors are securely fitted and located in the correct terminals, also referred to as checking for correct polarity.

Incorrect polarity would be easily identified during a simple visual inspection where wiring problems such as neutral and earth conductors were in the wrong connections. The correct way to put this problem right would be to safely isolate the electrical supply, verify the supply was the correct polarity and then refit the conductors into the proper terminals as indicated on the connection terminal.

If any damage, installation errors or defects are discovered during a visual inspection they must be made good before any electrical installation certificate can be issued.

Quite often during a visual inspection a plumber can come across burnt or scorched connections. If a conductor and its insulation have been scorched, this must be remedied before any certification is produced. The cause of the problem must be identified and often both the conductor and component will need to be replaced.

Loose wiring or reduction of conductor csa could cause high resistance at this point. However, if the inspection and testing of an installation prove to be satisfactory then a signed installation certificate, together with a schedule of inspections and a schedule of test results, should be given to the person ordering the electrical work (IET On-Site Guide). This person is often the customer but could be an agent, landlord or site representative.

Any visual inspection must always come before tests of electrical installations using instruments. A dead test should always be carried out before a live test therefore all tests should be carried out in their correct sequence.

▼ Table 5.5 Sequence of dead testing on an electrical installation

Sequence of dead testing	Test instrument
Continuity of earth conductors	Low reading ohmmeter
Continuity of final ring conductors	Low reading ohmmeter
Insulation resistance	Insulation resistance tester
Polarity	Low reading ohmmeter

According to the Regulation 4(3) of the Electricity at Works Regulations 1989, it is preferable that electrical supplies should be made dead before any work begins. The regulations do also mention that some work such as fault finding and testing may require the electrical equipment to be energised. It is essential that if live

testing is the only practical way to carry out this type of work then the person must be competent and fully trained to work in such a high risk environment. In addition they must only use approved testing equipment not neon screwdrivers or homemade test lamps. Finally, they must let others in the vicinity know what they are doing and protect the area with barriers and notices.

▼ Table 5.6 Sequence of live testing on an electrical installation

Live test sequence	Instrument
Polarity	GS38 approved test leads
Earth electrode resistance	Earth loop impedance tester
Earth fault loop impedance	Earth loop impedance tester
Prospective fault current	Prospective fault current tester
Functional testing	RCD tester

Once it is confirmed that it is safe to work on an installation, a live test can be carried out.

All types of electrical testing come with an element of risk to the person carrying it out. They must be aware that they are responsible for their own safety and for the safety of others who could be affected by their actions.

On heating circuits, the control system should have only one means of isolation but quite often it is assumed that a circuit is dead because isolation has been carried out from the fused switched spur connection unit.

After the safe isolation of gas supply for example, an engineer could open the case of a boiler and expect all of the terminals to be dead. At this stage it is essential to carry out a dead test to establish that there is no current present using a commonly used instrument known as a voltage indicating device. Sometimes there are surprise readings showing that there is a current present because there has been a rogue supply installed, such as a remote frost thermostat with its supply taken from a different circuit from that serving the heating system.

This stresses the importance of a visual inspection of wiring arrangements and of dead testing. To remedy this situation the frost thermostat would have to be re-installed as part of a fully controlled system with a single point of isolation in accordance with the manufacturer's instructions.

It is worth noting that the control systems within a boiler vary and the manufacturer's instructions and the boiler labels will indicate if the AC supply has been transformed to DC.

401

Routine checks and diagnostics

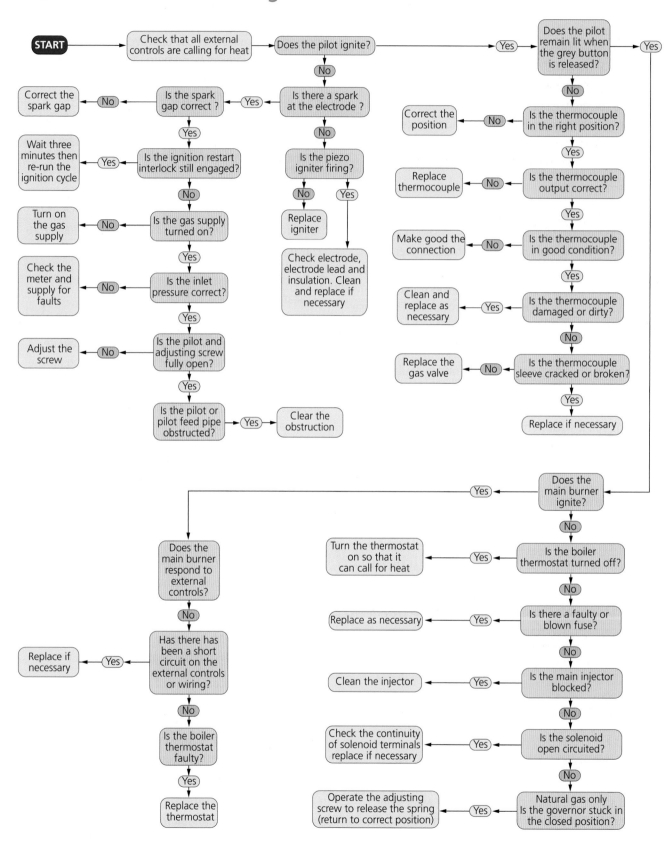

▲ Figure 5.111 Fault-finding flow chart

Basic sequence of operation for a central heating control system

Typically plumbers and heating engineers will work on domestic heating installations controlled by electrically controlled components designed to be operated by a programmer. In order to effectively diagnose problems and faults, it is important to be aware of the function of each component, so that the most common faults and problems on heating systems can often be remedied by following manufacturer's fault-finding sequences and flow charts. The simple sequence of operation follows a logical flow pattern once the programmer has been switched on. At each stage the component should be visually and manually checked and this will help with basic fault diagnosis.

By following this basic sequence of operation of control, fault finding can be carried out on most heating installations.

One of the most common problems found is errors in wiring, therefore it is advisable to always double check your work during electrical installation.

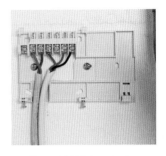

▲ Figure 5.113 A typical example of an exposed backplate connection block for a programmer

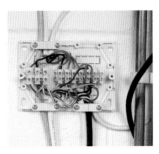

▲ Figure 5.114 A typical wiring centre is where all the conductors of a central heating control system meet. By following the manufacturer's wiring instructions, efficient fault finding can be carried out

When a programmer is calling for heat then the boiler should fire and the heating cycle should begin.

In Figure 5.115 the wires enter from the rear of the recessed metal box and are eventually connected to the terminals shown at the top of the thermostat. A rubber grommet should be fitted to the metal casing which houses the thermostat. This will prevent the abrasion of the cable insulation. Ensure that the earth wire is only used on the earth terminal and not as a conductor.

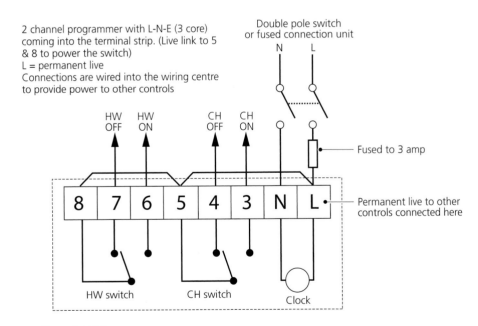

▲ Figure 5.112 Programmer

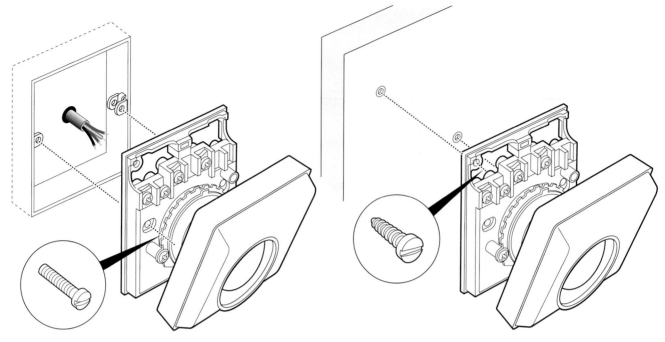

▲ Figure 5.115

Thermostat

If the boiler does not fire, the first place to check would be the thermostat to ensure that it is set above the existing ambient temperature of the room. A thermostat sensing element is comprised of two circular, flexible, metal plates which are welded together at the rims and encapsulate a liquid whose pressure changes greatly in response to small variations in temperature. In effect this dual diaphragm forms a 'bellow' which expands and contracts in sympathy with the ambient temperature changes. This movement then operates a snap action electrical switch which in turn controls the heating.

The operation of the thermostat can be checked when the circuit is isolated by attaching probes between the switching terminals of the thermostat and then moving the control dial on and off. By using an ohmmeter or multimeter set to ohms the switching process can be verified.

Zone valve

A typical problem with a heating system is the failure of a zone valve to operate. This will result in either no heating or no hot water. The first test is visual. If a zone valve does not open the lever mechanism will not move and the two most likely reasons are either the control

thermostat has failed or the servo motor on the zone valve has burnt out. If it is a new installation and the problem existed from the beginning then it could be a wiring problem.

Once a visual inspection of the wiring arrangements in accordance with the manufacturer's instructions has been confirmed then the next step is to check for power to the zone valve and this could involve live testing with a voltage indicator or test lamp. However, only fully trained and qualified persons can carry out such testing. A dead test using Ω resistance will also confirm if the motor has burnt out and an ohmmeter or calibrated multimeter could suffice for this task.

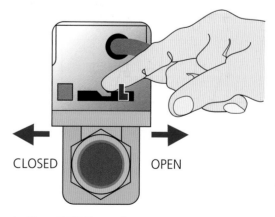

▲ Figure 5.116 Zone valve

Testing the resistance of a synchronous motor from a zone valve step-by-step

STEP 1 A synchronous motor is removed from a zone valve and tested with a multimeter set to Ω.

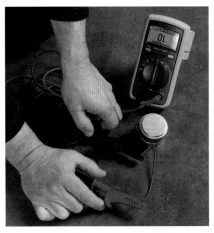

STEP 2 Motor showing a resistance reading.

STEP 3 Motor showing an infinity reading.

Because the motor is a resistive component there should be a reading when probes are connected to the leads attached to the motor.

If there is an infinity (OL) reading it means that the motor is no longer operational and should be replaced. The manufacturer's instructions provide expected Ωs readings for a working component.

KEY TERM

Resistance testing: the Ω scale is used for example to find out the level resistance of a coil in a motor, to ascertain whether it works or not, e.g. testing a motor on a zone valve.

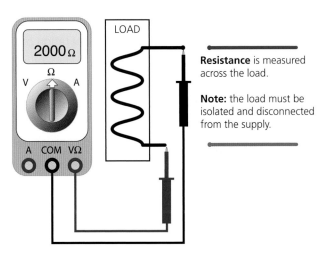

Resistance is measured across the load.

Note: the load must be isolated and disconnected from the supply.

▲ Figure 5.117 Resistance testing

An example of the sequence of switching which can be found in a *Honeywell* S-plan is shown in Figure 5.118 and works in the following way.

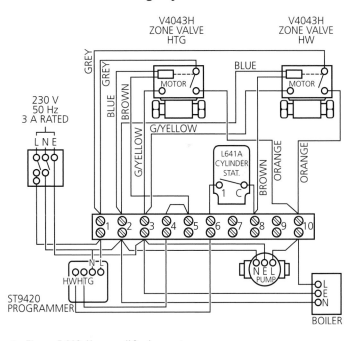

▲ Figure 5.118 *Honeywell* S-plan system

The heating circuit and the zone valve

The switched live (SL) from the room thermostat connects at the wiring centre with the brown wire to the zone valve. The signal from the brown wire will energise the motor, which in turn opens the valve and will rest when the lever arm makes contact with the micro-switch.

When open the micro-switch activates the permanent live grey wire and allows the permanent live to go down the orange wire where it connects at the wiring centre to send the signal to pump and boiler. The process is reversed when the room thermostat reaches the temperature set by the customer.

Therefore when the room thermostat reaches temperature the brown live supply to the motorised valve is cut. The valve then closes, due to a specially designed spring returning the mechanism, to the closed position. The switched live orange is then cut because the micro-switch is deactivated, as it is released by the returning lever arm. As a result the boiler and pump are turned off, but the grey wire still remains as the permanent live.

There are a range of zone valves available and some are power on types and others the power off type, and do not include a spring to return the mechanism to the OFF position. This type should not be fitted to an unvented cylinder flow pipe as isolation of the energy source must be the mechanical spring return type.

If the zone valve is open then a signal should be sent to the pump to operate it. If it is not working then it could be that the motor is defective or if the pump has not been in use for some time it may be jammed and this can be remedied by manual rotation of the rotation shaft. In the event that the pump is working then the last appliance to check will be the boiler.

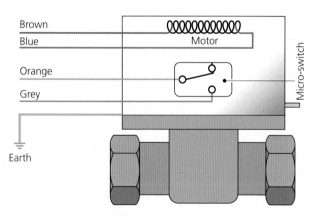

▲ Figure 5.119 2-port zone valve

Plumbers install a range of heating appliances and most of them require connection to an electrical supply.

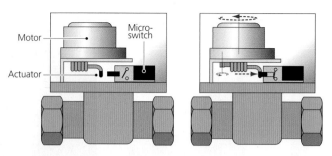

▲ Figure 5.120 Motorised zone valve

Boiler

The boiler display panel should indicate its mode of operation. For example, the boiler could have locked out and the reset button will have to be activated. As with all fault finding it is advisable to check the obvious first such as fuel and electrical supply or if the boiler thermostat is calling for heat.

If all the components are functioning correctly and the boiler still does not operate then another sequence of operations can be followed to help find out any problems occurring in the boiler.

A typical gas fired, fanned flued boiler has a range of electrical components, which often require testing and fault diagnosing during the course of a maintenance schedule. There are a range of test instruments required to safely and effectively enable accurate testing.

For example an ohmmeter could be used to determine the condition of a motor in a fan or in a motorised valve and by checking the results against manufacturer's instructions and standard testing procedures, accurate diagnosis can be achieved.

This is an example of a basic sequence of operation on a gas fired fanned flued boiler.

- Signal to boiler control panel from zone valve.
- Boiler thermostat calling for heat.
- Fan operates.
- Pressure switch senses pressure changes through the movement of the fan.
- Ignition process begins via control panel.
- Flame verified.
- Gas valve solenoid opens.
- Main burner ignites.

There are variations of sequences depending on the flame supervision system and this one can be applied to an appliance with flame rectification.

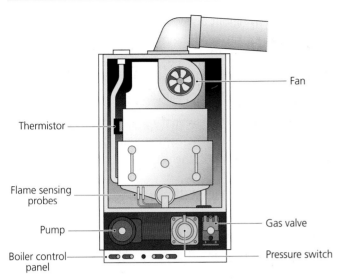

- Fan
- Thermistor
- Flame sensing probes
- Pump
- Gas valve
- Boiler control panel
- Pressure switch

▲ Figure 5.121 A standard gas fired fanned flued boiler with key components identified

Switches and resistors

The sequence of operations here represents a series of switches and resistors. A switch controls the flow of a current through a circuit or to a resistor and will be in either an open or closed position. A resistor converts electrical current into a form of energy. The energy could be heat, movement, magnetism or even light. A circulating pump is in fact a resistor as current is used to induce movement into a circulating impeller. With this information the fault-finding sequence can be identified as switches and resistors. For example the boiler thermostat is a **switch** while the fan motor is a **resistor** and the gas valve solenoid also a resistor.

KEY TERMS

Switch: component that breaks an electrical circuit by interrupting or diverting current.

Resistor: a passive thru-terminal electrical component that resists electrical current.

By following the basic sequence of operation for a boiler as shown above most basic faults can be detected. During the analysis of component specific faults, dead testing is recommended although sometimes live testing may be the only option. A visual survey will help detect some faults at the outset such as non-operational lights on the control panel or leads and hoses disconnected and scorch marks.

Boiler control panel

Many modern boilers have fault indicating panels which display a code which when used in conjunction with the manufacturer's fault-finding flow charts will identify the source of the fault.

▲ Figure 5.122 External boiler control section with operating temperature and fault displays to be used in conjunction with the manufacturer's fault-finding flow charts

This is the first step to assess the operation of the appliance as it is always good practice to operate an appliance within an installed system to evaluate if all the components are working properly before any work begins. Any problems should be reported and documented then discussed with your supervisor or if you have been authorised to do so, directly with the customer. Listening for excessive noise from components during operation such as humming, chattering or grinding will give some indication of the condition of the moving parts. Sometimes poor electrical connections will produce scorch marks and if the problem is not visible the fault can be identified by the smell of burning.

By referring to the manufacturer's instruction each LED light on the panel of a modern boiler will reveal its functional stage during operation. Often there is a digital display on the front of the boiler showing any fault or operational codes. If any faults are evident then direct reference to the manufacturer's fault-finding flow chart is required.

ACTIVITY

During routine maintenance of a heating installation listen to the boiler sequence of operation when there is a call for heat or hot water. Each sound will determine which component is operating. Use the manufacturer's flowchart to follow and confirm the sequences. After an initial discussion with a customer, it is good practice to operate the appliance before carrying out maintenance.

Working on an electrical appliance

Before carrying out any work on an electrical appliance it is important to read the manufacturer's instructions. The customer or the person ordering the work can impart only so much information to the engineer about the operation of the appliance.

It is essential that preliminary safety checks are carried out before checking the components on an electrical appliance as shown in the section on safe isolation.

▲ Figure 5.123 An engineer referring to manufacturer's flow charts

Boiler thermostat

A boiler thermostat typically consists of a liquid filled phial which is inserted into a socket on the boiler. This liquid in this phial expands and contracts in relation to the ambient heat. The capillary wire is connected onto liquid bellows which will open or close a switch when the liquid expands or contracts. The liquid is usually ethanol based but it used to be mercury on older appliances.

A dead test can be used to check if a boiler thermostat switch is functioning. Ensure safe isolation and disconnect the boiler thermostat from the circuit. By using the Ω scale, a test can be carried out to check operation of the thermostat switch.

Carrying out a dead test step-by-step

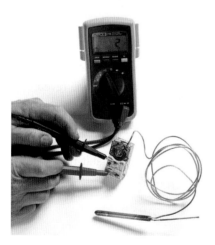

STEP 1 The switch is in the closed position as there is an Ω reading. Sometimes an audible buzzer is provided on meters to confirm the continuity of a circuit.

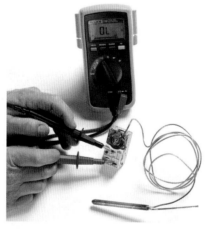

STEP 2 The switch is now in the open position as the meter is displaying the infinity (OL) sign.

By moving the thermostat dial and testing as shown above, the operation of the boiler thermostat switch can be verified.

STEP 3 The testing of an overheat thermostat can be carried out to check if the switch is operating correctly. Because there is an infinity reading (OL) it means that the switch is open and will require resetting to complete any circuit.

▲ Figure 5.124

A similar test can be carried out on an unvented cylinder temperature control which incorporates two phials, one for the standard operating thermostat and the other for the overheat thermostat.

The process of dead testing of a switch is shown in Figure 5.125.

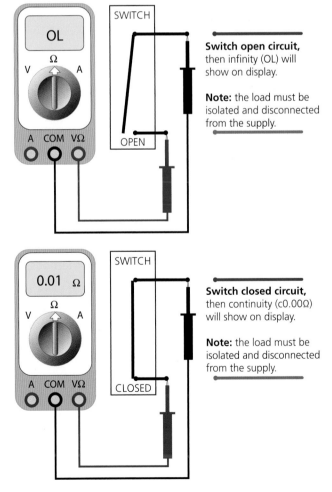

Switch open circuit, then infinity (OL) will show on display.

Note: the load must be isolated and disconnected from the supply.

Switch closed circuit, then continuity (c0.00Ω) will show on display.

Note: the load must be isolated and disconnected from the supply.

▲ Figure 5.125 Dead testing

Once the signal from the zone valve is received and the boiler thermostat calls for heat, the electrical signal is then relayed in conjunction with the control panel, to operate the fan.

The fan

As an advanced plumber you will invariably be involved in dealing with problems associated with appliance fans, and when qualified to do so you will need to be able to competently fault-find and repair them. In the illustration shown, the customer has reported that the boiler did not operate when they turned on the programmer. After carrying out safe electrical and fuel isolation, a basic visual inspection and referencing the manufacturer's flow chart, you could be led to the possible conclusion that the problem is a fan operational failure.

A continuity test while the appliance is safely isolated, known as a dead test, would confirm whether a wire or cable had integrity throughout is length. By using an ohmmeter or a multimeter, used on the ohms setting, the continuity of the wire can be tested. Often instruments have audible tone when used on the ohms setting and this is very useful when testing.

The fan is an essential part of the process of removal of products of combustion. If it does not operate because of poor electrical connections or motor failure, its operation will not be detected on the PCB and the ignition cycle will break down. When the fan operates the new movement of air is sensed by pulse tubes which are connected to a pressure operated switch. It is important that those hoses are properly connected and not kinked.

A continuity reading (closed circuit 0Ω) means that there is a short circuit and if the fan was connected it would blow a fuse.

When checking the performance of the fan motor, which is a resistive component, the meter should be set to the Ω scale. The test leads should then be connected across the live and neutral connections.

If there is a resistance recorded it would indicate that the component was operating satisfactorily.

When there is an open line (OL) infinity reading it means that the fan motor resistance winding is broken and therefore it will not turn the fan.

After confirming safe isolation, the switching function of an air pressure switch can be verified.

STEP 1 With appliance safely isolated remove the connection from the boiler circuit.

STEP 2 Resistance displayed on meter indicates that the fan motor is satisfactory.

STEP 3

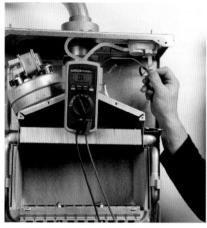

STEP 4

STEP 5

Air pressure switch

An air pressure switch serves a range of functions as its ON/OFF action sends signals to the PCB to indicate whether the fan is running or not. The action of the pressure switch can also prevent the boiler from firing if the heat exchanger primary flue way is blocked or obstructed. Therefore, it is essential it is tested as part of a boiler safety check.

This component consists of a diaphragm which responds to the pressure applied to its surface by the positive and negative movement of air generated by the fan via the pulse tubes. There the switch is fitted within the component which becomes activated when the diaphragm moves. This then sends a signal to the control panel to begin the ignition process.

The connections of the pulse tubes must be properly fitted and the electrical connections from the pressure operated switch should be checked for their condition and continuity.

There are two types of pressure switching methods, one which includes three wires and the other two. The process begins when the PCB checks the operation status of the switch to see if it is in the normally closed (NC) position which means that the fan is not running. A signal is then sent from the PCB to operate the fan and then it checks that the switch has moved to the normally open (NO) position and the boiler ignition process begins. The purpose of the sequence is to prevent the boiler from firing when the fan is not running.

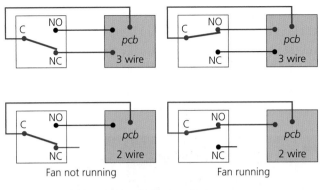

▲ Figure 5.126 Pressure switching methods

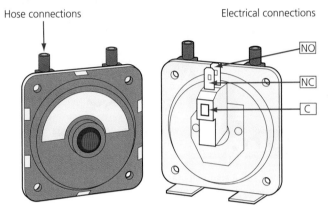

▲ Figure 5.127 Pressure switch, front and rear view

Sometimes engineers blow down one of the hoses and attach a multimeter across the terminals to test the operation of the switch. This is a logical approach but is not considered good practice as damage such as stretching the delicate diaphragm can occur. The pressure switch is only designed to deal with a limited range of pressures, typically 1 millibar. Some commercial pressure switches can be adjusted to perform in response to a wider range of pressures generated from a fan and this can be tested by attaching a monometer to the tube connection and then reading the pressure produced.

Electrical testing of the switch is therefore the correct way to proceed.

Sometimes live testing of a switch is necessary and the following drawings show the testing process.

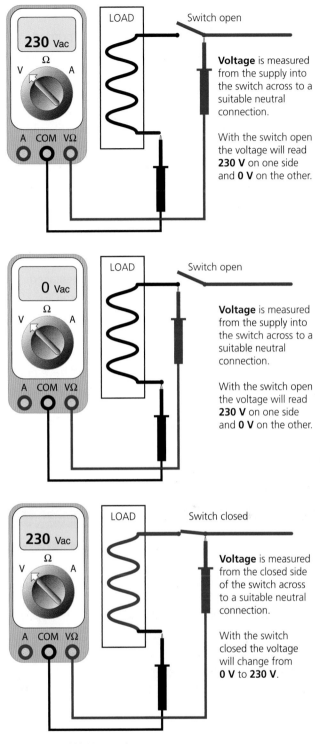

▲ Figure 5.128 Live testing

Printed circuit boards (PCB)

The control panel of a boiler is linked to the printed circuit board and is often a very expensive component. Therefore, great care should be taken when inspecting them. There are many components fitted to the PCB such as relays, rectifiers and transformers. Safe isolation is imperative and a visual inspection should precede any physical testing to check for obvious signs of distress such as scorching or burn marks. The receiving and sending of signals for components to operate a boiler mainly come from the PCB. Typical faults include loose connections and corrosion on the board. Sometimes the control panel can malfunction because of component failure on the board itself. This usually means that the whole unit must be replaced as modifications are not permitted. However PCBs are sometimes changed in error because the logical fault-finding sequence found in the manufacturer's instructions has not been followed and individual components remotely attached to the board have not been properly tested. Testing of the PCB is often limited to a continuity test on leads or removable fuses on the circuit board. Manufacturer's flow charts will help with diagnosing common problems on printed circuit boards and offer a parts list, with specific product reference numbers to enable accurate replacement.

▲ Figure 5.129 Printed circuit board

Ignition leads and probes

Once the control panel receives the signal from the pressure switch a spark generator on the printed circuit board will begin to send a high voltage signal, sometimes 15,000 V, through an insulated high tension lead to an electrode situated on the burner. Providing there is a gap of about 3–5 mm between the nearest metal components which is usually the gas pilot shroud, a spark will travel from the electrode to the shroud, then back to earth. Often the spark rate can be as high as eight sparks per second.

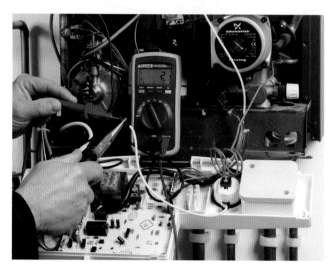

▲ Figure 5.130 A continuity test on a high tension lead from an electrode to the boiler PCB

KEY TERM

Continuity test: a test to ensure that a conductor has integrity along its whole length. For example it could be used to test a high tension lead to ensure its connection from the spark generator to the probe on the boiler burner bar is sound.

In this test the Ω range is used and there is a reading shown. Often multimeters have an audible device which indicates that there is continuity.

Problems and solutions

Sometimes a high tension lead can fail to send a high voltage signal to an electrode and the ignition sequence will fail because of this. The reason for such failure could include a poor connection on the printed board or at the electrode located on the burner manifold, where a part from a poor connection to the ceramic electrode casing caused itself to crack and arcing could occur at that point. Quite often a damaged or broken lead can fail to carry the signal along its whole length because it is arcing on another part of the boiler such as a damp area on the case.

A visual inspection of the lead and the connections can often reveal the problem. A continuity test to check the integrity of the lead can help diagnose the problem and a replacement lead can often provide a simple solution.

ACTIVITY

Inspect your test equipment to ensure that it is calibrated then carry out a simple continuity test while at work. Set your meter to Ω and see if your instrument has a continuity buzzer. Select a scrap piece of cable and apply the tips of your tester at each end. First of all take a Ω reading then try with the buzzer function. Once a system has been made electrically safe try the same test on either side of a fitted pump. If there are any brass push fitting connectors installed on the copper pipework, carry out the same test again and make a note of the readings.

▲ Figure 5.132 A solenoid on a gas valve

▲ Figure 5.133 A side view of a *Honeywell* multifunctional valve

A multimeter for example could be used to carry out a dead test on a solenoid with the meter set to read Ω.

▲ Figure 5.131 Spark ignition in action; the spark can be seen travelling from the tip of the electrode probe to the gas burner

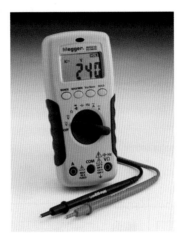
▲ Figure 5.134 A multimeter

Gas solenoid

Often there are two solenoids fitted on a multifunctional gas valve and during the ignition process which is generated from the control panel, a simultaneous signal is sent to open the pilot solenoid. When the pilot flame is ignited the conductance of the flame enables an electrical signal from the rectification device to travel back to the control panel, which in turn allows the main solenoid to open and the full flow of gas to travel to the burner and ignite. There are several variations on the ignition process and the one mentioned incorporates the intermittent pilot method.

A multifunctional gas valve will incorporate a solenoid valve, a flame supervision device and a pressure regulator. Once the solenoid is activated it will open and gas will flow to the burner. A solenoid can be tested when in the dead mode by checking the resistance on the coil to ensure it is operating properly.

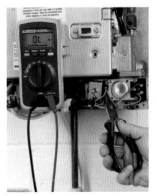
▲ Figure 5.135 Solenoid check

413

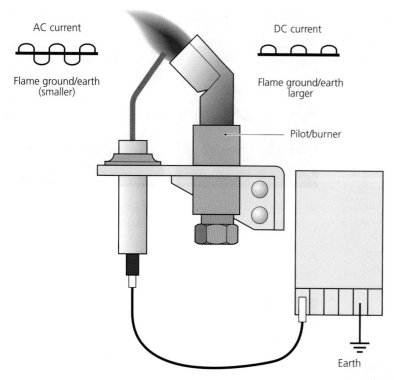

▲ Figure 5.136 Flame rectification

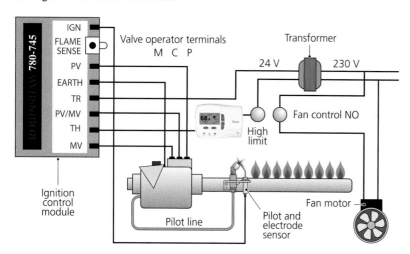

▲ Figure 5.137 Pilot ignition system

Flame rectification

Flame rectification is used as a flame supervision on many boiler designs. It works on the principle of utilising a flame as a conductor and passing a current through it. If there is a flame present then a signal can pass through it and back to a control box, which is when the boiler ignition sequence can begin. Therefore, no flame = no ignition.

The sequence of operation for the system shown in Figure 5.137:

- Thermostat calls for heat.
- An ignition controller sends spark to electrode to ignite gas.
- A controller utilises flame rectification to verify presence of the flame at pilot.
- An ignition controller sends signal to open gas valve solenoid.
- When room thermostat is satisfied the signal to the gas valve is terminated and the flame on the burner goes out.

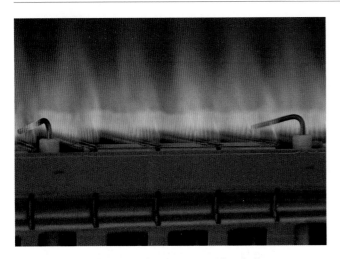

▲ Figure 5.138 Flame rectification on a modern boiler

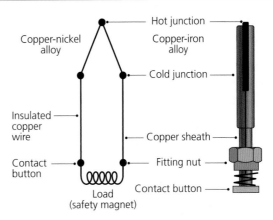

▲ Figure 5.139 Thermocouple

A thermocouple

A thermocouple is a component often found on a multifunctional gas valve and is located at the point where the pilot flame emerges from the outlet which is positioned next to a burner. From when the pilot flame initially envelopes the tip (12 mm area) of the thermocouple (hot junction), it takes about 20 seconds maximum to energise the contact button which is attached to the magnetic part of the flame supervision device, located on the gas valve. This sequence is known as the thermocouple pull in time.

Testing a thermocouple

The pilot flame produces a heat temperature of 650 ± 50°C at the hot junction. This in turn creates a small DC voltage across the hot and cold junction. The minimum voltage required to energise the magnet in the flame supervision device is **15 mV** DC, which means gas can flow via the solenoid to the burner when energised by a signal from the appliance thermostat. If the pilot flame is interrupted the magnet in the flame supervision device is de-energised and no gas can flow to the burner. The maximum time it takes to stop the flow of gas is 60 seconds maximum on domestic boilers and this is known as the drop out time.

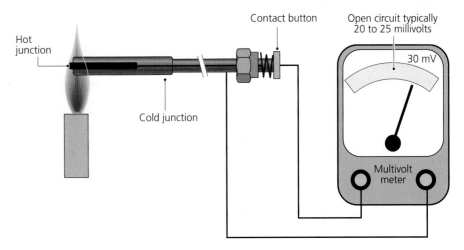

▲ Figure 5.140 Testing a thermocouple

Thermistors

Figure 5.142 shows an engineer testing a thermistor to see if the component is functioning. The space can be quite tight on a modern boiler but a well-designed appliance will always allow adequate access.

Many boilers use liquid expansion devices which respond to the temperature rise and fall of a boiler to operate control switches. The phial is located in a special pocket in a heat exchanger and is directly connected via a capillary tube to bellows which expand and contract to operate a calibrated electric thermostat switch which turns the gas solenoid on or off.

▲ Figure 5.141 Thermistors located on boiler pipework

▲ Figure 5.142 Testing thermistors in situation

A thermistor does a similar job to a thermostat but it works on a different principle. A thermistor is a variable resistor. The resistance of this component varies in relation to the rise and fall of temperature which makes it ideal for the efficient control of a modern boiler. Figure 5.141 shows the location of a thermistor on boiler pipework. It works in the following way: a low current of about 5 volts DC is sent from the PCB to the thermistor. This current then travels through the component and returns back to the PCB where the temperature of the resistance is measured. The resistance of the component relates to the temperature of the pipework where it is fitted. Depending on the amount of current that returns, the PCB can either

open or close the gas valve to increase or decrease the heat input into the appliance. The type of thermistors which are often found on combination boilers are called negative temperature co-efficient (NTC). The higher the temperature the lower the resistance that is created. By looking at a thermistor, it is impossible to know whether it is operating so a basic electrical test is carried out to confirm it is operating correctly.

To test this, use a multimeter or ohmmeter and set the instrument to the Ω scale. Carry out a dead test by attaching the probe to the two terminals of the component. There should be a reading showing resistance and by applying even a small amount of heat which can be generated from the fingers which are holding the thermistor, a change in the resistance reading should be evident. This indicates that it is working normally. If the reading remains constant then there is a possibility that it is defective and if there is an infinity reading this means that thermistor should be replaced.

Electrical resistance testing for a thermistor in this situation is usually at ambient room temperature, i.e. 25 °C, and at this temperature a common NTC may read 10,000 Ω.

▲ Figure 5.143 Resistances shown on meter

▲ Figure 5.144 OL on this test

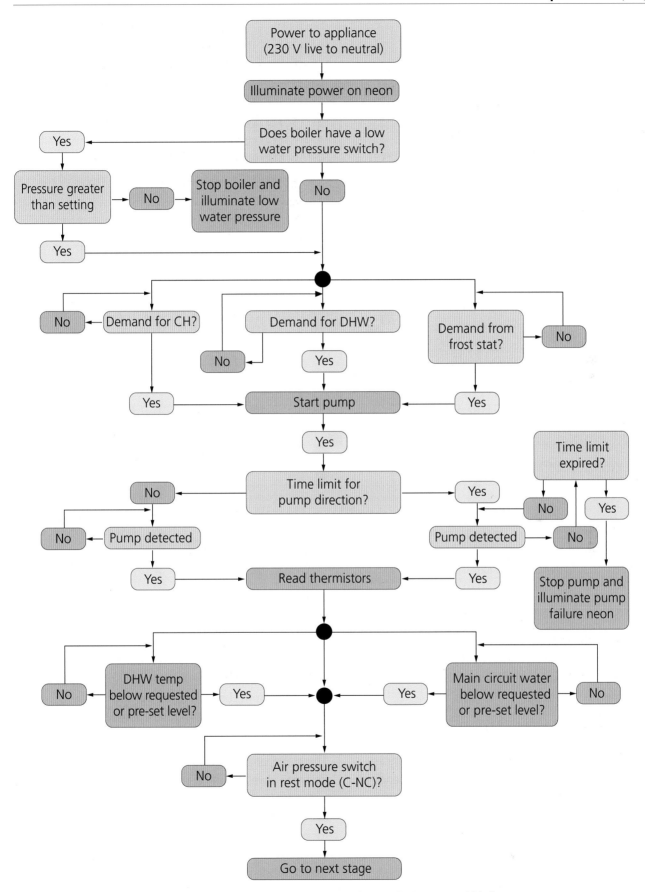

▲ Figure 5.145 A sample of a generic chart to show a logical sequence of typical faults on a combi boiler

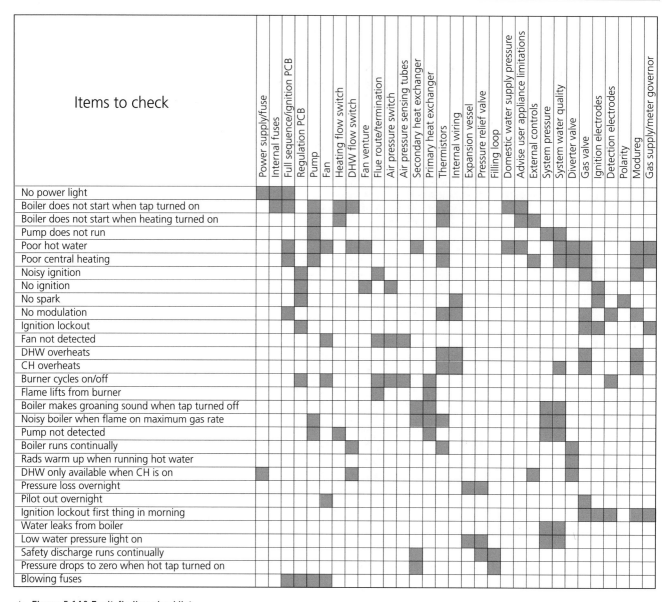

▲ Figure 5.146 Fault-finding checklist

When the chart in Figure 5.146 is used in conjunction with the manufacturer's flow chart then accurate fault-finding can take place and the correct component can be replaced or repaired.

The flow chart logic sequence is used for a combination boiler and will help an engineer identify any faults. This type of chart is commonly used on many appliances; logical sequences will be added or removed depending on the type of appliance.

New and alternative heat sources

The appliance in Figure 5.147 is a micro-combined heat and power unit and requires an electrical supply to energise the controls and pump which can be tested as part of a normal service by a suitably qualified plumber. In addition, it requires a connection to a special feed-in tariff (FiT) meter or smart meter to supply the energy generated from the stirling engine back into the electrical grid, which only a fully qualified electrician should undertake (PART P level A).

These appliances provide heating and hot water for a domestic dwelling and are controlled in the same way as a standard boiler but with the bonus of the production of about 3 kW of electrical energy. It is now quite common for new energy efficient appliances to incorporate more advanced electrical controlled systems, hence the need for the modern plumber to improve their knowledge and understanding of electrical installation, testing and fault finding.

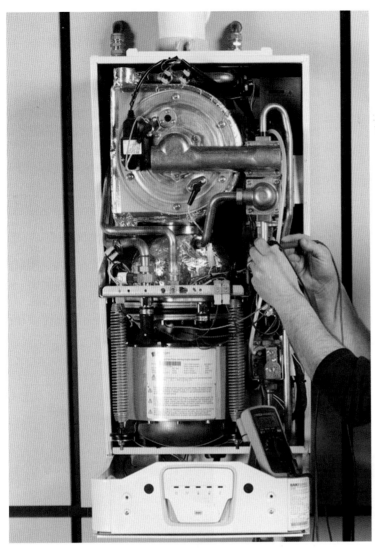

▲ Figure 5.147 There is a Stirling engine located beneath the heat exchanger

Other appliances such as biomass boilers are emerging on the market and even though some are manually fed with fuel they are electrically controlled in much the same way as an oil or gas boiler.

▲ Figure 5.148 Biomass boiler

Functional testing of an immersion heater installation using an insulation resistance tester

The **functional testing** of components typically includes the testing of an immersion heater which can fail due to the corrosive effects of electrolytic action. While in the operation mode and immersed in the water of the cylinder, such a failure can cause the RCBO on the radial circuit to trip out. These type of faults can be detected and diagnosed by the functional testing of components.

KEY TERM

Functional testing: a process carried out to check that components within an installation operate correctly. For example an immersion heater should be tested to ensure safe operation of heating element and thermostat.

▲ Figure 5.149 Electrical immersion damaged by corrosion

Immersion faults such as corrosion due to electrolytic action or work hardening will cause electrical failure and of course cannot be seen. This is why functional testing is important when examining the integrity of an immersed electrical element.

STEP 1 After safely isolating the supply to the immersion, remove the protective cover on the cylinder and proceed while referring to the manufacturer's instructions.

STEP 2 Carefully remove the phase (line) and neutral conductors from the immersion connection using an appropriate tool such as insulated long nose pliers. Always safely isolate before beginning any electrical testing.

STEP 3 Connect one test lead from an insulation resistance tester to a suitable earth point on the appliance.

→

STEP 4 Connect the next test lead to one of the element connections on the immersion.

STEP 5 Turn on the tester and set to 500 V.

STEP 6 Push the test button and hold for approximately three seconds.

STEP 7 There will be no fault if the reading shows OL (above 0.5 MΩ is also acceptable). Any reading below 0.5 MΩ should be investigated further.

When the test is finished turn off the tester and remove the leads.

An immersion removed from a cylinder was tested and the reading was below 0.5 MΩ.

▲ Figure 5.150 The reading on this immersion is 0.002 MΩ

▲ Figure 5.151 After a close visual inspection the fault shown above was found

Relays – switching and connections

Electrical relays are often found in heating components such as boilers and they are used for the remote control of switches and frequently for switching on a mains voltage circuit by means of a low voltage control. Both are comprised from an electromagnet and an armature which moves and in turn operates a switch. The armature is held away from the magnet by a spring when the current is off but when energised the armature is drawn onto the coil and the switch is then moved to the other position.

Many modern boilers use a solid state relay (SSR) or semiconductor switches, which have no moving parts and reduce the risk of sparks being created.

When a relay on a boiler receives a signal from a zone valve micro-switch the ignition sequence will then begin.

Isolation components for electrical appliances

Rocker switches

Sometimes there are very basic problems associated with a rocker plate (with or without CPC) with both single and double poles, which include incorrect polarity or loose connections. The rocking part of the switch can also fail. In addition, often when a switch is tightened onto a pattress that has not been fixed on a square surface then either part of the fitting can crack and this then creates the risk of exposed conductors.

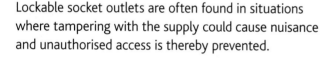

▲ Figure 5.152 Lockable socket outlet

▲ Figure 5.153 100 A main switch

Lockable socket outlets are often found in situations where tampering with the supply could cause nuisance and unauthorised access is thereby prevented.

Pull cord switches

Pull cord switches located on the ceiling are suitable for use within a bathroom zone. Their typical applications include electric showers and gas boiler installation within compartments that are located in restricted zones.

Common problems associated with pull cord switches are failing switching mechanisms and cord breakage. It is essential when installing one of these switches that sufficient slack is allowed in the length of the conductors for the components as any remedial or inspection work would prove difficult at a later date owing to the restriction of not being able to open the fitting fully. This is especially important when working on fittings located at a high level.

▲ Figure 5.154 Pull cord switches are typically used in bathrooms to avoid contact with electrical connections

INDUSTRY TIP

Where a pull cord switch fitted to a shower shows signs of burning insulation or terminals, it is likely that the wrong device has been installed and the pull cord is not correctly rated to switch the load correctly. Switches capable of switching the full load current must be installed, not isolators which are not switch rated.

Methods of correcting deficiencies in electrical components

Defective cables and broken connections

For example a plumber working on a gas appliance may find that the fan is not operating and as a result the ignition sequence is halted. The fan may not be operating because it is not receiving a signal from the boiler control box and this could be because there is a fault in that component but it could be because there is a broken conductor at the connection, which receives the signal to power the fan. The fan could be exposed to heat as well as movement and vibration when operating. Broken conductors could occur because of work hardening and even thermal movements or if there is a very tight radius on the cable or flex where it makes its connection.

▲ Figure 5.155 Electrical connections of a fan located on a gas boiler

The connections from the control box to junction box located at the fan are damaged. Although the white wire appears to be connected, it is in fact broken and therefore the fan cannot receive a signal to operate.

▲ Figure 5.156 The effects of a loose connection within a fused spur outlet

Although the cartridge fuse, found in some older consumer units could protect the circuit, it is possible to replace the wrong value cartridge fuse creating a dangerous situation and circuit protection could be compromised.

▲ Figure 5.157 Cartridge fuses of different ratings (1)

▲ Figure 5.158 Cartridge fuses of different ratings (2)

While safely isolating a system it is possible that re-wireable fuses can be discovered in older installations. They are slower to respond to overcurrent than MCBs as the weakest link is determined by the size of the fuse wire fitted. Although each fuse holder is colour coded and the amp limit visible, it is still possible to install the wrong fuse wire and put the safety of the circuit at risk.

A re-wireable fuse would be installed in an older consumer unit such as that shown in Figure 5.159. On some units MCBs can be installed instead or sometimes it is the engineer's decision to upgrade the consumer unit and only a fully qualified electrician should carry out this type of work.

Another example of broken connections in electrical systems is when an earth conductor is incorrectly used to carry a switched load as part of the control installation such as a *Honeywell* S- or Y-plan. Sometimes the wiring required for a room thermostat will require more than two conductors to operate the respective zone valves and often there is only PVC twin and earth available which is not suitable for this application and the earth conductor is incorrectly used. It has a smaller csa than the line and neutral conductor and although it could carry the load, it is bad practice and in any event is more susceptible to breaking because of over tightening. The loose sheath which should cover the exposed earth conductor can often hide the integrity and soundness of the connection.

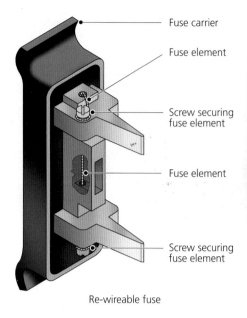

Fuse carrier

Fuse element

Screw securing fuse element

Fuse element

Screw securing fuse element

Re-wireable fuse

▲ Figure 5.159 Semi enclosed fuse to **BS 3036**

If electrical connections are properly fitted with the appropriate conductors in place, for example a power supply to an electric shower, and during the course of its operation one of those conductors is broken, it would follow that if this occurred to either the line or the neutral conductor the circuit would stop. If this happened to an earth conductor then a higher risk could be presented as the operation of the system protective devices such as MCBs could be compromised which could expose someone to a serious shock hazard.

Loose connections

High resistance faults often occur where conductors are connected in junction boxes or switches and connection itself is not tight. The reason for this could include poor workmanship, dirt or debris in the connection itself or the thermal movement of the conductor during its working cycle has resulted in it working loose. Vibration could also cause the loosening of conductors in a cold water booster pump installation during its working cycle or even acoustic transmissions from percussive water outlets creating water hammer. Good clipping and effective cable support will help reduce or eliminate these problems.
The effect of high resistance on a loose connection can lead to localised heating around a connection.

Loose connections where conductors are fitted cause high resistance and localised heat can occur. Extreme scorching or even fire can result when a high current circuit produces unwanted resistance at a loose connection resulting in a dissipation of hundreds of watts at that weak point. Components, cable and flex damage can also occur, which could result in the operation of the conductive protective device owing to the failure of the insulation, creating a short circuit. Fire is the worst outcome.

Another effect of loose connections can result in performance drop in different parts of the circuit with the outcome of appliance damage and even protective devices failing to operate.

Loose wiring caused this extreme effect on an unvented hot water cylinder immersion. There were also exposed conductors at the connector on the immersion heater.

Note: Once the thermostat was removed, the location of the fault is clear, with the damage originating from either a bad wiring connection to the thermostat and/or incorrect alignment of the thermostat to immersion element fasten connection. Both would cause a high resistance connection.

Inadequate earthing provision

Metal pipework on heating systems no longer requires bonding if the electrical system complies with current regulations. Older installations may have supplementary bonding. A typical example of supplementary bonding is shown in Figure 5.160.

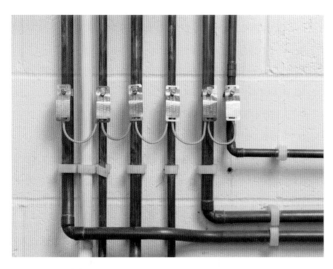

▲ Figure 5.160 6 mm² earth bonding conductor installed to pipework from a combination boiler on a central heating circuit

While working in installation it is possible to come across installations with no pipework earth bonding and exposed conductors. A qualified electrician can advise on the bonding required.

▲ Figure 5.161 Poor installation practice

Figure 5.162 shows an unorthodox and inadequate way to apply earth bonding to a gas pipe at a meter. The earth conductor shown is 10 mm² while the earth conductor on a 2.5 mm² twin and earth cable which supplies a socket outlet is circa 1.5 mm².

▲ Figure 5.162 Unorthodox and inadequate way to apply earth bonding to a gas pipe at a meter

In any event equipotential bonding at a gas meter should be connected not more than 600 mm from the outlet pipework and before the first tee of the metal pipework – whichever is closer.

SUMMARY

Electrical test equipment should become a standard part of the modern fully qualified plumber's tool kit as new technologies and renewable energies emerge and develop. Continued professional development will help to create a world class workforce capable of providing excellence in installation, servicing and fault finding within the diverse and evolving craft of plumbing. However, never carry out any electrical work which is beyond your level of competency and always leave installation work which requires certification to an electrician.

Test your knowledge

1 How much current would flow in a circuit having 10 Ω resistance and a 230 V supply?

   a 2.3 A

   b 23 A

   c 230 A

   d 2300 A

2 How much current would be drawn by a 3 kW, 230 V water heater?

   a 0.07 A

   b 0.76 A

   c 1.30 A

   d 13.0 A

3 Which non-statutory document give 'Requirements for Electrical Installations'?

   a GS38

   b BS 7671

   c HSWA

   d EWR

4 What does RCD stand for?

   a Reacting circuit device

   b Resistive current deactivator

   c Residual current device

   d Remote circuit deactivator

5 What is a common rating of circuit breaker used to protect a 230 V, 3 kW immersion heater radial circuit?

   a 3 A

   b 13 A

   c 16 A

   d 32 A

6 What class of electrical power tools require a connection to the earth of a socket outlet?

   a Class I

   b Class II

   c Class III

   d Class 0

7 What is a proving unit used for when undertaking safe and secure isolation of an electrical supply?

   a It proves the circuit is no longer live.

   b It proves the voltage indicator works.

   c It proves the circuit breaker is locked.

   d It proves the voltage is disconnected.

8 What wording should be on the label attached to a bonding clamp?

   a Live part: do not touch

   b Live connection: do not remove

   c Safety electrical connection: do not remove

   d Safety bonding connection: do not connect

9 What are the identification colours of the two conductors connected to the micro-switch within a zone valve?

   a Orange and grey

   b Blue and brown

   c Blue and orange

   d Brown and grey

10 What device is used to monitor ambient air temperatures in a heating system?

   a Cylinder thermostat

   b Zone valve

   c Room thermostat

   d Circulating pump

11 What colour conductors are classed as live?

   a Green

   b Yellow

   c Blue and brown

   d Green and yellow

12 One conductor has a voltage of 100 V and another has a voltage of 400 V.

What reading would a voltmeter give if the meter probes were connected to each conductor?

  a 100 V
  b 200 V
  c 300 V
  d 400 V

13 What is the frequency of the electrical supply in the UK?

  a 20 Hz
  b 50 Hz
  c 60 Hz
  d 100 Hz

14 What is the value of IP protection that is regarded as finger protection?

  a IP2X
  b IPX2
  c IP4X
  d IPX4

15 What type of drawing shows the location of electrical accessories in a building?

  a Plan drawing
  b Wiring diagram
  c Circuit diagram
  d Schematic drawing

16 What rating of circuit breaker should be used to protect a 230 V, 9 kW shower circuit?

  a 20 A
  b 32 A
  c 40 A
  d 50 A

17 What is the maximum fuse rating that can be used in a fused connection unit?

  a 1 A
  b 3 A
  c 5 A
  d 13 A

18 What is the purpose of a proving unit when undertaking the safe isolation procedure?

  a To check the lock is fixed and secure
  b To ensure the supply to the circuit is dead
  c To enable the use of multiple padlocks
  d To verify the function of the voltage indicator

19 Describe the safe isolation process.

20 Describe the three parts of a cable.

21 Explain what is meant by IP54 when considering the ingress protection of electrical equipment.

22 Describe a non-electrical method of testing if a zone valve operates correctly.

23 Explain what is meant by 'flame rectification'.

24 Explain the purpose of a split-way consumer unit.

25 Describe situations where it would not be appropriate to bond a metal water pipe using a main protective bonding conductor.

26 Describe the general requirements of HSE document GS38.

27 Describe what a motorised zone valve does in a central heating control system.

Answers can be found online at www.hoddereducation.co.uk/construction.

CHAPTER 6

ADVANCED ENVIRONMENTAL TECHNOLOGY SYSTEMS

Wherever you seem to look these days there are environmental technologies being installed and used commercially and in domestic properties. Governments and house builders are under pressure to include a variety of systems to help cut back carbon emissions that are contributing to the greenhouse effect, and therefore climate change. As plumbers we are looking to install systems that offer practical and sustainable alternatives to fossil fuels, systems that potentially offer heat and power to the property.

This is a growing market area that is here to stay. Plumbers need to have a good working understanding along with specialist training in these areas to offer the customer suitable 'green technology' solutions within a domestic property.

By the end of this chapter, you will have knowledge and understanding of the following:

- the legislation relating to micro-renewable energy
- the operating principles of micro-renewable energy and water conservation technology
- the installation requirements of micro-renewable energy and water conservation technologies
- the benefits and limitations of micro-renewable energy and water conservation technologies.

All of these learning outcomes will be addressed for each environmental technology system in turn as follows:

- solar thermal (hot water)
- ground source heat pumps
- air source heat pumps
- biomass systems
- combined heat and power (CHP)
- rainwater harvesting
- grey water reuse
- solar photovoltaic
- micro-wind
- micro-hydro.

1 THE LEGISLATION RELATING TO MICRO-RENEWABLE ENERGY AND WATER CONSERVATION TECHNOLOGIES

Any environmental technology system comes under a number of Acts, Building Regulations and British Standards. These are in place so that manufacturers and installers make sure the systems are safe, work efficiently and are maintained to the correct standards. Building Regulations form part of the law in the UK so must be followed, whereas British Standards offer more practical guidance on how to conform with the regulations.

Town and Country Planning Act 1990

Before any work that increases the size of a property, changes the use of a property or sometimes installing items externally you will need to apply for **planning permission**.

However certain work can be carried out to a property that does not require planning permission, but this is quite often concerning minor or internal work. This is known as permitted development and typically comes with some criteria that must be met. Special care needs to be taken over properties that are 'listed' buildings. So, it is always worth checking with the local authority prior to any work starting.

Building Regulations

The Climate Change and Sustainable Energy Act 2006 bought micro-renewable energy systems under the requirements of the Building Regulations. Even if planning permission is not required because the installation comes under the 'permitted development', there are still requirements to comply with the relevant Building Regulations as shown below.

Building Regulations are a **statutory** requirement and therefore must be followed. Each regulation is a relatively brief document and are currently divided into 14 sections, each accompanied by an Approved Document which offers guidance on complying with the regulation.

▼ Table 6.1 The 14 parts of the Building Regulations

England and Wales		Northern Ireland
A	Structure	D
B	Fire safety	E
C	Site preparation and resistance to contaminants and moisture	C
D	Toxic substances	None
E	Resistance to the passage of sound	G
F	Ventilation	K
G	Sanitation, hot water safety and water efficiency	P
H	Drainage and waste disposal	J & N
J	Combustion appliances and fuel storage systems	L
K	Protection for falling, collision and impact	H
L	Conservation of fuel and power	F1 & F2
M	Access to and use of buildings	R
N	Glazing – safety in relation to impact, opening and cleaning	V
P	Electrical safety	None

Compliance with the Building Regulations is required when installing renewable technologies, but not all will be applicable and different installations will have to comply with different Building Regulations. These are indicated in each section.

It should be noted that due to devolution of government in the UK, each country's government takes responsibility for Building Regulations, so there are differences between each country.

▼ Table 6.2 Building Regulations in England and Wales

Status	Source
Primary legislation	Building Act 1984
Secondary legislation	Building Regulations 2010
Guidance	Approved codes of practice

▼ Table 6.3 Building Regulations in Scotland

Status	Source
Primary legislation	Building (Scotland) Act 1984
Secondary legislation	Building (Scotland) Regulations 2004 (amended 2009)
Guidance	Technical guide books

▼ Table 6.4 Building Regulations in Northern Ireland

Status	Source
Primary legislation	Building Regulations (Northern Ireland) 1979 (amended 2009)
Secondary legislation	The Building Regulations (Northern Ireland) 2012
Guidance	Technical booklets

2 HEAT-PRODUCING MICRO-RENEWABLE ENERGY

Micro-renewable energy systems have become more popular in recent times in an effort to cut domestic carbon footprints. If energy can be produced without having to burn fossil fuels and discharge carbon dioxide into the atmosphere, it will only help to slow down global warming. Renewable energy systems tend to be installed into new build properties or properties undergoing a full renovation; this is due to the disruption that is caused whilst installation takes place.

Solar thermal (hot water) systems

Solar thermal hot water systems use **solar radiation** to heat water, either directly or indirectly. Key components of these systems include the following:

- solar collector
- differential temperature controller
- circulating pump
- hot water storage cylinder
- auxiliary heat source.

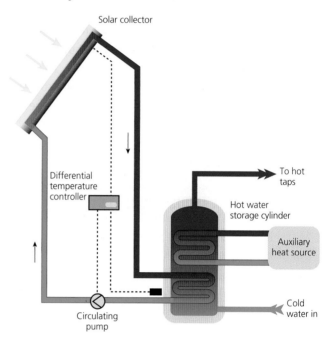

▲ Figure 6.1 Solar thermal system components

Solar collector

A solar thermal collector is designed to collect heat by absorbing heat radiation from the Sun. The heat energy from the Sun heats the heat transfer fluid contained in the system.

There are two types of solar thermal collectors that heat the collection fluid slightly differently.

- **Flat plate collectors** – these are less efficient, but also cheaper. The heat transfer fluid circulates around the collectors and is directly heated by the Sun. The collectors need to be well insulated to avoid heat loss.
- **Evacuated tube collector** – an evacuated tube is more efficient, but also more expensive. It consists of a specially coated, pressure-resistant, double-walled glass tube. The air is evacuated from the tube to aid the transfer of heat from the Sun to the pipe housed inside the tube. The pipe contains a heat-sensitive

liquid which vaporises. The warmed gas rises in the tube and the heat is focused on a small heat exchanger, which is plugged into the header tube. The header tube contains the heat transfer liquid which absorbs the heat from the evacuated tube. As the gas cools so the heat sensitive liquid condenses and runs down to the bottom of the evacuated tube, ready for the process to start again. The collectors must be mounted at a suitable angle to allow the vapour to rise and condense.

Differential temperature controller

The differential temperature controller (DTC) has sensors connected to the solar collector (high level) and the hot water cylinder (low level). It monitors the temperature at both points of the system. The DTC switches the circulating pump on when there is enough solar energy available and there is a demand for water to be heated. Once the hot water cylinder reaches the required temperature the DTC switches the circulating pump off.

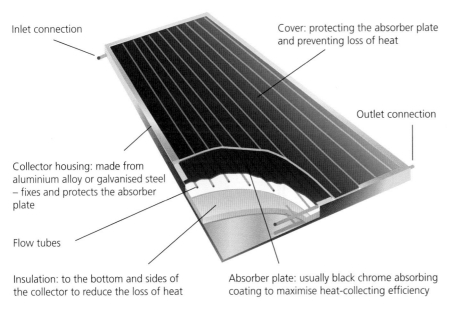

Inlet connection

Cover: protecting the absorber plate and preventing loss of heat

Outlet connection

Collector housing: made from aluminium alloy or galvanised steel – fixes and protects the absorber plate

Flow tubes

Insulation: to the bottom and sides of the collector to reduce the loss of heat

Absorber plate: usually black chrome absorbing coating to maximise heat-collecting efficiency

▲ Figure 6.2 Cutaway diagram of a flat plate collector

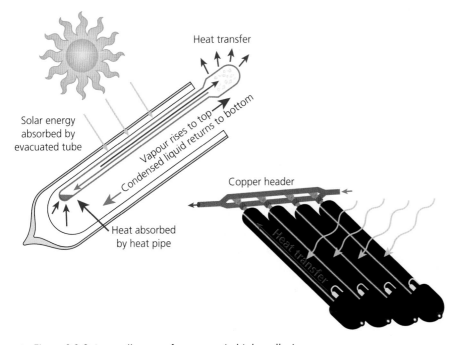

Heat transfer

Solar energy absorbed by evacuated tube

Vapour rises to top

Condensed liquid returns to bottom

Copper header

Heat absorbed by heat pipe

Heat transfer

▲ Figure 6.3 Cutaway diagram of an evacuated tube collector

Circulating pump

This is controlled by the DTC and circulates the system's heat transfer liquid around the solar hot water circuit. The circuit is a closed loop system between the solar collector and the hot water cylinder coil. The heat transfer fluid is normally water based and would normally contain **glycol** so that at night or during cold periods the system will not freeze. The circulating pump should operate only when there is either solar energy available or when there is demand for water to be heated.

> **KEY TERM**
>
> **Glycol:** a liquid anti-freeze which is odourless and colourless in its raw state.

> **INDUSTRY TIP**
>
> Always read the manufacturer's instructions regarding the percentage of glycol that is required in the system. If they are not followed the efficiency, performance and longevity of the system can be affected.

Hot water storage cylinder

This enables the transfer of heat from the solar collector circuit to the stored water. There are different configurations for the hot water cylinder:

- **Twin or multi-coil cylinder** – has at least two heat exchanger coils. The lower (primary) coil is for the solar heating circuit and the higher (secondary) coil for the auxiliary heating circuit. Cold water enters the base of the cylinder and is heated by the coils. If the solar coil cannot meet the demand for hot water then the boiler will heat the secondary coil.
- **Alternative use of one cylinder as a solar pre-heat cylinder** – the output which feeds a primary hot water cylinder.

The two arrangements that have been described are indirect systems, with the solar heating circuit forming a closed loop.

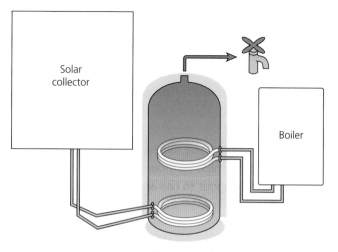

▲ Figure 6.4 Twin coil cylinder

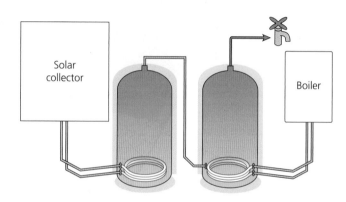

▲ Figure 6.5 Using two separate cylinders

Direct system

This is an alternative to the indirect system, where the domestic hot water that is stored in the cylinder is directly circulated through the solar collectors and is the same water that can be drawn off at the outlets. As this water is being drawn off, glycol cannot be used in the system, so these units can only be used where low temperatures will not affect the system.

Auxiliary heat source

Generally speaking, in the UK there will be times in the year when there will be insufficient solar energy available to provide adequate hot water to the property. On these occasions an auxiliary heat source will be required; this will normally be from a boiler. If a boiler is not present in the property, the use of an electric immersion heater is acceptable.

Location of solar thermal systems and building requirements

When deciding to install a solar thermal hot water system and whether a particular property is suitable it is important to consider several factors:

- **Orientation of the solar collectors** – the best direction for a solar collector to face is due south. However, as the Sun rises in the east and sets in the west, any location that has a roof facing east, south or west is suitable for mounting these collector panels on, although the efficiency of the system is reduced for any system not facing due south.
- **Tilt angle of the solar collectors** – at different times of the year, the elevation or height of the Sun relative to the horizon does vary – summer time being the highest and winter time being the lowest. Ideally, solar collectors should always be **perpendicular** to the path of the Sun's rays. It is generally not practical to change the tilt angle of the solar collector. A compromise angle has to be used. In the UK the angle is 35°. However, collectors will work, but less efficiently, from a vertical to horizontal position.
- **Shading of the solar collectors** – when positioning collectors, you must be aware of any structure, tree, chimney, aerial or other objects that could stand between the collector and the Sun's rays. This could mean a visit to the property at different times of the day to note any potential shadows. The Sun shines for a limited time each day, so any reduction in the amount of heat energy reaching the collector will reduce its ability to provide hot water to meet the demand.
- **The suitability of the structure for mounting the solar panel** – the structure, quite often the roof area, will need to be surveyed for the chosen mounting system for the collectors. Consideration will need to be given to the strength and condition of the structure and its suitability to the fixing type. The effect of wind and location must also be taken into consideration. The force exerted by the wind on the collectors can be quite noticeable which will affect the collectors and the fixings.
  If the installation is on a property which is jointly owned, such as flats, maisonettes and semi-detached, the ownership of the structure will need to be carefully considered and discussed with the customer.

The space or area needed to mount the collectors will be dependent on the demand for hot water. The number of outlets and people living in the property determines the demand.

- **Compatibility with the existing hot water system** – solar thermal systems provide stored hot water and do not provide instant hot water:
  - Properties using an under the sink or over the sink water heater and electric showers will not be suitable for a solar thermal system.
  - Properties using a combination boiler to provide hot water will not be suitable for a solar thermal system, unless substantial alterations are made to the existing system.

VALUES AND BEHAVIOURS

Always read the manufacturer's instructions concerning the installation factors for their system and discuss these with the customer to ensure they are clear on the proper operation of their system.

KEY TERM

Perpendicular: at an angle of 90° to a certain plane. In other words, the Sun's rays need to be at 90° to the collector where possible for maximum efficiency.

▼ Table 6.5 The effect of shading on collectors

Shading	% of sky blocked by obstacles	Reduction in output
Heavy	Up to 80%	50%
Significant	60–80%	35%
Modest	20–60%	20%
Little	Up to 20%	No real reduction

IMPROVE YOUR ENGLISH

A customer lives in a semi-detached property where the front of the house is north-east facing. It is very close to the coast line and has a roof angle of 45°. It is an older style property with slate tiles on the roof. Outline the discussion points that need to be raised with the customer.

Planning permission required for solar thermal systems

Permitted development applies where a solar thermal system is installed:

- on a property or block of flats
- on a building within the grounds of a property or block of flats
- as a standalone system in the grounds of a property or block of flats.

However, there are criteria to be met in each case.

- For a building mounted system:
  - the solar system must not protrude more than 200 mm above a wall or roof line
  - the solar system must not protrude above the highest point of the roof line (ridge), excluding the chimney.
- For a standalone system:
  - only one standalone system is allowed in the grounds
  - it must not exceed 4.0 m in height
  - it must not be installed within 5.0 m of the property boundary
  - it must not be more than 9.0 m² in area
  - no dimension can exceed 3.0 m in length.
- For both standalone and building mounted systems:
  - the system cannot be installed in the grounds of or be installed on a listed building or monument
  - if the system is to be installed within a conservation area or world heritage site then the system cannot be installed closer to a highway than the property is located.

In every other case planning permission will be required.

INDUSTRY TIP

It is always worth asking the local authority about the restrictions for these installations in your area.

ACTIVITY

Consider why is it important to consider the 'uplift forces' that wind can create before an installation takes place.

Compliance with Building Regulations

The Building Regulations that apply to solar thermal installations are listed in Table 6.6.

Other regulatory requirements applicable for solar thermal systems

- **BS 7671** The IET Wiring Regulations – this will cover the installation and maintenance of controls and circulators.
- Approved Document Part G3 Unvented hot water systems – this will cover the installation of the hot water cylinder and the temperature at which hot water must be stored.
- Water Supply (Water Fittings) Regulations 1999 – this will mainly cover the prevention of contamination where wholesome water is concerned.

▼ Table 6.6 Building Regulations applicable for solar thermal heating systems

Building Regulation	Title	Relevance
A	Structure	These systems put extra load forces on the structure, particularly the roof structure. Not only downward load from the weight but also uplift from the wind must be considered.
B	Fire safety	Where holes for pipes are made, this could reduce the fire resistance of the building fabric.
C	Resistance to contaminants and moisture	Where holes for pipes and fixings for collectors are made, this may reduce the moisture resistance of the building and allow the ingress of water (rainwater).
G	Sanitation, hot water safety and water efficiency	Hot water safety and water efficiency of the system installed.
L	Conservation of fuel and power	Energy efficiency of the system and the building it is installed in.
P	Electrical safety	The installation and testing of electrical controls and components.

Benefits and limitations of solar thermal systems

There are both advantages and disadvantages to using solar thermal systems. Advantages include:

- These systems reduce CO_2 emissions into the environment.
- They reduce energy costs for the customer.
- They are low-maintenance systems.
- They improve the energy rating of the building.

Disadvantages are as follows:

- The system may not be compatible with the existing hot water system.
- The system may not meet the demand for hot water in the winter period.
- There are high initial installation costs.
- The system requires an auxiliary heat source.

IMPROVE YOUR ENGLISH

Having an understanding of the advantages and disadvantages of available systems is necessary in order to adequately communicate these and offer professional advice to the customer so they can make an informed decision. Remember, you will need to be mindful of the technical terms you use which the customer may be unfamiliar with.

ACTIVITY

Use the internet to research the differences between a pressurised and a fully filled solar system.

Heat pumps

Heat pumps are now commonly being installed on to new build properties to increase the energy rating of the property. This again aligns with the reduction of the domestic carbon footprint. The pumps produce heat that can be effectively used within the property, which means less gas and electricity will be demanded by the property.

There are two main types of heat pump:

- ground source heat pump (GSHP)
- air source heat pump (ASHP).

They either extract the heat energy from the air or the ground which is replaced by the action of the Sun.

General working principle

A water pump moves water from a lower level to a higher level through the application of energy. As the name suggests, a heat pump moves heat from one location to another by the application of energy – in most cases electrical energy. Heat energy from the Sun exists in the air that surrounds us and in the ground below us – heat pumps extract this energy.

At absolute zero or 0 K (kelvin), there is no heat at all. This temperature equates to –273 °C, so even with an outside temperature of –10 °C there is a vast amount of free heat energy available.

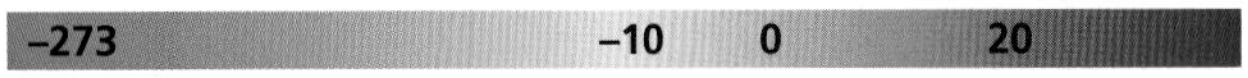

−273		−10	0		20

▲ Figure 6.6 Heat energy exists down to absolute zero or 0 K

Using a relatively small amount of energy that is stored in the air or the ground, this energy can be used to heat up our living accommodation, especially if linked to an underfloor heating system.

The general rule that must be understood with heat transfer is that heat moves from the warmer spaces to the cooler spaces.

A heat pump contains a **refrigerant**. The external air or ground is the heat source that gives up the heat energy. When the refrigerant is passed through the heat source it is cooler than the surroundings so it absorbs the heat.

KEY TERM

Refrigerant: a substance or mixture, usually a fluid, used in a heat pump and refrigeration cycle. In most cycles it undergoes phase transitions from a liquid to a gas and back again.

The compressor on the heat pump then compresses the refrigerant causing the gas to heat up. When the refrigerant is passed into the interior of the heat pump it is now hotter than the surroundings and gives up its heat to the cooler surrounding. The refrigerant is then allowed to expand, where it once again turns into a liquid; as it expands so it cools down and the cycle starts over again.

ACTIVITY

At home, inspect your fridge carefully and note how the tubes inside the fridge at the back are cold, but the tubes on the outside of the fridge at the back are hot.

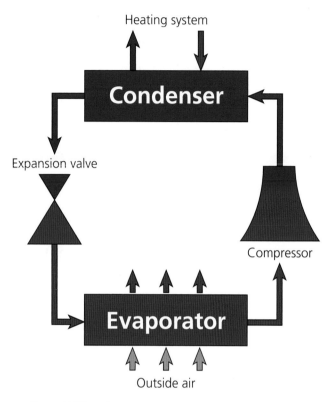

▲ Figure 6.7 The refrigeration process

The only energy that is required to drive the system is what is required for the compressor. The greater the difference in temperature between the refrigerant and the heat source (ground or air) the greater the efficiency of the heat pump.

KEY POINT

The greater the difference in temperature between the refrigerant and the heat source (ground or air) the greater the efficiency of the heat pump.

If the temperature difference between the heat source and the refrigerant is small, then the compressor will have to work much harder, making the system less efficient.

It is not uncommon for heat pumps to have efficiencies up to 300 per cent. For example, for an electrical input of 3 kW, the heat output should be around 9 kW.

IMPROVE YOUR MATHS

For an electrical input of 2.2 kW, what should be the expected return in heat output be?

To compare the heat savings of other appliances, see Figures 6.8–6.10.

Electricity input 1 kW  Heat output 1 kW (100% efficiency)

▲ Figure 6.8 The efficiency of an electrical panel heater

Gas input 1 kW  Heat output 0.95 kW (95% efficiency)

▲ Figure 6.9 The efficiency of an A rated condensing gas boiler

Electricity input 1kW (+ 2 kW free heat extracted from air) 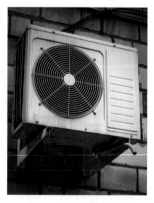 Heat output 3 kW (equates to 300% efficiency)

▲ Figure 6.10 The efficiency of an air source heat pump

The efficiency of a heat pump is measured in terms of the coefficient of performance (CoP). The coefficient of performance (CoP) is a ratio between the heat delivered and the power input of the compressor:

$$\text{CoP} = \frac{\text{Heat delivered}}{\text{Compressor power}}$$

> **KEY POINT**
> The higher the CoP value, the greater the efficiency.

The higher CoP values are generally achieved in the milder weather, because in the colder weather the compressor has to work harder to extract the heat.

Storing excess heat produced

Heat pumps are not able to provide instant heat, so therefore work better when run continuously. Start-stop operations will shorten the lifespan of the heat pump. A **buffer tank** is incorporated into the system so that when heat is not required in the system, the heat pump can 'dump' the heat into the vessel and thus keep running. When heat is needed, it can be drawn from the buffer tank. These buffer tanks can be used with both ground source and air source heat pumps.

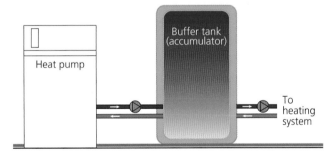

▲ Figure 6.11 Storing heat in a buffer tank

> **KEY TERM**
>
> **Buffer tank**: a large white storage vessel used for storing excess hot water until demanded.

Ground source heat pumps

A ground source heat pump (GSHP) extracts low temperature free heat from the ground, magnifying it to a higher temperature and then releasing it when required for space heating and water heating.

Key components of GSHPs include:
- heat collection loop and circulating pump
- heat pump
- heating system.

The collection of heat from the ground is accomplished by means of a pipe run containing a mixture of water and antifreeze, which is buried in the ground. This type of system is known as a 'closed loop system' There are three methods of burying the pipe used to collect the heat, each having advantages and disadvantages:
- horizontal loops
- vertical loops
- slinkies.

Horizontal loops

Piping is installed in horizontal trenches that are about 1.5–2.0 m deep. Horizontal loops require more piping than vertical loops (around 200 m of pipe for an average house), although a proper calculation will be required for every installation. This installation will require a large area to sink the pipework in.

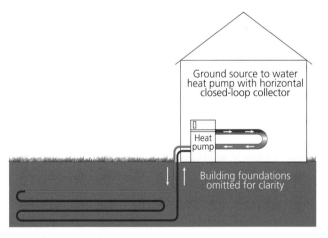

▲ Figure 6.12 Horizontal ground loops

Vertical loops

Most commercial installations use vertical loops. Holes are bored to a depth of 15–60 m deep, depending on soil conditions. Pipes are then inserted into the boreholes. This style of system needs far less land area, but the deeper the hole the greater the pump force required to circulate the system fluid. Access will need to be available for the drilling rigs.

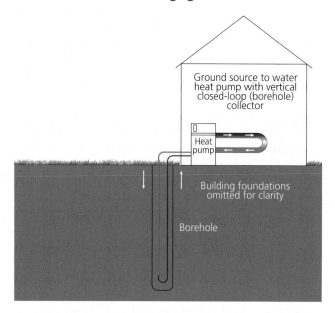

▲ Figure 6.13 Vertical ground loops

Slinkies

These coils are flattened, overlapping coils that are spread out in shallow trenches and then buried. They are able to concentrate the heat transfer in to a smaller area of land, which reduces the trench size required. A 10.0 m slinky trench will produce around 1 kW of heating load.

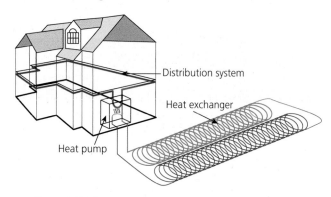

▲ Figure 6.14 Slinkies

▲ Figure 6.15 Slinkies installed in a trench

INDUSTRY TIP

It is obvious from Figure 6.15 that digging a trench creates a major disruption to existing properties, so they are best installed during the original build.

The water and antifreeze fluid is circulated around the pipework sunk in the ground by means of a circulating pump. The low-grade heat from the ground is passed over the heat exchanger which transfers the heat to the refrigerant gas. The refrigerant gas is then compressed (temperature increased) after which it passes over a second heat exchanger where the heat is transferred to the heating loop that feeds the radiators or underfloor heating.

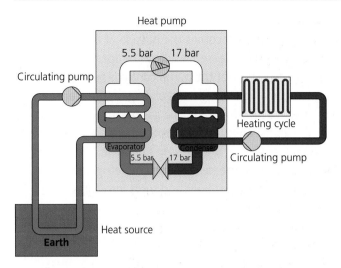

Heat pump

5.5 bar   17 bar

Circulating pump

Heating cycle

Evaporator   Condenser

5.5 bar   17 bar   Circulating pump

Heat source

Earth

▲ Figure 6.16 Ground source heat pump operating principle

The final heat output from a GSHP is lower than that of a standard gas boiler. A GSHP output temperature would normally be around 40 °C, whereas a gas boiler output temperature would be around 70–80 °C. This makes a GSHP very suitable to be installed on an underfloor heating system.

By reversing the refrigeration process, a GSHP could be used to provide cooling during the summer period.

Location and building requirements
The building that a GSHP supplies heat to will need to be well insulated; this will allow the GSHP to work effectively due to the low output temperatures.

Provision for machinery to access the property to either dig a trench or drill the borehole needs to be a major consideration and an important discussion point with the customer if the property is occupied.

Planning permission
The installation of a ground source heat pump is usually considered to be a permitted development and therefore will not require a planning application to be made. It is always worth checking with the local authority to find out if they have any criteria that need to be adhered to.

Compliance with Building Regulations
The Building Regulations that apply to the installation of a GSHP are listed in Table 6.7.

Other regulatory requirements applicable for GSHPs
- **BS 7671** The IET Wiring Regulations – this will cover the installation and maintenance of controls and circulators.
- F gas (fluorinated) requirements if working on refrigeration pipework – these are the refrigerant gases used in GSHPs which come under strict laws.

▼ Table 6.7 Building Regulations applicable for GSHPs

Building Regulation	Title	Relevance
A	Structure	Where heat pumps and components put additional load on the building and where holes are drilled to allow pipework to pass from outside to inside the property
B	Fire safety	Where holes for pipework may reduce the fire-resistant integrity of the building
C	Resistance to contaminants and moisture	Where holes for pipework may reduce the moisture resistant integrity of the building
E	Resistance to the passage of sound	Where holes for pipework may reduce the soundproof integrity of the building or cause a nuisance to nearby buildings
G	Sanitation, hot water safety and water efficiency	Hot water safety and water efficiency within the system
L	Conservation of fuel and power	Energy efficiency of the system and the building
P	Electrical safety	The installation and testing of electrical controls and components in the system

Benefits and limitations of GSHPs

As with solar thermal systems, there are both benefits and drawbacks to using GSHPs. Advantages include:

- high efficiency
- reduction in energy bills
- reduction in CO_2 emissions
- safer appliance as there is no combustion
- lower maintenance costs compared to combustion appliances
- long life span
- no requirement for any fuel storage
- offer cooling in the summer period
- more efficient than an air source heat pump.

Disadvantages are as follows:

- initial high installation costs
- potential large land areas required
- design and installation are complex
- unlikely to work efficiently with an existing heating system
- refrigerants could be harmful to the environment
- more expensive to install than an air source heat pump.

Air source heat pumps

An air source heat pump (ASHP) extracts the free heat from low temperature air and releases it where required for space heating and water heating.

▲ Figure 6.17 Air source heat pump

Key components include:

- heat pump containing a heat exchanger, compressor and expansion valve
- heating system.

An ASHP works in a similar way to a refrigerator, but the cooled area becomes the outside of the property and the heat is released inside the property.

- The pipework of the pump contains refrigerant that can be a liquid or gas depending on the stage of the cycle. The refrigerant as a gas passes through the heat exchanger (evaporator), where the low outside temperature is drawn across the heat exchanger using a fan. The heat warms the refrigerant – the liquid refrigerant boils and turns into a gas.
- The warmed refrigerant vapour flows to the compressor, where it is compressed, causing its temperature to rise.
- Following the compressor, the refrigerant passes through a second heat exchanger (condenser), where the refrigerant loses its heat to the heating system water, because at this stage the refrigerant is hotter than the system water. The heating system carries the heat away to heat the building.
- The cooled refrigerant has started to turn back into a liquid as it passes the expansion valve, where its temperature drops suddenly and completely turns into a liquid. It flows back to the evaporator heat exchanger to start the cycle once again.

There are two common types of ASHP in use today:

- Air-to-water – the type described above, which can provide both space heating and water heating.
- Air-to-air – this type is only suitable for space heating.

The final heat output from an ASHP is lower than that of a standard gas boiler. Ideally, an ASHP should be used in conjunction with an underfloor heating system or low temperature radiators.

Location and building requirements for ASHPs

When deciding whether an ASHP is suitable you must consider the following:

- The property must be well insulated.
- There must be space on the ground or on an outside wall to mount the unit.
- There will need to be space around the unit to allow good air flow.
- The ideal system to link an ASHP up to is an underfloor heating system.

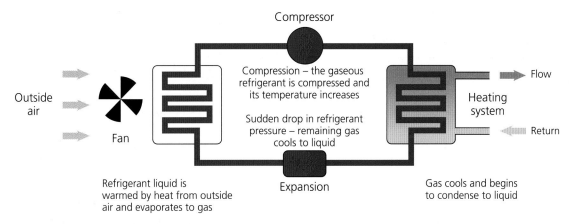

Compressor

Compression – the gaseous refrigerant is compressed and its temperature increases

Sudden drop in refrigerant pressure – remaining gas cools to liquid

Outside air

Fan

Flow

Heating system

Return

Refrigerant liquid is warmed by heat from outside air and evaporates to gas

Expansion

Gas cools and begins to condense to liquid

▲ Figure 6.18 Air source heat pump operating principle

- The payback period for an ASHP is shorter if it replaces an electric, coal or oil system, rather than a gas boiler.

> **INDUSTRY TIP**
>
> ASHPs are ideal for new builds where the insulative value of the property is greater and the likelihood of an underfloor heating system being installed is greater.

Planning permission required for ASHPs

A permitted development applies to an ASHP when:
- it is installed on a house or block of flats
- it is installed on a building within the grounds of a house or block of flats
- it is installed within the grounds of a house or block of flats.

There are however criteria to be met, mainly due to noise levels:
- The ASHP must comply with Microgeneration Certification Scheme (MCS) planning standards.
- Only one ASHP can be installed on the building or in the grounds.
- The volume of the outdoor compressor must not exceed 0.6 m³.

- The ASHP cannot be installed within 1.0 m of the property boundary.
- The ASHP cannot be installed on a pitched roof.
- If the unit is installed on a flat roof, it must not be within 1.0 m of the edge of the flat roof.
- If the unit faces a highway, it cannot be mounted above ground level.
- The unit cannot be installed on a designated monument.
- The unit cannot be installed on or in the grounds of a listed building.
- An ASHP cannot be installed within a conservation area or a world heritage site.

> **ACTIVITY**
>
> Use the internet to look up and detail the Microgeneration Certification Scheme (MCS) planning standards and what they outline by visiting: www.microgenerationcertification.org

> **INDUSTRY TIP**
>
> Older houses in the UK are generally not well insulated. Those with solid walls obviously cannot have cavity insulation, but internal insulation could be used. Loft insulation is important and there are government grants available.

Compliance with Building Regulations

▼ Table 6.8 Building Regulations applicable for ASHPs

Building Regulation	Title	Relevance
A	Structure	Where heat pumps and components put additional load on the roof and walls of the property
B	Fire safety	Where holes for pipework may reduce the fire-resistant integrity of the building
C	Resistance to contaminants and moisture	Where holes for pipework may reduce the moisture resistant integrity of the building
E	Resistance to the passage of sound	Where holes for pipework may reduce the soundproof integrity of the building or cause a nuisance to nearby buildings
G	Sanitation, hot water safety and water efficiency	Hot water safety and water efficiency within the system
L	Conservation of fuel and power	Energy efficiency of the system and the building
P	Electrical safety	The installation and testing of electrical controls and components in the system

Other regulations applicable for ASHPs

- **BS 7671** The IET Wiring Regulations – this will cover the installation and maintenance of controls and circulators.
- F gas (fluorinated) requirements if working on refrigeration pipework – these are the refrigerant gases used in ASHPs which come under strict laws.

Benefits and limitations of ASHPs

Again, there are both advantages and disadvantages to using ASHP systems. Advantages include:

- high efficiency
- reduction in energy bills
- reduction in CO_2 emissions
- the units are safe as no combustion takes place
- low maintenance compared to combustion appliances
- no requirement for fuel storage
- provide cooling in the summer period
- cheaper and easier to install than a GSHP.

Disadvantages are as follows:

- less efficient with existing heating systems
- less efficient than a GSHP
- high initial costs
- less efficient in winter
- the fans are noisy
- need to incorporate a defrost cycle to stop the heat exchanger freezing in winter.

Biomass systems

Biomass systems have actually been used for many years. Biomass is biological material that can be burnt and used to create heat. Biomass material can come from animal matter or plants that have recently been sourced, whereas fossil fuels have taken millions of years to source.

Both fossil fuels and biomass have to be burnt to produce heat which is used within the system, and both produce carbon dioxide. This is a greenhouse gas that is linked with global warming. The difference is that biomass material absorbs carbon dioxide when it grows, reducing the current carbon dioxide levels in the atmosphere. When burnt, the carbon dioxide is released into the atmosphere, the net result being no overall increase in carbon dioxide levels – carbon neutral.

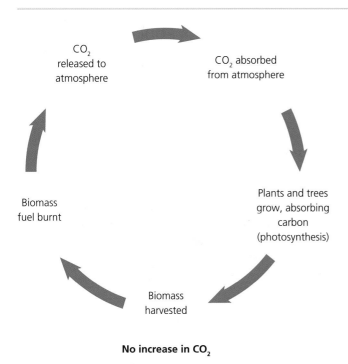

CO$_2$ released to atmosphere

CO$_2$ absorbed from atmosphere

Plants and trees grow, absorbing carbon (photosynthesis)

Biomass fuel burnt

Biomass harvested

No increase in CO$_2$

▲ Figure 6.19 The carbon cycle

Biomass fuel does not achieve the same heat output as fossil fuels, as fossil fuels are far more dense. However, biomass, if managed correctly, is sustainable, whereas fossil fuels are not.

Materials that can be turned into biomass:

- wood
- crops, such as elephant grass, reed grass and oil seed rape
- by-products – agricultural by-products like straw, grain husks and animal waste
- food waste – an estimated 35 per cent of food in the UK is wasted
- industrial waste.

> **KEY POINT**
>
> These materials require a large storage facility, which has cost implications.

Wood-related biomass material is the most popular and primary domestic biomass fuel. Wood is a sustainable material and trees that are managed correctly can be relatively fast growing. These can be managed on a 3–5-year crop rotation.

Woody biomass material is supplied in three different forms:

- small logs
- wood chips – mechanically shredded trees and branches
- wood pellets – formed from sawdust and shavings that are compressed into pellets.

The calorific value of woody biomass material is generally low. The greener (wetter) the wood, the lower the calorific value.

> **INDUSTRY TIP**
>
> Calorific value is the energy given off by burning. The drier the wood, the higher the value, and therefore the hotter the flame.

A woody biomass boiler can be as simple as a wood burner heating a room, which is manually fed wood by opening the doors; to an automated boiler which is fed pellets constantly. The feed to an automated boiler is done either by an auger drive (cork screw) or by suction. The boiler is monitored very carefully by the use of thermostats which balance the feed of biomass material, the air intake and the fan speed.

Location and building requirements

Whenever installing a biomass boiler, the following considerations must be made:

- space for the storage of biomass material
- good access for biomass deliveries
- biomass boiler may not be permitted in smokeless zones.

> **ACTIVITY**
>
> A barn conversion is to have a biomass boiler installed. The property is remote and is accessed via narrow country lanes. The deliveries are made by heavy goods vehicles. Outline the considerations that will need to be made before the installation takes place.

In the past it has been commonplace to burn coal and wood as a domestic heat source, but this caused many cities to have poor air quality and created smogs (fog and smoke). There were public outcries regarding air quality, so in 1956 the Clean Air Act was formed. It was subsequently replaced by the Clean Air Act 1993, which made it illegal to sell or burn unauthorised fuel in a smoke-controlled area, unless it was burnt in an 'exempt appliance'.

INDUSTRY TIP

Exempt appliances are able to burn smoky fuel without emitting smoke into the atmosphere.

Planning permission

Planning permission is not normally required to install a biomass boiler in a domestic property, if all the work is internal. If the installation requires external flues to be installed, it will normally be classed as permitted development as long as the flue is to the rear or side elevation and is not higher than 1.0 m above the roof ridge.

If the property is a listed building, checks will need to be made with the local authority for both internal and external work.

If the property is within a conservation area or in a world heritage site, checks will need to be made with the local authority for internal work, external work and any additional construction for storage.

Compliance with Building Regulations

The Building Regulations that apply to a biomass installation are given in Table 6.9.

Other regulations applicable for biomass systems

- The Water Supply (Water Fittings) Regulations 1999 apply. The main area of concern is the avoidance of cross contamination between system water and wholesome water; this is known as backflow prevention.
- **BS 7671** will apply to the installation's electrical supplies and controls.

Benefits and limitations of biomass systems

The advantages of installing a biomass system include:
- It is a carbon neutral system.
- It is a sustainable fuel source.
- When burnt the waste gases are low in nitrous oxide, with no sulphur dioxide.

INDUSTRY TIP

Nitrous oxide and sulphur dioxide are both greenhouse gases which are linked to global warming.

The disadvantages of installing a biomass system include:
- Transportation of the biomass fuel is expensive.
- Storage space is required for the fuel. (Pellets and chips are bulk delivered.)
- Control of heat is not instant (as it would be for a gas boiler). The fuel cannot be instantly removed to stop combustion.
- It requires a suitable flue system.

▼ Table 6.9 Building Regulations applicable for biomass installations

Building Regulation	Title	Relevance
A	Structure	Where components affect the load placed on the structure of the building or excavations are close to the building
B	Fire safety	Where holes for pipework reduce the fire-resisting integrity of the structure
C	Resistance to contaminants and moisture	Where holes for pipework reduce the moisture-resisting integrity of the structure
E	Resistance to the passage of sound	Where holes for pipework reduce the sound-resisting integrity of the structure
G	Sanitation, hot water safety and water efficiency	Water efficiency
L	Conservation of fuel and power	The efficiency of the appliance and system including insulation
P	Electrical safety	Installation of supply and control wiring for the system

Micro-combined heat and power systems

These systems are installed so that the fuel source used to supply heat and hot water to the property is also used to supply electricity at the same time. Generally, the fuel source is natural gas or LPG, but could also be biomass. The diagrams illustrate an old inefficient boiler, a new condensing boiler and a micro-combined heat and power (mCHP) unit.

Figure 6.20 shows that the old inefficient boiler is only 65 per cent efficient, as 35 per cent of the heat is lost out of the flue. The condensing boiler is a lot more efficient at 95 per cent. The mCHP unit will achieve the same efficiencies as the condensing boiler but in a different way. 80 per cent of the heat is used to provide heat and 15 per cent is used to generate power.

This type of system is known as 'heat-led', as the primary function (80 per cent) is to provide heat to the property, whilst the power generation is secondary (15 per cent). The unit will only produce electricity if there is a demand for heat. Most mCHP units can produce between 1 kW and 1.5 kW of electricity. These units are known as a carbon-reducing technology rather than a carbon-free technology.

Working principle of a mCHP

CHP units have been around for several years now, but more recently domestic versions have become available. The domestic versions are gas fired and use a Stirling engine to produce electricity.

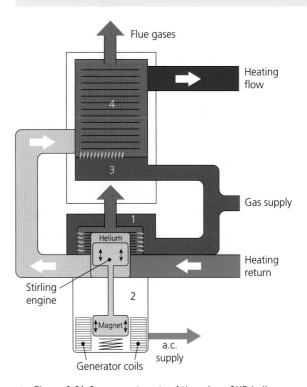

▲ Figure 6.21 Component parts of the micro-CHP boiler

Key components of the mCHP boiler:
- the engine burner
- the Stirling engine generator
- the supplementary burner
- the heat exchanger.

When there is a call for heat, the engine burner fires and starts the Stirling generator. The engine burner produces about 25 per cent of the full heat output.

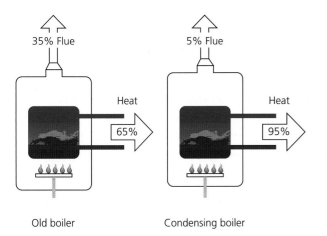
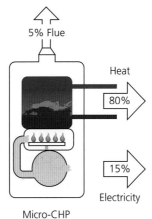

▲ Figure 6.20 The efficiency of different boilers

The burner preheats the heating system return water before it passes on to the main heat exchanger. Hot flue gases from the engine burner are passed across the heat exchanger to heat the heating system water even more. If there is more demand than supply, then the supplementary burner operates.

The Stirling engine within the unit uses the expansion and contraction of internal gases to operate the piston. The gases do not leave the engine and no explosive combustion takes place. Helium is often used as the internal gas. When the engine burner fires, the helium expands, forcing the piston downwards. The return water from the heating system passes across the engine causing the gas to contract. A spring mechanism in the engine returns the piston to the stop and the process starts again.

INDUSTRY TIP

When considering an mCHP installation it is important to consider the purchase costs, but also the running and maintenance costs of these systems.

Location and building requirements

To make an mCHP system viable the following criteria should be met:

- The building should have a high demand for space heating. The larger the building the greater the carbon savings.
- A building with good insulation is not normally suitable, as a well-insulated building is unlikely to have a high demand for space heating.

If these criteria are not met the system will be inefficient.

Planning permission

The same planning permissions for mCHP apply as with the biomass boiler discussed above.

Compliance with Building Regulations

The same Building Regulations will apply to mCHP systems as to a biomass installation as discussed above.

Other regulations applicable to mCHP

- The Gas Safety (Installation and Use) Regulations will apply to these installations, so a Gas Safe engineer must carry out the installation of the gas supply.

- The Water Supply (Water Fittings) Regulations will need to be referenced to avoid any contamination of wholesome water.
- **BS 7671** The IET Wiring Regulations will apply to the electrical supply and controls.
- G83 requirements will have to be followed when connecting the generator to the system.
- Microgeneration Certification Scheme will need to be referred to when electricity is generated.

Benefits and limitations of mCHP

The advantages of installing an mCHP system include:

- It has the ability to produce electricity and not be dependent on building direction or weather conditions.
- The system generates electricity while there is demand for heat.
- A feed in tariff is available (but there are limitations).
- It saves carbon over centrally generated electricity.
- It reduces the building's carbon footprint.

The disadvantages of installing an mCHP system include:

- The initial installation costs are high.
- It is not suitable for all properties (low demand for heat/well insulated).
- It has limited capacity for electrical generation.

3 WATER CONSERVATION TECHNOLOGIES

The population of the UK (and the world) is expanding and thus the demands on the water supply systems are increasing. In the UK, unlike many other countries, the water supplied to properties is suitable for drinking straight from the tap – potable water.

We use this water not only for drinking, but also for washing, bathing and watering our gardens.

Even in the UK, fresh water is a limited resource. Alongside the increase in demand, there is a great deal of pressure on this vital resource. Water conservation is one way of making sure demand does not exceed supply and shortages are avoided.

The two methods of water conservation that are commonly used in the UK are:

- rainwater harvesting
- grey water reuse.

The Code for Sustainable Homes sets a target for reducing average drinking water consumption from 150 litres per person per day to an optimum of 80 litres.

ACTIVITY

Access 'The Code for Sustainable Homes' and list the areas being highlighted for new builds. www.theccc.org.uk/wp-content/uploads/2019/02/UK-housing-Fit-for-the-future-CCC-2019.pdf

At present the adopted target is 103 litres; Part G sets the level at 125 l. Whichever target is used, the conclusion is that a reduction in consumption from 150 litres is vital.

VALUES AND BEHAVIOURS

Given the demands for water and the problems water extraction from rivers and lakes causes, it is essential that better use is made of the available water. More houses are being built and the people living in those houses are putting a demand on the water supply. This increased demand affects water levels in rivers and reservoirs and therefore puts pressure on the natural environment around them. Customers are becoming well informed about water conservation, so it is important to be able to discuss these technologies with them.

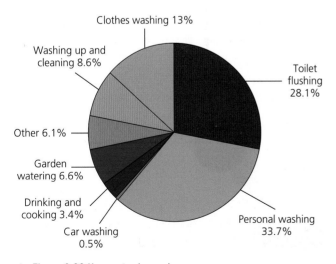

▲ Figure 6.22 How water is used

Looking at how water is used, you can see that on average the amount of water used for drinking or food preparation (potable) is estimated to be around 3.4 per cent of the total consumption. This outlines that there are obvious opportunities for water savings.

▼ Table 6.10 Types of water

Terminology	Meaning
Wholesome, potable water	Water that is suitable for human consumption. The water supplied from the local water authority to properties in the UK is known as WHOLESOME or POTABLE
Rainwater	Water captured from the roof line of a property when it rains, flowing into the gutter and downpipe to be collected
Grey water	Waste water from wash basins, showers, baths and washing machines
Blackwater	Sewage

Rainwater harvesting

Rainwater harvesting refers to the process of capturing and storing rainwater from the surface it falls on, rather than letting it run off into the drains. The harvesting of rainwater can result in sizeable reductions in wholesome water usage. Instead of refilling the WC with wholesome water, the WC could be refilled with rainwater. Every time a WC is fully flushed it uses 6 litres of water. Water authorities claim 1 litre of tap water costs around 1p. Over a period of time this adds up. This offers financial savings for the customer and carbon reductions in the environment.

If harvested water is correctly filtered, stored and used regularly, it will not remain in the storage tank for excessive periods of time. It can be used for:

- flushing toilets
- car washing
- garden watering
- supplying washing machines.

Harvested rainwater CANNOT be used for:

- drinking water
- washing dishes
- food preparation
- personal hygiene (baths and showers).

Rainwater is classified as fluid category 5, which is the highest risk category (see Book 1, Chapter 5 for coverage of this topic).

The process of rainwater harvesting step-by-step

The harvesting of rainwater follows the process of:

1 collection
2 filtration
3 storage
4 reuse.

Step 1: collection of rainwater (harvesting)

Rainwater can be collected from the roof line or from hard standing surfaces. From the roof line it flows off the tiles into the gutter and down the downpipe towards the storage tank. The amount of water harvested is governed by:

● the size of the roof area (or hard standing)
● annual rainfall for the area.

The roof material must be taken in to account when installing a harvesting system. Roofs covered in copper, lead, asbestos or bitumen may not be suitable to have a rainwater harvesting system installed due to the potential health risk. Also, rainwater that is harvested from hard standing, such as a driveway, could potentially be contaminated with oil.

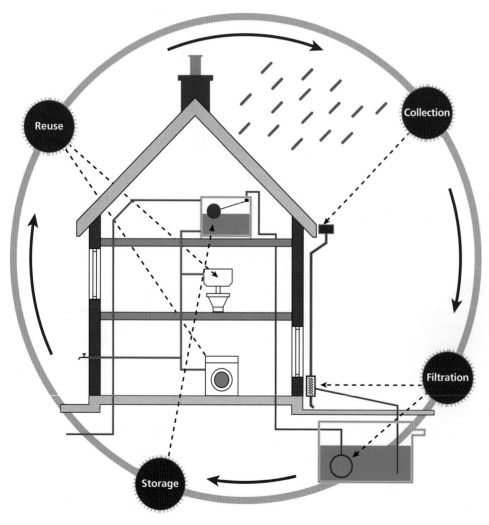

▲ Figure 6.23 The rainwater harvesting cycle

Step 2: filtration

As the rainwater passes down the system towards the storage tank, it passes through an in-line filter to remove debris such as leaves; this is known as a mechanical filter. These are flushed out. The efficiency of the filter will determine how much of the captured water ends up in storage. There may well be microscopic or even carbon filters to remove suspended particles. Manufacturers will normally expect over 90 per cent of captured water to enter the storage tank through the filters.

Step 3: storage

The storage tanks can be installed above ground or below ground and can vary in size from a small 'next to the house' tank to a large buried tank holding thousands of litres. Below-the-ground tanks will need excavation work carried out, which is disruptive, whilst above-the-ground tanks will need an area of ground to be sited. Whichever tank is used, it will need to be protected against freezing in winter time, heating up in summer which allows bacteria to grow, and direct sunlight which allows algae to grow. It will also need protection against contamination.

The size of the tank will be determined by the rainwater available and the demand for the harvested water. Different manufacturers offer various calculators to size the tank effectively.

A submersible pump is used to transport the water from the storage tank to the point of demand. The tank will also incorporate an overflow pipe which leads to the drainage system.

> ### INDUSTRY TIP
>
> If the property has a garden, a suggestion could be made to the customer to install a basic water butt fed from a downpipe.

Step 4: reuse of rainwater

Once the rainwater is stored in the tank, there are two types of system options available for the distribution of this reuse water within the property:

- **Indirect distribution** – in this system the reuse water is pumped to a storage cistern (header tank) at high level within the property, quite often in the loft. This in turn feeds the outlets under low pressure within the property.

The storage cistern in the loft MUST incorporate a backflow prevention air gap to meet with Water Regulations.

The arrangement of water control and overflow will ensure the air gap is maintained. The rainwater level control connects to the main control unit that operates the submersible pump, so that water is drawn from the main storage tank when required. At times, mainly in the summer period, when there is less rainfall, fresh water is introduced into the system via a wholesome water inlet controlled by means of a float operated valve.

- **Direct distribution** – in this system the reuse rainwater is pumped directly to the outlets on demand. At times of low rainwater, the control unit will introduce wholesome water to supply the outlets. The backflow prevention to meet the Water Regulations is incorporated in the control unit. This type of system uses more energy than the indirect distribution, but does not require the storage space.

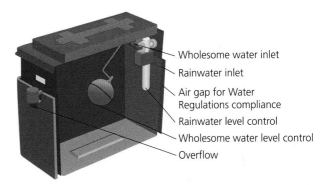

Wholesome water inlet
Rainwater inlet
Air gap for Water Regulations compliance
Rainwater level control
Wholesome water level control
Overflow

▲ Figure 6.24 Header tank with backflow protection

> ### KEY POINT
>
> Remember, the term backflow refers to the flow of water in a direction it is not intended to flow. Refer back to Book 1, Chapter 8.

Location and building requirements

When deciding on the suitability of a location for the installation, the following points will need to be considered:

- if there is enough rainfall in the area to meet the demand of the property
- if a suitable supply of wholesome water will be required for back up
- above ground tanks must avoid the risk of freezing, warming and sunlight
- if below ground, tanks need excavation work.

IMPROVE YOUR MATHS

Use the internet to locate current data about the average annual rainfall in your area along with the rainfall intensity.

For example, the Met Office provides maps of average annual rainfall across the UK: www.metoffice.gov.uk/binaries/content/assets/metofficegovuk/pdf/research/library-and-archive/library/publications/factsheets/factsheet_4-climate-of-the-british-isles.pdf

Using your findings, work out approximately how much rain could be collected from a 15 m² roof area.

Planning permission

Generally, planning permission is not required for the installation of a rainwater harvesting system if it does not alter the outside appearance of the property. However, it is always worth consulting with the local authority, especially if the system uses an above-the-ground tank or the property is listed.

Compliance with Building Regulations

The regulations outlined in Table 6.11 will apply.

Other regulations applicable for rainwater harvesting

The Water Supply (Water Fittings) Regulations 1999 apply to rainwater harvesting systems. The key area of concern is the potential cross contamination between rainwater and wholesome water. This is known as backflow prevention.

KEY POINT
Rainwater is classified a category 5 risk – severe.

The usual method of preventing backflow between rainwater and wholesome water is by the use of an AA air gap.

ACTIVITY
Refer to the Water Supply (Water Fittings) Regulations 1999 and use internet research to identify different types of air gaps.

Any pipework used to supply outlets with rainwater will need to be labelled to identify it from wholesome water. The outlets will also need to be labelled 'unsuitable for drinking' as displayed in Figure 6.25.

▲ Figure 6.25 Outlet label

▼ Table 6.11 Building Regulations applicable for rainwater harvesting systems

Building Regulation	Title	Relevance
A	Structure	Where the components affect the loadings placed on the structure or excavations are close to the building
B	Fire safety	Where holes for pipework reduce the fire-resisting integrity of the structure
C	Resistance to contaminants and moisture	Where holes for pipework reduce the moisture-resisting integrity of the structure
E	Resistance to the passage of sound	Where holes for pipework reduce the sound-resisting integrity of the structure
G	Sanitation, hot water safety and water efficiency	Water efficiency
H	Drainage and waste disposal	Where gutter and rainwater pipes are connected to the system
P	Electrical safety	Installation of supply and control wiring for electrical components

You must also adhere to the following regulations for rainwater harvesting:

- **BS 7671** will apply to the electrical wiring and controls – this will cover the installation and maintenance of controls and circulators.
- **BS 8515:2009** is the Code of Practice for rainwater harvesting systems – this outlines key design, installation and maintenance criteria for these systems.

Benefits and limitations of rainwater harvesting systems

The advantages of using rainwater harvesting systems include:

- a reduction in the use of wholesome water
- a reduction in water bills
- the water does not require any treatment prior to use
- rainwater harvesting is less complicated than grey water reuse.

Disadvantages are as follows:

- The quantity of water is limited to the collection area.
- The quantity of water is limited to the rainfall in the area.
- Initial installation costs are high.
- A water meter should be fitted.

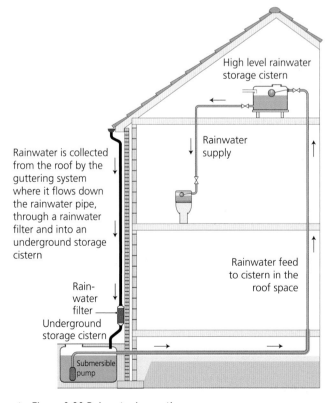

High level rainwater storage cistern

Rainwater supply

Rainwater is collected from the roof by the guttering system where it flows down the rainwater pipe, through a rainwater filter and into an underground storage cistern

Rainwater feed to cistern in the roof space

Rain-water filter

Underground storage cistern

Submersible pump

▲ Figure 6.26 Rainwater harvesting

Grey water reuse

Grey water is the waste water from baths, showers, basins and washing machines. It gets its name from the cloudy grey appearance of the waste water. Capturing and reusing this water for permitted uses, like flushing a WC, reduces the consumption of wholesome water.

> **KEY POINT**
> Whilst the waste from a kitchen sink can be classed as grey water, the waste from a kitchen sink is not often used because of the fats, oils, greases and food particles that it will contain.

The grey water collected will have the potential of being contaminated with human intestinal bacteria and viruses, skin particles and hair. It will also contain soap, detergents and cosmetic particles which could cause bacterial growth. Add to this the warm temperature of the grey water waste, it is the ideal conditions to encourage the growth of bacteria.

For these reasons grey water cannot be stored for more than a few hours. The less-polluted water from basins, showers and baths is favoured for grey water collection. Where greater supply of grey water is required, washing machine water can be collected.

> **HEALTH AND SAFETY**
> The capacity for storing grey water, a category 5 (high risk) water, is limited to a few hours given the pathogen content and potential for rapid growth of bacteria. Remember, some sources of grey water, such as basins and showers, are safer to harvest from than others, such as washing machines.

Grey water is classified as category 5 water, which is the high-risk category, under the Water Regulations. Grey water can pose a severe health risk due to the pathogen content. Untreated grey water deteriorates rapidly when stored, so all systems that store grey water will incorporate the appropriate level of treatment.

If grey water is filtered and stored correctly then it can be used for:

- flushing toilets
- car washing
- garden watering
- washing clothes (after additional treatment).

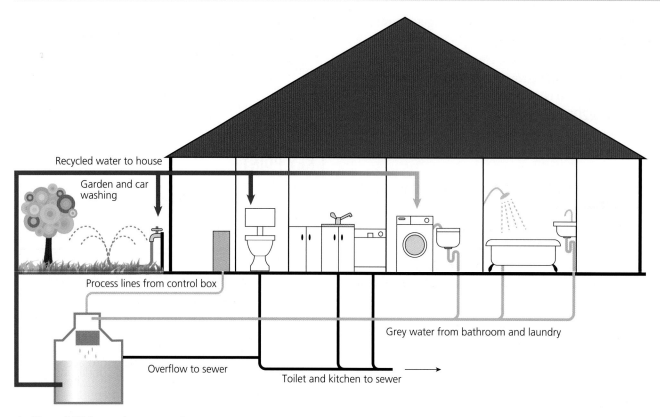

Recycled water to house

Garden and car washing

Process lines from control box

Grey water from bathroom and laundry

Overflow to sewer

Toilet and kitchen to sewer

▲ Figure 6.27 Grey water reuse system

Grey water cannot be used for:

- drinking water
- washing dishes
- food preparation or washing
- personal hygiene (baths and showers).

Types of grey water reuse systems

There are several differing types of grey water reuse systems that are available in the UK, but they all have similar features:

- storage tank for treated water
- pump
- distribution system (moving the stored water to the point of use)
- some form of treatment.

Direct reuse system

The grey water is collected from appliances and directly reused without treatment or storage and could be used to water the garden. Even so, grey water is not considered suitable for watering fruit or vegetables as there is a possibility of passing on pathogen and bacterial content.

Short retention system

Grey water from baths and showers is collected in a cleaning tank where it is treated; basic debris is removed and heavier particles are allowed to settle. The remaining water is transferred to a storage tank. The storage tanks are relatively small, being around 100 litres, which would be enough water for 18–20 flushes. If the stored water is not used within a short period, normally around 24 hours, the water is purged and the system cleaned, leaving a small amount of water for toilet flushing. This avoids the grey water deteriorating and beginning to smell. This type of system can result in water savings of 30 per cent.

This type of system would normally be fitted in a new build as it is more difficult to retrofit. This system is best suited for a self-contained bathroom and toilet, collecting and serving the water within the same room.

Physical and chemical system

This system uses a filter to remove the physical debris from the collected water. After the water has been filtered, chemical disinfectants such as chlorine or bromine are added to prevent any bacterial growth whilst the water is stored.

Biomechanical system

This type of system is the most advanced of the grey water reuse systems. It uses both biological and physical methods to treat the collected grey water. These systems vary in size, but on average are a similar size to a large refrigerator.

Grey water enters the system through the filter, where particles like hair and textiles are filtered out. The filter is monitored electronically and is flushed out periodically. The water then enters the main recycling chamber where organic matter is decomposed using bio-cultures. The grey water remains in this chamber for three hours before being pumped into the second chamber for further bio-culture treatment. Biological sediment settles at the bottom of both these chambers, which is removed. After a further three hours the grey water passes through a UV filter and into the final storage tank where it is ready for use.

When there is a demand for grey water, it is pumped to the point of use. At times when treated water availability is low, fresh water can be introduced into the system.

The grey water from this system can be used to wash clothing as well.

Biological system

This system uses some of the principles used by sewage treatment works. In this system, biological growth is encouraged rather than inhibited, by the introduction of oxygen into the waste water. The oxygen is introduced by means of a pump into the storage tank. Bacteria then 'digest' the organic matter contained in the grey water.

A more natural method of introducing oxygen is through the use of reed beds. In nature, reeds, which thrive in water logged conditions, transfer oxygen to their roots. The grey water is allowed to infiltrate through the reed bed. The added oxygen from the roots of the reeds and the naturally occurring bacteria remove the organic matter contained in the grey water. The disadvantage of reed beds is the land area required and the expertise required.

There are companies that make and offer smaller-scale systems, such as the green roof recycling system (GROW).

ACTIVITY

Research reed bed filtration systems to assess the range of systems available in the UK.

Location and building requirements

When considering the installation of a grey water reuse system, the following should be taken into account:
- a suitable supply of grey water to meet the demand
- suitability of location and availability of space to store the grey water
- storage tanks need to be located away from heat and direct sunlight to avoid algae growth
- location or storage needs to be protected from freezing
- supply of wholesome water is required
- access for excavation equipment may be required
- a water meter will need to be fitted on incoming mains.

Planning permission

Planning permission on the whole is not required for a grey water reuse system, although it is always worth checking with the local authority, especially if the system is above ground, in a designated area or the building is listed. If a building is required to house the grey water storage system, then planning permission will be required.

Compliance with Building Regulations

The Building Regulations that apply to grey water reuse systems are given in Table 6.12.

Other regulations applicable for grey water reuse systems

- The Water Supply (Water Fittings) Regulations 1999 apply. The main area of concern is the avoidance of cross contamination between grey water and wholesome water; this is known as backflow prevention. Grey water is classified as category 5, which is a high health risk. The usual method is by use of an air gap.
- Any pipework used to supply outlets and any outlet using grey water needs to be labelled to avoid any mistakes.
- The local authority must be informed prior to installing a grey water system.
- **BS 7671** will apply to the installation's electrical supplies and controls.

INDUSTRY TIP

Traditionally, water charges were based on the rateable value of the property. Nowadays a water meter monitors the use of water and will identify the reduction in wholesome water use.

▲ Figure 6.28 Grey water warning label

▼ Table 6.12 Building Regulations applicable for grey water reuse systems

Building Regulation	Title	Relevance
A	Structure	Where components affect the loading placed on the structure of the building or excavations are close to the building
B	Fire safety	Where holes for pipework reduce the fire-resisting integrity of the structure
C	Resistance to contaminants and moisture	Where holes for pipework reduce the moisture-resisting integrity of the structure
E	Resistance to the passage of sound	Where holes for pipework reduce the sound-resisting integrity of the structure
G	Sanitation, hot water safety and water efficiency	Water efficiency
H	Drainage and waste disposal	Where waste pipes are connected to the drainage system
P	Electrical safety	Installation of supply and control wiring for the system

Benefits and limitations of grey water reuse systems

The advantages of using grey water reuse systems include:

- a reduction in water bills
- a reduction on the demand for wholesome water
- a wide range of system designs
- potential to provide more reusable water than rainwater harvesting.

Disadvantages are as follows:

- long payback period
- difficult to integrate into an existing system
- only certain appliances can be supplied by grey water reuse
- potential cross contamination risk
- water meter will need to be fitted on the property supply
- the need for filtering and pumping contributes to the property carbon footprint.

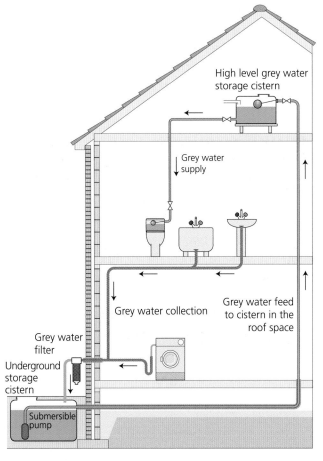

▲ Figure 6.29 Grey water reuse

4 ELECTRICITY-PRODUCING MICRO-RENEWABLE ENERGY TECHNOLOGIES

The electricity-producing micro-renewable energy technologies that will be discussed in this section are:

- solar photovoltaic
- micro-wind
- micro-hydro.

The major advantages of these technologies are that they do not use any of the planet's dwindling fossil fuel resources. They also do not produce any carbon dioxide (CO_2) when running.

With each of the electricity-producing micro-renewable energy technologies, two types of connection exist:

- on-grid or grid-tied – where the system is connected in parallel with the grid-supplied electricity
- off-grid – where the system is not connected to the grid but supplies electricity directly to current-using equipment or is used to charge batteries and then supplies electrical equipment via an inverter.

The batteries required for off-grid systems need to be deep-cycle type batteries, which are expensive to purchase. The other downside of using batteries to store electricity is that the batteries' life span may be as short as five years, after which the battery bank will require replacing.

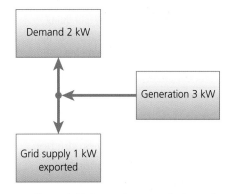

▲ Figure 6.30 Generation exceeds demand

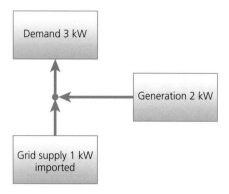

▲ Figure 6.31 Demand exceeds generation

With on-grid systems, any excess electricity generated is exported back to the grid. At times when the generation output is not sufficient to meet the demand, electricity is imported from the grid.

While the following sections will be focused primarily on on-grid or grid-tied systems, which are the most common type in use, an overview of the components required for off-grid systems is included to provide a complete explanation of the technology.

Solar photovoltaic (PV)

Solar photovoltaic (PV) is the conversion of light into electricity. Light is electromagnetic energy and, in the case of visible light, is electromagnetic energy that is visible to the human eye. The electromagnetic energy released by the Sun consists of a wide spectrum, most of which is not visible to the human eye and cannot be converted into electricity by PV modules.

Working principles

The basic element of photovoltaic energy production is the PV cell, which is made from semiconductor material. A semiconductor is a material with resistivity that sits between that of an insulator and a conductor. Whilst various semiconductor materials can be used in the making of PV cells, the most common material is silicon. Adding a small quantity of a different element

(an impurity) to the silicon, a process known as 'doping', produces n-type or p-type semiconductor material. Whether it is n-type (negative) or p-type (positive) semiconductor material is dependent on the element used to dope the silicon. Placing an n-type and a p-type semiconductor material together creates a p-n junction. This forms the basis of all semiconductors used in electronics.

When photons, which are particles of energy from the Sun, hit the surface of the PV cell they are absorbed by the p-type material. The additional energy provided by these photons allows electrons to overcome the bonds holding them and move within the semiconductor material, thus creating a potential difference or – in other words – generating a voltage.

Photovoltaic cells have an output voltage of 0.5 V, so a number of these are linked together to form modules with resulting higher voltage and power outputs. Modules are connected together in series to increase voltage. These are known as 'strings'. All the modules together are known as an 'array'. An array therefore can comprise a single string or multiple strings. The connection arrangements are determined by the size of the system and the choice of inverter. It should be noted that PV arrays can attain d.c. voltages of many hundreds of volts.

ACTIVITY

What is one danger associated with photovoltaic systems?

There are many arrangements for PV systems but they can be divided into two categories:
- off-grid systems, where the PV modules are used to charge batteries
- on-grid systems, where the PV modules are connected to the grid supply via an inverter.

The key components of an off-grid PV system are:
- PV modules
- a PV module mounting system
- d.c. cabling
- a charge controller
- a deep-discharge battery bank
- an inverter.

Other components, such as isolators, will also be required.

▲ Figure 6.32 Off-grid system components

On-grid systems where the PV modules are connected to the grid supply via an inverter

The key components of an on-grid PV system are:

- PV modules
- a PV module mounting system
- d.c. cabling
- an inverter
- a.c. cabling
- metering
- a connection to the grid.

Other components such as isolators will also be required.

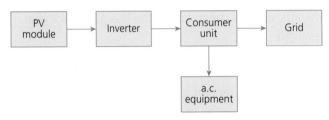

▲ Figure 6.33 On-grid system components

PV modules

A range of different types of module, of various efficiencies, is available. The performance of a PV module is expressed as an efficiency percentage: the higher the percentage the greater the efficiency.

- Monocrystalline modules range in efficiency, from 15 to 20 per cent.
- Polycrystalline modules range in efficiency, from 13 to 16 per cent, but are cheaper to purchase than monocrystalline modules.
- Amorphous film ranges in efficiency, from 5 to 7 per cent. Amorphous film is low efficiency but is flexible, so it can be formed into curves and is ideal for surfaces that are not flat.

Whilst efficiencies may appear low, the maximum theoretical efficiency that can be obtained with a single junction silicon cell is only 34 per cent.

PV module mounting system

Photovoltaic modules can be fitted as on-roof systems, in-roof systems or ground-mount systems.

- On-roof systems are the common method employed for retrofit systems. Various different mounting systems exist for securing the modules to the roof structure. Most consist of aluminium rails, which are fixed to the roof structure by means of roof hooks. Mounting systems also exist for fitting PV modules to flat roofs. Checks will need to be made to ensure that the existing roof structure can withstand the additional weight and also the uplift forces that will be exerted on the PV array by the wind.
- In in-roof systems the modules replace the roof tiles. The modules used are specially designed to interlock, to ensure that the roof structure is watertight. The modules are fixed directly to the roof structure. Several different systems are on the market, from single-tile size to large panels that replace a whole section of roof tiles. In-roof systems cost more than on-roof systems but are more aesthetically pleasing. In-roof systems are generally only suitable for new-build projects or where the roof is to be retiled.
- Ground-mount systems and pole-mount systems are available for free-standing PV arrays.

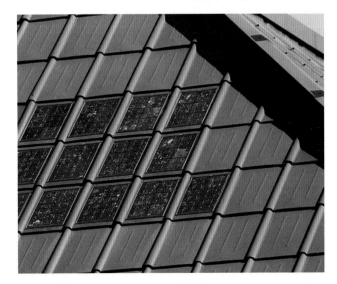

▲ Figure 6.34 In-roof mounting system

ACTIVITY

With regards to a PV installation, who would normally be responsible for mounting the roof brackets and panels and testing and connecting the electrical system?

Tracking systems are the ultimate in PV mounting systems. They are computer-controlled motorised mounting systems that change both **azimuth** and tilt to track the Sun as it passes across the sky. Ideally the modules should face due south, but any direction between east and west will give acceptable outputs.

KEY TERM

Azimuth: refers to the angle that the panel direction diverges from facing due south.

Inverter

The inverter's primary function is to convert the d.c. input to a 230 V a.c. 50 Hz output, and synchronise it with the mains supply frequency. The inverter also ensures that, in the event of mains supply failure, the PV system does not create a danger by continuing to feed power onto the grid. The inverter must be matched to the PV array with regard to power and d.c. input voltage, to avoid damage to the inverter and to ensure that it works efficiently. Both d.c. and a.c. isolators will be fitted to the inverter, to allow it to be isolated for maintenance purposes.

ACTIVITY

What is the purpose of an inverter?

Metering

A generation meter is installed on the system to record the number of units generated, so that the feed-in tariff can be claimed.

Connection to the grid

Connection to the grid within domestic premises is made via a spare way in the consumer unit and a 16 A overcurrent protective device. An isolator is fitted at the intake position to provide emergency switching, so that the PV system can easily be isolated from the grid.

Location and building requirements

When deciding on the suitability of a location or building for the installation of PV, the following considerations should be taken into account.

Adequate roof space available

The roof space available determines the maximum size of PV array that can be installed. In the UK, all calculations are based on 1,000 Wp (watts peak) of the Sun's radiation on 1 m² so, if the array uses modules with a 15 per cent efficiency, each 1 kWp of array will require approximately 7 m² of roof space. The greater the efficiency of the modules, the less roof space that is required.

The orientation (azimuth) of the PV array

The optimum direction for the solar collectors to face is due south; however, as the Sun rises in the east and sets in the west, any location with a roof facing east, south or west is suitable for mounting a PV array, but the efficiency of the system will be reduced for any system not facing due south.

▲ Figure 6.35 Ideal orientation is south

The tilt of the PV array

Throughout the year, the height of the Sun relative to the horizon changes from its lowest in December through to its highest in June. As it is generally not practical to vary the tilt angle throughout the year, the optimum tilt for the PV array in the UK is between 30 and 40°; however, the modules will work outside the optimum tilt range and will even work if vertical or horizontal, but they will be less efficient.

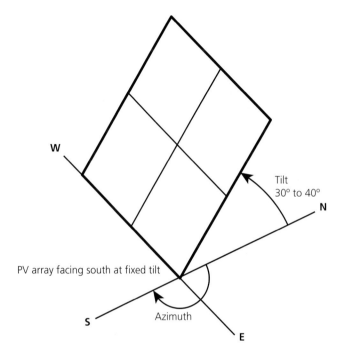

▲ Figure 6.36 Ideal tilt is 30–40°

Shading of the PV array

Any structure, tree, chimney, aerial or other object that stands between the PV array (collector) and the Sun will prevent some of the Sun's energy from reaching the collector. The Sun shines for a limited time and any reduction in the amount of sunlight landing on the collector will reduces its ability to produce electricity.

▲ Figure 6.37 Shading of PV system

Location within the UK

The location within the UK will determine how much sunshine will fall, annually, on the PV array and, in turn, this will determine the amount of electricity that can be generated. For example, a location in Brighton will generate more electricity than one in Newcastle, purely because Brighton receives more sunshine.

The suitability of the structure for mounting the solar collector

The structure has to be assessed for its suitability for fixing the chosen mounting system. Consideration needs to be given to the strength of the structure, the suitability of fixings and the condition of the structure. Consideration also needs to be given to the effect known as 'wind uplift', an upward force exerted by the wind on the module and mounting system. The strength of the PV array fixings and the fixings holding the roof members to the building structure must be great enough to allow for wind uplift.

In the case of roof-mounted systems on flats and other shared properties, consideration must also be given to the ownership of the structure on which the proposed system is to be installed.

A suitable place to mount the inverter

The inverter is usually mounted either in the loft space or at the mains position.

Connection to the grid

A spare way within the consumer unit will need to be available for connection of the PV system. If one is not available then the consumer unit may need to be changed.

Planning permission

Permitted development applies where a PV system is installed:

- on a dwelling house or block of flats
- on a building within the grounds of a dwelling house or block of flats
- as a stand-alone system in the grounds of a dwelling house or block of flats.

However, there are criteria to be met in each case.

For building-mounted systems:

- the PV system must not protrude more than 200mm from the wall or the roof slope
- the PV system must not protrude past the highest point of the roof (the ridgeline), excluding the chimney.

For stand-alone systems the following criteria must be met.

- Only one stand-alone system is allowed in the grounds.
- The array must not exceed 4 m in height.
- The array must not be installed within 5 m of the boundary of the grounds.
- The array must not exceed 9 m² in area.
- No dimension of the array may exceed 3 m in length.

For both stand-alone and building-mounted systems the following criteria must be met.

- The system must not be installed in the grounds or on a building within the grounds of a listed building or a scheduled monument.
- If the dwelling is in a conservation area or a World Heritage Site, then the array must not be closer to a highway than the house or block of flats.
- In every other case, planning permission will be required.

ACTIVITY

What documentation should be completed by the electrical installer after testing the new PV installation?

▼ Table 6.13 Building Regulations applicable for solar photovoltaic systems

Building Regulation	Title	Relevance
A	Structure	The PV modules will impose both downward force and wind uplift stresses on the roof structure
B	Fire safety	The passage of cables through the building fabric could reduce the fire-resisting integrity of the structure
C	Resistance to contaminants and moisture	The fixing brackets for on-roof systems and the passage of cables through the building fabric could reduce the moisture-resisting integrity of the structure
E	Resistance to the passage of sound	The passage of cables through the building fabric could reduce the sound-resisting properties of the structure
L	Conservation of fuel and power	The efficiency of the system and the building overall
P	Electrical safety	The installation of the components and wiring system

Compliance with Building Regulations

The regulations outlined in Table 6.13 will apply.

Other regulatory requirements applicable for solar photovoltaic systems

- **BS 7671** The IET Wiring Regulations will apply to the PV installation.
- G83 requirements will apply to on-grid systems up to 3.68 kW per phase; above this size the requirements of G59 will need to be complied with. (G83 and G59 are both specific Engineering Recommendations that cover part of the installation and commissioning of PV systems.)
- Microgeneration Certification Scheme requirements will apply.

INDUSTRY TIP

All systems will require some penetration of the building fabric, be it the roof or walls, depending on the building type and construction. You should be able to describe the methods of making good for all building fabrics.

Benefits and limitations of solar photovoltaic systems

Advantages of solar photovoltaic systems include the following:

- They can be fitted to most buildings.
- There is a feed-in tariff available for electricity generated, regardless of whether it is used on site or exported to the grid.
- Excess electricity can be sold back to the distribution network operator (DNO).
- There is a reduction in electricity imported.
- It uses zero carbon technology.
- It improves energy performance certificate ratings.
- There is a reasonable payback period on the initial investment.

Disadvantages include the following:

- Initial cost is high.
- The system size is dependent on available, suitable roof area.
- It requires a relatively large array to offset installation costs.
- It gives variable output that is dependent on the amount of sunshine available. Lowest output is at times of greatest requirement, such as at night and in the winter. Savings need to be considered over the whole year.
- There is an aesthetic impact (on the appearance of the building).

Micro-wind

Wind turbines harness energy from the wind and turn it into electricity. The UK is an ideal location for the installation of wind turbines, as about 40 per cent of Europe's wind energy passes over the UK. A micro-wind turbine installed on a suitable site could easily generate more power than would be consumed on site.

Working principles

The wind passing the rotor blades of a turbine causes it to turn. The hub is connected by a low-speed shaft to a gearbox. The gearbox output is connected to a high-speed shaft that drives a generator which, in turn, produces electricity. Turbines are available as either horizontal-axis wind turbines (HAWT) or vertical-axis wind turbines (VAWT).

A HAWT has a tailfin to turn the turbine so that it is facing in the correct direction to make the most of the available wind. The gearbox and generator will also be mounted in the horizontal plane.

Vertical-axis wind turbines, of which there are many different designs, will work with wind blowing from any direction and therefore do not require a tailfin. A VAWT also has a gearbox and generator.

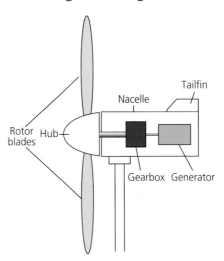

▲ Figure 6.38 Horizontal-axis wind turbine

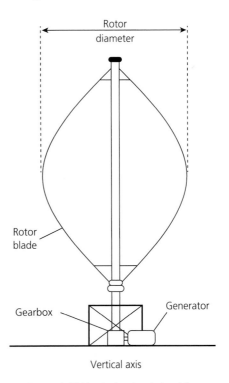

▲ Figure 6.39 Vertical-axis wind turbine

The two types of micro-wind turbines suitable for domestic installation are:

- pole-mounted, free-standing wind turbines
- building-mounted wind turbines, which are generally smaller than pole-mounted turbines.

Micro-wind generation systems fall into two basic categories:

- on-grid (grid-tied), which is connected in parallel with the grid supply via an inverter
- off-grid, which charge batteries to store electricity for later use.

The output from a micro-wind turbine is wild alternating current (a.c.). 'Wild' refers to the fact that the output varies in both voltage and frequency. The output is connected to a system controller, which rectifies the output to d.c.

In the case of an on-grid system the d.c. output from the system controller is connected to an inverter which converts d.c. to a.c. at 230 V 50 Hz, for connection to the grid supply via a generation meter and the consumer unit.

With off-grid systems the output from the controller is used to charge batteries so that the output is stored for when it is needed. The output from the batteries then feeds an inverter so that 230 V a.c. equipment can be connected.

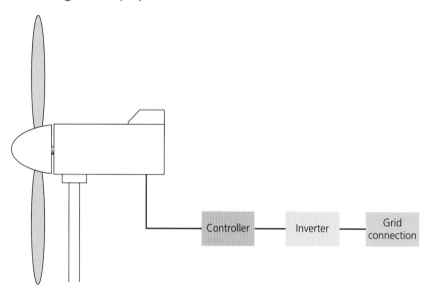

▲ Figure 6.40 Block diagram of an on-grid micro-wind system

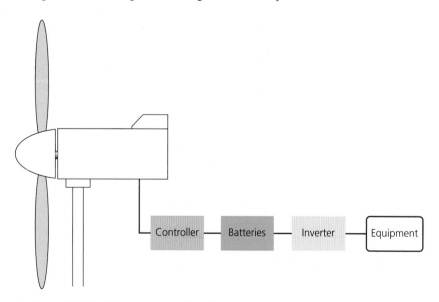

▲ Figure 6.41 Block diagram of an off-grid micro-wind system

Location and building requirements

When considering the installation of a micro-wind turbine, it is important to consider the location or building requirements, including:

- the average wind speed on the site
- any obstructions and turbulence
- the height at which the turbine can be mounted
- turbine noise, vibration, flicker.

Wind speed

Wind is not constant, so the average wind speed on a site, measured in metres per second (m/s), is a prime consideration when deciding on a location's suitability for the installation of a micro-wind turbine.

Wind speed needs to be a minimum of 5 m/s for a wind turbine to generate electricity. Manufacturers of wind turbines provide power curves for their turbines, which show the output of the turbine at different wind speeds. Most micro-wind turbines will achieve their maximum output when the wind speed is around 10 m/s.

Obstructions and turbulence

For a wind turbine to work efficiently, a smooth flow of air needs to pass across the turbine blades.

The ideal site for a wind turbine would be at the top of a gentle slope. As the wind passes up the slope it gains speed, resulting in a higher output from the turbine.

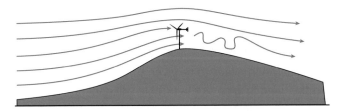

▲ Figure 6.42 A suitable site for a micro-wind turbine

Figure 6.43 illustrates the effect on the wind when a wind turbine is poorly sited. The wind passing over the turbine blades is disturbed and thus the efficiency is reduced.

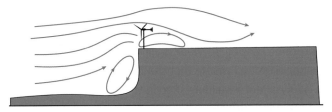

▲ Figure 6.43 An unsuitable site for a micro-wind turbine

Any obstacles, such as trees or tall buildings, will affect the wind passing over the turbine blades.

Where an obstacle is upwind of the wind turbine, in the direction of the prevailing wind, the wind turbine should be sited at a minimum distance of 10 × the height of the obstacle away from the obstacle. In the case of an obstacle that is 10 m in height, this would mean that the wind turbine should be sited a minimum of 10 × 10 m away, which is 100 m from the obstacle.

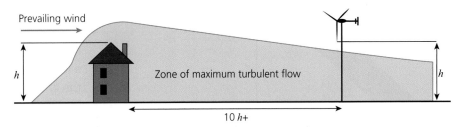

▲ Figure 6.44 Placement of micro-wind turbine to avoid obstacles

Turbine mounting height

Generally, the higher a wind turbine is mounted, the better. The minimum recommended height is 6–7 m but, ideally, it should be mounted at a height of 9–12 m. As a wind turbine has moving parts, consideration needs to be given to access for maintenance. Where an obstacle lies upwind of the turbine, the bottom edge of the blade should be above the height of the obstacle.

Turbine noise

Consideration needs to be given to buildings sited close to the wind turbine as the turbine will generate noise in use.

Turbine vibration

Consideration needs to be given to vibration when the wind turbine is building mounted. It may be necessary to consult a structural engineer.

Shadow flicker

Shadow flicker is the result of the rotating blades of a turbine passing between a viewer and the Sun. It is important to ensure that shadow flicker does not unduly affect a building sited in the shadow-flicker zone of the wind turbine.

The distance of the shadow-flicker zone from the turbine will be at its greatest when the Sun is at its lowest in the sky.

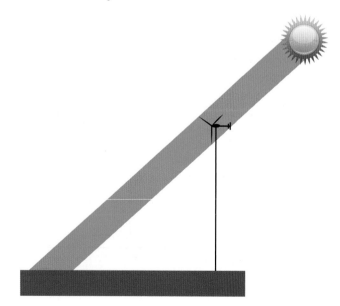

▲ Figure 6.45 The area affected by shadow flicker

Planning permission

Whilst permitted development exists for the installation of wind turbines, it is severely restricted, so, in the majority of installations, a planning application will be required. The permitted development criteria are detailed below.

Permitted development applies where a wind turbine is installed:
- on a detached dwelling house
- on a detached building within the grounds of a dwelling house or block of flats
- as a stand-alone system in the grounds of a dwelling house or block of flats.

It is important to note that permitted development for building-mounted wind turbines only applies to detached premises. It does not apply to semi-detached houses or flats.

Even with detached buildings or stand-alone turbines there are criteria to be met.
- The wind turbine must comply with the MCS planning standards, or equivalent.
- Only one wind turbine may be installed on the building or within the grounds of the building.
- An air source heat pump may not be installed on the building or within the grounds of the building.
- The highest part of the wind turbine (normally the blades) must not protrude more than 3 m above the ridge line of the building or be more than 15 m in height.
- The lowest part of the blades of the wind turbine must be a minimum of 5 m from ground level.
- The wind turbine must be a minimum of 5 m from the boundary of the premises.
- The wind turbine cannot be installed on or within:
  - land that is safeguarded land (usually designated for military or aeronautical reasons)
  - a site that is designated as a scheduled monument
  - a listed building
  - the grounds of a listed building
  - land within a national park
  - an area of outstanding natural beauty
  - the Broads (wetlands and inland waterways in Norfolk and Suffolk).

- The wind turbine cannot be installed on the roof or wall of a building that fronts a highway, if that building is within a conservation area.

The following conditions also apply.
- The blades must be made of non-reflective material.
- The wind turbine should be sited so as to minimise its effect on the external appearance of the building.

Compliance with Building Regulations

The regulations outlined in Table 6.14 will apply.

▼ Table 6.14 Building Regulations applicable for micro-wind systems

Building Regulation	Title	Relevance
A	Structure	A wind turbine mounted on a building will exert additional structural load, as well as forces, due to its operation
B	Fire safety	Cable entries and fixings may reduce the fire-resisting integrity of the building structure
C	Resistance to contaminants and moisture	Cable entries and fixings may reduce the moisture-resisting integrity of the building fabric
E	Resistance to the passage of sound	Cable entries may reduce the sound-resisting integrity of the building fabric
L	Conservation of fuel and power	The efficiency of the system and the building
P	Electrical safety	Installation of wiring and components

Other regulatory requirements applicable for micro-wind systems

- For on-grid systems, the requirements of the Distribution Network Operator (DNO) will apply.
- **BS 7671** The IET Wiring Regulations will apply to the installation of micro-wind turbines.

ACTIVITY

What is meant by a 'feed-in tariff'?

Benefits and limitations of micro-wind turbines

Advantages of micro-wind turbines include the following:
- They can be very effective on a suitable site as the UK has 40 per cent of Europe's wind resources.
- There are no carbon dioxide emissions.
- They produce most energy in winter, when consumer demand is at its maximum.
- A feed-in tariff is available.
- This can be a very effective technology where mains electricity does not exist.

Disadvantages include the following:
- Initial costs are high.
- The requirements of the site are onerous.
- Planning can be onerous.
- Performance is variable and is dependent on wind availability.
- Micro-wind turbines cause noise, vibration and flicker.

Micro-hydro-electric

All rivers flow downhill. This movement of water from a higher level to a lower level is a source of free kinetic energy that hydro-electric generation harnesses. Water passing across or through a turbine can be used to turn a generator and thus produce electricity. Given the right location, micro-hydro-electric is the most constant and reliable source of all the micro-generation technologies and is the most likely of the technologies to meet all of the energy needs of the consumer.

As with the other micro-generation technologies, there are two possible system arrangements for micro-hydro schemes: on-grid and off-grid systems.

Working principles

Whilst it is possible to place generators directly into the water stream, it is more likely that the water will be diverted from the main stream or river, through the turbine, and back into the stream or river at a lower level. Apart from the work involved with the turbines and generators, there is also a large amount of civil engineering and construction work to be carried out to route the water to where it is needed.

The main components of the water course construction are:

- intake – the point where a portion of the river's water is diverted from the main stream
- the canal that connects the intake to the forebay
- the forebay, which holds a reservoir of water that ensures that the penstock is pressurised at all times and allows surges in demand to be catered for
- the penstock, which is pipework taking water from the forebay to the turbines
- the powerhouse, which is the building housing the turbine and the generator
- the tailrace, which is the outlet that takes the water exiting the turbines and returns it to the main stream of the river.

See Figure 6.46.

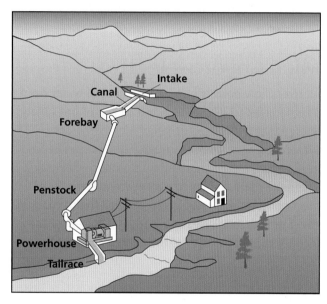

▲ Figure 6.46 The component parts of a micro-hydro system

INDUSTRY TIP

The idea of water turbines sounds great but of course few people live near a water course that is suitable for this application. PV and solar thermal systems are more adaptable for different locations.

To ascertain the suitability of the water source for hydro-electric generation, it is necessary to consider the head and the flow of the water source.

Head

The head is the vertical height difference between the proposed inlet position and the proposed outlet. This measurement is known as 'gross head'.

Head height is generally classified as:
- low head – below 10 m
- medium head – 10–50 m
- high head – above 50 m.

There is no absolute definition for each classification. The Environment Agency, for example, classifies low head as below 4 m. Some manufacturers specify high head as above 300 m.

Net head

This is used in calculations of potential power generation and takes into account losses due to friction, as the water passes through the penstock.

Flow

This is the amount of water flowing through the water course and is measured in cubic metres per second (m³/s).

▲ Figure 6.47 The meanings of 'head' and 'flow'

Turbines

There are many different types of turbine but they fall into two primary design groups, each of which is better suited to a particular type of water supply.

Impulse turbine

In an impulse turbine, the turbine wheel or runner operates in air, with water jets driving the runner. The water from the penstock is focused on the blades by means of a nozzle. The velocity of the water is increased but the water pressure remains the same so there is no requirement to enclose the runner in a pressure casing. Impulse turbines are used with high-head water sources.

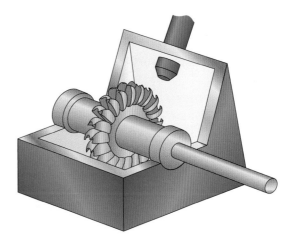

▲ Figure 6.49 Pelton turbine

Turgo

This is similar to the Pelton but the water jet is designed to hit the runner at an angle and from one side of the turbine. The water enters at one side of the runner and exits at the other, allowing the Turgo turbine to be smaller than the Pelton for the same power output. This type of turbine is used with water sources with medium or high heads of water.

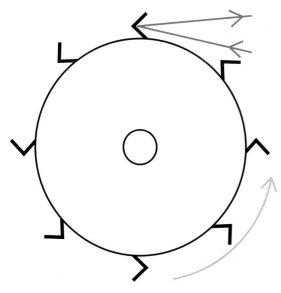

▲ Figure 6.48 Impulse turbine

Examples of impulse turbines are described below.

Pelton

This consists of a wheel with bucket-type vanes set around the rim. The water jet hits the vane and turns the runner. The water gives up most of its energy and falls into a discharge channel below. A multi-jet Pelton turbine is also available. This type of turbine is used with water sources with medium or high heads of water.

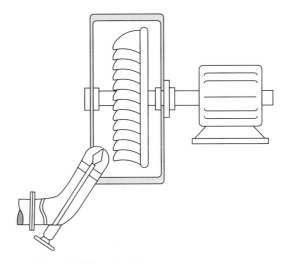

▲ Figure 6.50 Turgo turbine

Cross-flow or Banki

With this type of turbine the runner consists of two end-plates with slats, set at an angle, joining the two discs, much like a water wheel. Water passes through the slats, turning the runner and then exiting from below. This type of turbine is used with water sources with low or medium heads of water.

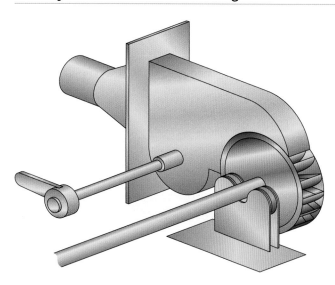

▲ Figure 6.51 Cross-flow or Banki turbine

Reaction turbine

In the reaction turbine, the runners are fully immersed in water and are enclosed in a pressure casing. Water passes through the turbine, causing the runner blades to turn or react.

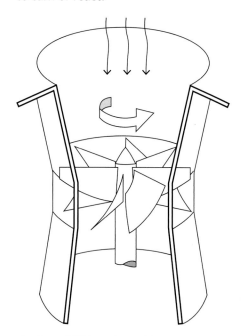

▲ Figure 6.52 Reaction turbine

Examples of reaction turbines are described below.

Francis wheel

Water enters the turbine housing and passes through the runner, causing it to turn. This type of turbine is used with water sources with low heads of water.

Kaplan (propeller)

This works like a boat propeller in reverse. Water passing the angled blades turns the runner. This type of turbine is used with water sources with low heads of water.

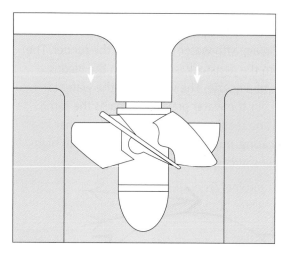

▲ Figure 6.53 Kaplan or propeller turbine

Reverse Archimedes' screw

The Archimedes' screw consists of a helical screw thread, which was originally designed so that turning the screw – usually by hand – would draw water up the thread to a higher level. In the case of hydro-electric turbines, water flows down the screw, hence reverse, turning the screw, which is connected to the generator. This type of turbine is particularly suited to low-head operations but its major feature is that, due to its design, it is 'fish-friendly' and fish are able to pass through it, so it may be the only option if a hydro-electric generator is to be fitted on a river that is environmentally sensitive.

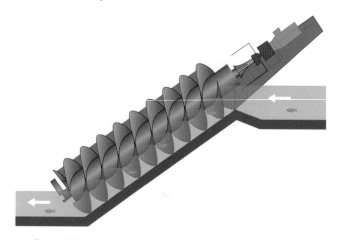

▲ Figure 6.54 Reverse Archimedes' screw

Location building requirements

When considering the installation of a micro-hydro turbine, the following location or building requirements should be taken into account.

- The location will require a suitable water source with:
  - a minimum head of 1.5 m
  - a minimum flow rate of 100 litres/second.

The water source should not be subject to seasonal variation that will take the water supply outside of the above parameters.

- The location has to be suitable to allow construction of:
  - the water inlet
  - the turbine/generator building
  - the water outlet or tailrace.

Planning permission

Planning permission will be required.

A micro-hydro scheme will have an impact on:

- the landscape and visual amenity
- nature conservation
- the water regime.

The planning application will need to be accompanied by an environmental statement detailing any environmental impact and what measures will be taken to minimise these. The environmental statement typically covers:

- flora and fauna
- noise levels
- traffic
- land use
- archaeology
- recreation
- landscape
- air and water quality.

Compliance with Building Regulations

The regulations outlined in Table 6.15 will apply.

▼ Table 6.15 Building Regulations applicable for micro-hydro-electric systems

Building Regulation	Title	Relevance
A	Structure	If any part of the system is housed in or connected to the building, then structural considerations will need to be taken into account
B	Fire safety	Where cables pass through the building fabric they may reduce the fire-resisting properties of the building fabric
C	Resistance to contaminants and moisture	Where cables pass through the building fabric they may reduce the moisture-resisting properties of the building fabric
E	Resistance to the passage of sound	Where cables pass through the building fabric they may reduce the sound-resisting properties of the building fabric
P	Electrical safety	Installation of components and cables

Other regulatory requirements applicable for micro-hydro-electric systems

- **BS 7671** The IET Wiring Regulations
- G83 requirements for grid-tied systems
- Microgeneration Certification Scheme requirements
- Environment Agency requirements

In England and Wales, all waterways of any size are controlled by the Environment Agency. To remove water from these waterways, even though it may be returned – as in the case of a hydro-electric system – will usually require permission and a licence.

There are three types of licence that may apply to a hydro-electric system.

- An abstraction licence will be required if water is diverted away from the main water course. The major concern will be the impact that the project has on fish migration, as the majority of turbines are not fish-friendly. This requirement may affect the choice of turbine. It may mean that fish screens are required over water inlets or, where the turbine is in the main channel of water, a fish pass around the turbine may need to be constructed.
- An impoundment licence – an impoundment is any construction that changes the flow of water, so if changes or additions are made to sluices, weirs, etc that control the flow within the main stream of water, an impoundment licence will be required.
- A land drainage licence will be required for any changes made to the main channel of water.

An Environment Site Audit (ESA) will be required as part of the initial assessment process. The ESA covers:

- water resources
- conservation
- chemical and physical water quality
- biological water quality
- fisheries
- managing flood risk
- navigation of the waterway.

INDUSTRY TIP

Micro-hydro systems are no good on rivers that suffer seasonal droughts or are subject to winter flooding or freezing, unless preventative measures are taken to protect the generator and control gear.

Benefits and limitations of micro-hydro-electric systems

Advantages of micro-hydro-electric systems include the following:

- There are no on-site carbon emissions.
- Large amounts of electricity are output, usually more than required for a single dwelling. The surplus can be sold.

- A feed-in tariff is available.
- There is a reasonable payback period.
- It is an excellent system where no mains electricity exists.
- It is not dependent on weather conditions or building orientation.

Disadvantages include the following:

- It requires a high head or fast flow of water on the property.
- It requires planning permission, which can be onerous.
- Environment Agency permission is required for water extraction.
- It may require strengthening of the grid for grid-tied systems.
- Initial costs are high.

SUMMARY

With depleting fossil fuel resources and pressure being put on households to reduce their carbon footprint, plumbers have a vital role to play. It is vital that you are knowledgeable and up-to-date with current technologies so you are able to offer a customer sound advice and successfully install their system. You will never be an expert in every renewable technology, but being aware of each technology and being a specialist in one or two means that you play an important part in this conservation drive being implemented by governments all over the world. This is only going to be a growing area for plumbers to be trained in.

Test your knowledge

1 A system that uses a fluid to capture the heat from the Sun is known as:

   a Photovoltaic

   b Solar voltaic

   c Solar thermal

   d Solar thermistic

2 What is a buffer tank is used for?

   a Holding water pressure until demand is placed in the system

   b Holding hot water until there is a demand

   c Holding glycol to prevent corrosion

   d Holding cold water to replace demanded hot water

3 Which of the following is a limitation of a ground source heat pump?

   a Can only be used on mains water

   b Only be used on a bungalow

   c Has potential large land area requirements

   d Needs to be installed at least 500 m away from a water source

4 Which one of the following is a benefit of an ASHP?

   a Cheap to install

   b High efficiency

   c No noise

   d No restrictions on location

5 The air source heat pump works on the principle of:

   a Cold heat change

   b Electromagnetic induction

   c The refrigeration cycle

   d The Otto cycle

6 Which of the following areas CANNOT be supplied by harvested rainwater?

   a Flushing toilets

   b Food preparation

   c Garden watering

   d Supplying washing machines

7 Which one of the following is NOT a style of pipework loop for a ground source heat pump?

   a Vertical loop   c Slinky

   b Horizontal loop   d Rounded loop

8 If you work on refrigeration pipework, what regulation must be followed?

   a F Gas Regulations

   b H Gas Regulations

   c B Gas Regulations

   d M Gas Regulations

9 Prior to installing a solar thermal system on a roof line, which one of the following is NOT a location consideration that must be considered?

   a Orientation of collectors

   b Tilt of collectors

   c Style of the roof tiles

   d Suitability of structure

10 Which one of the following can grey water be used for?

   a Food preparation

   b Cold water supply to a bath

   c Hospital cleaning

   d Washing clothes (after additional processes)

11 Before any micro-renewable and water conservation technology is installed, what must be applied for from the local authority?

   a Planning permission

   b The Building Control officer to visit the site

   c Detailed plans to be drawn up

   d Recommendations on positioning the technology

12 Which Approved Document for England and Wales covers the 'Resistance to the passage of sound'?

   a K

   b E

   c M

   d D

13 Which of the following would be a crucial component in a solar thermal installation?

a Condenser

b Ground loop

c Water filter

d Evacuated tube

14 Which Building Regulation must be complied with when installing a solar PV system?

a K

b D

c L

d M

15 Which phrase is correct in relation to a heat source (air or ground) installation?

a The smaller the temperature difference between the refrigerant and the heat source, the greater the efficiency

b The larger the temperature difference between the refrigerant and the heat source, the greater the efficiency

c The smaller the temperature difference between the refrigerant and the heat source, the lower the efficiency

d The larger the temperature difference between the refrigerant and the heat source, the lower the efficiency

16 When installing an air source heat pump, what item can store additional heat until required by the system?

a Hot water cylinder

b Expansion vessel

c Accumulator

d Condenser

17 Why is the net result of a biomass system 'carbon neutral'?

a It does not release any carbon dioxide when burnt

b The carbon dioxide that is produced is retained within the system

c The biomass material absorbs carbon dioxide when it grows

d The carbon (ash) left over is put back into the ground

18 Which item is a central component of a micro-CHP boiler?

a Stirling engine

b Compressor

c Evaporator

d Flat plate collector

19 Approximately what percentage of used water within a domestic property is used to flush toilets and could be saved if a rainwater harvesting system were installed?

a 12%

b 17%

c 28%

d 33%

20 Which of the following fuels is carbon neutral?

a Natural gas

b Peat

c Oil seed rape

d Coal

21 Why is it important to have a high coefficient of performance?

22 List and outline the Building Regulations that a grey water system comes under.

Regulation	Title	Relevance

23 Describe why a twin coil cylinder should be used on a solar thermal system.

24 Describe the working principle of a solar thermal collector.

25 Outline why Building Regulations Part E is important to consider when installing an ASHP.

26 Complete the table below of the Building Regulations that need to be followed for the different environmental technologies.

Regulation for England and Wales	Content
A	
B	
C	
D	
E	
F	
G	
H	
J	
K	
L	
M	
N	
P	

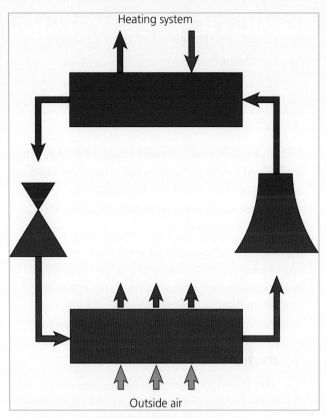

Heating system

Outside air

27 Describe why 'shading' is a consideration when designing a solar system.

28 Complete the following diagram for the refrigeration process, identifying the following parts:
- Expansion valve
- Compressor
- Condenser
- Evaporator

29 State four benefits and four limitations for a grey water reuse system.

Answers can be found online at www.hoddereducation.co.uk/construction.

CHAPTER 7

DOMESTIC FUEL SYSTEMS

For hundreds of years humans relied on solid fuel in the form of wood and coal to heat their homes. Then, in the 1850s, gas in the form of coal gas was used to heat and light dwellings and factories. This was followed soon after by oil.

These fuels, coal, gas and oil, are known as hydrocarbons and, because of the way they were formed millions of years ago, they are very carbon rich. When they are combusted, they produce copious amounts of CO_2 which has systematically altered the Earth's climate and this has led to the phenomenon known as global warming.

Now, less than 300 years later, hydrocarbon fuels are all but depleted and the damage to the climate they have caused practically irreparable. With gas and oil reserves set to last only 50 years and much of the coal left below the Earth's surface unreachable, we have to look for alternative forms of energy for our heat and light.

This chapter will investigate the types of fuels used in the appliances we install and identify the reasons that certain fuels are chosen. We will also take a look at how these fuels are supplied and stored, and the physics of the combustion of fuels. Finally, we will study the construction of flues and chimneys and how we test them to ensure the safety of the customer.

The learning outcomes for this chapter are:
- factors affecting fuel selection
- combustion processes of fuel supply systems
- principles of chimney/flue systems.

1 FACTORS AFFECTING FUEL SELECTION

The heating appliances that we install are fuelled by a selection of energy sources, some that have been around for many years and some that are relatively new technology. In this first section, we will investigate these fuels both old and new. We will learn where they come from and the consequences of using them.

Identify the types of fuels used in appliances

There are five categories of fuels:
- natural gas
- liquid petroleum gas (LPG)
- oil
- solid fuel
- sustainable, low carbon fuels.

Natural gas

Natural gas is a combustible mixture of hydrocarbon gases and is probably the most widely used hydrocarbon fuel on Earth. It is colourless and odourless in its purest form and, when it is combusted, it releases a vast amount of energy with fewer emissions than many other common fossil fuels. Natural gas is naturally occurring and is usually found during the extraction of oil from deep below the Earth's surface but it can also be found near coal formations and seams.

▼ Table 7.1 The composition of gas

Gas	Chemical symbol	Percentage (%)
Methane	CH_4	70–90
Ethane	C_2H_6	0–5
Propane	C_3H_8	0–20
Butane	C_4H_{10}	0–5
Hydrogen sulphide	H_2S	0–5
Nitrogen	N_2	0–5
Carbon dioxide	CO_2	0–8
Water vapour	H_2O	

Natural gas is composed primarily of five combustible gases, two inert gases and water vapour.

The distinctive 'rotten eggs' smell that natural gas has is added to the gas when it is cleaned of the impurities and **naphtha** it contains at the refinery. The smell is a chemical called mercaptan and is added to aid the detection of gas leaks.

Natural gas is lighter than air, having a specific gravity of 0.6–0.8.

Natural gas is available in most cities, towns and villages through a national grid of underground pipes, with only the most isolated of places not connected to the grid.

The **calorific value** (CV) of gas is usually 37.8–43 MJ/m³ depending on where the gas was extracted from.

Most natural gas used in the UK comes from the North Sea, but other sources include Russia and the Middle East.

KEY TERMS

Naphtha: a waxy oil deposit that is present in natural gas in its unrefined state. It is removed and later reused in other products such as cosmetics.

Calorific value: the amount of energy stored in the gas in its uncombusted state. It is the amount of energy released when the gas is combusted. It is measured in megajoules per cubic metre or MJ/m³.

Liquid petroleum gas (LPG)

Liquid petroleum gas, like natural gas, is a fossil hydrocarbon fuel that is closely linked to oil. About two thirds of all LPG used is extracted directly from oil wells; the rest is extracted during the manufacture of petroleum from crude oil.

There are many types of LPG but, generally, only two types of LPG are used commercially. These are propane and butane. These gases share common elements but in different quantities and these are reflected in their chemical symbols:

- propane – three atoms of carbon and eight atoms of hydrogen (C_3H_8)

- butane – four atoms of carbon and 10 atoms of hydrogen (C_4H_{10})
- iso-butane is butane that has the same elements, but these are connected in a slightly different way.

▼ Table 7.2 The composition of LPG fuels

LPG attribute	Propane	Butane
Chemical formula	C_3H_8	C_4H_{10}
Energy content: MJ/m³	95.8	111.4
Energy content: MJ/kg	49.58	47.39
Boiling temp: C°	–42	–4
Pressure @ 21°C: kPa	858.7	215.1
Flame temp: C°	1967	1970
Gas volume: m³/kg	0.540	0.405
Relative density: H_2O	0.51	0.58
Relative density: air	1.53	2.00
L per kg	1.96	1.724
kg per L	0.51	0.58
Specific gravity @ 25°C	1.55	2.07
Density @ 15°C: kg/m³	1.899	2.544

Both of these compounds are heavier than air in their gaseous form, with propane having a specific gravity of 1.5 and butane having a specific gravity of 2.0. In liquid form both are thinner than water, butane having a relative density of 0.58 and propane, 0.51.

When LPG gas is subjected to high pressure, it turns into a liquid, but it also takes up less space than the gas. One litre of LPG in its liquid state makes 274 litres of LPG gas. This means that one cylinder of LPG liquid is equivalent to 274 cylinders of LPG gas.

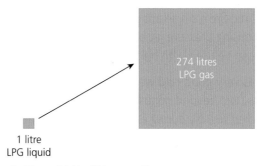

274 litres
LPG gas

1 litre
LPG liquid

▲ Figure 7.1 Liquid to gas ratio

▼ Table 7.3 The uses of LPG fuels

Compound	Uses
Butane (C_4H_{10})	Used for portable supplies, such as camping equipment, boats and barbecues. Not much use for plumbing or heating installation as it boils (turns from a liquid to a gas) at −4 °C.
Propane (C_3H_8)	Has a very low boiling point at −42 °C. Can be used in domestic situations as an alternative to natural gas where the mains gas supply is not available. Many appliances are available for use with propane, including boilers, cookers, fires and water heaters.
Iso-butane (C_4H_{10})	Used as a refrigerant in domestic refrigerators and fridge/freezers.

Environmentally, LPG is relatively clean when compared to other fuels such as coal or oil, creating far less air pollution in the form of soot and carbon particulates, sulphur and carbon dioxide and, therefore, adds less to global warming than might be realised.

Cost, however, is an issue, since LPG is much more expensive than conventional natural gas.

Fuel oil (kerosene grade C2, 28 second viscosity oil to **BS 2869:2017**)

A simple definition for fuel oil is a liquid by-product of crude oil, which is produced during petroleum refining. There are two main categories under which it is classified. One is distillate oils, such as diesel fuel, and the others are residual oils, which includes heating kerosene. It is distillate fuel oil that is generally used for home heating.

Around 95 per cent of domestic boilers burning fuel oil in domestic properties use kerosene, which is also known generically as C2 grade 28 second viscosity oil. This is the preferred oil fuel grade for domestic heating, due to its clean combustion. Modern oil central heating boilers only require a single annual service, if being used with an atomising pressure jet burner. It is the only oil grade that can be used with balanced or low-level flues.

Kerosene has very good cold weather characteristics and remains fluid beyond minus 40 °C although it does tend to thicken slightly during extremely cold weather.

Kerosene is a high carbon fuel and is clear or very pale yellow in colour. Newer boilers have a label inside the casing with information on nozzle size and pump pressure that show that the boiler has been set up to use kerosene. It may also reference the British Standard for kerosene **BS 2869** grade C2.

▲ Figure 7.2 A domestic kerosene oil tank for oil fired heating

Solid fuel (coal, coke and peat)

There are three main types of solid fuel. These are:

- coal
- coke
- peat.

Coal

This is a fossil fuel created from the remains of plants that lived and died between 100 and 400 million years ago when large areas of the Earth were covered with huge swamps and forest bogs.

The energy that we get from coal comes from the energy that the plants absorbed from the Sun millions of years ago. The process is called photosynthesis. When plants die, this stored energy is usually released during the decaying process, but when coal is formed the process is interrupted preventing the release of the trapped solar energy.

As the Earth's climate evolved and the vegetation died, a thick layer of rotting vegetation built up that was covered with water, silt and mud, stopping the decaying process. The weight of the water and the top layer of mud compressed the partially decayed vegetation under heat and pressure, squeezing out the remaining oxygen leaving rich hydrocarbon deposits. What once had been plants gradually fossilised into a combustible carbon-rich rock we call coal.

Types of coal

Coal is classified into four main types, depending on the amount of carbon, oxygen, and hydrogen present. The higher the carbon content, the more energy the coal contains.

▼ Table 7.4 The different types of commercially available coal

Coal type	Heat content (kWh/Kg)	Carbon content (%)	Description
Lignite	2.2–5.5	25–35	The lowest type of coal, lignite is crumbly and has high moisture content. Most lignite is used to produce electricity.
Sub-bituminous	5.5–8.3	35–45	Typically contains less heating value than bituminous coal but more moisture.
Bituminous	7–10	45–86	Formed by added heat and pressure on lignite. Made of many tiny layers, bituminous coal looks smooth and sometimes shiny. It has two to three times the heating value of lignite. Bituminous coal is used to generate electricity and is an important fuel for the steel and iron industries.
Anthracite	10	86–97	Created where additional pressure combined with very high temperature inside the Earth. It is deep black and looks almost metallic due to its glossy surface.

Coal is still used for central heating boilers both domestic and industrial and for steam and electricity generation.

▲ Figure 7.3 Open cast coal mine

Coke

Coke is produced by heating coal in coke ovens to around 1000 °C. During this process, the coal gives off methane gas and coal tar, both of which can be cleaned and reused. Coke burns cleanly and without a flame and gives out a lot of heat but it has to be mixed with coal as it will not burn by itself.

Coke is a smokeless fuel that is valued in industry because it has a calorific (heat) value higher than any form of natural coal. It is widely used in steel making and in certain chemical processes but can also be used in many domestic boilers and room heaters.

Peat

Peat is an organic material that forms over hundreds of thousands of years from the decay of plant material in the absence of oxygen, in boggy waterlogged ground. This encourages the growth of moss which forms the basis of the peat. As the plants die, they do not decompose. Instead, the organic material slowly accumulates as peat because of the lack of oxygen in the bog. Peat is a poor-quality fossil fuel which is easily cut and dried.

Peat has a high carbon content, but much less than coal, with large amounts of ash produced during combustion.

It is used in many domestic fires, room heaters and peat-burning stoves.

Sustainable, low-carbon fuels

Low carbon can be classified as fuels made from renewable sources:

- **Solar thermal** – solar thermal technology utilises the heat from the Sun to generate domestic hot water supply to offset the water heating demand from other sources such as electricity or gas.
- **Solid fuel (biomass)** – the term biomass can be used to describe many different types of solid and liquid fuels. It is defined as any plant or animal matter used directly as a fuel or that has been converted into other fuel types before combustion. When used as a heating fuel, it is generally solid biomass including wood pellets, vegetal waste (including wood waste and crops used for energy production), animal materials/wastes and other solid biomass.
- **Heat pumps** – a heat pump is an electrical device with reversible heating and cooling capability. It extracts heat from one medium at a low temperature (the source of heat) and transfers it to another at a high temperature (called the heat sink), cooling the first and warming the second. They

work in the same way as a refrigerator, moving heat from one place to another. Heat pumps can provide space heating, cooling, water heating and air heat recovery. There are several different types:

- ground source heat pumps
- air source heat pumps
- water source heat pumps
- geo-thermal heat pumps.

- **Combined heat and power (CHP)** – combined heat and power is a plant where electricity is generated and the excess heat generated is used for heating. It is used primarily for district heating systems, but micro-CHP has also been developed for domestic properties.

- **Combined cooling, heat and power (CCHP)** – very similar to CHP (see above), combined cooling, heat and power uses the excess heat from electricity generation to achieve additional building heating or cooling.

- **District heating** – this uses the principle of CHP to supply low-carbon, high-efficiency heat to high-density properties, businesses and public buildings. This system is used extensively in mainland Europe to good effect. In the UK, there are more than 17,000 district heating systems installed, mainly in city areas. District heating relies on high efficiency and areas with a large and dense population. There is one highly efficient heat source: heat from this source is transferred through plate heat exchangers into the primary pipe

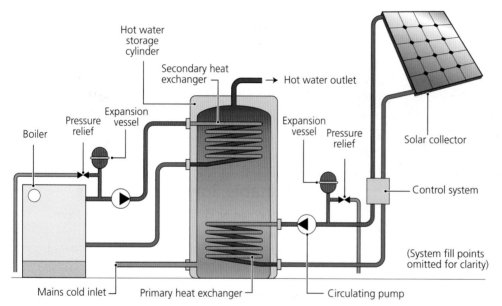

▲ Figure 7.4 Solar thermal system

▲ Figure 7.5 Biomass wood pellets

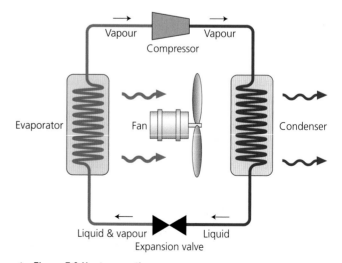

▲ Figure 7.6 Heat pump theory

network, which is highly insulated and transports the heat to individual properties. Each property has its own heat interface unit (HIU). This interface becomes the 'boiler' for each property, which allows the heat to enter the systems within the property.

INDUSTRY TIP

Look at the following clip to learn more: www.youtube.com/watch?v=MhJVsSkxg7s

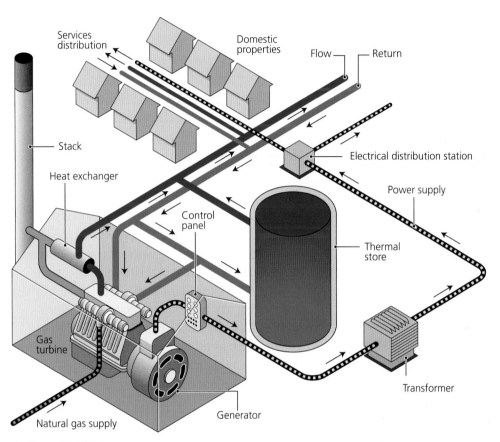

▲ Figure 7.7 A CHP system

Factors that affect the selection of fuels

There are many reasons why the fuels used in domestic appliances are chosen:

- **Availability** – the availability of fuels plays a big part when choosing the right fuel for an installation. For example, for most of the UK, natural gas is available piped to the home with no interruption of supply. However, in many rural areas, piped gas supply is many miles away. In this instance, like other fuels such as coal and oil, gas is delivered by suppliers and the customer is dependent on a regular fuel delivery. Whilst in most cases this does not pose a problem, in times of inclement weather, deliveries may be interrupted or cancelled, leaving the customer with no means of heat or cooking.

- **Appliance type** – the type of fuel available will dictate the type of appliance used and vice-versa. Some appliances may be duel-fuel types, where two types of fuel may be used in the same appliance. It must be remembered that gas appliances must be purchased in line with gas type available. A natural gas appliance CANNOT be used on an LPG supply. Similarly, a coal fired boiler will have solid fuels recommended by the manufacturer and must not be used with other types of solid fuel.

- **Fuel storage requirements** – with exception of natural gas, all fuels require storage space within the boundary of the property. With solid fuels and biomass, these can take up considerable space. Biomass also needs to be kept reasonably dry. Where oil and LPG are concerned, space may not be the issue. However, safe distances from the property to store the fuel may be dictated either by legislation or manufacturer's instructions.

- **Environmental considerations** – most fuels used in hot water and heating systems directly or indirectly create waste products that are harmful to the atmosphere. These may be by direct pollution, such as soot and sulphur emissions from coal and oil combustion, or saturation of the atmosphere by carbon dioxide (CO_2). Some gases released by fossil fuel combustion, such as nitrogen dioxide (NO_2) are extremely toxic in large quantities.

  With solid fuels, there is an added environmental problem in the form of ash and clinker left over from the combustion process that requires careful consideration and disposal. Consultation and advice should be sought from the fuel supplier and local authority as to the recommended disposal methods.

- **Smoke control legislation** – under the Clean Air Act of 1993, local authorities may declare that a district is a smoke control area. It is an offence under this Act to emit smoke from a chimney from a boiler or furnace located within an area designated a smoke control area. In Greater London, this act is used to control the emissions, not just from oil and solid fuel boilers but also biomass appliances that may also emit other noxious fumes, fly ash particulates and low levels of ozone gas.

  In some instances, certain appliances and smokeless type fuels may be exempt from the Clean Air Act. In these circumstances, advice should be sought from the Department for the Environment and Rural Affairs (DEFRA).

- **Cost** – this is a major factor when choosing the right fuel. Heating oil prices fluctuate widely, depending on the price of crude oil, whilst LPG prices remain consistently high. By far the cheapest of the fossil fuels is natural gas.

- **Client preference** – in towns and cities, the choice of fuel for heating appliances is limited. Natural gas is the preferred fuel chosen by customers for both heating and cooking appliances simply because it is readily available. Solid fuel, in the form of smokeless fuel, is still used in some areas. In rural settings, heating oil is preferred. LPG is expensive, and this is often the reason that this fuel is rejected. Many new build properties are actively seeking greener alternatives to fossil fuels, with heat pumps and electric boilers being chosen because of their very low carbon emissions.

Sources of information for fuel supply installation

Boilers, cookers, room heaters and fires require a supply of fuel, whether that fuel is piped directly to the door or delivered by a tanker. Each fuel has specific supply and storage requirements that must comply with certain documents:

- **Regulations** – there are certain regulations that fuel supply systems must comply with to maintain the safety of the property where the appliances are installed and the safety of the building occupants. Solid fuel and oil systems such as coal, coke, biomass and heating oils are simple to understand, since the fuel is readily visible. However, gaseous fuel systems such as those for natural gas and LPG tend to be much more stringent, since these fuels are at pressure and cannot be seen. Regulations include:
  - The Gas Safety (Installation and Use) Regulations 1998
  - Approved Document J of the Building Regulations – Combustion appliances and fuel storage systems (2010 edition incorporating 2010 and 2013 amendments)
  - **BS 7671** The IET Wiring Regulations.
- **British Standards** – there are many British Standards and European Standards that give recommendations when installing fuel systems. There is a comprehensive list of British Standards in Approved Document J of the Building Regulations – Combustion appliances and fuel storage systems.
- **Manufacturers' instructions** – manufacturers of appliances and components will often give advice about the installation of the fuel system to the appliance. These may sometimes conflict with the regulations and British Standards. In this instance, the manufacturer's instructions must always be followed.
- **Guidance notes** – guidance notes are produced by regulatory bodies and professional associations to

assist in the compliance of the regulations. Many guidance notes are produced by the Health and Safety Executive. They should be read in conjunction with the regulations and manufacturer's instructions.

Regulatory bodies that govern the installation of fuel systems

Before we investigate the regulatory bodies concerned with the installation of fuel systems and appliances, we must first understand what a regulatory body is.

A **regulatory body** is an organisation set up by the Government to monitor, control and guide various sectors within industry. Its aims are to protect consumers and to educate and guide installers in the ways of good practice. Occasionally, it may be necessary for a regulatory body to prosecute, in the interests of public safety, those installers who refuse to comply with regulations. In the plumbing and heating industry, it is compulsory to belong to the regulatory bodies if you engage in the installation of either gas, oil or solid fuel appliance and fuel supply systems.

KEY TERM

Regulatory body: an organisation set up by the Government to monitor, control and guide various sectors within industry.

In the plumbing and heating industry, there are three regulatory bodies:
- **Gas Safe** – Gas Safe is the UK registration body for the installation, maintenance and repair of gas installations and appliances. By law, ALL operatives engaging in domestic natural gas and LPG installations MUST be registered with Gas Safe and must hold various qualifications within the gas industry.
- **OFTEC** – OFTEC is the registration body for the installation and maintenance of oil fired heating appliances and fuel systems. Registration is

voluntary, but being a member is considered good practice. OFTEC registration means that installers are able to self-certify installations without the need for local authority intervention and inspection. OFTEC also administer recognised and authorised training courses for installers.
- **HETAS** – this is the official body that is recognised by the UK Government for approving solid fuel and biomass domestic heating systems, fuels and appliances. HETAS also manage a register of approved, competent installers and servicing businesses and oversee HETAS registered training courses.

INDUSTRY TIP

More information can be found at:
www.gassaferegister.co.uk
www.oftec.co.uk/technicians/industry-news-and-training/training-for-heating-technicians
www.hetas.co.uk/professionals/training-courses

▲ Figure 7.8 OFTEC logo

▲ Figure 7.9 HETAS logo

Storage requirements for fuels

In this section, we will investigate the methods of safely storing:

- coal
- oil
- LPG
- biomass.

Storing coal

The Solid Fuel Association recommend that coal should be stored outside of any dwelling in a purpose made bunker to protect the fuel from damage. There are a number of recommendations of how coal should be stored:

- Coal may be stored inside or outside the property.
- Coal should be covered to reduce the contaminants that can enter the fuel.
- A smooth, hard floor is important as it allows easy shovelling of the fuel.
- If the fuel is stored in a coal bunker, a slight slope on the base of a coal bunker prevents water from collecting inside the bunker. Keeping the fuel dry makes it easier to combust.
- The area around the coal bunker should be well lit to ensure safe bagging and shovelling.
- Good ventilation of the bunker helps to prevent a build of moisture, allowing the fuel to stay dry.

INDUSTRY TIP

Unlike other fuel sources, there are no special rules, regulations or restrictions when it comes to storing fuel, other than storing it away from the heating appliance or boiler.

Storing fuel oil

The following information is intended as a general guide as the regulations regarding oil storage may vary slightly, depending on the location of the installation.

Oil storage tank specifications

Generally, oil storage tanks up to 3500 litres capacity, supplying oil to a single domestic property, can be made of either plastic or steel. The actual size for any given installation will depend on the individual requirements. Any tank installed should conform to the following specifications:

- OFS T100 for plastic storage tanks
- OFS T200 for steel storage tanks
- Safety and environmental standards for fuel storage sites (HSE).

Oil tanks should be inspected annually as part of the heating system regular servicing. Oil tanks have a useful working life of around 20 years and using a tank beyond this time carries the risk of failure.

Protection of the environment

Some tank installations require a secondary containment system, known as a bund, to counteract the risk of pollution from oil spillage. This may be achieved by using an integrally bunded oil tank with secondary oil containment built in or building an oil impermeable containment wall around the tank installation. These are generally required where the tank is close to a river or water source. The bund must be capable of holding 110 per cent of the oil tank

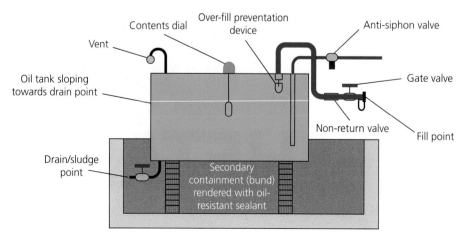

▲ Figure 7.10 An oil tank with an oil proof bund wall

contents. Usually, a standard risk assessment is required by a registered oil installer to ascertain if a bunded installation is required.

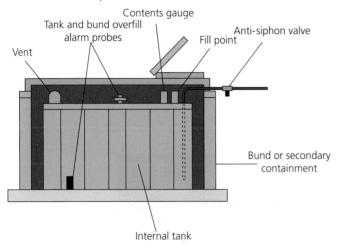

▲ Figure 7.11 Secondary containment tank

The location of fuel oil tanks

The siting of oil tanks must comply with fire separation distances to protect the fuel oil from a fire or heat source that may occur within the building itself. It is very unlikely that any fire would occur within the tank itself. The regulations state that fuel oil tanks should be sited:

- 1.8 m from non-fire rated eaves of a building
- 1.8 m from a non-fire rated building or structure such as a garden shed
- 1.8 m from openings such as doors or windows in a fire rated building or structure such as a brick-built house or garage
- 1.8 m from oil fired appliance flue terminals
- 760 mm from a non-fire rated boundary such as a wooden boundary fence
- 600 mm from any trellis or foliage that does not form part of the boundary.

If any of these requirements cannot be met, then a fire protection barrier with at least a 30-minute fire rating must be provided. A minimum separation space of 100 mm is required between the tank and any fire-resistant barrier unless the tank manufacturer specifies a larger distance.

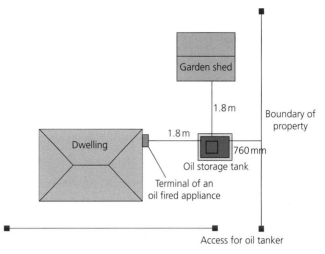

▲ Figure 7.12 Siting oil tanks

Storage of liquid petroleum gas

It should be remembered that LPG is heavier than air and will 'search' for the lowest position if a leak occurs and, although LPG has a distinctive smell, this will not be apparent until a person is at the same level as the low-lying gas.

> **KEY POINT**
>
> Above all else, LPG is extremely flammable and explosive, and the siting of any LPG storage tanks must comply with certain recommendations and any gas installation is subject to the Gas Safety (Installation and Use) Regulations 1998.

Siting the LPG storage tank

LPG storage tanks can either be sited above or below ground. Below-ground tanks are subject to ground conditions and the proximity of the water table.

According to Health and Safety Executive recommendations, there should be a minimum separation distance between the LPG storage tank and any building, boundary line or fixed source of ignition. These distances are shown in the table below. There should be no drains or gullies in the vicinity of the tank, unless those drains are protected by a water trap to prevent the gas from entering the drainage system.

▼ Table 7.5 The distances from buildings and structures for LPG storage tanks

Maximum LPG capacity			Minimum separation distances		
Of any single vessel in a group			Of all vessels in a group		
LPG capacity (tonnes)	Typical water capacity (litres)	LPG capacity (tonnes)	From buildings, boundary, property line or fixed source of ignition		Between vessels (m)
			Without fire wall (m)	With a fire wall (m)	
0.05 to 0.25	150 to 500	0.8	2.5	0.3	1
> 0.25 to 1.1	> 500 to 2500	3.5	3	1.5	1
> 1.1 to 4	> 2500 to 9000	12.5	7.5	4	1

Ventilation and conditions around the LPG storage tank

There should be plenty of room around the tank to allow good air circulation so that pockets of the heavier-than-air gas cannot build up around it should a leak occur. The area should also be kept free of rubbish and weeds and any grass should be kept short.

Protection against impact

Tanks and their associated pipework should not be located in areas where motor traffic is likely. However, if this is unavoidable, then a suitable protection barrier should be installed in the form of either bollards or crash barriers. A security fence is not suitable since this is unlikely to offer the required protection.

> **KEY POINT**
>
> Further guidance on location and spacing for vessels and requirements concerning fire wall provision reference is available in the Liquid Gas UK COP 1 Part 1 and buried vessels Liquid Gas UK COP 1 Part 4: www.liquidgasuk.org/codes/cops

The LPG gas cylinder option

It is often a good idea to start using LPG with a LPG cylinder installation until the exact usage of the installation is known. Large bulk storage tank installations only become viable when usage exceeds 2000 litres to 2500 litres per year. The average bulk storage tank user uses around 2300 litres per year.

An LPG gas cylinder installation typically uses 47 kg propane gas cylinders located at the dwelling in a lockable cabinet. This type of cylinder installation usually uses either 2 × 47 kg cylinders or 4 × 47 kg cylinders.

▲ Figure 7.13 LPG gas cylinder

Storage requirements for biomass fuels

The storage requirements of the various types of biomass fuels can influence a client's decision because key points, such as site access, space requirements and even aesthetics of the storage vessel itself need careful consideration before the installation begins. Storage considerations for biomass fuels such as wood chips or pellets should be considered early on in any biomass system design.

There are many storage options for biomass and all of them need to be watertight. Water ingress can severely affect biomass fuel quality and, as a consequence, the operation of the biomass boiler. Wood pellets, for example, that have a low moisture content will expand if they get wet and this can also damage the wood store itself.

- **Container or hook bin** – wood chips can be delivered in a container, often called a hook bin, where the fuel is delivered in a container that forms the fuel storage, which connects directly to the fuel extraction system. However, these are quite expensive because at least two bins are required.
- **Covered shed** – these are relatively cheap and easy to install. Fuel delivery is quite straightforward. For large stores, the use of manual handling equipment, such as a front-end loader or mechanical grab, is recommended.
- **Hoppers** – the hopper is a chute with extra storage capacity. They are relatively inexpensive to install. The hopper has a 'V'-shaped floor, sloped at approximately 40°. This allows the fuel to fall directly onto the boiler feed screw located at the base of the floor.
- **Silos** – these are purpose-made rigid structures that are relatively inexpensive to install but may require special delivery equipment to maintain the biomass supply.
- **Flexible silos** – these are prefabricated, collapsible structures designed specifically for smaller installations where access may be restricted, such as in a confined space or a roof space. The fuel delivery system is usually where the fuel is blown into the hopper. This system uses two hoses: one to blow in the fuel and the other to extract any dust.
- **Underground bunker** – underground bunkers are ideal for larger installations with easy access for tipper-truck delivery. The feasibility of an underground bunker will depend upon such factors as ground type, water table and cost.

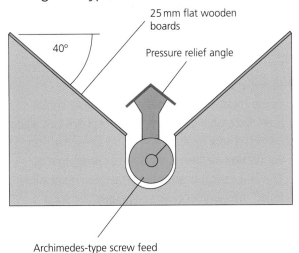

▲ Figure 7.14 A biomass hopper floor

▲ Figure 7.15 Biomass silo

HEALTH AND SAFETY

Safety masks should be worn when moving wood chips or wood pellets as the dust can pose a health risk. Dust can also pose a significant explosion risk if the area is not ventilated sufficiently.

Considerations that could affect the storage requirements of fuels

When considering the type of fuel system to be used in a dwelling, there are several factors that need special consideration:

- space for fuel storage
- delivery requirements
- safety
- weather conditions
- distribution
- proximity to dwelling.

Space for fuel storage

Space for fuel storage is a major factor when deciding which fuel system to use. Most fuels require specific distances in which to site storage vessels, tanks or silos. This may take the form of environmental concerns, as with heating oil, or explosion or fire risks, as with LPG. Where biomass is concerned, it may be the sheer mass of the fuel that is problematic.

Delivery requirements

The transportation and delivery requirements for domestic fuels differ according to the fuel, as described below.

- **Heating oil** – most oil tankers carrying domestic heating oil carry 45 m of hose. This is suitable for most installations. However, extra-long hoses can be requested. Consumers should remember to measure around any corners or obstacles when stipulating the oil tank distance from the access point.
- **LPG** – LPG bulk deliveries are usually delivered in mini-LPG tankers. These are 2.6 m wide and require access 2.75 m wide with a minimum access road width of at least 3 m. It is a requirement that a line of sight is maintained between the storage vessel and the tanker, with a maximum hose length of 40 m, to ensure the safe delivery of LPG to the bulk storage tank.
- **Coal/smokeless fuels** – solid fuels such as coal and smokeless fuels are delivered in sealed 25 kg bags. Deliveries are arranged as required.
- **Biomass** – in October 2015, rules to support sustainable fuels for the domestic Renewable Heat Initiative (RHI) came into effect for all biomass heating systems. The domestic RHI scheme aims to support homeowners and landlords who have invested in renewable heating technologies. This includes biomass, heat pumps and solar thermal panels. The idea behind the RHI scheme is to reward those people that stick to the RHI rules regarding sustainable supplies of fuel by paying them a tariff per kWh. Payments are made every three months for a period of seven years.
- **Access for biomass fuel deliveries** – biomass pellets can be blown up to 30 m via hoses, but this distance often causes problems such as clogging of the hose and break-up of the fuel. It is recommended that deliveries of biomass should be within a 20 m limit of the fuel store. A lorry of around 2 m wide will need to be able to gain access to the property.

▲ Figure 7.16 A typical small LPG tanker for domestic deliveries

Safety

All fuels, by their very nature, are flammable and some are even explosive. With this in mind, the storage of fuels should be considered with care. Here are some points to consider:

- **Confined spaces** – solid fuels, such as biomass and coal, are kept in confined spaces. There are several problems with this:
  - **Fire** – although rare, bunker and fuel store fires can occur, especially where the store is directly connected to the boiler room. Generally, biomass wood chips are too wet to ignite but if they begin to de-compost they can self-ignite. Liquid fuels, such as kerosene, do not usually combust unless they are either atomised or vaporised, but they can become dangerous near excess heat because the vaporisation process begins at a relatively low temperature of around 65 °C.
  - **Explosion** – LPG, because it is heavier than air, settles at low level. In the event of a leak, the build-up of gas may not be noticed, despite the fact that a chemical called mercaptan is added to make the gas detectable by smell.
    Some fuels, such as coal and biomass, create dust. Excessive dust in the atmosphere can also be extremely explosive. A good air-extraction ventilation system is vital in confined spaces. The HSE recommends building in an 'explosion relief' into any storage space used for solid fuels that create dust. This can be a plywood panel in a bunker or silo that creates a weak spot to release the explosive energy. The HSE produces a fact sheet, HSG103 Safe handling

of combustible dust, which is available from its website at: www.hse.gov.uk

- **Carbon monoxide build-up** – for any confined space close to the place of combustion of a fossil fuel, combustion problems may lead to a build-up of carbon monoxide (CO), which is highly toxic. An audible CO alarm installation is recommended in fuel storage facilities.
- **Slips, trips and falls** – fuel stores of all kinds are dangerous places. Build-up or spillages of fuel create slip, trip and fall hazards. Some hazards may be limited by fuel store design. However, where solid fuel and biomass are concerned, the fuel storage space height may be high and so safety nets and harnesses should be considered.
- **Fuel delivery** – fuels are delivered to properties by either tanker (heating oil, LPG, biomass) or flat-bed truck (solid fuels – coal, coke, etc.). Care should be exercised while fuel deliveries are taking place. Follow the recommendations of the fuel delivery driver.
- **Personal hygiene** – there should be no reason for the fuel itself to be handled. However, in the event that contact with the fuel must be made, always wear appropriate PPE, such as overalls, gloves, hard hat, goggles and respirator (especially in dusty environments).

Weather conditions

The prevailing weather can have a severe effect on the storage of fuels. Bad weather, such as wind, rain, hail and snow, is often a cause for late deliveries and even cancellations of fuel deliveries, especially in rural areas. In almost all cases, fuel is delivered by large tanker or flat-bed vehicles that find it next to impossible to negotiate small, narrow roads when the weather conditions are poor. While the weather can be unpredictable in the UK, good planning of fuel deliveries can reduce the impacts of bad weather. Ordering more when severe weather is forecast can often mean the difference between running out of fuel and keeping the heating on.

Similarly, bad weather can render some fuels, such as wood chip and wood pellet biomass, almost unusable. Coal and coke too suffer from the negative effect of excessive rain, whereby the fuel can become too wet to burn effectively. Wood pellets swell from the effects of the rain and these then clog fuel delivery

to the fuel bed of the boiler. Wood chip biomass can begin to de-compost if it gets too damp and this, paradoxically, can cause the fuel to heat internally and spontaneously combust.

> **KEY POINT**
> It is vital that fuels are kept dry and that they are delivered in good condition for optimum combustion efficiency to occur.

Distribution

The distribution of fuels becomes a vital consideration, especially the further outside a major town or city you live. Natural gas coverage in the UK through the national grid stands at around 7000 km of pipelines, but there are still many rural areas that are too far away from the grid for a supply to be economically viable. In these cases, other fuel supplies have to be considered.

By far the most viable fuel in rural areas is domestic heating oil, otherwise known as C2 grade, 28 second viscosity kerosene. Distribution of this still vital fuel is nationwide. However, kerosene poses an environmental risk if leakage occurs, especially where the installation lies close to a watercourse, river or stream or where the water table is high.

LPG distribution is also very comprehensive, with most areas in the UK reachable by tanker. However, there are certain restrictions with LPG that do not exist with heating oil, such as that the delivery driver must have line of sight to the LPG storage tank at all times during delivery of the liquid gas. LPG is also very expensive as a domestic heating fuel.

Coal and coke solid fuels continue to be readily available all over the UK, although many areas now forbid the use of these fuels because of the environmental pollution they release. If solid fuel is to be used, then local authority advice should be sought.

The use of biomass in rural and suburban areas is permitted under the permitted development legislation, which came into force in 2008. However, some areas, especially suburban districts, may put restrictions on its use if they lie within a smoke control zone. Outside of these zones, there are no major restrictions other than a requirement not to emit 'dark smoke'. In most cases, domestic biomass does not fall into this category.

Proximity to dwelling

The installation of fuel storage and its requirements with regard to the proximity of the dwelling are covered elsewhere in this chapter.

2 COMBUSTION PROCESSES OF FUEL SUPPLY SYSTEMS

Combustion is an exothermic chemical reaction in which a fuel reacts violently with oxygen to produce heat and light.

The combustion processes

For the combustion process to take place, there must be fuel, oxygen and ignition present. The fuel can be a solid such as wood, a liquid such as petrol, or a gas such as propane. Oxygen is known as an oxidiser or an oxidising agent. To create combustion or fire, we a need third element in the form of heat or an ignition source. These three elements, fuel, oxygen and ignition combine into what is known as the fire triangle.

KEY POINT

All three elements, fuel, oxygen and ignition, need to be in place for combustion to happen. Take any of the three away and combustion will not occur.

▲ Figure 7.17 The fire triangle

The main constituents of complete and incomplete combustion

Combustion can take two forms:

- **Complete combustion** – complete combustion occurs when there is enough oxygen for the maximum heat to be produced during the combustion process. Complete combustion of hydrocarbons produces CO_2 and water vapour. This can be expressed as follows:

$$CH_4 + 2O_2 \rightarrow 2H_2O + CO_2, \text{ or}$$

$$\text{hydrocarbon} + \text{oxygen} \rightarrow \text{water} + \text{carbon dioxide}$$

- **Incomplete combustion** – incomplete combustion occurs when there is not enough oxygen present to allow the combustion process to turn all of the carbon present in the gas into CO_2. When this occurs, the products of combustion are different than those produced for complete combustion:

$$O_2 + CH_4 \rightarrow H_2O + CO$$

In this instance, there is not enough oxygen (O_2) to allow proper combustion to take place and so carbon monoxide (CO) is produced.

Flame picture

The flames produced by appliances can be a good indication of whether complete combustion is taking place within the appliance and at the burner:

- **Complete combustion** – produces a stable blue flame with a bright blue inner cone and a darker blue outer cone. The flame is attached to the burner and is stable. It can also be called a neutral flame.
- **Incomplete combustion** – largely due to lack of combustion air supply, incomplete combustion produces an overly large flame that has no form or shape that is predominantly yellow in colour. The flame appears to be without direction and floppy. It is also known as a carburising flame as it produces large amounts of carbon soot.

Flame pictures

When gas and oxygen are burned in a mixture, a chemical reaction takes place and a flame is produced. This happens when a mixture of gas and air is heated to its ignition temperature.

There are two main types of flame relevant to the combustion process and they are **pre-aerated** and **post-aerated**.

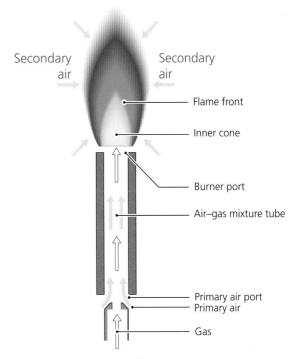

Pre-aerated flame

▲ Figure 7.18 Four major components make up a pre-aerated burner. They are the primary air port, injector, mixing tube and the burner incorporating the flame ports

KEY TERMS

Pre: means 'before'.

Pre-aerated flame: air is entrained in the mixing tube before ignition.

Post: means 'after'.

Post-aerated flame: air is drawn for combustion from the surrounding air once the flame is lit, often resulting in a loose yellow, floppy flame.

A pre-aerated flame is the most common type used in domestic gas appliances. This is where some of the air required for combustion is pre-mixed in a tube before it is ignited at the outlet such as the burner. The air that is provided is called primary air.

A pre-aerated flame is smaller than a post-aerated flame but both give off the same amount of heat. The pre-aerated flame burns with a well-defined inner cone inside the outer mantle.

A post-aerated flame takes its air for combustion from the surrounding atmosphere after the gas leaves a pipe or tube and has a luminous appearance. In other words, all of the air for the combustion process is provided once a gas leaves the burner ports.

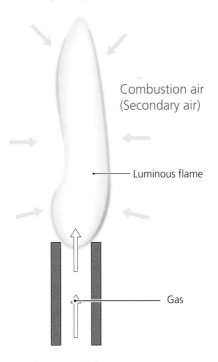

Post-aerated flame

▲ Figure 7.19 The post-aerated flame which receives secondary air only for combustion will result in a loose yellow, floppy flame. Similarly, a pre-aerated flame with the primary air port blocked will also produce the same result

Causes of incomplete combustion

Incomplete combustion can be catastrophic for the appliance. Incomplete combustion can lead to an appliance sooting up, causing restricted flow of the products of combustion through the appliance combustion chamber. This can lead to the appliance overheating and scorching taking place with the result that carbon monoxide (CO) is released into the room where the appliance is installed. Incomplete combustion is caused by:

- **Lack of oxygen** – fossil fuels require oxygen at specific ratios for complete combustion to take place. For natural gas, for instance, this ratio is 2:1 or 2 atoms of oxygen for every atom of natural gas. If there is not enough oxygen for complete combustion, then incomplete combustion occurs with the result that carbon monoxide (CO) is produced.
- **Too much fuel** – appliances are designed to burn a specific quantity of fuel for the given supply of air. If there is more fuel than there is air supply, then incomplete combustion will occur because the air supply available is inadequate to combust all of the fuel completely.
- **Vitiated air** – the phrase vitiated air simply means air that has been made impure. This can occur when there are many people in a room all breathing out carbon dioxide. If the supply of fresh air entering the room is insufficient to maintain the O_2/CO_2 balance, then the air is made impure by the extra amount of CO_2 being exhaled. Vitiated air is usually a problem with open flued appliances where the combustion air is taken from the room where the appliance is installed.
- **Flame impingement** – usually occurs when the flames inside an appliance touch a colder surface, such as the heat exchanger in the combustion chamber. The cold surface causes the flame to momentarily cool down and this prevents complete combustion of the fuel.

KEY TERM

Vitiated air: the word 'vitiated' simply means 'to be made impure', so 'vitiated air' is air that has been made impure.

INDUSTRY TIP

Flame impingement is also known as flame chilling.

The signs of incomplete combustion

When complete combustion occurs, the flame picture (or the colour of the flame) indicates the state of the flame. There are many signs that incomplete combustion of a fuel is occurring within an appliance. If any of the below are detected, they should be investigated and rectified as a matter of urgency:

- **Yellow, floppy flame** – these are usually much larger than normal burner flames and appear to be without direction. In effect, the flame is searching for oxygen with which to combust. The flame produced is yellow because the CO_2 within the flame is being combusted with the fuel in the absence of oxygen. This effectively reacts to produce carbon monoxide and carbon soot deposits.
- **Sooting** – occurs because of the lack of oxygen in the combustion process. The soot produced is black carbon soot that has a powdery quality. It will quickly coat the inside of a heat exchanger, affecting the burner and further restricting air supply. This leads to further sooting. Eventually the heat exchanger will become blocked, preventing the combustion gases from escaping out of the building through the flue or chimney.
- **Staining around the appliance** – in many instances, staining occurs around the appliance when incomplete combustion takes place. This appears as a dark brown scorch or stain on the appliance and/or the wall or fireplace where the appliance is installed. Any evidence of staining must be investigated and the cause identified.

Carbon monoxide: its effects and how to prevent CO poisoning

Carbon monoxide is a silent killer. It has no smell and gives no sign or warning that it is present. Less than 1 per cent of carbon monoxide within a room or space is enough to kill everyone in it. Yet it is preventable.

By using correct installation practices and regular appliance service and maintenance, CO poisoning due to faulty appliances can be prevented and eliminated.

In this section, we will investigate carbon monoxide poisoning, its effects on humans, the symptoms and the ways of warning of its presence.

What is carbon monoxide poisoning?

Carbon monoxide is a poisonous gas. It has no smell, it has no taste. Breathing in carbon monoxide enables the gas to enter the bloodstream, where it mixes with the body's **haemoglobin**.

KEY TERM

Haemoglobin: the part of the red blood cell that carries vital oxygen to the brain. When CO enters the bloodstream, the haemoglobin becomes carboxyhaemoglobin, which effectively blocks the red blood cells from carrying the oxygen to the brain and this causes the body's cells and tissue to fail and die.

The symptoms of CO poisoning

The symptoms of CO poisoning are often mistaken for flu and even food poisoning, especially at low exposure levels. Symptoms typically include:

- a tension-type headache
- a redness to the cheeks – the look of being flushed (without the fever)
- dizziness
- nausea and sickness
- tiredness and confusion
- severe stomach cramp
- shortness of breath.

In most cases, the symptoms become progressively worse with prolonged exposure. If the exposure is severe, then further symptoms may present themselves:

- impaired mental state or personality changes
- **vertigo**
- ataxia, or loss of physical co-ordination caused by the underlying damage being caused to the brain
- chest pain
- seizures
- eventually, loss of consciousness
- death.

KEY TERM

Vertigo: the feeling that the sufferer or objects around them are moving when they are not; feels like a spinning or swaying movement.

HEALTH AND SAFETY

The best way of treating someone who is suspected of having CO poisoning is to remove them to clean oxygenated air as soon as possible. However, this may present problems if the room is full of CO. The overriding concern is not to become a victim yourself. If a high concentration of CO is present, then the time that can be spent within the environment is very limited indeed. In heavily saturated areas death can occur in a matter of minutes.

Remember! Around 10–15 per cent of people who have survived CO poisoning suffer long-term health issues as a result.

The purpose of CO detectors (to **BS EN 50291: 2018** Gas detectors. Electrical apparatus for the detection of carbon monoxide in domestic premises)

The purpose of a CO detector is to warn of the presence of small amounts of carbon monoxide in a space or room. Different types of CO alarm work by different methods:

- **Biometric sensor** – a gel changes its colour when it absorbs carbon monoxide. The colour change triggers the alarm.
- **Metal oxide semiconductor** – this is based on a silica chip detecting carbon monoxide. When CO is detected, the chip's electrical resistance is lowered, and this sounds the alarm.
- **Electrochemical sensor** – here, electrodes are immersed in a chemical. Changes in electrical current are detected when CO is present, which activates the alarm.

The requirements for ventilation

Combustion air requirement

Oxygen is required for the combustion process. The amount of oxygen needed for complete combustion depends on the type of fuel. For natural gas (CH_4), the oxygen to fuel ratio for complete combustion to occur is 2:1.

Since the atmosphere of Earth is not 100 per cent oxygen, the actual amount of air needed for complete combustion of natural gas is much more than the 2:1 ratio. The atmosphere of Earth consists of:

- nitrogen: 78.08 per cent
- oxygen: 20.95 per cent
- argon: 0.93 per cent
- carbon dioxide: 0.04 per cent.

For calculation purposes, we can say that our atmosphere is only 20 per cent oxygen. Since a ratio of 2:1 of oxygen is required and oxygen makes up only one fifth of the atmosphere, for complete combustion of natural gas to occur, the ratio of air to natural gas for complete combustion is 10:1.

A ratio of 2:1 can be expressed as follows:

$$2O_2 + CH_4$$

This means two atoms of O_2 plus one atom of CH_4. This combination when combusted produces:

$$CO_2 + 2H_2O$$

Or one atom of CO_2 and two atoms of H_2O.

So, complete combustion can be expressed as the following:

$$2O_2 + CH_4 \rightarrow CO_2 + 2H_2O$$

Cooling air requirement

Some appliances, such as room sealed, or balanced flue appliances do not require a separate provision for combustion air. The combustion air is taken directly from outside via the flue arrangement. However, some cooling air may be required to prevent overheating of the appliance or the space where the appliance is installed. This can either be taken directly to the outside via airbricks and ventilators or to another suitable internal space via unrestricted grilles.

The methods of supplying ventilation

Ventilation is necessary to remove stale air from buildings and replace it with fresh air. Ventilation in buildings helps to:

- moderate internal temperatures
- control internal humidity
- replenish necessary oxygen
- reduce condensation, odours, dust, bacteria and carbon dioxide
- create air movement – this improves thermal comfort.

There are two methods of supplying ventilation:

- natural
- mechanical.

> **INDUSTRY TIP**
>
> Ventilation of a building is subject of Building Regulations Approved Document F – Ventilation 2010 with 2013 amendments.

Natural ventilation for combustion air

All open-flued appliances that burn fossil fuels will need to replace the air in the room where the appliance is installed, simply because it is the air in the room that is used during the combustion process. In many cases this means the installation of air vents/grills direct to the outside air.

Where an appliance is fitted in a cupboard or compartment, then air vents/grilles are required to be fitted either direct to the outside air or to the inside room where the cupboard or compartment is located. However, if air vents draw air from another room, then the room must draw air direct from the outside.

The air provided through grilles and air vents performs two tasks:

- It provides air needed for the combustion of the fossil fuel.
- Ventilation is required for the correct operation of the flue.

In some cases, where the appliance is fuelled by natural gas and is below 7 kW input, an air vent may not be required. In this instance, the air is provided by 'adventitious air' where the air that naturally infiltrates the room through draughts, under doors and through cracks is sufficient to allow replacement of the air used for combustion.

Locating air vents and grilles

These should be placed where the occupants are not tempted to cover them due to noise or draughts. Draught discomfort can be avoided by placing the air vents close to the appliance, such as floor vents or by drawing supplementary air from intermediate spaces such as halls, or by ensuring a good mix of incoming air. Air vents MUST NOT be placed within a fireplace opening.

When installing air vents, it must be remembered that the cavity in the wall must be sealed to the air vent opening and any fly/insect screens must be removed from the grille. It must also be remembered that air vents need to be properly sized in accordance with the input rating of the appliance and the type of fuel being combusted to give the correct amount of air required.

Supply of air for combustion to gas fired appliances

An air supply for gas fired appliances should be provided in accordance with the following British Standards:

- **BS 5871.3:2005** Specification for the installation and maintenance of gas fires, convector heaters, fire/back boilers and decorative fuel effect gas appliances. Decorative fuel effect gas appliances of heat input not exceeding 20 kW (second and third family gases)
- **BS 5871.2:2005** Specification for the installation and maintenance of gas fires, convector heaters, fire/back boilers and decorative fuel effect gas appliances. Inset live fuel effect gas fires of heat input not exceeding 15 kW, and fire/back boilers (second and third family gases)
- **BS 5440.2:2009** Flueing and ventilation for gas appliances of rated input not exceeding 70 kW net (first, second and third family gases). Specification for the installation and maintenance of ventilation provision for gas appliances.

Mechanical ventilation for combustion air

Combustion air can be provided mechanically through the use of fans, provided a positive pressure zone is maintained in the room where the boiler is installed. It is the operation of the fan blowing combustion air into the room that creates a positive pressure which is necessary to ensure that the products of combustion do not back spill into the room. Negative pressure zones must be avoided. Generally, mechanical ventilation methods require the use of an interlock, which is an electrical switch that shuts down the gas appliance should the ventilation fan fail. These systems are not usually specified for domestic properties and are more likely to be installed in commercial boiler rooms.

Installation practices for ventilation

Ventilation is required to provide combustion air where open flued and flueless appliances, such as open fires, gas fires, gas boilers and gas cookers are installed. It should be remembered that ventilation needs to be:

- **Adequately sized** – ventilation should be sized to provide sufficient replacement air to supplement the air used during the combustion process.
- **Continuous size** – it should be of continuous size throughout the thickness of the wall.
- **Sleeved** – it must be sleeved through the wall so that any cavities in the wall construction are sealed off from the ventilation opening. This is to prevent gas entering the cavity in the event of a gas escape.
- **Permanently open** – ventilation openings must be permanently open to the outside and must not be open/closed type vents.
- **Fly screen removed** – fly screens must be removed to prevent them from becoming blocked with dust, spiders' webs, etc.
- **Correctly positioned** – ventilation must be correctly positioned in accordance with the appliance manufacturer's instructions, Building Regulations Approved Document F – Ventilation 2010 with 2013 amendments and British Standard **BS 5440 Part 2** Specification for the installation and maintenance of ventilation provision for gas appliances.

Ventilation categories

Ventilation is required for all the categories below but is calculated in a different way for each individual situation:
- flueless appliances
- open-flued appliances
- open-flued appliances located in a compartment
- room-sealed appliances located in a compartment.

Adventitious ventilation

Buildings admit draughts from the outside and this is called 'adventitious ventilation' which equates to about 35 cm^2 of free air. When air for combustion is calculated, 5 cm^2 is required for each kW of gas used, which in effect means the first 7 kW of the input rating of a gas appliance can be deducted. It is worth noting that because of the advancement of draught reduction in many new buildings, adventitious allowance may be reduced because of highly effective draught proofing incorporated into the building structure or design.

IMPROVE YOUR MATHS

A 16 kW open-flued boiler is installed in a room.

$$16 \text{ kW} - 7 \text{ kW} = 9 \text{ kW}$$

Ventilation for 9 kW = 9 kW × 5 cm^2 = 45 cm^2

A vent providing 45 cm^2 of free air should therefore be located on an outside wall to provide air supply for combustion.

KEY POINT

A vent designed to provide air for combustion should be marked with the amount of free air it provides.

If an open-flued appliance is installed in a compartment then adventitious air is never deducted from the calculation.

Similarly the same rule applies to flueless appliances such as cookers or to open-flued decorative fuel effect (DFE) gas fires, unless the DFE manufacturer states otherwise.

Decorative fuel effect gas appliances are covered in **BS 5871 Part 3** which states that they require 100 cm^2 of ventilation unless manufacturers' instructions differ.

A tumble dryer also usually requires 100 cm^2 and this depends on the room size where it is located. A fire/back boiler unit (BBU) will require a minimum of 100 cm^2 unless the manufacturer's instructions state otherwise.

Gross and net

When working out your ventilation requirements, calculate either in gross or net. Gross is the higher heating value which takes into account the latent heat of the vaporisation of the water in the combustion process, whereas net assumes that the latent heat of the water in the fuel has not been recovered. Therefore, when calculating in net, allow 5 cm^2 per kilowatt input and for gross, allow 4.5 cm^2 per kilowatt input.
- To convert gross to net, divide by 1.11.
- To convert net to gross, multiply by 1.11.

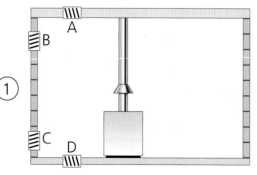

In a room

One vent required in either position A, B, C or D direct to outside air.
(5 cm^2/kW of rated heat input above 7 kW)

▲ Figure 7.20 Ventilation options for open-flued appliances in a room or internal space

Compartment ventilation

A compartment is an enclosure designed to house a gas appliance. This will require high- and low-level ventilation unless the manufacturer's instructions state otherwise. The purpose of vents located on an **appliance compartment** is to provide air for complete combustion, to enable the correct operation of the flue and for appliance cooling. There is no adventitious air allowance in compartments.

KEY TERM

Appliance compartment: an enclosure specifically designed or adapted to house one or more gas appliances.

If an appliance is to be installed in a compartment then the following rules must be adhered to:

- Both of the air vents should communicate with the same room or space.
- Vents must be located at both the lowest and highest practicable locations within the compartment.
- Both of the air vents must be on the same wall to the outside.
- If a compartment contains two or more appliances, the aggregate maximum input rating of both of them must be added together when calculating vent sizes.
- Range-rated appliances must be calculated using their maximum input rating.
- There should be 75 mm clearances around the front, the sides, and above and below an appliance situated in a compartment unless the manufacturer's instructions state otherwise.
- An appliance compartment should never be used for storage purposes because of the risk of fire and blocking of air vents.

Small rooms such as cloakrooms and WCs are not usually considered as compartments but just like small appliance compartments they can be susceptible to vitiation which is caused by a downdraught. In addition heat loss from appliances can cause high ambient temperatures to occur. Therefore an engineer should carefully assess such installations and decide whether or not to treat them as compartments and install the required ventilation.

KEY POINT

If combustion air for open-flued appliances located in compartments is drawn from two different external walls, cross-flow of ventilation can occur which can result in unsatisfactory burner and/or flue performance.

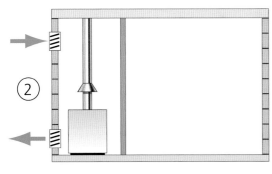

In a compartment

Ventilated direct to outside air 10 cm^2/kW of rated heat input at low level and 5 cm^2/kW of rated heat input at high level

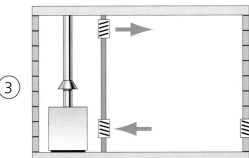

In a compartment

Ventilated via a room or internal space 20 cm^2/kW of rated heat input at low level and 10 cm^2/kW of rated heat input at high level. The room is ventilated as in Figure 7.20

▲ Figure 7.21 Compartment ventilation for open-flued appliances

IMPROVE YOUR MATHS

If we calculate the ventilation requirement for an open-flued appliance of 15 kW input installed in a compartment with vents facing externally, we would use the formula of multiplying the lower vent by 10 cm² times the kilowatt input of the appliance and the higher level multiplied by 5 cm².

Therefore:

$10 \text{ cm}^2 \times 15 \text{ kW} = 150 \text{ cm}^2$ at the lower level

$5 \text{ cm}^2 \times 15 \text{ kW} = 75 \text{ cm}^2$ at the higher level

If the vents in the compartment face internally then the following equation applies: input rating of the appliance times 20 cm² for the lower vent and 10 cm² for the higher vent.

Therefore:

$20 \text{ cm}^2 \times 15 \text{ kW} = 300 \text{ cm}^2$ at the lower level

$10 \text{ cm}^2 \times 15 \text{ kW} = 150 \text{ cm}^2$ at the higher level

Ventilation in series for open-flued appliances

Sometimes open-flued appliances are located some distance away from an outside wall and in order to provide air for combustion it will be necessary to install ventilation via another wall between the appliance and the outer wall. This method is called installing air vents 'in series' and the following rules must be adhered to.

- When venting through one wall to an appliance then the internal vent will need to be the same size as the one located on the outer wall.
- If venting through more than one room the internal vents need to be 50 per cent bigger than the external vent.
- If air is provided from the outside air as shown in Figure 7.22 then both internal and external vents remain the same size.

> **KEY POINT**
>
> There is never any allowance for adventitious air for appliances installed in compartments.

In Figure 7.23, the external air vent number 1 remains the same but both internal vents 2 and 3 are 50 per cent bigger than the external vent.

In Figure 7.24, vents A and B are sized for the compartment as normal in accordance with the maximum input rating of the appliance. However, both internal vents 2 and 3 are still required to be 50 per cent bigger than vent 1.

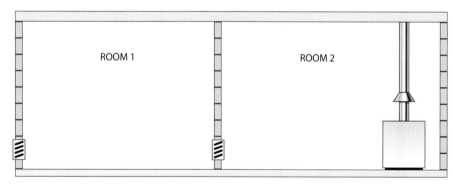

▲ Figure 7.22 Venting through one internal wall

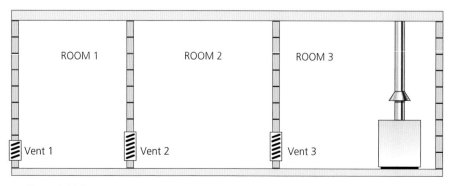

▲ Figure 7.23 Venting through two internal walls

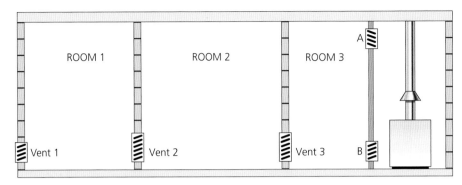

▲ Figure 7.24 Venting through two internal walls and into a compartment

> **KEY POINT**
>
> A low-level vent in an internal wall should be located no more than 450 mm above floor level in order to reduce the spread of smoke in the event of a fire.

Ventilation calculations for room-sealed appliances

Room-sealed appliances do not require air for combustion as they receive this air from the outside via the special ducts in the flue which are balanced to atmospheric pressure and the appliance itself is sealed from the effects of the room – hence the term room-sealed. The reason the air is needed is for cooling and to keep the appliance operating at its designed temperature.

When room-sealed appliances are installed in compartments, ventilation is required for cooling only and is provided from vents located at both high and low level at the greatest possible vertical distance apart to encourage a convective flow.

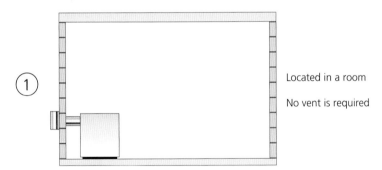

(1)

Located in a room

No vent is required

▲ Figure 7.25 Room-sealed appliances do not require vents

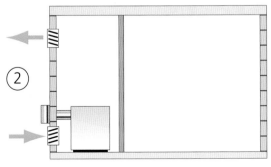

(2)

Compartment

If the appliance is ventilated direct to outside air then 5 cm² kW of free air ventilation is required both at high and low level

▲ Figure 7.26 Compartment ventilation for room-sealed appliances to outside air

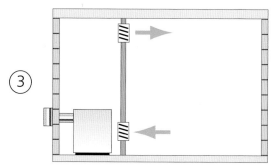

Compartment

Ventilated via a room or internal space 10 cm²/kW of rated heat input at both high and low level

▲ Figure 7.27 Compartment ventilation for room-sealed appliances via room/internal space

Many manufacturers now state that ventilation is not required for compartment installations because of the low surface temperature of their room-sealed boilers. The correct operating temperature of an appliance relates directly to its performance and the following information is essential to maintain optimum performance.

It is important for the purposes of cooling not to exceed air temperatures of:

- 25 °C up to 100 mm from finished floor level (FFL)
- 32 °C at 1.5 m above FFL
- 40 °C at 100 mm below and up to ceiling level.

Vents and grilles

- Ventilation must allow free air to pass and vents should not be closable.
- The air cannot be taken from bath or shower rooms.
- Vents should not incorporate any gauzes or screens.
- Air vent openings should be not be larger than 10 mm and no smaller than 5 mm.
- Air vents located externally should be located so that they will not become blocked.
- Air vents located on internal walls should not be fitted any higher than 450 mm.
- An air vent should never penetrate a protected shaft or stairway.
- Vents can be made from a range of materials such as terracotta, plastic, brass and aluminium.

Plastic vents are installed through a wall using a core drill. The free area for combustion is printed on the vent.

Sometimes a more aesthetically pleasing style of vent may be required to complement a design within a dwelling, but in any event the free air rating of the fitting will still be shown on its surface.

An engineer should always verify the free air admitted through a vent and this should preferably be marked on the vent. While inspecting a vent it can sometimes be seen that the space between two vents has become blocked either by debris or even intentionally by an owner to prevent unwanted draughts. Therefore the removal of the outer cover of the vent may be required.

As a practical solution to the problem of customers blocking vents, an innovative device which incorporates baffles to reduce the effect of draughts has been developed. However, tests have shown that the vent may not always produce the required amount of air for combustion. Therefore a spillage test should always be carried out when one of these devices is installed.

The brown plastic terracotta part will be located on the external wall to blend with brickwork; the free air volume of the device is then protected with ducting which incorporates a baffle system and the device eventually terminates with a white plastic vent inside the building.

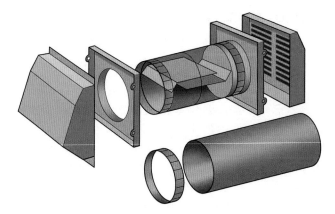

▲ Figure 7.28 A circular option incorporating a baffle system to fit a cored hole through a wall

High- and low-level vents should be located at the greatest vertical distance apart.

If a duct comes from high level, the duct should terminate below the level of the burner. It is not acceptable to duct from low to high level.

On louvre doors the total free area should be equal to that of the calculated high- and low-level vents.

A typical louvre door requires the gap between each of the louvres to be measured individually by the engineer to calculate the free air that can pass through them.

Vents for compartments

Compartments containing gas appliances should be labelled accordingly.

> **IMPORTANT**
> **DO NOT BLOCK THIS VENT.**
> **DO NOT USE FOR STORAGE.**

▲ Figure 7.29 Typical example of label to be fixed on a cupboard

The purpose of vents located on an appliance compartment is to provide air for complete combustion, to enable the correct operation of the flue and for appliance cooling.

External vents need to be more heavy duty than internal vents given that they are for use on exterior walls where strength and security are important.

An external vent can be constructed from a range of materials and be of various sizes which are sufficient for the requirements of the appliance or appliances installed in the compartment.

Ducting requirements

A duct which connects two air vents should be no longer than 3 m in length and should have no more than two 90° bends in the design. Ducting size should be increased by 50 per cent for each extra part or section after every 3 m. Flexible ducting should not be used as this could cause sagging low points and the potential for condensation to form.

> **KEY POINT**
> Ducting runs, where unavoidable, should be kept to a minimum of not more than 3 m in length.

Hit-and-miss type vents should never be used as a means of ventilation for gas appliances because they can be adjusted to close all ventilation and, even if it were permanently left open, the gauze, which is never acceptable as a vent component, could become blocked, thereby reducing any air passing through into the room.

> **KEY POINT**
> Hit-and-miss vents can be opened or closed therefore they are not suitable for the combustion ventilation of gas appliances.

Flueless ventilation requirements

When setting up flueless ventilation:
- Ventilation must be provided directly from outside and it should be possible to open a window or a hinged panel in the same room in accordance with the Building Regulations.
- Ventilation can only be conveyed from one room to another if it is ducted.
- There must be sufficient room volume and there are restrictions on maximum input ratings of a flueless appliance.
- Smaller rooms will need increased purpose-provided ventilation.
- Manufacturers' instructions must always be adhered to.
- Flueless space heaters are constrained by W/m³ of room volume and individual appliance rated input. Manufacturers' instructions will give guidance on specific appliance installation requirements.

▼ Table 7.6 Flueless ventilation guidance (selected section)

Type of appliance	Maximum rated input limit (net)	Room volume (m³)	Permanent vent size (cm³)	Openable window or equivalent also required
Domestic oven, hotplate, grill or any combination thereof	None	< 5	100	Yes
		5 to 10	50	
		> 10	Nil	
Instantaneous hot water heater	11 kW	< 5	Installation not permitted	Yes
		5 to 10	100	
		> 10 to 20	50	
		> 20	Nil	

If the room or internal space containing these appliances has a door which opens directly to the outside, then no permanant opening is required.

The full guide to the minimum permanent opening free area for flueless appliances can be found on Table 6 of **BS 5440 Part 2:2009**.

Flueless ventilation calculation

If, for example, a flueless water heater of 10 kW net were located in a room measuring 2 m × 2 m × 2.4 m, the total volume of the room would be 9.6 m³. By referring to the flueless ventilation guidance it can be seen that if the room was between 5 and 10 m³, then 100 cm² ventilation would be required, as well as an openable window or equivalent such as a hinged panel. If a domestic cooker were located in the same room it would require a permanent vent and 50 cm² with an openable window or equivalent. Details, guidance and ventilation requirements for other flueless appliances are given in Table 6 in **BS 5440 Part 2:2009**.

Intumescent air vents

Intumescent air vents are special types of air vents that are designed to close in the event of a fire to stop the spread of smoke and fumes. It is important to check the vents to ensure the correct free air space. **BS 5440 Part 2:2009** describes them as an assembly specified for preventing the spread of fire, consisting of a metal louvre or grille with an intumescent block secured behind it which incorporates a latticework of holes to provide continuous ventilation but which will expand and close in the event of extreme heat build-up such as in a fire. When fitted to doors the assembly usually has a louvre or grille on both sides.

KEY TERM

Intumescent air vents: an intumescent vent contains a substance which swells when exposed to heat and blocks the free air opening which will help prevent the spread of smoke in a fire.

Multi-appliance installations

Where a room or an **internal space** contains more than one gas appliance then the air vent free airs should be calculated from the greatest of the following:

- the aggregate maximum rated heat input of all flueless space heating appliances
- the aggregate maximum rated heat input of all open-flue space heating appliances*
- the greatest maximum rated heat input of any other type of appliance in the same area, for example, this could be an oil boiler.

* There is an exception to the second point when there is a situation where the interconnecting wall between two rooms has been removed and, as result, the room contains only two similar chimneys each fitted with a similar gas fire of an individual rating less than 7 kW. In this situation an air vent may not be required.

KEY TERM

Internal space: an indoor space not classified as a room because it is either a hall, passageway, stairway or landing.

IMPROVE YOUR MATHS

A room or an internal space contains:

- three gas appliances with a volume of 9 m³
- an open-flued boiler with an input of 25 kW gross
- a flueless water heater with an input of 11 kW net
- a balanced flue cooker of 28 kW input net.

We must first work out the individual ventilation requirements of each appliance.

Open-flued boiler:

$25 \div 1.11 = 22.52$ (converting gross to net)

$22.52 - 7 = 15.52$ (removing allowance for adventitious air)

$15.52 \times 5 = 77.6$ cm² (multiplying the final figure by the factor of 5 cm² to find out air required)

By referring to the flueless ventilation guidance it can be seen that the flueless water heater will require 100 cm² plus an openable window.

A balanced flue cooker of 28 kW input will require no ventilation.

Therefore a 100 cm² vent would be the correct size with the addition of an openable window in this instance.

Effects of fans

Warm air heater fans, fans in flues of open-flued appliances, ceiling (paddle) fans, room extractor fans, externally ducted tumble dryers, cooker hoods and fans used for extracting radon gas all can potentially reduce the ambient pressure to an appliance and therefore adversely affect the operation of a flue; they should therefore be operated during spillage testing. The spillage test will involve first testing with the fan off and then testing with any fan on to assess the performance of the appliance. Where applicable the fan to be tested should operate in both directions. An engineer must make an engineering judgement based on specific testing procedures.

BS 5440 Part 2 suggests that that 50 cm² extra free air space added to the existing air space will usually solve any problems with spillage from gas appliances that is related to ventilation problems. Once the extra ventilation has been provided then the appliance should be tested again for spillage.

> **KEY POINT**
>
> When a fan is deemed to be causing the spillage of an open-flued appliance, **BS 5440 Part 2** suggests that an additional 50 cm² be added, then the system tested again.

Radon gas

In areas where radon gas has been identified as a problem, ventilation should not be taken from the space below the ground floor level, for example by use of a floor vent.

No design for ventilation for gas appliances should in any way interfere with the remedial measures that may already be in place to prevent radon gas entering the habitable part of a building.

When testing for spillage, any extraction systems should be turned on and running.

Passive stack ventilation

Passive stack ventilation is a ventilation system using ducts from the ceiling of a room to terminals on the roof. It operates by incorporating the principles of the natural stack effect. This means movement of air due to difference in temperature between inside and outside and the effect of the wind passing over the roof of the dwelling. This system is becoming popular in new dwellings but under no circumstances should it be used for ventilating gas appliances.

3 PRINCIPLES OF CHIMNEYS AND FLUE SYSTEMS

Flues and chimneys are required to safely remove the products of combustion from burning fossil fuels to the outside air, where they can do no harm.

The operating principles of chimney/flue systems

The operating principles of a flue/chimney are to:
- remove combustion products to the outside
- draw in air for combustion

Chimney stacks and flues work on the principle that the hot flue gases rise because they are less dense than the cooler air outside of the stack. This creates a pressure difference, which is known as updraught. The movement within the stack expels the hot flue gases to the outside whilst drawing in air for combustion into the appliance.

KEY POINT

These two factors affect the amount of updraught created by the chimney:
- the hotter the gases, the stronger the updraught
- the taller the chimney, the stronger the updraught at a given temperature difference.

Table 7.7 illustrates the relationship between heat and height and how they work in conjunction to produce updraught.

The types of chimney/flue systems

All central heating appliances need a flue to remove the products of combustion safely to the outside. The basic concept is to produce an updraught, whether by natural means or by the use of a fan, to eject the fumes away from the building.

Flue types are divided into three basic categories:
- A – flueless
- B – open-flued
- C – room-sealed.

These categories are then further divided by the addition of a second number which identifies if the flue is natural draught or has a fan, and if the fan is located upstream or downstream of a heat exchanger. Manufacturers' instructions and appliance data plates will give details on what type of flue or flue variations appliances are designed for.

▼ Table 7.7 The relationship between heat and height

		Chimney height in metres (m)							
	°C	3	5	6	8	9	10	12	
This is the average temperature difference between the flue gases in the chimney and the outside air.	500	26	39	52	65	78	92	105	The figures in red are the pressure differences between the inside of the chimney and the outside air measured in pascals (Pa).
	400	24	36	48	60	73	85	97	
	300	21	32	43	54	64	75	86	
	200	18	26	35	44	52	61	70	
	100	11	17	23	28	34	39	45	
	50	4	5	7	9	11	13	14	
	20	2	2	3	4	5	5	6	

▼ Table 7.8 Flue types

Flue type	Category letter with first digit	Flue of chimney design	Natural draught identified by the second digit	Fan downstream of heat exchanger identified by the second digit	Fan upstream of heat exchanger identified by the second digit
Flueless	A	Not applicable	A_1	A_2	A_3
Open-flued	B_1	With draught diverter	B_{11}	B_{12}	B_{13}
	B_2	With draught diverter	B_{21}	B_{22}	B_{23}

→

▼ Table 7.8 Flue types (continued)

Flue type	Category letter with first digit	Flue of chimney design	Natural draught identified by the second digit	Fan downstream of heat exchanger identified by the second digit	Fan upstream of heat exchanger identified by the second digit
Room-sealed	C_1	Horizontal balanced flued inlet with air ducts to outside air	C_{11}	C_{12}	C_{13}
	C_2	Inlet and outlet ducts connect to a common duct system in SE-duct for multi-appliance connections	C_{21}	C_{22}	C_{23}
	C_3	Vertical – balanced flue and inlet ducts to outside	C_{31}	C_{32}	C_{33}
	C_4	Inlet and outlet connections to U-duct system for multi-appliance system	C_{41}	C_{42}	C_{43}
	C_5	Unbalanced flue or inlet air ducted system	C_{51}	C_{52}	C_{53}
	C_6	Appliance purchased with a flue or air inlet ducts	C_{61}	C_{62}	C_{63}
	C_7	Vertical flue to an outlet which takes its air from a loft space	C_{71}	C_{72}	C_{73}
	C_8	Flue connected to a common duct system which takes an air supply from outside making it an unbalanced system	C_{81}	C_{82}	C_{83}

Open flues

The open flue is the simplest of all flues. Because heat rises, it relies on the heat of the flue gases to create an updraught. There are two different types:

- natural draught
- forced draught.

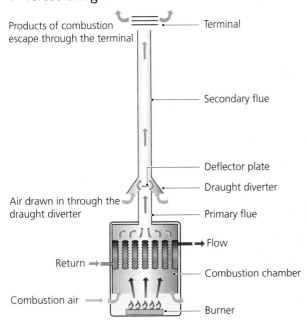

▲ Figure 7.30 The operation of an open flue

Products of combustion escape through the terminal — Terminal

— Secondary flue

— Deflector plate
— Draught diverter

Air drawn in through the draught diverter

— Primary flue

— Flow

Return →

— Combustion chamber

Combustion air →

— Burner

With a boiler having this type of flue, air for combustion is taken from the room in which the boiler is located. The products of combustion are removed by natural draught vertically to atmosphere, through a suitable terminal. The room must have a route, for combustion air, direct from outside. This is usually supplied through an air brick on an outside wall. All natural, draught, open flue appliances work in this way. The material from which the flue is made, however, will differ depending on the type of fuel used.

Occasionally, an open flue may be forced draught. This is where a purpose designed fan is positioned either before the combustion chamber or close to the primary flue. The fan helps to create a positive updraught by blowing the products of combustion up the flue. Forced draught open flues are not suitable for all open flue types and it will depend upon the boiler manufacturer and the boiler/flue design.

The flue type for an open-flued system can be one of the following:

- rigid single-walled
- rigid double-walled (twin-walled) with or without insulation
- flexible single- or double-walled (twin-walled).

Single-walled flues

Single-walled flues are usually connected by a socket and a spigot with the socket facing uppermost. They can be made from stainless steel or vitreous enamel, and older flues were made from cement and even asbestos. Single-walled flues are not suitable for external installations or even in uninsulated loft spaces as cold temperatures affect the flue performance.

> **KEY POINT**
>
> Any open flue under 10" requires a flue terminal.

Double-walled flues

A double-walled or (twin-walled) metal chimney should be installed with the internal socket facing uppermost. Whenever bayonet joints are utilised then the full twist movement process should be applied to ensure that the joint is complete and secure. They provide thermal protection during the conveyance of flue gases. This is achieved by the air gap between the inner and outer walls. Some flues have insulation in this space.

Whenever a double-walled metal chimney is connected to an appliance or chimney fittings and components, then an appropriate adaptor should be used. Similarly, when the connection of different makes of metal chimneys are carried out, it is recommended that the chimney manufacturer's recommended adaptor is used. With all applications of flue connection, it is essential that the appliance manufacturer's instructions be followed. Double-walled metal chimneys are mainly made from stainless steel or zalutite outer shells and stainless steel or aluminum inner shells. To avoid condensation no external run of twin-walled flue pipe insulated with an air gap should exceed 3 m.

When installing flues, fittings and components, unless the manufacturer specifically gives permission and details, then metal chimney components or fittings should not be cut. Each individual section must be examined before assembly is completed and any sections that have damaged joints or other internal damage should not be used.

Never improvise a connection or adapt a flue. Always consult the manufacturer's instructions. Connections to primary flues and secondary flues can be made up with proprietary components such as heat-resistant rope and fire cement, for example, and these should always be checked for integrity when testing and inspecting the operation of an appliance. Creative ideas can lead to leaks and fires on gas installations.

Flexible flue liners

A flexible flue liner is fitted to the primary flue outlet of a fire/back boiler unit (BBU) and from there it travels within an existing chimney to a terminal. At the point where it enters the base of the chimney a register plate must be installed to prevent secondary flue pull. In the same way, it must be sealed at the top of the chimney with a sealing plate where it connects with the terminal. A typical way of sealing the **annular space** between the chimney and the flexible flue liner is with the use of mineral wool.

> **KEY TERM**
>
> **Annular space:** the required 25 mm gap between any hot surface of a flue and any combustible materials when travelling through a floor in a dwelling.

BS 5440 Part 1 states that under normal operating conditions, a correctly installed metallic liner conforming to **BS EN 1856.1:2003**, **BS EN 1856.2:2004** and **BS 715:2005** should operate safely for at least the operational lifespan of an appliance, which is normally 10 to 15 years.

Half-round terminals are not fit for use as flue terminals because they are there purely to ventilate a decommissioned stack in order to prevent condensation. When carrying out a flue flow test, an engineer should always check to ensure that the flue terminal is not restricted or impeded by any adjacent obstruction such as a TV aerial.

> **KEY POINT**
>
> Aerials can affect the performance of a flue in the same way as a tree that is located too close to a terminal or even a wind turbine that operates nearby.

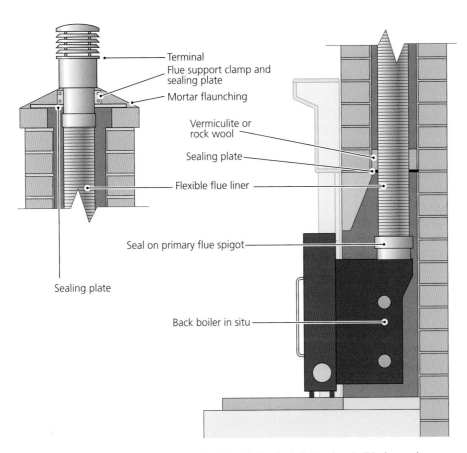

▲ Figure 7.31 A sectional view of a typical back boiler installation in a builder's opening

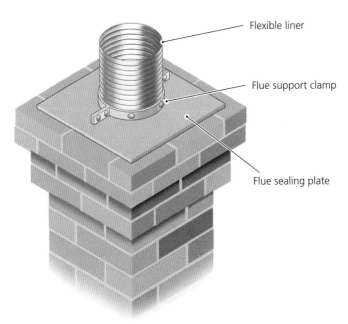

▲ Figure 7.32 When installing a flexible flue within a traditional chimney, the chimney pot must often be removed and a sealing plate installed to prevent secondary flue pull

Flue terminals

The design of flue terminals has changed over the years and many of the older styles are no longer fit for use. Terminals should always be fitted to flue outlets up to 170 mm diameter. Above that dimension they are not considered necessary unless there is a risk of birds nesting.

Builder's opening

If a solid fuel appliance was previously installed in a builder's opening then the chimney should be cleaned and inspected by a Gas Safe-registered chimney sweep and any flue damper removed or permanently fixed in the fully open position. The base of a traditional class I chimney is called the builder's opening and this is where the products of combustion are discharged via the spigot of the space heater as shown in Figure 7.33. There is a lintel at the top of the fireplace and a chairbrick at the rear which sometimes requires removing if it impedes the flow of the products of combustion from the spigot, as there should be no obstruction or surface within 50 mm. The appliance should have a non-combustible hearth at the base of a thickness no less than 12 mm.

KEY POINT

Any installation which has previously used a solid fuel appliance must have the chimney swept and tested if an open flued space heater is to be installed. Any obstructions such as flue dampers should be either removed or fixed in the fully open position.

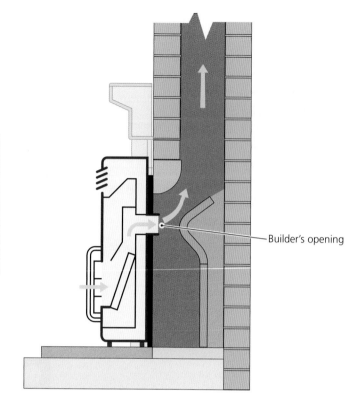

Builder's opening

▲ Figure 7.33 Builder's opening with a gas fired space heater installed

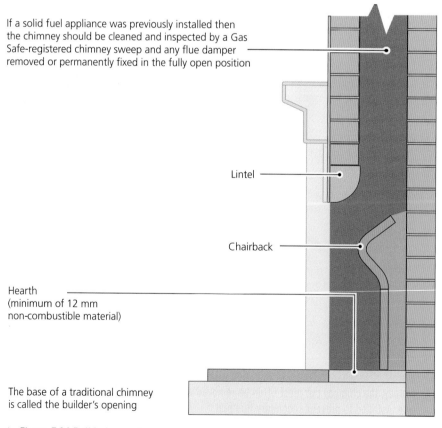

If a solid fuel appliance was previously installed then the chimney should be cleaned and inspected by a Gas Safe-registered chimney sweep and any flue damper removed or permanently fixed in the fully open position

Lintel

Chairback

Hearth (minimum of 12 mm non-combustible material)

The base of a traditional chimney is called the builder's opening

▲ Figure 7.34 Builder's opening

Void volume (catchment space)

At the base of the builder's opening is the area where the appliance is located known as the 'void volume'. The dimensions of the area behind the appliance are critical when determining the suitability of an installation and Table 7.9 gives guidance on different types of application determined on individual type and previous usage of a chimney.

The void volume is determined by the length × width of the opening × the distance below the appliance spigot. The spigot of the appliance must not be less than 50 mm from any obstruction and must protrude at least 12 mm past the closure plate. In any event the manufacturer's instructions will give specific installation details which could differ from the British Standard.

▼ Table 7.9

Chimney type/circumstance	Void volume (litres)	Spigot height (mm)
Unlined brick	12	250
Lined brick – new/unused/used with gas	2	75
Lined brick – previously solid fuel/oil	12	250
Flue block – new/unused/used with gas	2	75
Flue block – previously solid fuel/oil	12	250

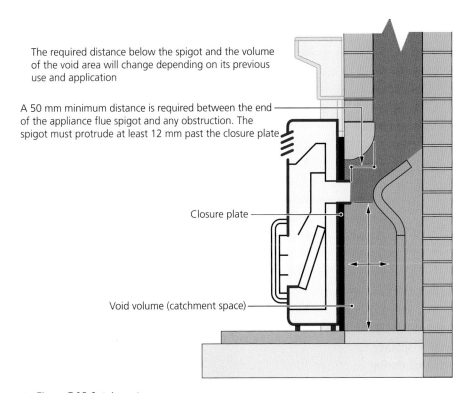

The required distance below the spigot and the volume of the void area will change depending on its previous use and application

A 50 mm minimum distance is required between the end of the appliance flue spigot and any obstruction. The spigot must protrude at least 12 mm past the closure plate

Closure plate

Void volume (catchment space)

▲ Figure 7.35 Catchment area

Terminal guards

Flue terminal guards should be installed on balanced flue appliances if the underside flue is less than 2 m from the ground. There should be a minimum of a 50 mm space between the guard and any hot surface of the flue terminal. The purpose of this guard is to protect anyone from coming in contact with the hot surface.

BS 5440 Part 1:2008 gives precise information about the location of flues in respect to openings to a building. While this installation might have been acceptable when it was installed, it may not comply with the new standard. An engineer must test and assess the performance of a flue to ensure that it operates without causing a risk to anyone. If the appliance operates without POCs entering a building then it could be considered 'not to current standard' (NCS). **Engineering judgement** is essential and in some instances this installation could be classified as 'at risk' (AR). If POCs did enter the building then the appliance would be 'immediately dangerous' (ID).

KEY TERM

Engineering judgement: this is a technical decision which is based on the competence of a person who has an appropriate combination of technical education, training and practical experience in the specific field of work. Competence in specific areas of gas work is verified by assessments of an engineer's theoretical and practical knowledge at an independent nationally approved ACS gas centre, and then registration with the HSE approved Gas Safe register.

On open flues where there is evidence that a chimney is used for nesting by birds, squirrels or other wildlife, or if any such problem is known of in the vicinity, then a suitable guard or terminal should be fitted to the chimney to prevent entry of these creatures. This is especially important in areas where birds such as jackdaws are known to roost.

A chimney should be inspected and reinforced if required before fitting a guard to ensure that it can support such a fitting. Once a terminal guard has been installed the appliance should be checked for spillage to ensure that POCs are being effectively cleared during the combustion process.

The Brewer Birdguard as shown in Figure 7.36 is suitable for gas, oil or solid fuel systems. It prevents the entry of birds and can be installed in chimney pots with a diameter of 150–250 mm.

Square birdguards are designed to fit an internal pot size of 6–10" square and have a strap fixing suitable for gas, oil or solid fuel systems. This guard protects from birds nesting and the entry of rain and other debris.

Both of the terminal guards have versions that are suitable for gas and are designed to **BS 5871:2005** requirements.

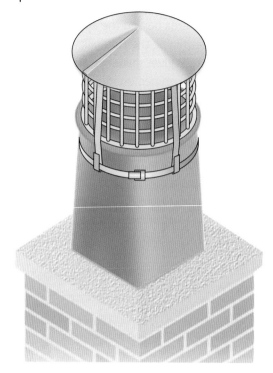

▲ Figure 7.36 Brewer Birdguard

▲ Figure 7.37 Square birdguard

Flue blocks

As with a conventional flue, there are several components that make up a flue block system. There are some key points to consider to enable safe and effective operation of such a flue. For example, the minimum cross-sectional area of new chimneys should be no less than 16,500 mm² and any angle build into an offset should be no greater than 30°.

Flue blocks are designed to save space within living areas and are built into the fabric of the building. They form part of the structure and are staggered and bonded into a standard pattern.

When a gas fire or a fire/back boiler unit is to be connected to the chimney then the chimney manufacturer's starter or recess block(s) that are appropriate for that particular appliance type must be fitted at the base of the chimney along with a lintel or cover block. Sometimes an engineer must deal with the temperature transfer from an appliance to the building next to it and a **cooler plate** located at the rear of the appliances could help solve this problem. In addition flue block chimneys should not be directly faced with plaster, or plaster cracking might occur.

KEY TERM

Cooler plate: a device used behind an appliance in a flue block installation to prevent the unwanted transfer of heat from the appliance.

BS 5440 Part 1:2008 explains that flue blocks are more resistive to the flow of flue gases than metal chimneys of the same cross-sectional area. In addition, any mortar extrusions where joints are made will increase resistance even more. Internally extruded mortar should be removed and coring should be carried out through the erection to remove all extrusions and droppings. Previously engineers used a small canvas bag of sand which was lowered on a rope from the top of the chimney to remove any excess mortar.

When connecting a metal chimney component to a flue block chimney, the manufacturer's transfer block should be used. Any metal chimney component connected to such a block should not project into the flue such that it restricts the cross-sectional area of the flue. When connecting a gas fire to the base of a flue block chimney there should be a debris collection space below the spigot.

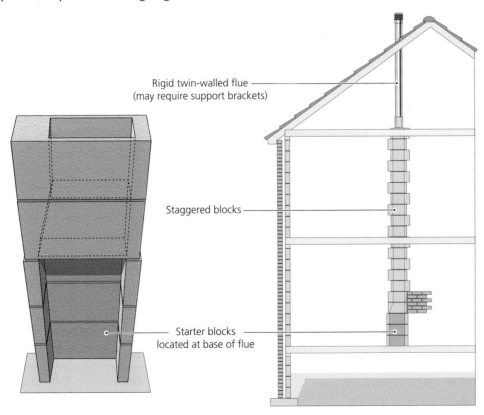

Rigid twin-walled flue
(may require support brackets)

Staggered blocks

Starter blocks
located at base of flue

▲ Figure 7.38 Flue blocks

Ridge terminals

Often flues terminate at ridge terminals which are located at the highest part of the roof. It is important that they are located at 1500 mm away from any adjacent structure which is higher than the ridge vent position. In addition if the vent has openings on all four sides, it is essential that they should be positioned so that they are a minimum of 300 mm from other such ridge terminals. It is important that ridge terminals are located so that the connected flue pipe will be at least 50 mm from any combustible materials such as roofing timbers.

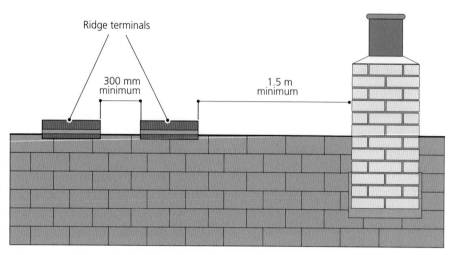

▲ Figure 7.39 Siting for ridge terminals

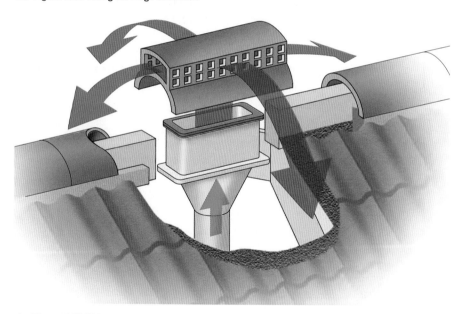

▲ Figure 7.40 Ridge terminal

Room sealed (balanced) flues

This boiler draws its air for combustion directly from outside through the same flue assembly used to discharge the flue products. This boiler is inherently safer than an 'open flue' type, as there is no direct route for flue products to spill back into the room. There are two basic types:

- natural draught
- fan assisted (forced draught).

Natural draught

Natural draught room sealed appliances have been around for many years and there are still many thousands in existence. The basic principle is very simple – both the combustion air (fresh air in) and the products of combustion (flue gases out) are situated in the same position outside the building. The products of combustion are evacuated from the boiler through a duct that runs through the combustion air duct, one inside the other.

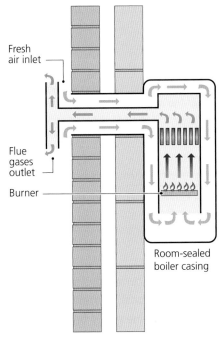

▲ Figure 7.41 The operation of a natural draught room sealed boiler

Fan assisted (forced draught)

Fan-assisted room sealed appliances work in the same way as their natural draught cousins, with the products of combustion outlet positioned in the same place (generally) as the combustion air intake but there are two distinct differences:

- The process is aided by a fan, which ensures the positive and safe evacuation of all combustion products and any unburnt gas which may escape.
- The flue terminal is circular, much smaller and can be positioned in many more places than its predecessors.

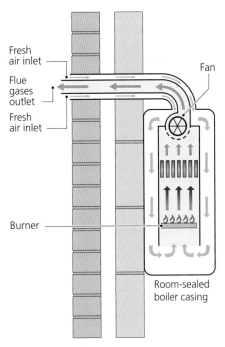

▲ Figure 7.42 The operation of a fan-assisted room sealed boiler

There are two very different versions of the fan-assisted room sealed boiler. These are:

- The fan positioned on the combustion products outlet from the heat exchanger – creates a negative pressure within the casing.
- The fan positioned on the fresh air inlet blowing a mixture of gas and air to the burner – creates a positive pressure within the boiler casing. All modern condensing boilers use this principle.

The components within chimney/flue systems

Open flues contain several different parts. Each of these does a specific job (see Table 7.10).

▼ Table 7.10 The functions of open flue system components

Primary flue	The section of flue which connects the draught diverter (if fitted) to the appliance. It is here where the greatest updraught is created. On open flued appliances, the primary flue must be at least 600 mm in length and have no bends or changes of direction. In some open flue appliances, the primary flue and draught diverter are integrated into the appliance.
Draught diverter	Often called the draught break, it allows pressure fluctuations in the flue system to be moderated. The draught diverter has several functions: ● It diverts down-draught away from the burner, preventing blow-back across the burner flame. This can often be a problem in windy conditions or in flues that have excessive height. ● It helps to dilute the flue gases, thus reducing carbon dioxide content. ● It helps to reduce the pull of the secondary flue. ● It helps in preventing the loss of flame stability at the burner.
Secondary flue	The part of the flue that carries the products of combustion towards the outside. This can either be installed outside or inside the property, depending on the position of the appliance within the property. Secondary flues should: ● be as straight as possible and take the shortest possible route ● must not contain any 90° bends ● changes of direction should be kept to a minimum angle of 45° ● be constructed of twin-wall insulated to reduce heat loss and to prevent condensation within the flue.
Terminal	The part where the flue gases evacuate the property to the outside. Flues up to 170 mm require a terminal. The terminal's job is to: ● help the flue gases discharge to atmosphere ● prevent wind, rain, leaves, birds, etc. from penetrating the flue.

The effects of layout on chimney/flue systems

The way a flue is designed and installed greatly affects its ability to perform the function of dispersing the products of combustion safely away from the building to the outside environment. There are several factors that need careful consideration.

Flue route and bends in flues

As mentioned earlier, the route a flue takes should be as straight as possible, travelling vertically throughout its length. Bends in flues should be avoided as these have the effect of creating frictional resistance, which slows down the velocity of the flue gases, but where this is not possible should be kept to a minimum angle of 45°.

Effective height

The effective height of a flue is different than the physical, measured height of the flue. The effective height takes into consideration the frictional resistances encountered as the products of combustion flow upwards through the flue towards the terminal.

IMPROVE YOUR MATHS

Effective height must be calculated using this calculation:

$$He = Ha \times (Ki + Ko) \div [(Ki + Ko) - KeHa + \Sigma K]$$

Where:

He = equivalent height

Ha = vertical height

Ki = inlet resistance from appliance

Ko = outlet resistance from flue

Ke = resistance factor per m run of selected flue

ΣK = sum total of resistance due to total pipe length (including any direction changes) and fittings

To be able to complete the calculation, reference must be made to the following tables to determine the resistance factors Ki, Ko and Ke.

→

▼ Table 7.11 Inlet and outlet resistance factors

Appliance	Internal resistance (Ki)	Flue diameter (mm)	External resistance (Ko)
Appliances excluding gas fires:			
100 mm spigot	2.5	100	2.5
125 mm spigot	1.0	125	1.0
150 mm spigot	0.48	150	0.48
Gas fire	3.0		
Gas fire and back boiler unit	2.0		

▼ Table 7.12 Chimney components resistance factors

Component	Internal size (mm)	Resistance factor (Ke)	Component	Internal size (mm)	Resistance factor (Ke)
Pipe (per metre run)	100	0.78	45° bend	100	0.61
	125	0.25		125	0.25
	150	0.12		150	0.12
Brick chimney (per metre run)	213 × 213	0.02		197 × 67	0.3
				231 × 65	0.22
				317 × 63	0.13
Pre-cast blocks (per metre run)	317 × 63	0.35	Terminal at ridge	100	2.5
	231 × 65	0.65		125	1.0
	197 × 67	0.85		150	0.48
	200 × 75	0.6	Terminal not at ridge	100	0.6
	183 × 90	0.45		125	0.25
	140 × 102	0.6		150	0.12
90° bend	100	1.22	Raking block (per block)	Any	0.3
	125	0.5	Transfer block	Any	0.85
	150	0.24			

To make the calculation easier, the information is entered into the table below:

1	2	3	4	5	6	7	8	9	10	11
Ki	Ko	1 + 2	Ke	Ha	4 × 5	3 − 6	∑K	7 + 8	3 ÷ 9	5 × 10

- In column 1, enter the internal resistance of the appliance (Ki) from Table 7.11.
- In column 2, enter the external resistance of the flue (Ko) from Table 7.11.
- In column 3, enter the total of column 1 + column 2.
- In column 4, enter the resistance factor per m run of flue pipe (Ke) from Table 7.12.
- In column 5, enter the total measured height vertically in metres, above the draught diverter.

- In column 6, enter the sum of column 4 × column 5.
- In column 7, enter the sum of column 3 − column 6.
- In column 8, enter the total of all resistance factors (∑K) due to flue pipe and fittings. These are taken from Table 7.12.
- In column 9, enter the sum of column 7 + column 8.
- In column 10, enter the sum of column 3 ÷ column 9.
- In column 11, enter the equivalent height. This is calculated by entering the sum of column 5 × column 10.

→

So, how does the calculation work?

Look at Figure 7.43. It shows the installation of a 25 kW input open flued gas appliance installed on a 125 mm twin wall flue pipe system. Extract the information for the calculation table from the drawing.

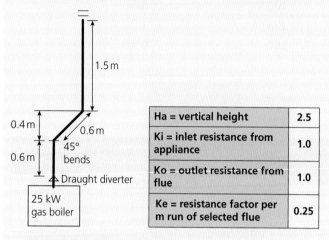

Ha = vertical height	2.5
Ki = inlet resistance from appliance	1.0
Ko = outlet resistance from flue	1.0
Ke = resistance factor per m run of selected flue	0.25

▲ Figure 7.43 Installation of a 25 kW input open flued gas appliance installed on a 125 mm twin wall flue pipe system

1 The internal resistance of the appliance from Table 7.11 = 1.0 (Ki)

2 The external resistance of the flue from Table 7.11 = 1.0 (Ko)

3 Column 1 + column 2 = 2

4 The resistance factor of the flue pipe 125 mm dia. from Table 7.12 = 0.25 (Ke)

5 The height of the flue vertically = 2.5 m (Ha)

6 Column 4 × column 5 = 0.625

7 Column 3 – column 6 = 1.375

8 The total sum of the resistance of pipe and fittings (ΣK) is as follows:

2 × 45° bends @ 0.25 = 0.5

1 × terminal @ 0.25 = 0.25

2.5 m flue pipe @ 0.25/m = 0.625

Total = 0.5 + 0.25 + 0.625 = 1.375. This is entered in column 8.

9 Column 7 + column 8 = 2.75

10 Column 3 ÷ column 9 = 0.727

11 Column 5 × column 10 = 1.817

1	2	3	4	5	6	7	8	9	10	11
Ki	Ko	1 + 2	Ke	Ha	4 × 5	3 – 6	ΣK	7 + 8	3 ÷ 9	5 × 10
1.0	1.0	2	0.25	2.5	0.625	1.375	1.375	2.75	0.727	1.817

The equivalent height in this instance is 1.817 m. To check whether this height is sufficient, it must be compared to the figure in Table 7.13.

The appliance calculated was an open flue gas boiler of 25 kW input. The table tells us that the minimum height of the flue must be at least 1 m. Therefore, the calculated flue equivalent height suggests that the flue design equivalent height at 1.817 m is adequate.

▼ Table 7.13 Minimum equivalent flue height and size

Appliance	Minimum height (m)	Minimum diameter (mm)	Minimum area (mm²)
Gas fire connected to pre-cast block system	2	125	12,000
Other gas fires and gas fire/back boiler unit	2.4		
All other appliances <70 kW net input (excluding DFE and ILFE gas fires)	1	Same diameter of flue spigot	
Appliances of 70 kW–3.5 MW net input	3		

IMPROVE YOUR MATHS

Look at the drawing in Figure 7.44. It shows the installation of a 35 kW input open flued gas appliance installed on a 125 mm twin wall flue pipe system. Extract the information for the calculation table from the drawing.

Using the formula in the previous example, calculate the effective height of the flue in the drawing. Enter the information on to the table provided and state whether the height is sufficient enough to operate correctly.

A blank calculation table is provided below for you to enter the information.

1	2	3	4	5	6	7	8	9	10	11
Ki	Ko	1+2	Ke	Ha	4×5	3−6	ΣK	7+8	3÷9	5×10

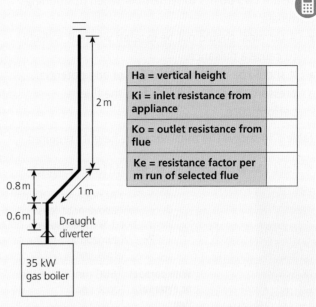

Ha = vertical height	
Ki = inlet resistance from appliance	
Ko = outlet resistance from flue	
Ke = resistance factor per m run of selected flue	

▲ Figure 7.44 Installation layout of an open boiler flue

Flue temperature, condensation and cooling

For a flue to work efficiently, the temperature inside the flue needs to remain as constant as possible. When a flue is cold, the products of combustion move slowly. This can create condensation and, frequently, the spillage of flue gases around the draught diverter until the flue warms sufficiently for the velocity to increase. It takes around 5 to 10 minutes after first lighting for the flue to warm to a velocity that is adequate to evacuate the products of combustion safely away from the building.

As the velocity of the flue increases, frictional resistance inside the flue is created. This leads to turbulence. Frictional resistance and turbulence cause a slowdown in velocity until a maximum velocity is reached.

Once the flue is warmed, the loss of heat from inside the flue must be kept to a minimum. Loss of heat leads to cooling of the flue, which creates condensation, a rapid decrease in velocity and a lack of updraught.

KEY POINT

These three factors – frictional resistance, turbulence and cooling – can result in the loss of updraught and the spillage of the products of combustion into the room where the appliance is installed.

Termination

The position of the terminal is crucial for the efficient evacuation of the products of combustion from the building. It is vital that the terminal position does not affect the velocity of the flue. When cooling occurs, the flue gases may be seen as a plume of gas slowly discharging from the terminal. A velocity of 6 m/s is required for the safe evacuation of combustion gases.

Terminals need to have an opening that is equal to twice the cross-sectional area of the flue itself.

There are two points to remember when positioning terminals:

- A terminal should not be located where it is likely to cause a nuisance.
- It should be positioned outside of any potential pressure zone that could affect the flue performance.

Table 7.14 shows the minimum flue heights required.

▼ Table 7.14 Minimum flue heights required for termination

	Pitched roof		At ridge, or 600 mm above or at least 1.5 m measured horizontally to the roof line
<70 kW net input	Flat roof	With parapet or external flue	600 mm above the roof line*
		Without parapet and providing internal routed flue	250 mm above the roof line*
>70 kW net input	Pitched or flat		1000 mm above the roof line**

*If within 1.5 m of a nearby structure the flue must be raised to 600 mm above that point.

**If within 2.5 m of a nearby structure the flue must be raised to 1000 mm above that point.

The layout and features of chimney and flue construction

Chimney and flue construction depend on the materials that they are constructed from. All flues and chimneys are designed to evacuate the products of combustion, whether that is coal smoke or gas fumes, to the atmosphere. There are many different types.

Brick and masonry chimney construction (class 1 flue)

Typically, chimneys can be classified into two categories:

- chimneys designed for use with fireplaces
- chimneys designed for use with appliances.

There are differences between these types, but both share the same basic functions – to remove the products of combustion safely to atmosphere.

When designing chimneys, the geographical location must be considered. Much of the United Kingdom has weather which is classified as severe or very severe exposure to wind and driving rain, and the construction of the chimney should be such that ensures long-lasting, trouble-free life, in adverse conditions.

Brick chimneys constructed before 1966 were built to accept coal fires and were generally 225 mm × 225 mm (9 in × 9 in) in dimension. Since 1966, chimneys were built with a 175 mm diameter or 200 mm × 200 mm square clay, pumice or concrete flue linings to **BS EN 1857:2010**. There is a specific reason for this. Brick chimneys without linings are generally considered too large for gas fires. The size of the flue means that combustion gases cool rapidly as they reach the top of the chimney, which causes the velocity to slow down to the point where the water vapour in the gas products of combustion condenses on to the inside of the chimney. This causes the chimney to deteriorate at the mortar joints. By the inclusion of a clay, pumice or concrete flue lining, chimneys can be used for both solid fuel fires and gas fires without deterioration of the chimney structure.

Masonry chimneys should be constructed of frost-resistant bricks above the roof line with sulphur resistant mortar used in the joints. Below the roof, the stack may be constructed of normal bricks and mortar. Several points should be remembered:

- Where chimneys penetrate the roof, weatherings will be required to stop the rain from penetrating the building below.
- Because chimneys are in exposed positions, a damp-proof course (DPC) may be required to be built into the brickwork to stop downwards saturation of rain.
- A coping stone made of precast concrete must be bedded on the top of the chimney to, again, help in preventing rain penetration of the structure.

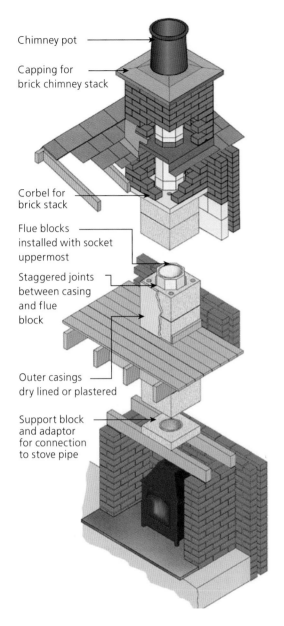

Chimney pot

Capping for brick chimney stack

Corbel for brick stack

Flue blocks installed with socket uppermost

Staggered joints between casing and flue block

Outer casings dry lined or plastered

Support block and adaptor for connection to stove pipe

Free-standing stove in a recess

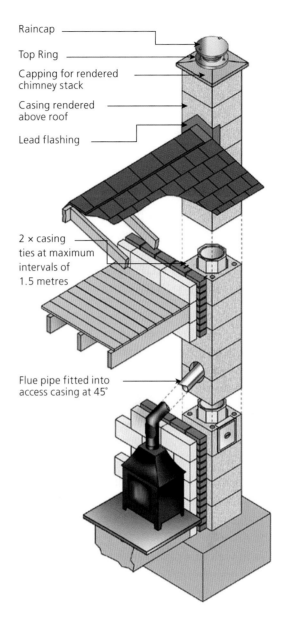

Raincap

Top Ring

Capping for rendered chimney stack

Casing rendered above roof

Lead flashing

2 × casing ties at maximum intervals of 1.5 metres

Flue pipe fitted into access casing at 45°

Free-standing stove with external chimney and pre-formed flue entry kit

▲ Figure 7.45 Cross section of a brick-built chimney

Pre-cast flue blocks (class 2 flue) gas flue blocks certified to **BS EN 1858:2008**

Pre-cast flue blocks are manufactured from pre-cast concrete and are designed to be built and bonded into the house wall construction. This can be either a party wall between dwellings or the inner leaf of an external wall. In the roof space, the flue converts to a normal 125 mm twin wall flue pipe. It is usually distinguishable by either a metal terminal on the roof or a terminal at the ridge of the roof (known as a ridge terminal). The main problem with a pre-cast concrete flue is that the 'pull' on the flue can be quite poor because the flue blocks are very shallow and do not give a good flue combustion products extraction rate.

Pre-cast flue blocks are designed with a socket and spigot system, which provides a concrete to concrete joint that is sealed with a special heat resisting sealant.

Early pre-cast concrete flues were jointed using normal cement mortar, which often squeezed out into the flue itself. This created a partial blockage in an already restricted flue space causing more problems with the flue extraction rate.

An important point to remember about this kind of flue is that not all gas fires can be installed on the flue system and the fire must be checked to ensure that it can be fitted safely to the flue.

Rigid metallic flue systems
BS EN 1856.1:2009

There are two types of rigid metal flue systems:

- **Single wall flue pipes** – in most cases, only suitable as linings to existing chimneys. However, some single wall flue pipes can be installed directly on to an appliance, such as Aga cookers and stoves. Single wall flue pipes are generally made from high grade stainless steel, especially when used as flue liners that line existing flues. This is so they do not corrode in the highly corrosive sulphur-enriched surroundings of existing chimneys. Flue pipes that connect to the appliance are usually manufactured from enamelled steel (usually either coloured black or white).

- **Twin wall flue pipes** – either made from stainless steel (for externally installed flues) or galvanised steel (for internally installed flues). They are twin walled to create an insulating barrier to guard against excessive condensation that occurs with single walled flues.

  Twin wall flues have an interlocking socket and spigot jointing system. Bends are multi-angled that swivel to the desired angle. Special appliance connectors are required for connecting to the appliance.

  The correct specification of flue pipe must be installed with the type of fuel used on the appliance. Different fuels have a different type of twin wall flue.

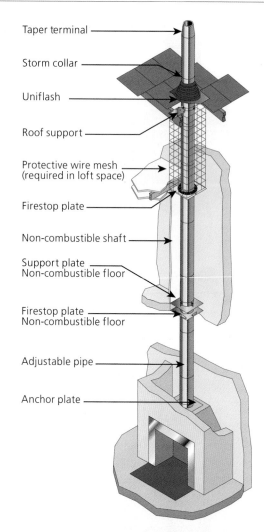

Taper terminal
Storm collar
Uniflash
Roof support
Protective wire mesh (required in loft space)
Firestop plate
Non-combustible shaft
Support plate
Non-combustible floor
Firestop plate
Non-combustible floor
Adjustable pipe
Anchor plate

▲ Figure 7.46 Twin walled flue pipe system

▲ Figure 7.47 Twin walled flue pipe

Flexible metallic liners (types and suitability) BS EN 1856.2

Flexible flue liners are made from stainless steel. They are used specifically to line existing chimneys. There are two specific types:

- **Single skin liners** – specifically used for gas appliances such as gas fires and gas boilers. They must never be directly connected to the appliance when the appliance and the flue liner are exposed. They MUST be connected to a flue pipe. Single skin flue liners must not be used for solid fuel appliances.
- **Double skin liners** – manufactured from two layers of stainless steel, and specifically manufactured for use on wood and multifuel stoves. Again, these must not be connected directly to the appliance when the appliance and liner are exposed.

Flexible liners are threaded down existing chimneys but only after the chimney has been swept to remove any existing soot and debris. The chimney must be sealed around the flue liner at the top and bottom effectively creating a twin walled flue. This helps prevent excessive condensation.

Installation requirements for chimney/flue systems from relevant documents

The construction and installation of flues and chimneys is covered by several documents:

- **Approved Document J of the Building Regulations** – this document covers the application and construction of chimneys and flues and should be used in conjunction with the relevant British Standards. Together these documents dictate the mandatory requirements to work to for a chimney/flue to function safely.
- **British Standards** – the following British Standards contain information regarding the design, installation and termination methods of chimneys and flues:
  - **BS EN 13502:2002** Specification for clay flue linings and flue terminals
  - **BS EN 1443:2019** Chimneys. General Requirements
  - **BS 5440.1:2008** Installation and maintenance of flues and ventilation for gas appliances of

rated input not exceeding 70 kW net (1st, 2nd and 3rd family gases). Specification for installation and maintenance of flues
  - **BS 5871.3:2005** Specification for installation of gas fires, convector heaters, fire/back boilers and decorative fuel effect gas appliances. decorative fuel effect gas appliances of heat input not exceeding 20 kW (2nd and 3rd family gases)
  - **BS EN 15287.1:2007+A1:2010** Chimneys. Design, installation and commissioning of chimneys. Chimneys for non-room sealed heating appliances
- **Manufacturer's instructions** – these should be consulted when designing and installing flues, chimneys and appliances. In some cases, manufacturer's instructions may contradict the regulations or British Standards. In these instances, the manufacturer's instructions must take precedence over all other documents.

Inspection and testing procedures for chimney/flue systems

Inspection and testing of flues and chimneys ensures that they continue to work correctly and safely. There are several different tests that must be employed both to the flue/chimney system and the appliance itself.

Visual inspection (BS 5440 Part 1)

All chimneys and flues should be visually inspected before an appliance is installed and during the appliance annual service. There are several points to remember:

- The flue/chimney should have no obstructions and only serve one room or appliance.
- If a gas appliance is fitted, then a terminal should be fitted that conforms to **BS 5440 Part 1**. If the chimney serves a coal/solid fuel fire, a terminal is not recommended but an appropriate chimney pot is.
- Dampers and restrictor plates must be removed or fastened in the permanently open position.
- The catchment space should be of sufficient size for the appliance installed and be free of debris and sealed from the surrounding structure.

- Any staining or signs of spillage around the opening of the chimney/flue or appliance MUST be investigated.
- The flue/chimney should be inspected outside to ensure that it is in good condition and that there is no deterioration of the structure. This is especially important when inspecting brick chimneys. Worn and deteriorating brickwork can leak fumes.
- Where the chimney/flue passes through a roof space, it should be inspected to ensure the structure's integrity and the condition of the joints.

When a visual inspection has been completed, a flue flow test (also called a smoke test) can be performed.

Flue flow (BS 5440 Part 1)

A flue flow test shows the 'pull' or extraction efficiency of the flue or chimney. It also highlights any structural problems that the flue might have. To correctly perform a flue flow test, the following should be observed:

1 Check that there is adequate combustion air supply to the room where the appliance is installed in line with the appliance manufacturer's instructions.
2 Close all doors and windows in the room where the appliance is going to be installed.
3 If the appliance has a closure plate, such as a gas fire, the closure plate should be in position while the flue flow test is carried out.
4 Check the flue flow by lighting a smoke pellet in the chimney opening. These should burn for at least 30 seconds and produce at least 5 m³. A successful test will show smoke emerging from the chimney stack where the appliance is to be installed. Check carefully to ensure that the smoke only emerges from one terminal and that no smoke is detected emerging from the chimney brickwork joints. Smoke drift can occur whereby the smoke emerges from more than one terminal. This means that the chimney is deteriorating from the inside, which can lead to down draught of fumes into other rooms. Any smoke from anywhere other than the correct chimney terminal indicates the chimney has failed and must be repaired before any appliance is installed.
5 Check that no smoke enters the room during a smoke test. This could indicate down draught. If smoke is detected, the chimney should be warmed for 5 to 10 minutes and then retested.

▲ Figure 7.48 Flue flow smoke pellets

Chimneys can fail flue flow tests because of weather conditions. Severe winds, the proximity of trees and other buildings can all influence the pressure zones around the chimney/flue and affect the results of a flue flow test.

Assuming that the chimney/flue is in good condition the appliance can be fitted and a spillage test performed.

Spillage test (BS 5440 Part 1)

A spillage test is performed using a smoke match to ensure that the products of combustion are evacuating the building through the flue/chimney directly to the atmosphere. The manufacturer's instructions must be consulted to ensure that the smoke match is placed in the correct position on the appliance. The procedure, again, depends on the appliance, but generally:

1 All doors and windows must be closed to the room where the appliance is installed.
2 If the property has any extraction fans, such as cooker hoods, bathroom fans or tumble dryers, these should be put into operation and the doors

between the extraction fan and the appliance left open. If the extraction fan has several settings, it MUST be operated at full extraction rate.

3 Light the appliance and leave it running for between 5 and 10 minutes.

4 In the position dictated by the manufacturer's instructions, place the smoke match and observe the behaviour of the smoke. The smoke should be pulled into the appliance and not pushed back into the room as this would indicate spillage of fumes into the room. Any spillage detected means that the test has failed. In this instance, leave the appliance running for a further 10 minutes and re-test. If spillage is still obvious, shut the appliance down, and disconnect until the problem has been cured. A warning label must be attached until such time as the appliance is operating correctly.

▲ Figure 7.49 Smoke matches for spillage testing

Flue gas analysis

Gas Safe recommends that all domestic gas appliances are tested using a flue gas analyser to ensure that the appliances are operating safely and to maximum efficiency. Flue gas analysers are more than capable of detecting many different gases and calculating the exact concentrations of elements, such as oxygen, nitrogen, CO_2 and CO, present in the products of combustion.

Flue gas analysis is performed by inserting a probe into the flue of the appliance. A sample of the products of combustion are then taken and the levels of CO, CO_2, CO/CO_2 ratio, oxygen and the flue gas temperature calculated. For most appliances, it is the CO/CO_2 ratio that determines correct combustion of the gas and it is this that determines whether a service is required or not.

▲ Figure 7.50 Flue gas analyser

Where a reading is given that shows poor combustion, this could be the result of a number of problems, including:

● poor maintenance
● defective combustion surfaces such as burners or injectors
● incorrectly set gas rate or pressure.

SUMMARY

Fossil fuels, and more specifically coal and coal related products, have fallen out of favour over recent years because of the damage fossil fuel combustion and the resulting CO_2 is causing to the climate of planet Earth. Yet, as far as the 'home' is concerned, natural gas continues to be the fuel of choice for home heating and cooking. Similarly, natural gas still has the largest fuel usage in the generation of electricity in the UK.

In both of these uses – electricity generation and home heating/cooking – natural gas looks set to be the leading fuel for many years to come until a viable renewable, cheaper and less polluting alternative becomes available.

Test your knowledge

1 What is the typical content of methane within natural gas?

   a 5%

   b 20%

   c 70%

   d 95%

2 What is the calorific value of propane?

   a 39 MJ/m³

   b 58.6 MJ/m³

   c 95.8 MJ/m³

   d 111.4 MJ/m³

3 Which grade of fuel oil is most suitable for use with domestic oil burners?

   a C2

   b D2

   c E2

   d F2

4 Which coal type has the highest heat content (kW/kg)?

   a Lignite

   b Anthracite

   c Bituminous

   d Sub bituminous

5 Referring to section 1 of this chapter, at what distance should an oil storage tank for a heating appliance be positioned away from a wooden fence if the tank is not concealed within a fire-rated barrier?

   a 600 mm

   b 760 mm

   c 1.2 m

   d 1.8 m

6 Select the correct formula for the complete combustion of natural gas:

   a $C_4H_{10} + O_2 = CO_2 + CO$

   b $C_4H_{10} + 2O_2 = CO_2 + CO$

   c $CH_4 + O_2 = CO + H_2O$

   d $CH_4 + 2O_2 = CO_2 + 2H_2O$

7 The section of an open flue which provides the initial draught is known as what?

   a Draught diverter

   b Terminal

   c Primary flue

   d Secondary flue

8 What is the minimum angle of change in direction within an open flue?

   a 45°

   b 75°

   c 90°

   d 120°

9 An appliance which takes its air directly from outside through the same flue assembly used to discharge products of combustion and does not rely on a fan is known as:

   a Room sealed natural draught

   b Room sealed forced draught

   c Open flued natural draught

   d Open flued induced draught

10 Which of the following is most likely to indicate incomplete combustion of a high efficiency gas appliance?

   a Condensation from the drain

   b A loose yellow flame

   c A smell of gas

   d Neat blue flame

11 Which of the following is an advantage of using coke as a fuel?

   a The carbon content is low

   b The calorific value is high

   c Coke is a sustainable fuel

   d It is easily produced

12 Which of the following is not a factor that needs to be considered when selecting a solid fuel type?

   a Availability

   b Storage requirements

   c Cost

   d Network connection

13 Which of the following is the regulatory body for oil fired boiler installation and maintenance?

  a Gas Safe

  b HETAS

  c OFTEC

  d NICEIC

14 Which of the following fuel types does NOT require any storage provision?

  a Oil

  b Natural gas

  c LPG

  d Biomass

15 There are three elements to the fire triangle to form combustion. If there is fuel and heat available, what is the third element that is required?

  a Oxygen

  b Nitrogen

  c Hydrogen

  d Methane

16 In a natural gas appliance, what colour flame proves complete combustion?

  a Yellow

  b Red

  c Orange

  d Blue

17 With a natural gas appliance, what is the oxygen to fuel ratio required for complete combustion?

  a 4:1

  b 3:1

  c 2:1

  d 1:1

18 A building admits draughts from the outside, which is called 'adventitious ventilation'. What kilowattage can be deducted for adventitious ventilation?

  a 35 kW

  b 7 kW

  c 5 kW

  d 15 kW

19 When an open flued gas appliance has been installed, how much ventilation is required for each 1 kW rating above the adventitious ventilation?

  a 5 cm^3

  b 10 cm^3

  c 15 cm^3

  d 20 cm^3

20 Which test would you use a smoke match on?

  a Flue flow test

  b Visual inspection

  c Spillage test

  d Flue gas analysis test

21 What does air consist of? State the percentages of each part.

22 When mechanical means are used to supply air for combustion or remove products of combustion within a gas fired appliance, what must be provided should the ventilation fail?

23 Describe the purpose of the draught diverter.

24 Determine the minimum height of a flue terminating above a flat roof with a parapet if the appliance is rated at 28 kW and not within 1.5 m of another structure.

25 Explain the purpose of a spillage test.

26 What are the five categories of domestic fuels?

27 Describe how a heat pump works.

28 What are the regulatory bodies for the following fuel types?

  ● Solid fuels

  ● Oil

  ● Gas

29 Describe how a fan assisted (forced draught) room sealed appliance works.

Answers can be found online at www.hoddereducation.co.uk/construction.

CHAPTER 8

ADDITIONAL TOPICS

In this chapter we will look at the effectiveness of relationships between the plumber and the client, the plumber and their suppliers and the plumbing team with other on-site trades, to enable systems to be installed quickly, efficiently and with minimal problems.

We will also look at how you can plan your career development, setting goals to help realise your plans.

1 PLAN WORK SCHEDULES FOR A SYSTEM INSTALLATION

Identify other trades involved in the installation process

- **Carpenters/joiners** – The wood trades provide a vital function on site during the initial building phase, fitting door and window frames, floor joists and roof trusses. During the second phase they will fix internal doors, skirting boards, architraves, etc.
- **Electricians** – Install and test all electrical installation work on site, including power, lighting, fire and smoke alarms and security systems, usually running the cables in trunking or conduits for neatness.
- **Gas fitters** – Install natural gas lines in domestic properties and in commercial or industrial buildings. On some sites they may also install large appliances and pipelines.
- **Plasterers** – Responsible for wall and ceiling finishing, dry lining and external rendering, if required, using a mixture of both modern and traditional techniques.
- **Painters and decorators** – Responsible for wall and ceiling finishing, including painting skirting boards, architraves and any specialist decorating such as murals, frescos, etc.
- **Tilers** – Responsible for internal and external tiling of walls and floors and specialist tiling such as swimming pools and wet rooms.

Describe effective working relationships between trades

Communication between the various trades on site

We have now identified some of the trades that have a direct involvement with plumbing installations whether it is an electrician completing the wiring to a central heating system or a tiler completing the tiling to a bathroom prior to the installation of a wash basin or WC.

The diverse nature of the construction industry brings together these individual, very different trades with the sole aim of successfully completing the request of the customer, on time and to an acceptable standard.

Plumbing installations on a construction site rely on the cooperation, communication and coordination between various trades to ensure that the installation is completed as smoothly as possible. Consideration of customer requirements is vital and can often mean that specifications and plans have to be altered and amended to suit their wishes. This can involve negotiation between the trades to accommodate the alterations to planned schedules of work and timings. It must be remembered, however, that there are no set rules and different companies and construction sites will approach these problems in a number of ways.

More often than not, a schedule of work will have been drawn up which highlights the timings of such items like first fix and second fix operations and the trades that are involved. Careful planning and verbal communication is paramount to ensure that the schedules do not go astray.

Communication between the company and the customer

Communication between the company and the customer takes place at every stage of the contract from the initial contact to customer care at the contract completion. Written communication can take the form of:

- **Quotations and estimates** – Both of these are written prices as to how much the work will cost to complete. A quotation is a fixed price and cannot vary. An estimate, by comparison, is not a fixed price but can go up or down if the estimate is not accurate or the work is completed ahead of schedule. Most contractors opt for estimates because of this flexibility.
- **Invoices/statements** – Documents that are issued at the end of any contract as a demand for final payment. Invoices and statements can be from the supplier to the contractor for payment for materials supplied or from the contractor to the customer for services rendered. Usually a period of time is allowed for the payment to be made.
- **Statutory cancellation rights** – A number of laws give the customer the legal right to cancel contracts after the customer signs a contract. There is usually no penalty for cancellation providing the cancellation is confirmed in writing within a specific timeframe. Most cancellation periods start when the customer receives notification of their right to cancel up to seven days before work commences.

Communication between the employer and the employee

One of the key points about running a successful business is the relationship between the employer and the employee. Businesses are successful when the management and staff work together, are motivated and engage in constructive dialogue.

Whereas in the past, pay and working conditions varied from employee to employee and the employer had the power to 'hire and fire' as they saw fit, today employers and employees are actively encouraged to engage in discussions about matters across the whole spectrum of a business including their respective rights.

Types of communication

There are a number of ways that companies communicate with customers, staff and suppliers, and other companies, such as:

- Written communication
  - Letters
  - Emails
  - Faxes
  - Text messaging
- Verbally (should always be backed up with written confirmation to prevent confusion)
  - Face to face
  - Via the telephone

Written communication

Letters are an official method of communication and are usually easier to understand than verbal communication. Good written communication can help towards the success of any company by portraying a professional image and building goodwill. Official company business should always be in written form, usually on company headed paper and should have a clear layout. The content of the letter must be well written, using good English, correct grammar and be divided into logical paragraphs. Examples of business letters are sales letters, information letters, general enquiry and problem-solving letters.

Emails have emerged as a hugely popular form of communication because of the speed with which information can be transferred. As with letters, they should be well written and laid out, using correct grammar and spelling to convey professionalism, whether the recipient is a client, customer or colleague.

Faxes are another useful form of communication for businesses. They are used mainly for conveying documents such as orders, invoices, statements and contracts where the recipient may wish to see an authorising signature. Again, the basic rules apply with regard to layout, grammar and content. Remember always to use a cover page that is appropriate for your company. This is an external communication that reflects the business and company image.

Messaging is used a lot these days as an easy, convenient and cheap way to pass on information. This is an informal method of communication that may use simple text messaging or a smartphone app such as WhatsApp. This method of communication can be used to update a customer about your potential arrival time if you are delayed, for example, but should not be used for any formal information.

Oral communication

The spoken word is, more often than not, our main method of communication, especially in a work context. In order to present a professional image and communicate effectively, you must consider what you are saying, your tone of voice, your body language, and the response of your listener.

Describe the elements of a plumbing system installation schedule for a domestic dwelling

Plumbing installations have a finite timescale in which they must be completed. An estimate for a complete plumbing installation in, say, a new-build property that includes hot and cold water supplies, central heating installations and sanitation pipework will include a timeline that the installer will need to work to if the company is to make a profit from the installation. In many cases, the installation can be plotted on a schedule of work or a Gantt chart so that time is allotted to each phase of the installation. By using a Gantt chart, material deliveries can be planned for a certain day and staff loadings for the job can be calculated.

Staffing an installation is a delicate matter as too few plumbers on site will quickly lead to a job falling behind schedule. Similarly, too many plumbers on a job can sometimes have the same result as the presence of more installers leads to greater organisational problems. There is often a fine line between getting enough people on site and making a profit.

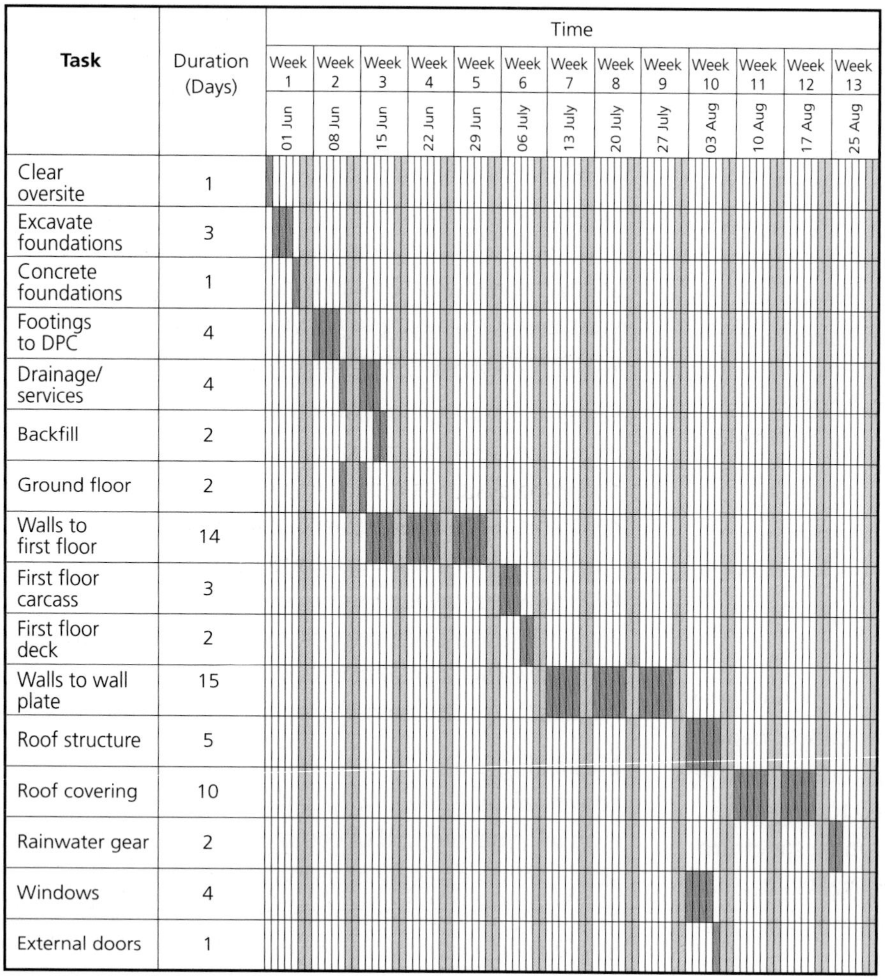

Task	Duration (Days)	Week 1 01 Jun	Week 2 08 Jun	Week 3 15 Jun	Week 4 22 Jun	Week 5 29 Jun	Week 6 06 July	Week 7 13 July	Week 8 20 July	Week 9 27 July	Week 10 03 Aug	Week 11 10 Aug	Week 12 17 Aug	Week 13 25 Aug
Clear oversite	1	■												
Excavate foundations	3	■												
Concrete foundations	1	■												
Footings to DPC	4		■											
Drainage/ services	4			■										
Backfill	2			■										
Ground floor	2			■										
Walls to first floor	14				■	■								
First floor carcass	3						■							
First floor deck	2						■							
Walls to wall plate	15							■	■	■				
Roof structure	5										■			
Roof covering	10											■	■	
Rainwater gear	2												■	
Windows	4										■			
External doors	1										■			

▲ Figure 8.1 A Gantt chart

Explain the sequence of work in a domestic dwelling plumbing system installation

A well-functioning plumbing system that meets or exceeds the customer's requirements is the result of a number of important aspects:

- good design
- good planning
- good installation
- correct commissioning and setting up procedures.

Planning a plumbing installation

Planning a plumbing installation involves:

- designing the system
- coordinating the availability of staff to undertake the installation
- ordering and coordinating the delivery of the materials
- installing the first fix
- installing the second fix
- filling and commissioning the system
- completing the benchmarking paperwork
- handover to customer
- removing all scrap and unused materials from site.

Planning a plumbing installation is often completed using a Gantt chart, as described in the previous section.

Designing a plumbing installation

Designing a plumbing installation will involve taking measurements from site drawings or visiting the site in person and taking measurements from the building so that pipework sizes, heat losses and heat emitter outputs, flow rates, hot water temperatures etc. can be calculated. The layout of a building is instrumental in how we design the systems for it. In many cases, the position of appliances, such as bathroom suites, has already been dictated by the architect's drawings and our job is to design a functioning system based upon these predetermined positions. Where central heating is concerned, positioning components, such as radiators, may be a little more flexible and consultation with the customer is needed to ensure that the position of radiators, the boiler and so on is satisfactory. It is here that the designer/installer will get a feel for where pipework runs can be installed and a decision made about the system type and the materials that will be used. Many installers now favour polybutylene pipe over copper tubes and fittings because of the benefits it offers in installation time.

Once the design is completed, an estimated cost of the installation can be prepared.

Ordering and storing the materials

When a customer has accepted an estimate, ordering of the materials can take place. Most plumbing companies shop around for the best deals on boilers and radiators and will not be dependent on one sole supplier. Alternatively, a company might have favourable contract rates with its supplier which will supply all the items needed, including tubes and fittings, at discounted rates.

The materials, obviously, must arrive early either before the job is started or on the day that the installation is to begin. A phased delivery is often the best method to use, as delivery of key items and appliances can be planned to coincide with the progression of the installation.

Materials that arrive on site must be stored in a secure and safe lock-up to prevent theft and to ensure that a check can be made of the materials in stock at any one time. Fragile materials such as sanitary ware should be kept separate and stored so as to prevent breakages.

Installation planning

Once the system has been designed in accordance with the customer's wishes, the installation planning can take place. Installation can be divided into five separate and distinct phases:

- **First fix** – Usually the first fix phase is where the installer will get their first look at the property. They will walk the job and plan the pipework routes, marking any floorboards that require lifting. On new build installations, the plumber will arrive before any ceilings are fixed and often before the upper floors are down. Marking and notching/drilling of the joists will take place in accordance with the Building Regulations and the pipework for the hot and cold water, the central heating, any gas pipework and sanitary pipework, waste pipes and so on will

527

be installed. At this stage, because there are no appliances or components installed, the plumber will position the pipework tails to where the appliances will eventually be fitted, using the working drawings of the building to position the pipework correctly. Any pipework that is to be positioned behind plasterboard walls, such as droppers for central heating behind the dot and dab plasterboards, will be installed. This phase of pipework is often called 'carcassing'. Once the carcassing has been completed, it must be fully pressure tested according to **BS EN 806**, which is 1.1 times the maximum design pressure.

- **Second fix** – The second fix takes place after all of the internal work, such as fitting plastering, skirting boards and internal doors, has been done. Where bathroom suites are installed, the bath is fitted first so that the tiler can tile around the bath and any areas where the wash basin and WCs are to be fitted. These can then be completed once the tiler has finished. Boilers, radiators and any central heating electrical controls can be installed and any hot water storage vessels, cold water cisterns fitted and connected. Once the second fix has been completed, commissioning can begin.

- **Commissioning and testing** – Commissioning and testing procedures depend upon the system being commissioned. It is at this point that the system is filled up with water to full operating pressures and the systems are run for the first time. Any leaks must be cured and flow rates and pressures checked to ensure that the installation meets the design specification. Central heating systems can be balanced and the temperatures checked against the design specification. Benchmarking the system can take place during this stage of the installation.

- **Snagging** – Snagging is the term used to describe the curing of minor problems that have emerged during the commissioning and testing process.

- **Signing off** – Commonly called 'handover', this is where we present the customer with all of the system documentation, including benchmarking certificates, Building Regulation compliance certificates, manufacturers' instructions and commissioning documentation. This is often presented in a system folder together with any

emergency contact details for use in the event of a problem. The customer should be instructed in the use of all system controls and shown where isolation points for the water, gas and electricity are. Any system servicing requirements, such as annual boiler servicing, should also be pointed out.

Describe difficulties that may arise when supervising system installations

Conflicts in the workplace

When people work together in groups, there will be occasions when individuals disagree and conflicts occur. Whether these disagreements become full-blown feuds or instead fuel creative problem-solving is, in large part, up to the person in charge. Conflicts can occur for many reasons, such as:

- unfair working conditions
- unfair pay structures
- clash of personalities
- language differences
- attitudes towards ethnic differences.

It is important to deal with workplace conflicts quickly and effectively, as if left unchecked they can affect morale, motivation and productivity, and potentially cause stress and even serious accidents. Conflicts may occur between:

- **employer and employee** – may need union involvement or some form of mediation
- **two or more employees** – will need employer intervention
- **customer and employer** – may need intervention by a professional body
- **customer and employee** – will need employer intervention.

Dealing with workplace conflicts

There are several ways that your employer may deal with disagreements. They should:

- Identify the problem. Make sure everyone involved knows exactly what the issue is, and why they are arguing. Talking through the problem helps everyone to understand that there is a problem, and what the issues are.

- Allow every person involved to clarify their perspectives and opinions about the problem. They should make sure that everyone has an opportunity to express their opinion. They may even establish a time limit for each person to state their case. All participants should feel safe and supported.
- Identify and clarify the ideal end result from each person's point of view.
- Work out what can reasonably be done to achieve each person's objectives.
- Find an area of compromise to see if there is some part of the issue on which everyone agrees. If not, they may try to identify long-term goals that mean something to all parties.

Informal counselling is one method that helps managers and supervisors to address and manage conflict in the workplace. This may be in the form of:

- meetings
- negotiation/mediation sessions
- other dispute-resolving methods.

It is important that employees know that there is someone to go to if a conflict develops. If an employee has a conflict with another member of staff, then they should first discuss the problem with their immediate supervisor. In extreme cases where the matter cannot be resolved, then mediation or union involvement may be required.

In the plumbing industry, workplace conflicts can usually be resolved by the Joint Industry Board (JIB).

The effects of poor communication at work

The effects of poor communications can be extremely harmful to both businesses and personnel. If poor communication exists then goals will not be achieved and this could develop into problems within the company. It can lead to demotivation of the workforce and the business will not function as a cohesive unit. The effects are obviously negative:

- Employees become mistrustful of management and, often, of each other.
- Employees argue and reject their manager's opinions and input.
- Employees file more grievances related to performance issues.

- Employees don't keep their manager informed and avoid talking to management.
- Employees do their best to hide their deficiencies or performance problems.
- Employees refuse to take responsibility.

Poor communication in the workplace can disrupt the organisation and cause strained employee relations and lower productivity which can often result in the following issues:

- Time may be lost as instructions may be misunderstood and jobs may have to be repeated.
- Frustration may develop, as people are not sure of what to do or how to carry out a task.
- Materials may be wasted.
- People may feel left out if communication is not open and effective.
- Messages may be misinterpreted or misunderstood causing bad feelings.
- People's safety may be put at risk.

All of these problems will eventually filter down to existing and potential customers, and when that happens, customer confidence will disappear leading to a possible collapse of the company.

Problems arising from the delivery of materials

Occasionally, problems can arise with the suppliers that deliver plumbing equipment, appliances and materials to site:

- **Resource shortages** – Lack of materials is a big contributing factor when jobs and contracts are not completed on time. When delivery dates are missed, it has a knock-on effect:
  - Operatives are left standing idle.
  - Jobs get behind on time.
  - Completion dates are missed.
  - Customers become annoyed at the lack of progress.
- **Poor quality components** – Many of the components, fittings and appliances used in the plumbing industry are mass-produced. Occasionally, fittings and appliances are delivered to site that have not undergone quality checks and arrive not fit for purpose. This can cost time and money in seeking replacements. Common problems include the following:

- Appliances such as boilers arrive with faulty components that are only discovered when the appliance is commissioned.
- Delays occur because bathroom suites often arrive with damage that has occurred during transit or poor quality components or parts missing.
- Fittings occasionally arrive either of the wrong type or the wrong size.

There are a number of *dos* and *don'ts* to observe when parts, appliances, components and equipment are delivered to site:

- DO check all materials that are delivered whilst the delivery driver is present. Any items that are found that are incorrect can be sent back with the driver and replacements requested immediately.
- DO check ALL items and not just the large ones.
- DO count all items and tick them off against the delivery note.
- DO check all sanitary items for damage. It is very difficult to request replacements after the delivery driver has left site.
- DO keep the delivery note in a safe place.
- DO NOT sign for anything that hasn't been checked against the delivery note.
- DO sign the advice note 'unchecked' if you do not have the time to check all items whilst the delivery driver is present.

Describe handover procedures

When the system has been tested and commissioned, it can then be handed over to the customer. The customer will require all documentation regarding the installation and this should be presented to the customer in a file, which should contain:

- all manufacturers' installation, operation and servicing manuals for the boilers, heat emitters and any other external controls such as motorised zone valves, pumps and temperature/timing controls fitted to the installation
- the commissioning records and certificates
- the Building Regulations Compliance certificate
- an 'as fitted' drawing showing the position of all isolation valves, drain-off valves, strainers, etc. and all electrical controls.

The customer must be shown around the system and shown the operating principles of any controls, time clocks and thermostats. Emergency isolation points on the system should be pointed out and a demonstration of the correct isolation procedure in the event of an emergency. Explain to the customer how the systems work and ask if they have any questions. Finally, point out the need for regular servicing of the appliances and leave emergency contact numbers.

2 PLAN FOR CAREERS IN BUILDING SERVICES ENGINEERING

Career planning helps realise your ambitions. A plan helps you focus on what you should be doing to start a new career or progress in the career you are in. Planning needs time and careful consideration to make your career happen rather than you letting your career happen to you.

In order to plan a career you will need to be able to identify the support available to you.

Resources to help you include:

- the internet (career guidance sites, and industry sites of employers you might want to apply to)
- professional bodies/organisations such as the National Careers Service
- educational support and guidance, such as your college careers advisor
- role models – people in the industry that you admire and could ask for advice on how they've got to where they are
- networking – through attending industry events or asking to be introduced to people who might be able to help you through advice or contacts within companies that are recruiting
- job centres and recruitment agencies.

The internet is a logical and easy place to look for sources of information to support career planning. However, resources found on the internet should always be read carefully in the light of the original purpose of the website or blog and the actual benefit you can glean from them. For example, many websites relate just to overseas employment or specific sectors. Before relying on such information, make sure it is relevant to your needs.

The most useful sites for career planning are generally those provided by UK Government departments or agencies sponsored to promote careerawareness. The National Careers Service, at http://nationalcareersservice.direct.gov.uk, provides general advice and career-planning information.

Once you have determined your career path, the National Careers Service website can provide useful information in a general context, including information on government support and guidance. Depending on circumstances, individuals may be able to seek government funding support for certain retraining.

Specific requirements and qualifications need to be researched from relevant trade organisations, competent person registration schemes, etc. There are many UK Accreditation Service (UKAS) accredited awarding bodies that provide support and guidance to trainees and to those already qualified wishing to keep themselves up to date and maintain their own continuing professional development (CPD). If you are already qualified in a specific trade or profession, your professional institution will have a recognised development programme, criteria for meeting their requirements, a mapping process and access to mentors so you can complete the process.

Elements of career planning

There is a vast amount of information designed to support you through the stages of planning your career and seeking a job. This includes:

- goal setting
- curriculum vitae (CV)
- personal statements
- covering letters
- SMART targets
- SWOT analysis.

Goal setting

A goal is an outcome that an individual or organisation is trying to reach. It is likely to be quite general and long range; it can look quite idealistic and doesn't include all the practical details of exactly what is required and when it is required by. An example of a goal could be to have a complete career change within the next five years. This may be not fully detailed but that goal is the starting point in the planning process.

When setting your goal, make sure you can answer the four questions that begin with W: What, Where, Who, and When?

- What position do I want?
- Where will I find relevant information?
- Who will I contact?
- When will I make contact?

▲ Figure 8.2 A candidate in discussion with a career officer

You should also know the difference between a specific goal and a vague goal. For example, there is a big difference between saying, 'I would like a job' and 'I would like to be an apprentice plumber'. You may also like to set a goal of contacting a certain number of companies a week. This is then measurable.

So, if you want to attain anything of significance, you must sit down and define what you really want, put it in writing, develop a real plan, and lay out the guidelines for completion. There's no better way to accomplish a really strong desire than a well-written plan.

Set out your goals

Break your goals into short term, medium term and long term. The short- and medium-term goals can be thought of as stepping stones to the long-term goal.

For example:

In one year's time I want to have …

In five years' time I want to be …

In ten years' time I want to be able to …

ACTIVITY

Define your own short-, medium- and long-term goals.

Curriculum vitae (CV)

How you write your CV and covering letter is up to you, but there are some basic rules to follow if you want to create the best impression.

You should include a summary of your educational and academic background, and skills you have demonstrated. You should also include any interests, hobbies and work experience.

KEY POINT

A curriculum vitae (CV) can be literally translated as 'course of life'. This provides an overview of a person's experience and other qualifications. A CV is typically the first item that a potential employer encounters regarding a job seeker. CVs are often used to screen applicants in order to shortlist them for an interview.

Most employers would expect to receive a CV along with a covering letter outlining your suitability for the position that you are applying for.

Personal statement

A personal statement is a short summary of your key skills and experience that you should put at the top of your CV. It is vital to spend time getting this right, as many employers will often use this statement to decide whether or not to read the rest of your CV. The best advice is to keep it short: your personal statement should be just a few lines or bullet points, around 50–100 words.

Covering letter

A covering letter accompanies a CV (and/or completed application form). It is an opportunity to highlight what is in your CV and to provide any real examples to support your ability to do the job.

SMART targets

SMART targeting is an acronym for the five steps of Specific, Measurable, Attainable, Relevant and Time-based targets and is one of the most effective tools used to achieve a desired goal.

▼ Table 8.1 SMART targets can be a useful tool to help you in your career path

S Specific	What is the task to be done?
M Measurable	What evidence could be used to show if and how well the task has been done?
A Attainable	Is the task possible?
R Relevant	Why is this target important?
T Time-based	Are review dates built in to check progress?

SWOT analysis

When planning your career, the first thing to look at is what you can already do. Ask yourself, 'What am I already good at or do I have an aptitude for?' It may be that you already have a number of skills and qualifications that are transferrable to a new career. A useful self-analysis tool is SWOT analysis.

SWOT is an acronym for Strengths, Weaknesses, Opportunities, Threats. It is a planning tool used to understand the strengths, weaknesses, opportunities and threats involved in a project or in a business. It involves identifying the internal and external factors that are either supportive or unfavourable to achieving that objective.

Different roles in building services engineering

Building services engineering is made up of four key industries:

- electrotechnical
- plumbing and domestic heating
- heating and ventilating
- air conditioning and refrigeration.

There are many optional career pathways within the building services engineering sector, including:

- installation electrician
- heating and ventilation service and maintenance engineer
- plumber
- heating and ventilation installation engineer
- refrigeration engineer
- maintenance electrician ductwork installer
- air-conditioning engineer.

Typical roles within building services

Plumber

Plumbing and domestic heating is a responsive and continually developing industry. The plumber is responsible for the installation of complex cold and hot water systems, sanitation, heating systems and domestic fuel burning appliances such as gas, oil or solid fuel boilers. In recent years, environmental technologies have been integrated within the industry and the modern tradesperson now undertakes a huge variety of jobs, including:

- installing and maintaining central heating systems, hot and cold water systems and drainage systems
- installing, commissioning and maintaining solar water heating, rainwater harvesting or grey water recycling systems
- installing and maintaining gas, oil and solid fuel appliances, including biomass.

Installation electrician

An installation electrician is responsible for the installation of power, lighting, fire protection, security and structured cabling. They may also maintain modern electrical systems and the equipment they serve to ensure effective and efficient operation.

Heating and ventilation engineer

A heating and ventilation engineer installs complex heating equipment and pipework systems to exact design specifications within large buildings such as office blocks, hospitals, schools, etc.

Ductwork installer

A ductwork installer is responsible for the installation of complex ductwork and ventilation systems to exact design specifications within large buildings.

Refrigeration engineer

A refrigeration engineer may install, service and maintain systems and equipment that control and maintain the quality, temperature and humidity of air within modern buildings. Another aspect of the job may be to install, service and maintain refrigeration and environmental technology systems throughout the UK in places such as supermarkets, hospitals, food processing and research establishments.

These are just a few of the many roles within the building services industry which may lead to different pathways in your chosen career. For example, you may progress into a supervisory role either within the same company or a different company. You may wish to become involved in the design and estimating of large-scale projects. There are many people who have started apprenticeships and progressed to running their own businesses or you may wish to go into education and train others. The opportunities available within the industry are many and varied and all are attainable.

Types of work in building services engineering

The industry itself is also varied in its different forms of work, which include:

- contract work
- consultancy
- subcontraction
- casual labour.

Contract work

This means you are providing a service/labour to another company under terms specified within an agreement. It could mean you don't work regularly for an employer but are self-employed.

Consultancy

Once a lot of experience and qualifications have been attained, there are opportunities for consulting work. A consultant is a professional who provides expert advice in a particular area, for example, in:

- hot water
- cold water
- central heating
- underfloor heating
- renewables.

Subcontractor

A subcontractor is a person or company that is hired by a main contractor to perform a specific task as part of an overall project and is normally paid for the services by the main contractor. A building company that doesn't have its own plumber may subcontract the plumbing work out.

Casual labour

This can be part-time, piece or temporary work, which means someone looks for and accepts any type of work within their skill sector. They are not part of the permanent workforce.

3 THE REQUIREMENTS TO BECOME A QUALIFIED OPERATIVE IN BUILDING SERVICES ENGINEERING

Becoming highly skilled in your own craft will be your first priority but over time you may wish to explore some form of supervisory responsibilities.

Opportunities for progression within the sector include: supervisor, manager, business owner, sideways moves to different crafts, assessor/trainer, designer, surveyor, estimator, apprenticeship, engineer and director.

To progress in any trade you will need to achieve your qualifications and gain experience alongside the qualification. It is important to have thought about your career path and to have set yourself attainable goals in the short, medium and long term.

As you progress, you may need to become a 'competent' person. This means you have been trained, tested and passed in a certain area of work. A certificate is given that states you are competent to carry out work in that area.

Competent persons

Competent Person Schemes were introduced by the Government to enable companies to self-certify plumbing and heating works that fall under the scope of Building Regulations. Self-certification provides a much more cost-effective route compared with the alternative of notifying work through local building control bodies.

> **KEY POINT**
>
> Competent Person Schemes apply the principles of self-certification and are based on giving people who are competent in their field the ability to self-certify that their work complies with the Building Regulations without the need to submit a building notice and thus incurring local authority inspection costs or fees.

Areas for competency include:

- gas
- electrics
- unvented water G3
- cold water
- oil
- solid fuel
- environmental technologies
- health and safety.

Gas

The Gas Safe Register is the official gas registration body for the United Kingdom, the Isle of Man and Guernsey, appointed by the relevant health and safety authority for each area. By law, all gas engineers who are actively working on gas installations must be on the Gas Safe Register. Registration must be renewed every five years.

Electrics

The NICEIC (National Inspection Council for Electrical Installation Contracting) Competent Person Scheme (CPS) allows registered installers who are competent in their field to self-certify certain types of building work as compliant with the requirements of the Building Regulations in England and Wales.

Unvented hot water G3

The Approved Building Regulations Document G (Section G3) requires that an unvented hot water system with a capacity of more than 15 litres should be installed and commissioned by a competent person who is a member of the Competent Person Scheme. This allows the self-certification of certain types of work and provides exemption from notification under the Building Regulations.

> **KEY POINT**
>
> It is a legal requirement to be registered with the appropriate body prior to carrying out any work. If the requirements of the scheme are not met the operative may face fines, imprisonment or loss of licence to practise.

Cold water

The approved plumber has been given a very useful concession in that he/she may start work without notification or prior consent on certain types of work, provided he/she issues the customer (and for some types of work the water supplier) with a certificate of compliance when the work is completed. This can save up to ten days of waiting for the water supplier's consent and reduces the paperwork of notification.

Oil

OFTEC (the Oil Firing Technical Association) was formed in 1991 and replaced the former Domestic Oil Burner Equipment Testing Association (DOBETA), an organisation formed after the enforced split of Shell-Mex and BP in the early 1970s.

OFTEC also administers a Competent Person Scheme and encourages those working in the oil firing industry to become registered. Becoming registered with OFTEC allows installers to self-certify installation work without the need to have it checked by local authority building control (where applicable).

Solid fuel

HETAS (the Heating Equipment Testing and Approval Scheme) is the official body recognised by the Government to approve biomass and solid fuel domestic heating appliances, fuels and services including the registration of competent installers. HETAS administers a Competent Person Scheme which will allow competent persons to self-certify installations work.

Environmental technologies

Competence in renewable technologies is becoming more important as environmental issues come to the forefront. Training in these areas is offered under the 'environmental' heading but more specific training is offered by independent manufacturers of these technologies. These include both environmental care and renewable energies.

Health and safety

The Construction Skills Certification Scheme (and the resultant CSCS card) proves competency in site health and safety, and is a requirement for any site work. Working on a building site requires that an operative is in possession of a current Competent Person Scheme card.

SUMMARY

In this chapter we have investigated the aspects of good system installation that you will need to have at least a basic knowledge of, as well as discussing good communication techniques.

We have also looked at ways in which you can take the next step towards getting a job in the building services sector.

Test your knowledge

1 Which three methods of communication are all written forms?

2 What is the difference between a quotation and an estimate?

3 What are 'statutory cancellation rights'?

4 A well-functioning plumbing system that meets or exceeds the customer's requirements is the result of a number of important aspects. What are they?

5 Identify two sources of information that could help you with career planning.

6 What is the purpose of a personal statement?

7 Identify four items that could be included in a CV.

8 In SWOT analysis, what do the letters SWOT stand for?

9 What are SMART targets?

10 Name three job roles in building services engineering.

11 What is a subcontractor?

12 How often must an operative renew their Gas Safe registration?

13 What regulations do people need to comply with under the Competent Person Scheme?

14 An approved plumber may start work without notification or prior consent on certain types of plumbing systems, provided they issue which type of certificate?

Answers can be found online at www.hoddereducation.co.uk/construction.

Practice synoptic assessments

Practice assessment 1

A customer requires an upgrade from an old floor-mounted open-flued boiler to a wall-mounted room-sealed boiler.

1 Describe to the customer the differences between the two boilers, and outline any safety advantages.

2 From the diagram of the property and the old radiator sizes, work out the approximate boiler size required.

3 Complete a method statement for the installation of the new boiler.

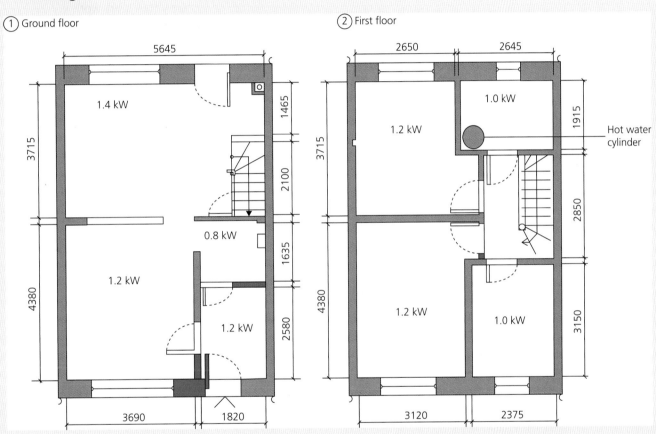

① Ground floor

② First floor

Hot water cylinder

Practice assessment 2

You are installing a new bathroom in a new build house. The owner is looking around and sees that you are installing a thermostatic mixing valve under the bath. They ask what it is and why you have to install it.

1 Explain how Building Regulations Part G outlines safety temperatures for domestic appliances.

2 Complete the table by writing in all the correct temperatures next to the descriptions.

Situation	Temperature
Legionella growth	
According to Building Regulations Part G, hot water must **never** exceed	
Unvented hot water cylinder thermostat	
Unvented hot water cylinder high-limit thermostat	
Unvented hot water cylinder temperature/pressure release valve	
Maximum temperature for domestic water at the outlet of a TMV to a batch, according to Part G	
Domestic bath hot water temperature at the outlet of a TMV	
Maximum temperature for communal or public showers, according to Part G	
Commercial basin, according to the code of practice for safe water temperatures	
Commercial bidet, according to the code of practice for safe water temperatures	
Temperature the hot water should reach at an outlet within 30 seconds, according to Part G	
Maximum temperature of a cold water storage cistern, according to the Water Regulations	
Maximum temperature of a cold water storage cistern, according to the BS EN 806	

Glossary

AC: alternating current flows in *both* directions.

Air change rate: a measure of how many times the air within a defined space (normally a room or a house) is replaced per hour, usually through natural ventilation.

Annular space: the required 25 mm gap between any hot surface of a flue and any combustible materials when travelling through a floor in a dwelling.

Appliance compartment: an enclosure specifically designed or adapted to house one or more gas appliances.

Approved plumber: a plumber that has undertaken specific Water Regulations training and is recognised as competent by the water undertaker.

Aquifer: a type of rock that holds water like a sponge.

Azimuth: refers to the angle that the panel direction diverges from facing due south.

Backflow loop: prevents backflow.

Bernoulli's principle: states that when a pipe is suddenly reduced in size, the velocity of the water increases but the pressure decreases. The principle can also work in reverse. If a pipe suddenly increases in size, then the velocity will decrease but the pressure will increase slightly.

Blue water corrosion: occurs in copper plumbing systems causing a blue-green colouration to the water.

Body language: movements and postures which communicate attitudes and feelings.

Boiler interlock: this is NOT a single control. It is a combination of several controls working in conjunction to ensure that the boiler does not fire unless it is required. It is key in ensuring good system efficiency and saving energy.

Bonding: a term used to describe the connection of extraneous conductive parts to the earthing system.

Boyle's law: One of two gas laws that determine the characteristics of a gas.

Buffer tank: a large white storage vessel used for storing excess hot water until demanded.

Calorific value: the amount of energy stored in the gas in its uncombusted state. It is the amount of energy released when the gas is combusted. It is measured in megajoules per cubic metre or MJ/m^3.

CENELEC: commonly used to refer to Comité Européen de Normalisation Électrotechnique, which is the European Committee for Electrotechnical Standardisation.

Competency: the degree to which a person has the ability to complete something.

Conductor: this is the part of a cable which current passes through. In most cables, this is made of copper and should have a low resistance.

Continuity test: a test to ensure that a conductor has integrity along its whole length. For example, it could be used to test a high-tension lead to ensure its connection from the spark generator to the probe on the boiler burner bar is sound.

Corrosion: the breaking down or destruction of a material, especially a metal, through chemical reactions.

Cryptosporidium: a gastrointestinal bacteria that affects humans and cattle and presents itself as severe diarrhoea. It usually affects children between the ages of 1 and 5, but it can affect anyone and the symptoms can be very severe in people with low immune systems.

Cutting off the supply: depending on the equipment and the circumstances, this may be no more than normal functional switching (on/off) or emergency switching by means of a stop button or a trip switch.

DC: direct current is chemically produced and is an electrical current in which electrons flow in a single direction.

De-zincification resistant (DZR): a type of brass that resists electrolytic corrosion.

Dead leg: when a hot tap is opened, a certain amount of cold water is usually drawn off and allowed to run to drain before hot water arrives at the tap. This wasted, cold water is known as a dead leg. Under the Water Regulations, dead legs must be restricted (see Table 2.3). If this is not possible, then secondary circulation is required.

Diagnostic: concerned with identifying problems.

Dead testing: testing carried out on electrical components or parts of an installation when the electrical supply is disconnected.

Downstream: in water systems, downstream means travelling away from the point of supply.

Duty holder: any person or organisation holding a legal duty under the Health and Safety at Work etc. Act 1974.

Duty of care: in British law, this is a moral and legal obligation imposed on an organisation or an individual, which necessitates that a standard of reasonable care is adhered to. If the standard of care is not met, then the acts are considered to be negligent and damages may be claimed for in a court of law.

Dynamic pressure: also known as 'running pressure', this is the water pressure when outlets are open and water is flowing.

Earth: earth with a capital E represents the potential of the ground we stand on.

Economical: good value for money.

Effective roof area: different to the actual size of the roof area. In effect it is the plan view area of the roof.

Electric shock: where a current flows through the human body and causes an accident or injury as a result.

Electrolytic corrosion: a process of accelerated corrosion between two or more differing metals when placed in an electrolytic environment.

Engineering judgement: this is a technical decision which is based on the competence of a person who has an appropriate combination of technical education, training and practical experience in the specific field of work. Competence in specific areas of gas work is verified by assessments of an engineer's theoretical and practical knowledge at an independent nationally approved ACS gas centre, and then registration with the HSE approved Gas Safe register.

Equal potential: where the voltage between any two parts is within safe touch voltage levels, usually 50 V AC but dependent on the location.

Equilibrium: in perfect balance (e.g., the pressure is balanced both sides of the valve). When referring to drainage systems this relates to keeping the air pressure even within the system, so that any negative pressure or pressure fluctuation does not cause any trap seal loss and therefore the ingress of foul air into the property.

Extraneous conductive parts: metallic parts of a building structure or services that have a low resistance path to the general mass of earth but do not form part of the electrical system.

From every source of electrical energy: many accidents occur due to a failure to isolate all sources of supply to or within equipment (for example, control and auxiliary supplies, uninterruptable power supply (UPS) systems or parallel circuit arrangements giving rise to back feeds).

Functional testing: a process carried out to check that components within an installation operate correctly. For example, an immersion heater should be tested to ensure safe operation of heating element and thermostat.

Glycol: a liquid anti-freeze which is odourless and colourless in its raw state.

Guidance Note GS38: electrical test equipment for use by electricians (published by the Health and Safety Executive, HSE) was written as a guideline to good practice when using test equipment on circuits operating at voltages >50 V AC or >120 V DC or where tests use these voltages. It is intended to be followed, in order to reduce the risk of danger and injury when performing electrical tests.

Haemoglobin: the part of the red blood cell that carries vital oxygen to the brain. When CO enters the bloodstream, the haemoglobin becomes carboxyhaemoglobin, which effectively blocks the red blood cells from carrying the oxygen to the brain and this causes the body's cells and tissue to fail and die.

Head: the pressure exerted by a column of water under gravity.

Insulation: the material that covers the conductor, and should have a high resistance stopping current flow. Insulation is intended to stop current from leaking from one conductor into another conductor or person which could in turn cause an electric shock.

Integral filling loop: a filling loop that is designed and installed as part of the boiler by the manufacturer.

Interconnected: connected together to form one cistern.

Internal space: an indoor space not classified as a room because it is either a hall, passageway, stairway or landing.

Intumescent air vents: an intumescent vent contains a substance which swells when exposed to heat and blocks the free air opening which will help prevent the spread of smoke in a fire.

IP2X: meaning that there is no hole in the barrier or enclosure greater than 12.5 mm in diameter, which provides 'finger protection', meaning no person can insert their finger and touch live parts.

IP4X: meaning that there is no hole in the barrier or enclosure greater than 1 mm in diameter stopping parts from falling into the enclosure.

Isolation: this means the disconnection and separation of the electrical equipment from every source of electrical energy in such a way that this disconnection and separation is secure.

Legible: readable.

Line: the conductor, having brown coloured insulation, which is normally connected to terminals marked L.

Live conductor: a conductor intended to be energised in normal service, and therefore includes a neutral conductor.

Live testing: a test is carried out when components are live.

Loading unit: a number or a factor, which is allocated to an appliance. It relates to the flow rate at the terminal fitting, the length of time in use and the frequency of use.

Mandatory: required by law or regulation.

Mass flow rate: the mass of a substance (e.g. kilograms) which passes in a unit of time.

Multi-storey building: a building having more than three floors.

Naphtha: a waxy oil deposit that is present in natural gas in its unrefined state. It is removed and later reused in other products such as cosmetics.

National standard: based on International Standards produced by the International Electrotechnical Commission (IEC), member nations create their own versions specific to their needs. Other CENELEC countries use the term 'rules' rather than 'regulations'. For example, the national wiring standard in the Republic of Ireland is the National Rules for Electrical Installations (ET101).

Neutral: the conductor, having blue coloured insulation, which is normally connected to terminals marked N.

Overheat protection: when water cannot circulate or the thermostats have been satisfied and the motorised valves have closed, the boiler will continue to heat up for a short period even though the burner has shut down. This is because of the latent heat in the boiler casing. If the boiler overheats, the high-limit thermostat will activate, and the boiler will fail to operate when it is next required. A pump-overrun circuit, which is fitted to most modern boilers, will ensure that the pump continues to run when the boiler has shut down to dissipate any latent heat. If the motorised valves are closed, the automatic bypass valve opens from the pump pressure to allow water circulation, allowing the excess heat to dissipate, keeping the boiler temperature below high-limit shut-down.

Pathogen: a germ or bacteria.

Performance test: carried out on a sanitary system to ensure that after simultaneous operation of appliances connected to the same soil stack, the trap depths remaining should be at least 25 mm deep.

Perpendicular: at an angle of 90° to a certain plane. In other words, the Sun's rays need to be at 90° to the collector where possible for maximum efficiency.

Planning permission: official permission from the local authority allowing a new build, alteration or addition to an existing building to be made.

Post: means 'after'.

Post-aerated flame: air is drawn for combustion from the surrounding air once the flame is lit, often resulting in a loose yellow, floppy flame.

Potable: pronounced poe-table, from the French word 'potable' meaning drinkable.

Pre: means 'before'.

Pre-aerated flame: air is entrained in the mixing tube before ignition.

Pryolised: when a material begins to decompose due to elevated temperatures.

PVCu (unplasticised poly vinyl chloride): a common material used in rainwater guttering and pipework systems.

Ratio: 1:600 means 1 mm fall for every 600 mm length of gutter.

Refrigerant: a substance or mixture, usually a fluid, used in a heat pump and refrigeration cycle. In most cycles it undergoes phase transitions from a liquid to a gas and back again.

Regulatory body: an organisation set up by the Government to monitor, control and guide various sectors within industry.

Residual current device (RCD): a sensitive device which trips, cutting current from a circuit, should a very small fault occur between any live conductor and earth. They are intended to give maximum protection against electric shock and can be identified by the fact they look like circuit breakers but have a test button located on them.

Resistance testing: the Ω scale is used, for example, to find out the level resistance of a coil in a motor, to ascertain whether it works or not (e.g. testing a motor on a zone valve).

Resistor: a passive thru-terminal electrical component that resists electrical current.

Reverse osmosis: method of purifying water.

Rodding point: a place where the drain or section of drain can be accessed to clear any blockages.

Schmutzdecke: a layer of mud that is saturated with friendly, water cleansing bacteria.

Secure: security can best be achieved by locking off with a safety lock (such as a lock with a unique key). The posting of a warning notice also serves to alert others to the isolation.

Solar radiation: radiant energy emitted by the Sun.

Standard English: use of English following correct spelling and grammar.

Static pressure: this is the water pressure when no flow is occurring. This is always greater than the dynamic pressure.

Statutory: the law, and therefore there are serious consequences if they are not followed.

Stratification: describes how the temperature of the water varies with its depth. The nearer the water is to the top of the cistern, the warmer it will be. The deeper the water, the colder it will be. This tends to occur in layers, whereby there is a marked temperature difference from one layer to the next. The result is that water quality can vary, the warmer water near the top being more susceptible to biological growth such as *Legionella pneumophila* (Legionnaires' disease).

Sub-station transformer: a piece of equipment which is owned by the electricity distribution network operator (DNO) and is used to step down large distribution voltages of 11000 V to 230 V for supplies into houses. Sub-stations are sometimes located behind panel fences and can serve up to 100 houses or more depending on its size. Sometimes they are located on poles where they serve one or two houses in more rural locations. In cities and large towns, they are normally located in brick or concrete structures.

Supplementary bonding: where a bonding conductor is installed either between pipes or from a socket outlet or other accessory to a pipe. Unlike MPB, supplementary bonding doesn't come from the MET. The minimum csa of cable permitted for supplementary bonding is 4 mm² where the cable is in free air. Supplementary bonding is sometimes also called cross-bonding.

Switch: component that breaks an electrical circuit by interrupting or diverting current.

Thermal cycling: heating and cooling of metal (in this case) which causes expansion and contraction and, eventually, the loosening of terminals.

Tanking: a process used to ensure that a wet room area installation is completely leak free.

Thermal shock: the rapid cooling or heating of a substance that can lead to failure of the material.

Tone of voice: a way of sounding to express meaning or emotion. For example, your tone of voice can communicate confidence and conviction, assuring customers that you are knowledgeable and capable.

Transpose: to rearrange the information to determine a different part of the formula.

Turbidity: refers to how clear or cloudy the water is due to the amount of total suspended solids it contains. The greater the amount of total suspended solids (TSS) in the water, the cloudier it will appear. Cloudy water can therefore be said to be turbid.

Type B circuit breakers: the most sensitive type of CB and should be the types used to protect circuits in domestic type installations. Other types are type C intended for motors and transformers and type D for very specialised machines such as welding equipment or medical equipment.

Upstream: in water systems, upstream means travelling toward the point of supply.

Verbal: the spoken word. Any verbal communication should always be backed up with written confirmation to verify any agreements and clarify any details to prevent confusion.

Verifiable: able to be checked.

Vertigo: the feeling that the sufferer or objects around them are moving when they are not; feels like a spinning or swaying movement.

Vitiated air: the word 'vitiated' simply means 'to be made impure', so 'vitiated air' is air that has been made impure.

Water table: the point where the earth below ground becomes saturated with water causing water to pool.

Water undertaker: a water authority or company that supplies clean, cold wholesome water under Section 67 of the Water Act 1991.

Index

Note: page numbers in **bold** indicate location of key term definitions.